METHODS IN MOLECULAR BIOLOGY

Series Editor
John M. Walker
School of Life and Medical Sciences,
University of Hertfordshire, Hatfield,
Hertfordshire AL10 9AB, UK

For further volumes:
http://www.springer.com/series/7651

RNA Detection

Methods and Protocols

Edited by

Imre Gaspar

Developmental Biology Unit, EMBL Heidelberg, Heidelberg, Germany

Humana Press

Editor
Imre Gaspar
Developmental Biology Unit
EMBL Heidelberg
Heidelberg, Germany

ISSN 1064-3745 ISSN 1940-6029 (electronic)
Methods in Molecular Biology
ISBN 978-1-4939-7212-8 ISBN 978-1-4939-7213-5 (eBook)
DOI 10.1007/978-1-4939-7213-5

Library of Congress Control Number: 2017957614

Cover Caption: Localization of *oskar* mRNA (in cyan, labelled with a set of FIT probes) to the posterior pole of a developing *Drosophila* oocyte (on the right). *oskar* mRNA is transcribed in the nurse cells (on the left) and transported through cytoplasmic bridges into the oocyte by cytoplasmic dynein. In the oocyte, localization of *oskar* to the posterior pole is mediated by kinesin heavy chain (in red, labelled with mKate2 fluorescent protein). Note, that *oskar* mRNA is exclusively expressed in the female germ-line (nurse cells and the oocyte) and it is absent from the somatic follicular epithelium (on the right, marked by kinesin) surrounding the oocyte.

Printed on acid-free paper

This Humana Press imprint is published by Springer Nature
The registered company is Springer Science+Business Media LLC New York
The registered company address is: 233 Spring Street, New York, NY 10013, U.S.A.

Preface

In the last couple of years, research based on high-throughput assays revealed that the RNA world is much more complex than initially anticipated entangling virtually all areas of cell and developmental biology. Semiautomated visual screens demonstrated that large fraction of the transcriptome is distributed nonuniformly within the cell, suggesting the presence of underlying active localization mechanisms. On the other hand, the nonbiased capture of RNA interactome showed that 8–10% of the total proteome could directly bind (m)RNA, including hundreds of novel RNA binding proteins, such as enzymes of fundamental biosynthesis pathways, components of the cytoskeleton, the endocytosis, and secretory pathways. Some of these novel RNA-binding proteins harbor low-complexity domains, making them capable of spontaneously self-assembling into higher-order structures both in vitro and in vivo, dynamically forming RNA-containing membraneless organelles, such as Cajal bodies, nuclear speckles, and paraspeckles, RNA and stress granules, nuage or germ granules. In specimen previously considered homogeneous including tumorous malformations, quantitative RNA imaging and correlative high-content imaging coupled with single cell transcriptome analysis demonstrated a high degree of heterogeneity which directly impacts prognosis and possible therapy. As revealed by in vivo functional assays of single-molecule sensitivity, this heterogeneity is mainly due to the stochastic nature of the underlying biological processes, such as transcription and translation.

The advances in RNA biology render advanced RNA detection and visualization tools invaluable to cell and developmental biologists as well as to medical researchers and practicing clinicians. Although the amount of technology development in the last couple of years renders it impossible to cover every possible aspects of RNA detection, this volume aims to introduce the various concepts and the methods of detecting RNA in biological material in a variety of model systems. The detailed protocols and the tips and tricks of the presented assays will allow the optimization and the adaptation of these methods to address different biological questions of RNA, and hopefully, this volume of the MiMB series becomes a useful everyday companion of every novel or experienced scientists of the expanding RNA world.

Heidelberg, Germany *Imre Gaspar*

Contents

Preface . . . . *v*
Contributors . . . . *xi*

1 The Secret Life of RNA: Lessons from Emerging Methodologies 1
Caroline Medioni and Florence Besse

2 Quantification of 2′-*O*-Me Residues in RNA Using Next-Generation Sequencing (Illumina RiboMethSeq Protocol) 29
Lilia Ayadi, Yuri Motorin, and Virginie Marchand

3 Identifying the m^{6}A Methylome by Affinity Purification and Sequencing 49
Phillip J. Hsu and Chuan He

4 PARIS: Psoralen Analysis of RNA Interactions and Structures with High Throughput and Resolution 59
Zhipeng Lu, Jing Gong, and Qiangfeng Cliff Zhang

5 Axon-TRAP-RiboTag: Affinity Purification of Translated mRNAs from Neuronal Axons in Mouse In Vivo 85
Toshiaki Shigeoka, Jane Jung, Christine E. Holt, and Hosung Jung

6 LCM-Seq: A Method for Spatial Transcriptomic Profiling Using Laser Capture Microdissection Coupled with PolyA-Based RNA Sequencing 95
Susanne Nichterwitz, Julio Aguila Benitez, Rein Hoogstraaten, Qiaolin Deng, and Eva Hedlund

7 Spatial Transcriptomics: Constructing a Single-Cell Resolution Transcriptome-Wide Expression Atlas 111
Kaia Achim, Hernando Martínez Vergara, and Jean-Baptiste Pettit

8 Single mRNA Molecule Detection in *Drosophila* 127
Shawn C. Little and Thomas Gregor

9 Detection and Automated Analysis of Single Transcripts at Subcellular Resolution in Zebrafish Embryos 143
L. Carine Stapel, Coleman Broaddus, and Nadine L. Vastenhouw

10 Super-Resolution Single Molecule FISH at the *Drosophila* Neuromuscular Junction 163
Joshua S. Titlow, Lu Yang, Richard M. Parton, Ana Palanca, and Ilan Davis

11 Detection of mRNA and Associated Molecules by ISH-IEM on Frozen Sections 177
Catherine Rabouille

12 Hybridization Chain Reaction for Direct mRNA Detection Without Nucleic Acid Purification 187
Yao Xu and Zhi Zheng

13 In Situ Detection of MicroRNA Expression with RNAscope Probes 197
Viravuth P. Yin

14 Padlock Probes to Detect Single Nucleotide Polymorphisms 209
Tomasz Krzywkowski and Mats Nilsson

15 Quantifying Gene Expression in Living Cells with Ratiometric Bimolecular Beacons 231
Yantao Yang, Mingming Chen, Christopher J. Krueger, Andrew Tsourkas, and Antony K. Chen

16 Optimizing Molecular Beacons for Intracellular Analysis of RNA 243
Mingming Chen, Yantao Yang, Christopher J. Krueger, and Antony K. Chen

17 Live Imaging of Nuclear RNPs in Mammalian Complex Tissue with ECHO-liveFISH 259
Dan Ohtan Wang

18 In Vivo Visualization and Function Probing of Transport mRNPs Using Injected FIT Probes 273
Jasmine Chamiolo, Imre Gaspar, Anne Ephrussi, and Oliver Seitz

19 Visualizing RNA in Live Bacterial Cells Using Fluorophore- and Quencher-Binding Aptamers 289
Murat Sunbul, Ankita Arora, and Andres Jäschke

20 Method for Imaging Live-Cell RNA Using an RNA Aptamer and a Fluorescent Probe 305
Shin-ichi Sato, Kenji Yatsuzuka, Yousuke Katsuda, and Motonari Uesugi

21 RNA Live Imaging in the Model Microorganism *Ustilago maydis*............ 319
Sabrina Zander, Kira Müntjes, and Michael Feldbrügge

22 Real-Time Fluorescence Imaging of Single-Molecule Endogenous Noncoding RNA in Living Cells 337
Hideaki Yoshimura and Takeaki Ozawa

23 Live Imaging of mRNA Synthesis in *Drosophila*.................................. 349
Hernan G. Garcia and Thomas Gregor

24 Imaging Newly Transcribed RNA in Cells by Using a Clickable Azide-Modified UTP Analog 359
Anupam A. Sawant, Sanjeev Galande, and Seergazhi G. Srivatsan

25 Detection of the First Round of Translation: The TRICK Assay 373
Franka Voigt, Jan Eglinger, and Jeffrey A. Chao

26 Imaging Translation Dynamics of Single mRNA Molecules in Live Cells 385
Suzan Ruijtenberg, Tim A. Hoek, Xiaowei Yan, and Marvin E. Tanenbaum

27 Systematic Detection of Poly(A)+ RNA-Interacting Proteins and Their Differential Binding.. 405
Miha Milek and Markus Landthaler

28 Isolation and Characterization of Endogenous RNPs from Brain Tissues . . . 419
Rico Schieweck, Foong yee Ang, Renate Fritzsche, and Michael A. Kiebler

29 Individual Nucleotide Resolution UV Cross-Linking and Immunoprecipitation (iCLIP) to Determine Protein–RNA Interactions . . . 427
Christopher R. Sibley

30 RNA Tagging: Preparation of High-Throughput Sequencing Libraries . . . 455
Christopher P. Lapointe and Marvin Wickens

31 RAP-MS: A Method to Identify Proteins that Interact Directly with a Specific RNA Molecule in Cells . . . 473
Colleen A. McHugh and Mitchell Guttman

Erratum to: Super-Resolution Single Molecule FISH at the *Drosophila* Neuromuscular Junction . . . E1

Index . . . *489*

Contributors

KAIA ACHIM • *Developmental Biology Unit, European Molecular Biology Laboratory, Heidelberg, Germany*
FOONG YEE ANG • *Cell Biology, Biomedical Center, Medical Faculty, LMU Munich, Planegg-Martinsried, Germany*
ANKITA ARORA • *Institute of Pharmacy and Molecular Biotechnology, Heidelberg University, Heidelberg, Germany*
LILIA AYADI • *IMoPA UMR7365 CNRS-Lorraine University, BioPole Lorraine University, Vandoeuvre-les-Nancy, France*
JULIO AGUILA BENITEZ • *Department of Neuroscience, Karolinska Institutet, Stockholm, Sweden; Department of Cell and Molecular Biology, Karolinska Institutet, Stockholm, Sweden*
FLORENCE BESSE • *Université Côte d'Azur, CNRS, Inserm, iBV, Nice, France*
COLEMAN BROADDUS • *Center for Systems Biology Dresden, Max Planck Institute of Molecular Cell Biology and Genetics, Dresden, Germany*
JASMINE CHAMIOLO • *Department of Chemistry, Humboldt University Berlin, Berlin, Germany*
JEFFREY A. CHAO • *Friedrich Miescher Institute for Biomedical Research, Basel, Switzerland*
ANTONY K. CHEN • *Department of Biomedical Engineering, College of Engineering, Peking University, Haidian District, Beijing, China*
MINGMING CHEN • *Department of Biomedical Engineering, College of Engineering, Peking University, Haidian District, Beijing, China; Peking-Tsinghua Center for Life Sciences, Peking University, Beijing, China; Academy for Advanced Interdisciplinary Studies, Peking University, Beijing, China*
ILAN DAVIS • *Department of Biochemistry, University of Oxford, Oxford, UK*
QIAOLIN DENG • *Department of Cell and Molecular Biology, Karolinska Institutet, Stockholm, Sweden*
JAN EGLINGER • *Friedrich Miescher Institute for Biomedical Research, Basel, Switzerland*
ANNE EPHRUSSI • *European Molecular Biology Laboratory (EMBL) Heidelberg, Heidelberg, Germany*
MICHAEL FELDBRüGGE • *Institute for Microbiology, Heinrich-Heine University Düsseldorf, Cluster of Excellence on Plant Sciences (CEPLAS), Düsseldorf, Germany*
RENATE FRITZSCHE • *Cell Biology, Biomedical Center, Medical Faculty, LMU Munich, Planegg-Martinsried, Germany*
SANJEEV GALANDE • *Center of Excellence in Epigenetics, Indian Institute of Science Education and Research, Pune, Pune, India*
HERNAN G. GARCIA • *Department of Molecular and Cell Biology, University of California, Berkeley, CA, USA; Department of Physics and Biophysics Graduate Group, University of California, Berkeley, CA, USA*
IMRE GASPAR • *European Molecular Biology Laboratory (EMBL) Heidelberg, Heidelberg, Germany*
JING GONG • *MOE Key Laboratory of Bioinformatics, Beijing Advanced Innovation Center for Structural Biology, Center for Synthetic and Systems Biology, Tsinghua-Peking Joint*

Center for Life Sciences, School of Life Sciences, Tsinghua University, Haidian, Beijing, China

THOMAS GREGOR • *Department of Physics, Lewis-Sigler Institute for Integrative Genomics, Princeton University, Princeton, NJ, USA*

MITCHELL GUTTMAN • *Division of Biology and Biological Engineering, California Institute of Technology, Pasadena, CA, USA*

CHUAN HE • *Department of Chemistry, Institute for Biophysical Dynamics, The University of Chicago, Chicago, IL, USA; Howard Hughes Medical Institute, The University of Chicago, Chicago, IL, USA; Department of Biochemistry and Molecular Biology, The University of Chicago, Chicago, IL, USA*

EVA HEDLUND • *Department of Neuroscience, Karolinska Institutet, Stockholm, Sweden*

TIM A. HOEK • *Hubrecht Institute, KNAW and University Medical Center Utrecht, Utrecht, The Netherlands*

CHRISTINE E. HOLT • *Department of Physiology, Development and Neuroscience, University of Cambridge, Cambridge, UK*

REIN HOOGSTRAATEN • *Department of Neuroscience, Karolinska Institutet, Stockholm, Sweden; Department of Translational Neuroscience, Brain Center Rudolf Magnus, Utrecht, Netherlands*

PHILLIP J. HSU • *Department of Chemistry, Institute for Biophysical Dynamics, The University of Chicago, Chicago, IL, USA; Howard Hughes Medical Institute, The University of Chicago, Chicago, IL, USA; Committee on Immunology, The University of Chicago, Chicago, IL, USA*

ANDRES JÄSCHKE • *Institute of Pharmacy and Molecular Biotechnology, Heidelberg University, Heidelberg, Germany*

HOSUNG JUNG • *Department of Anatomy, Brain Research Institute, and Brain Korea 21 PLUS Project for Medical Science, Yonsei University College of Medicine, Seoul, Republic of Korea*

JANE JUNG • *Department of Anatomy, Brain Research Institute, and Brain Korea 21 PLUS Project for Medical Science, Yonsei University College of Medicine, Seoul, Republic of Korea*

YOUSUKE KATSUDA • *Institute for Integrated Cell-Material Sciences (WPI-iCeMS), Kyoto University, Kyoto, Japan*

MICHAEL A. KIEBLER • *Cell Biology, Biomedical Center, Medical Faculty, LMU Munich, Planegg-Martinsried, Germany*

CHRISTOPHER J. KRUEGER • *Department of Biomedical Engineering, College of Engineering, Peking University, Haidian District, Beijing, China; Wallace H Coulter Department of Biomedical Engineering, Georgia Institute of Technology, Atlanta, GA, USA*

TOMASZ KRZYWKOWSKI • *Science for Life Laboratory, Department of Biochemistry and Biophysics, Stockholm University, Solna, Sweden*

MARKUS LANDTHALER • *RNA Biology and Posttranscriptional Regulation, Max Delbrück Center for Molecular Medicine Berlin, Berlin Institute for Molecular Systems Biology, Berlin, Germany*

CHRISTOPHER P. LAPOINTE • *Department of Biochemistry, University of Wisconsin-Madison, Madison, WI, USA*

SHAWN C. LITTLE • *Department of Cell and Developmental Biology, University of Pennsylvania Perelman School of Medicine, Philadelphia, PA, USA*

ZHIPENG LU • *Department of Dermatology, Center for Personal Dynamic Regulomes, Stanford University, Stanford, CA, USA*

Virginie Marchand • *IMoPA UMR7365 CNRS-Lorraine University, BioPole Lorraine University, Vandoeuvre-les-Nancy, France; Next-Generation Sequencing Core Facility, FR3209 BMCT, CNRS-Lorraine University, BioPole Lorraine University, Vandoeuvre-les-Nancy, France*

Colleen A. McHugh • *Division of Biology and Biological Engineering, California Institute of Technology, Pasadena, CA, USA*

Caroline Medioni • *Université Côte d'Azur, CNRS, Inserm, iBV, Nice, France*

Miha Milek • *RNA Biology and Posttranscriptional Regulation, Max Delbrück Center for Molecular Medicine Berlin, Berlin Institute for Molecular Systems Biology, Berlin, Germany*

Yuri Motorin • *IMoPA UMR7365 CNRS-Lorraine University, BioPole Lorraine University, Vandoeuvre-les-Nancy, France; Next-Generation Sequencing Core Facility, FR3209 BMCT, CNRS-Lorraine University, BioPole Lorraine University, Vandoeuvre-les-Nancy, France*

Kira Müntjes • *Heinrich-Heine University Düsseldorf, Institute for Microbiology, Cluster of Excellence on Plant Sciences (CEPLAS), Düsseldorf, Germany*

Susanne Nichterwitz • *Department of Neuroscience, Karolinska Institutet, Stockholm, Sweden*

Mats Nilsson • *Science for Life Laboratory, Department of Biochemistry and Biophysics, Stockholm University, Solna, Sweden*

Takeaki Ozawa • *Department of Chemistry, School of Science, The University of Tokyo, Tokyo, Japan*

Ana Palanca • *Department of Biochemistry, University of Oxford, Oxford, UK*

Richard M. Parton • *Department of Biochemistry, University of Oxford, Oxford, UK*

Jean-Baptiste Pettit • *European Bioinformatics Institute (EMBL-EBI), European Molecular Biology Laboratory, Hinxton, Cambridgeshire, UK*

Catherine Rabouille • *Hubrecht Institute of the KNAW/University Medical Centre Utrecht, Utrecht, The Netherlands; Department of Cell Biology, University Medical Center Utrecht, Utrecht, The Netherlands; Department of Cell Biology, University Medical Center Groningen, Groningen, The Netherlands*

Suzan Ruijtenberg • *Hubrecht Institute, KNAW and University Medical Center Utrecht, Utrecht, The Netherlands*

Shin-ichi Sato • *Institute for Integrated Cell-Material Sciences (WPI-iCeMS), Kyoto University, Kyoto, Japan*

Anupam A. Sawant • *Department of Chemistry, Indian Institute of Science Education and Research, Pune, Pune, India*

Rico Schieweck • *Cell Biology, Biomedical Center, Medical Faculty, LMU Munich, Planegg-Martinsried, Germany*

Oliver Seitz • *Department of Chemistry, Humboldt University Berlin, Berlin, Germany*

Toshiaki Shigeoka • *Department of Physiology, Development and Neuroscience, University of Cambridge, Cambridge, UK*

Christopher R. Sibley • *Division of Brain Sciences, Department of Medicine, Imperial College London, London, UK*

Seergazhi G. Srivatsan • *Department of Chemistry, Indian Institute of Science Education and Research, Pune, Pune, India*

L. Carine Stapel • *Max Planck Institute of Molecular Cell Biology and Genetics, Dresden, Germany*

MURAT SUNBUL • *Institute of Pharmacy and Molecular Biotechnology, Heidelberg University, Heidelberg, Germany*

MARVIN E. TANENBAUM • *Hubrecht Institute, KNAW and University Medical Center Utrecht, Utrecht, The Netherlands*

JOSHUA S. TITLOW • *Department of Biochemistry, University of Oxford, Oxford, UK*

ANDREW TSOURKAS • *Department of Biomedical Engineering, University of Pennsylvania, Philadelphia, PA, USA*

MOTONARI UESUGI • *Institute for Integrated Cell-Material Sciences (WPI-iCeMS), Kyoto University, Kyoto, Japan; Institute for Chemical Research, Kyoto University, Kyoto, Japan*

NADINE L. VASTENHOUW • *Max Planck Institute of Molecular Cell Biology and Genetics, Dresden, Germany*

HERNANDO MARTÍNEZ VERGARA • *Developmental Biology Unit, European Molecular Biology Laboratory, Heidelberg, Germany*

FRANKA VOIGT • *Friedrich Miescher Institute for Biomedical Research, Basel, Switzerland*

DAN OHTAN WANG • *Institute for Integrated Cell-Material Sciences (WPI-iCeMS), Kyoto University, Kyoto, Japan; The Keihanshin Consortium for Fostering the Next Generation of Global Leaders in Research (K-CONNEX), Kyoto, Japan*

MARVIN WICKENS • *Department of Biochemistry, University of Wisconsin-Madison, Madison, WI, USA*

YAO XU • *Department of Biochemistry and Molecular Biology, Institute of Basic Medical Sciences, Chinese Academy of Medical Sciences and School of Basic Medicine, Peking Union Medical College, Beijing, China*

XIAOWEI YAN • *Department of Cellular and Molecular Pharmacology, University of California, San Francisco, San Francisco, CA, USA*

LU YANG • *Department of Biochemistry, University of Oxford, Oxford, UK*

YANTAO YANG • *Department of Biomedical Engineering, College of Engineering, Peking University, Haidian District, Beijing, China*

KENJI YATSUZUKA • *Institute for Chemical Research, Kyoto University, Kyoto, Japan*

VIRAVUTH P. YIN • *Mount Desert Island Biological Laboratory, Kathryn W. Davis Center for Regenerative Biology and Medicine, Salisbury Cove, ME, USA*

HIDEAKI YOSHIMURA • *Department of Chemistry, School of Science, The University of Tokyo, Tokyo, Japan*

SABRINA ZANDER • *Heinrich-Heine University Düsseldorf, , Institute for Microbiology, Cluster of Excellence on Plant Sciences (CEPLAS), Düsseldorf, Germany*

QIANGFENG CLIFF ZHANG • *MOE Key Laboratory of Bioinformatics, Beijing Advanced Innovation Center for Structural Biology, Center for Synthetic and Systems Biology, Tsinghua-Peking Joint Center for Life Sciences, School of Life Sciences, Tsinghua University, Haidian, Beijing, China*

ZHI ZHENG • *Department of Biochemistry and Molecular Biology, Institute of Basic Medical Sciences, Chinese Academy of Medical Sciences and School of Basic Medicine, Peking Union Medical College, Beijing, China*

Chapter 1

The Secret Life of RNA: Lessons from Emerging Methodologies

Caroline Medioni and Florence Besse

Abstract

The last past decade has witnessed a revolution in our appreciation of transcriptome complexity and regulation. This remarkable expansion in our knowledge largely originates from the advent of high-throughput methodologies, and the consecutive discovery that up to 90% of eukaryotic genomes are transcribed, thus generating an unanticipated large range of noncoding RNAs (Hangauer et al., 15(4):112, 2014). Besides leading to the identification of new noncoding RNA species, transcriptome-wide studies have uncovered novel layers of posttranscriptional regulatory mechanisms controlling RNA processing, maturation or translation, and each contributing to the precise and dynamic regulation of gene expression. Remarkably, the development of systems-level studies has been accompanied by tremendous progress in the visualization of individual RNA molecules in single cells, such that it is now possible to image RNA species with a single-molecule resolution from birth to translation or decay. Monitoring quantitatively, with unprecedented spatiotemporal resolution, the fate of individual molecules has been key to understanding the molecular mechanisms underlying the different steps of RNA regulation. This has also revealed biologically relevant, intracellular and intercellular heterogeneities in RNA distribution or regulation. More recently, the convergence of imaging and high-throughput technologies has led to the emergence of spatially resolved transcriptomic techniques that provide a means to perform large-scale analyses while preserving spatial information. By generating transcriptome-wide data on single-cell RNA content, or even subcellular RNA distribution, these methodologies are opening avenues to a wide range of network-level studies at the cell and organ-level, and promise to strongly improve disease diagnostic and treatment.

In this introductory chapter, we highlight how recently developed technologies aiming at detecting and visualizing RNA molecules have contributed to the emergence of entirely new research fields, and to dramatic progress in our understanding of gene expression regulation.

Key words RNA detection, Transcriptomics, RNA structure, RNA localization, In vivo RNA imaging, Transcription, Translation, Ribonucleoprotein complexes, Interactome

1 Uncovering RNA Regulation via Large Scale Approaches

The advent of deep RNA sequencing (RNA-seq) technologies and their applications to various transcriptomes has revealed the unexpected complexity of eukaryotic RNA repertoires, composed of a plethora of noncoding species and a large number of alternative

Imre Gaspar (ed.), *RNA Detection: Methods and Protocols*, Methods in Molecular Biology, vol. 1649, DOI 10.1007/978-1-4939-7213-5_1, © Springer Science+Business Media LLC 2018

transcripts [1]. Recently, implementations of deep sequencing techniques adapted to comprehensive analysis of 3′ ends, specific RNA modifications or detection of translation events, have greatly enhanced our ability to interrogate gene regulation at the posttranscriptional level, leading to the discovery of new tunable regulatory layers. Furthermore, emerging technologies providing an up-to-the-single-cell spatial resolution have paved the way for spatially resolved transcriptomics, a new field that integrates both RNA profiling of defined cell types and retrieval of positional information. These approaches have a broad spectrum of applications in the study of regulatory networks underlying developmental processes or disease progression [2].

1.1 Multilevel Regulation of RNA Processing Revealed by Transcriptomic Methods

Posttranscriptional processing of RNAs is a several-step process that does not end with splicing of intronic regions. 3′ end sequencing, indeed, has revealed that alternative cleavage and polyadenylation of RNAs (APA) is pervasive in all eukaryotes examined so far, and that up to 70% of human genes use APA to generate transcripts that differ in the length of their 3′ UTRs [3, 4]. While the biological functions of APA remain to be demonstrated at a global scale, transcriptomic analyses have shown that this process is tightly regulated in response to differentiation programs as well as external signals [4]. For example, a widespread shift toward usage of proximal poly(A) sites has been associated with increased cell proliferation [3, 5]. Furthermore, although promoter-distal poly(A) isoforms tend to be enriched in neuronal tissues [6], changes in proximal/distal 3′UTR ratios are observed for specific groups of genes in response to neuronal activity [7]. Development of novel techniques tailored to transcriptome-wide detection of nucleoside modifications has also revealed the prevalence and the diversity of RNA posttranscriptional modifications, giving birth to the expanding field of epitranscriptomics [8, 9]. Interestingly, large-scale mappings of modifications such as A-to-I editing, nucleoside methylation (m^6A, m^5C, m^1A) or pseudo-uridinylation (Ψ) have shown that modifications are enriched at specific transcript locations, suggesting mark-specific functions. m^6A, for example, preferentially decorates the stop codon vicinity and large internal exons, while m^1A clusters around the AUG start codon and is associated with enhanced translation [10, 11]. Further highlighting potential regulatory functions of posttranscriptional RNA modifications, RNA marks are dynamically regulated in response to differentiation programs or environmental stimuli, and conserved across evolution [8, 9]. Although the impact of RNA modifications on dynamic regulation of gene expression still remains largely unclear, large-scale analyses are now paving the way to a better understanding of the role of the epitranscriptome.

1.2 Not Just Sequence: Transcriptome-Wide Capture of High-Order Structures

mRNAs or noncoding RNAs (ncRNAs) are not linear single-stranded molecules, but rather adopt 3D structures that are essential for their processing and function, yet not trivial to capture. To get a global idea about the RNA structurome landscape, Chang and coworkers implemented a method that identifies flexible single-stranded bases in RNAs for all four nucleotides, in living cells. Profiling mRNA structure in mouse embryonic stem cells, they found that m^6A methylation induces characteristic RNA structures that may be relevant for the control of gene expression [12]. More recently, Rouskin and coworkers developed a mutational profiling approach that, instead of generating population-average structures, provides multiple structural features at a single-molecule resolution [13]. This enables detailed studies of structural heterogeneity, in particular isoform-specific RNA structures, in cellular environments. As an alternative approach, three groups introduced cross-linking-based high-throughput strategies that capture RNA–RNA duplexes in cells, and identify the corresponding sequence pairs [14–16]. With these methods, both intramolecular and intermolecular base-pairing interactions can be mapped, giving insights into internal RNA structural conformations and higher-order structures respectively. Strikingly, the first applications of these methods have revealed intermolecular interactions involving all major classes of RNA, such as ncRNA–ncRNA interactions, ncRNA–mRNA interactions as well as mRNA–mRNA interactions. Furthermore, they have highlighted the preponderance of long-range, often conserved and dynamic internal interactions within and between 5′ and 3′UTRs and coding sequences [15]. Such interactions might be particularly relevant, as efficiently translated mRNAs tend to exhibit long-range end-to-end interactions, which supports the previously proposed circularization model for ribosome recycling and efficient mRNA translation [14]. In contrast, poorly translated mRNAs tend to contain clusters of short-range interactions near the beginning of the transcript, consistent with translation inhibition by structured elements in 5′ UTRs [14]. By providing a global view on how transcript structural organization can impact gene regulation, these methods have opened the door for functional studies of the conformational changes occurring in response to various conditions.

1.3 Large-Scale Spatiotemporal Mapping of Translation Profiles

As revealed by the limited correlation between mRNA and protein levels [17], translational control is an essential and regulated step in determining levels of protein expression. With the development of ribosomal profiling methods, in which deep sequencing is used to comprehensively map and quantify ribosome footprints, it has become possible to get instantaneous and sensitive detection of translation events [18]. Notably, ribosome profiling not only enables dynamic transcriptome-wide measurements of translational rates under various conditions, but also provides detailed information about the identity of translation products [19]. This has led to the discovery of a large number of unanticipated footprints that fall outside canonical coding regions, and correspond to translated short ORFs (sORFs) found in previously annotated noncoding RNAs, regulatory ORFs (such as uORFs) that contribute to translational regulation of downstream ORFs, or alternative start or stop sites generating extended protein isoforms [19–21]. Interestingly, recent implementations of the ribosome profiling approach now allow monitoring translational events localized to specific subcellular compartments or specific cell types. In proximity-specific ribosome profiling, for example, purification of ribosomal subunits that are biotinylated locally by the restricted activity of the BirA biotin ligase is performed prior to ribosome profiling. Using this technique, Weissman and coworkers were able to monitor translation at two distinct entry points to the ER and at the mitochondrial membrane, thus providing detailed insights into cotranslational translocation of proteins into these organelles [22, 23]. In translating ribosome affinity capture (TRAP), purification of a tagged ribosomal subunit expressed cell-type specifically is coupled to RNA-seq to profile the entire translated mRNA complement of defined cell populations. This method has enabled precise and dynamic profiling of specific neuronal cell types in mammalian brains, providing insights into the molecular changes underlying both neuronal cell differentiation programs and differential responses to specific drugs [24, 25]. Together, the versatility of developed translation profiling strategies makes it now possible to explore with unprecedented spatiotemporal resolution changes in both conventional and unconventional translational events.

1.4 Toward Spatially Resolved Transcriptomics

Until recently, most of our knowledge about transcriptome-wide regulations was derived from bulk assays applied to entire cell populations or tissues. Ensemble averaging methodologies, however, prevent the analysis of intracellular dynamics and mask biologically meaningful cellular heterogeneities. Recent single-cell RNA-seq technologies developed to overcome these limits now allow profiling of single-cell transcriptional landscapes [26]. Although limited in sensitivity, these fourth generation sequencing techniques can quantify intrapopulation heterogeneity and enable studies of cell states at very high resolution. Single-cell RNA-seq, for example, has been successfully used to deconvolve heterogeneous cell populations, and identify novel and/or rare cell types in complex tissues such as intestine, spleen, or brain [26–31]. It is also commonly used to study cell state transitions and to map cell trajectories over the course of dynamic processes such as differentiation or response to external stimulation. Detailed analyses of cell trajectories have led to the discovery of previously masked intermediate differentiation states, as well as key signaling pathways and regulators triggering switches in cellular state or fate [26, 32–34].

A major caveat of most single-cell isolation procedures, however, is that information about cell original spatial environment is lost during cell isolation. Thus, computational methods have recently been developed to infer the initial 3D position of isolated cells from their transcriptomic profiles using reference gene expression maps obtained by in situ hybridizations [35, 36]. Alternatively, approaches in which RNA is captured from tissue sections have been developed [37, 38]. In the zebrafish embryo, for example, sequencing of serial consecutive sections along different axes was used in combination with image reconstruction to generate 3D gene expression atlases at different developmental stages [38]. In mouse brains, the so-called spatial transcriptomics method has been used to visualize RNA distribution. In this method, histological sections are deposited on arrays that capture and label RNAs according to their position [39]. Together, single-cell and spatially resolved transcriptomic approaches now allow detailed and dynamics studies of gene regulatory networks. By enabling precise monitoring of disease progression and by revealing heterogeneities in tumor samples, they also have a profound impact on disease prognosis and definition of optimal therapeutic strategies [39–41].

2 Single-Molecule Approaches for Quantitative and Subcellular Analyses of RNAs

Single-molecule approaches have recently emerged as a powerful means to resolve individual RNA molecules within individual cells, and thus to overcome the limits of large-scale averaging analyses. Single-molecule FISH (smFISH) methods, in particular, now enable absolute quantification of transcript copy number as well as subcellular visualization of single RNA molecules in cultured cells or tissues. Strikingly, the high resolution and fidelity of these approaches have revealed the prevalence of subcellular RNA localization, and led to the discovery of a previously masked, but biologically relevant, cell-to-cell variability in gene expression.

2.1 Detecting Single RNA Molecules

2.1.1 From Conventional FISH to smFISH Methods

Conventional FISH methods, in which long antisense probes recruit enzymes that catalyze fluorogenic reactions, have been used in a wide range of cell types and organisms to qualitatively assess RNA distribution and abundance. These methods, while very sensitive, generate a strong experimental variability that prevents signal calibration and quantification. Over the last past 10 years, different approaches have been developed to detect single RNA molecules with photonic microscopy systems [42, 43]. These approaches have aimed on one hand at enhancing individual signal brightness and on the other hand at improving signal-to-noise ratio.

The first group of methodologies, pioneered by the Singer group [44] and further implemented by the Tyagi and Van

Oudenaarden groups [45], is based on the hybridization along the target RNA of multiple short oligonucleotide probes, each labeled with one or several fluorophores. The collective fluorescence arising from the binding of such arrayed probes generates a strong and localized signal detectable as a single diffraction-limited spot. Importantly, automatic detection and quantification of individual fluorescent spots obtained with these smFISH techniques generated numbers of molecules similar to those obtained by RT-QPCR [45]. The second group of methods uses signal amplification as a means to overcome the limited sensitivity of probes with direct fluorescence encountered in particular when working with RNA of small size. In hybridization chain reaction (HCR)-based methods, for example, target-specific probes trigger the self-assembly of metastable fluorescent RNA hairpins into large amplification polymers, resulting in a 200-fold increase in signal brightness [46, 47]. In branched DNA (bDNA) FISH, target specific probes create a landing platform for amplifier DNA molecules that in turn capture multiple labeled probes, resulting in enhanced brightness and signal-to-noise ratio [48]. Consistent with the robustness of this method, a very good correlation was obtained at the transcriptome-wide level between mean bDNA spot count per cell and transcript abundance measured with RNA-seq [49].

Thus, single-molecule approaches are providing tools for quantitative and spatially resolved analyses, opening the doors to detailed mechanistic studies of RNA regulatory processes in a cellular context.

2.1.2 Quantifying Transcript Copy Numbers

Because they provide unique means to accurately count copy numbers in individual cells, smFISH methods have been used to derive absolute measure of mRNA synthesis, nuclear export or decay. In mammalian cells and *Drosophila* embryos, for example, precise count of reporter or endogenous transcripts revealed both large cell-to-cell variations in transcript numbers, and poor correlations between nascent transcription and cellular transcript levels, revealing that transcription occurs in burst [50, 51]. In yeast, smFISH-based analyses showed that the stability switch observed for two mRNAs exhibiting mitosis-dependent decay depends on promoter activity rather than *cis*-regulatory sequences [52].

Quantification of absolute copy numbers has also provided opportunities to implement mathematical models for complex gene expression programs, and in particular to understand the establishment and interpretation of morphogen gradients. In *Drosophila* embryo, for example, Bicoid morphogen gradient could be accurately modeled by incorporating smFISH quantitative data about the distribution of individual *bicoid* mRNA molecules [53]. In *C. elegans* vulva induction model, measurements of EGF-induced gene expression at single-mRNA resolution, combined with mathematical modeling, revealed that downstream gene expression is not controlled exclusively by the external gradient, but also by dynamic changes in the sensitivity of induced cells [54].

2.1.3 Spatially Resolving Individual RNA Molecules

Being able to spatially resolve individual RNA molecules made it possible to precisely dissect the molecular processes underlying various aspects of RNA regulation. By positioning probe sets along the *β-actin* transcription unit, for example, Singer and coworkers were able to estimate transcription initiation and termination rates in response to serum activation [44]. To study the coupling between transcription and splicing, Tyagi and coworkers made use of two sets of probes targeting respectively an intronic sequence and the 3′UTR of a reporter mRNA. They showed that RNA binding splicing regulators can induce posttranscriptional splicing of specific introns [55]. Spatially detecting single molecules of different RNA species also provides a unique means to compare regulatory properties and establish correlation that would be masked by bulk assays. Indeed, quantification of nascent transcripts produced by loci located in different chromosomal contexts revealed the existence of chromosome-specific transcriptional regulations [56]. Furthermore, comparison of the transcriptional frequency of individual alleles within the same nucleus showed that the bursts in transcription observed for independent alleles do not correlate in default state [57, 58], but get coordinated in response to signaling pathways [57]. Finally, the development of methods via which transcripts with single nucleotide changes can be discriminated has provided a means to detect somatic mutations in patient samples, and thus to improve molecular disease diagnostics [59, 60]. Padlock probe-based methods, which rely on target-dependent circularization and amplification of probes [59, 61], have for example been used to detect point mutations in a frequently activated oncogene, and to study intratumor heterogeneity [62].

Combining single-molecule labeling with super-resolution imaging techniques is now the ongoing challenge, and promises to provide insights into the precise molecular and cellular interactions of RNA molecules with their environment [63].

2.1.4 Toward Systems Level Analyses

High throughput has classically been a limitation of image-based methodologies. However, recent progress in automatic image capture and processing, as well as combinatorial labeling of RNA molecules, has provided means to work at the transcriptomic scale in individual cells. By analyzing both the copy number and the subcellular distribution of about a thousand mRNAs in independent cultured HeLa cells, for example, Pelkmans and coworkers were able to cluster transcripts into functionally related groups using extracted features [49]. Strikingly, such clustering analyses revealed that spatial patterns of individual mRNAs were more powerful at identifying functionally relevant signatures than were mean spot counts. In their study, however, different mRNAs were imaged in different cells, preventing an analysis of covariations in gene expression levels. To overcome this limit, multiplexing FISH techniques enabling the simultaneous detection of hundreds to thousands of transcripts have recently been developed [64–66]. These methods have led to the discovery of gene clusters with substantially correlated expression patterns, as well as to functional predictions for unannotated genes belonging to these clusters. Ultimately, the objective of spatially resolved transcriptomic methods is to obtain exact cartographies of the entire transcriptomes of individual cells. As a first step toward this goal, in situ sequencing methods have been implemented over the last past 5 years [2, 67, 68]. In the FISSEQ next-generation fluorescent in situ RNA sequencing approach, for example, 3D in situ RNA-seq libraries cross-linked to the cellular protein matrix are created and sequenced using SOLiD partition sequencing [68]. Notably, such approaches not only provide quantitative and spatial information about RNA abundance and localization, but can also be used to monitor the behavior of alternatively spliced variants [68], or to visualize intratumoral heterogeneity in patient samples [62, 69]. Application of multiplexing and in situ sequencing methods to complex tissues and organisms is now the next step [46, 47, 66, 68, 70, 71], and should provide information on network-level regulatory processes at play during cell differentiation and disease progression [41].

2.2 Revealing Cell-to-Cell Heterogeneity in RNA Content

By enabling highly accurate measurements of individual RNA copy numbers, smFISH methods have revealed a previously underestimated cell-to-cell variability, with differences in transcript levels reaching up to 50% between genetically identical cells [51, 72–74]. While cell-to-cell variability may be a strategy used by unicellular organisms to improve the chances that a clonal population adapts to variable conditions, it seems not optimal for carrying out the precise programs underlying the early development or the complex tissue homeostasis characteristic of multicellular organisms [74]. Thus, this observation raises questions about how organisms cope with such a variability, but also about the origin of the observed fluctuations. Gene expression variability has been proposed to arise from both intrinsic sources (such as the inherent randomness of biochemical reactions) and extrinsic sources (such as variations in cell fitness or local environment). To determine whether cell-to-cell variability is stochastic, or rather determined by contextual parameters that may influence mRNA homeostasis, Pelkmans and coworkers compiled for millions of isolated mammalian cells both transcript count and a multivariate set of 183 features that quantify cellular phenotypic state as well as population context [72]. Strikingly, they uncovered that relating contextual features to regulatory state predicts the vast majority of the measurable variance, and thus that heterogeneities in cell morphometry or microenvironment are the dominant source of cell-to-cell variability in this system.

How is such a predictability compatible with the transcriptional noise observed in a wide range of organisms, and caused by stochastic bursts of transcription followed by periods of promoter quiescence [51, 56–58, 73, 75, 76]? Increasing evidence suggests that buffering mechanisms exist to reduce noise [74]. Nuclear retention of mRNAs, for example, has been shown to efficiently dampen fluctuations in transcriptional activity [72, 77], indicating that cellular compartmentalization provides a global means to confine transcriptional noise to the nucleus, without affecting steady-state levels. As proposed in the context of homeostatic liver tissue [73] or developing organisms [58], spatiotemporal averaging can also overcome molecular noise and reconcile highly pulsatile transcription with precise cytoplasmic accumulation. Indeed, after averaging over active loci and over long timescale (such as few hours of development) the contribution of intrinsic noise strongly decreases, and gene expression regulation becomes limited by extrinsic factors. In this context, constructing gene regulatory networks that minimize such an extrinsic variability is key, and appears to be a strategy adopted by both unicellular [78] and multicellular organisms [58].

2.3 The Prevalence of RNA Subcellular Localization

High-content, microscopy-based, smFISH experiments performed in cultured mammalian cells have provided transcriptome-level spatial information about the subcellular distribution of transcripts [49, 64]. These studies revealed that transcripts exhibit striking localization patterns, ranging from perinuclear or peripheral accumulations to more polarized accumulations. Complementary FISH analyses performed in differentiated cells, at the tissue-level, have further shown that virtually all the transcripts examined exhibited subcellular localization in some cell type, at some stage of *Drosophila* development [79–81]. Indeed, 661 of the 726 expressed transcripts (91%) analyzed in third instar larval tissues were localized in at least one cell type, the most common localization pattern being clustering within cytoplasmic foci [81]. Interestingly, subcellular RNA localization appears to be the norm rather than the exception for both coding and noncoding RNAs, as the vast majority of analyzed long ncRNAs were subcellularly localized during embryogenesis. Furthermore, comparison of subcellular localization across entire developmental programs, or between cell types, revealed that the capacity of RNAs to localize appears to depend both on developmental stage and cell type [79, 81]. By comparing the gene architecture of transcripts exhibiting subcellular localization versus homogenous distribution in the *Drosophila* ovary, Jambor and coworkers additionally found that subcellularly localized RNAs derive from genes with statistically longer and conserved noncoding regions, consistent with the importance of *cis*-regulatory sequences in controlling RNA fate [79]. The discovery of the prevalence of RNA localization raises the question of its global functional importance. While various examples have shown that the targeting of mRNAs to specific subcellular destinations provides a reservoir for local translation and onsite accumulation of the corresponding proteins [82, 83], more recent work combining global transcriptomics and proteomics analyses in invasive cells has revealed little correlation between the relative accumulation of mRNAs and proteins in cell protrusions [84]. This may reflect the need for translational activation of localized mRNAs in response to external signals, as shown extensively in neuronal cells. Alternatively, these results raise the intriguing possibility that mRNA targeting, by keeping transcripts away from their site of translation in the cell body, may also be used as a means to globally suppress translation. A systematic assessment of the accumulation pattern and the expression levels of proteins produced from localized mRNAs under various conditions should help getting a more comprehensive view on this regulatory process.

3 Live Imaging Approaches for Dynamic Analyses of RNAs

Having access to the temporal dimension is essential to precisely study posttranscriptional regulatory mechanisms. Besides classical injection of exogenous fluorescently labeled RNAs, many methodologies have been recently developed to visualize RNA dynamics in living cells or organisms, ranging from hybridization with fluorogenic probes to RNA tagging systems [42, 43]. These tools, when combined with the latest microscopy systems, allow live imaging of single molecules and precise dissection of all RNA regulatory steps, from transcription to translation. By providing unprecedented spatiotemporal resolution, they are also particularly useful to unravel the in vivo mechanisms involved in subcellular RNA targeting.

3.1 RNA Detection in Living Samples

3.1.1 Detecting Endogenous RNAs with Live FISH Methods

Live FISH methods, in which injected or transfected labeled antisense probes hybridize to target RNAs, have been implemented to monitor endogenous RNAs in real-time, reaching a close to single-molecule resolution. As working on living samples is incompatible with hybridization under denaturing conditions, or with washes removing unbound probes, several strategies have been developed to increase probe brightness and reduce background signals. Signal amplification is a first strategy adopted to produce the bright and photostable fluorescence required for live imaging. HCR-mediated signal amplification, for instance, was used to image low abundance RNAs such as miRNAs in living mammalian cells [85]. Alternatively, multiply labeled tetravalent MTRIP probes were developed, and used in particular to quantify viral RNA production and characterize individual viral particles in real-time [86, 87]. Designing probes that only fluoresce upon association with target RNAs is another strategy adopted to minimize background signals due to unbound probes. Molecular Beacons, for example, consist in oligonucleotides flanked by both a fluorophore and a quencher; they are designed such that fluorescence is quenched in the unbound state and unquenched upon binding to target RNA [88]. Latest generation beacons, optimized to overcome the instability and nuclear retention problems associated with the original molecules, have been successfully used in living cells. In primary cortical neurons, they enabled the dynamic study of axonal mRNA transport, and of the role of the RNA binding Protein TDP43 in this process [89, 90]. Two alternative methods, both using DNA intercalating dyes of the thiazole orange family to produce probes whose fluorescence dramatically increase upon binding, have been developed to study the spatiotemporal dynamics of RNAs in living cells or tissues. DNA FIT probes were used to track *oskar* mRNA molecules transported to the posterior pole of living *Drosophila* oocyte [91, 92], while ECHO-Fish probes were successfully used to dynamically monitor single RNA intranuclear foci in vertebrate cells [93].

3.1.2 Aptamer-Based Imaging of RNAs

In aptamer-based tagging approaches, RNA motifs that bind cognate molecules with high affinity are used to tag RNAs of interest. Spinach aptamers, for example, bind to and activate the fluorescence of DFHBI, a membrane permeable fluorogen compound analogous to GFP [94]. With the optimization of Spinach into brighter and more stable variants, and the further development of novel light-up aptamers such as RNA Mango, it is now possible to follow the dynamics of abundant RNAs in living organisms ranging from bacteria or yeast to human cells [95–100]. In a second group of approaches, RNAs of interest are tagged with stem-loop structures selectively recognized by coexpressed fluorescently tagged phage coat proteins. First developed by Singer and coworkers to study the transport of *Ash1* mRNA in living yeast [101], the MS2 stem loops/MCP-GFP binary system has since then been extensively applied to various cell types and whole organisms such as *Drosophila*, *Xenopus*, zebrafish, and mouse [102–105]. Interestingly, orthogonal phage coat protein–RNA tethering systems such as PCP/PP7 [106] or λN/BoxB [107] have been implemented, enabling both differential intramolecular labeling and simultaneous imaging of several RNA species. Of note, however, adding relatively long RNA tags may affect the regulation of RNAs under analysis. Furthermore, most of the studies performed to date rely on reporter RNAs expressed from engineered constructs. A notable exception has been provided by the Singer group, which generated a transgenic mice expressing MS2-tagged *β-actin* mRNA from the endogenous locus to dynamically analyze *β-actin* subcellular localization [105]. With the development of CRISPR techniques, endogenous tagging of RNAs should become standard in the forthcoming years.

3.1.3 Engineering Fluorescent Proteins for Recognition of Endogenous RNAs

To overcome the limits of monitoring genetically modified RNAs, fluorescent RNA binding proteins (RBPs) designed to detect endogenous RNAs have been generated. For example, fusion proteins between a fluorescent molecule and two Pum-HD RNA-binding domains engineered to each recognize specific eight base sequences present in target RNAs were designed to reveal mRNA dynamics within living mammalian cells [108, 109]. Another interesting approach is the RNA targeting cas9 (Rcas9) method that has recently emerged as a new method for tracking endogenous RNAs within living cells [110]. Here, the PAM sequence is provided by a separate oligonucleotide (PAMmer) that hybridizes on the target RNA, generating a landing platform for fluorescent nuclease-inactive Cas9 proteins. Remarkably, RCas9 enabled the tracking of *β-actin* mRNA trafficking to stress granules in living human cells without altering RNA or encoded protein levels [110]. Efforts to implement this method in vivo, in whole organisms, are underway.

3.2 Kinetic Dissection of RNA Regulatory Processes: From Transcription to Translation

The concomitant improvement of RNA tagging methods and imaging system sensitivity has led to a considerable increase in signal-to-noise ratio which, when combined with optimized single-particle tracking algorithms and mathematical modeling, enables the kinetic dissection of single RNA regulatory steps.

By tagging reporter transcripts with PP7 at the 5′ or 3′ ends, Singer and coworkers were for example able to differentially analyze transcription initiation, elongation and termination steps [106]. This revealed that gene firing rate is directly determined by the search times of rate-limiting trans-activating factors. Dynamic monitoring of transcription has also been performed in the context of entire *Drosophila* embryos [111, 112], revealing that the Bicoid transcription factor is not required for transcription initiation, but rather for persistence of transcriptional activity [112]. Interestingly, combining orthogonal tagging with dual color imaging allowed to simultaneously follow the transcription of independent RNAs, such as sense and antisense transcripts produced from the same locus [113] or allelic copies of the same gene [76], but also to perform dual labeling of a given transcript and follow its maturation. By differentially tagging intronic and exonic sequences of the same reporter pre-mRNA using PP7 (or λN) and MS2 stem loops, different groups were able to measure splicing kinetics of *β-globin* reporter genes. Carmo-Fonseca and coworkers, for example, demonstrated that splicing rate depends on splice site strength, but also on intron length, such that it is limited by the rate of transcription by RNA pol II [114]. Furthermore, Larson and coworkers showed that *β-globin* terminal intron splicing occurs stochastically before and after transcript release, thus indicating there is no checkpoint controlling the sequence of events [115].

Dual color imaging has also been key in dynamically studying nucleocytoplasmic export, a poorly studied yet active and selective step of RNA trafficking. By coimaging with high spatiotemporal resolution nuclear pore components and reporter mRNAs, the Singer and Shav-Tal groups were able to resolve individual transient steps of the nuclear export process. They show that the rate-limiting step for mRNA export is in fact not the transition through the nuclear pore itself, but rather access to nuclear pores, a process relying on nucleoplasmic diffusion [116, 117].

Until recently, live imaging of translation was prevented by the limited signal produced by single nascent proteins, and the background produced by already translated polypeptides. These limits have recently been overcome by the development of novel single-molecule imaging approaches, in which the 3′UTR of reporter mRNAs is tagged with PP7 or MS2 stem loops, and the 5′ ends of their ORFs with arrays of short peptide epitopes (SunTag or Flag/HA epitopes) recognized with high affinity by genetically-encoded fluorescent antibodies [118–121]. With these approaches, single translated mRNAs are visualized as bright colabeled punctae, and can be imaged over hours, providing precise measurements of the rates of translation initiation and translocation, or ribosome numbers [120, 121]. Interestingly, the Singer and Tanenbaum groups were able to show using the SunTag approach that translation, like transcription, occurs in burst, with "on" behaviors interposed by long periods of no translation [119, 120].

3.3 Unraveling Spatiotemporal Control of RNA Localization and Translation

3.3.1 Transporting RNAs to Specific Destinations

While asymmetric localization of endogenous transcripts had been observed since the early 80s [83], first line of evidence for cytoplasmic mRNA transport came from pioneer experiments, where exogenous fluorescently tagged RNAs were injected in *Drosophila* embryos and *Xenopus* oocytes [122, 123]. As revealed by live imaging of injected RNAs, and subsequently of in vivo-produced MS2-tagged transcripts, mRNAs undergo complex motions that are characterized either by directed motion or by passive diffusion [82, 83]. Diffusion of localizing mRNAs has been observed in primary fibroblasts, where the accumulation of endogenous *MS2-β-actin* mRNA at the leading edge appears to be mediated mainly by diffusion and trapping [124]. Directed transport of mRNAs relies on different mechanisms: it is mostly characterized by fast biased bidirectional motion along cytoskeletal elements, and directly implies molecular motors such as kinesins, dyneins, and myosins. Strikingly, live imaging analyses have shown that large net mRNA displacement at the population level does not necessarily involve strong biases at the single-molecule level. Indeed, tracking of individual MS2-tagged *oskar* mRNAs in *Drosophila* oocytes revealed a relatively small excess of kinesin-dependent mRNA movements toward the posterior pole [125].

A complex regulation of molecular motors has been described in different studies [104, 125–128], and is responsible for directed targeting of mRNAs to their precise final destination. By following in vivo the transport of MS2-tagged *Vg1* RNAs localizing to the vegetal pole of *Xenopus* oocytes, Mowry and coworkers observed distinct transport kinetics and directionality in different regions of the oocyte. While dynein is responsible for the unidirectional RNA transport characteristics of the upper vegetal cytoplasm, kinesin-1 is required for the bidirectional transport observed in the lower vegetal cytoplasm [104]. A tight temporal coordination in motor activities is also very important for the coupling between transport and anchoring at destination. As revealed by quantitative imaging in *Drosophila* oocytes, for example, a strong interplay between kinesin and dynein, and between the actin and microtubule cytoskeletons, is required for posterior accumulation of mRNA-containing germ granules [128].

How are these molecular motors recruited to actively transported mRNAs? As shown by Bullock and coworkers, RNA binding proteins (RBPs) associating with localizing elements play a key role in this process. Indeed, a specific structure found in mRNAs localizing apically in *Drosophila* embryos was shown to trigger RBP-mediated recruitment of dynein and directed transport [126]. Interestingly, the recruitment of molecular motors by RBPs can be induced by external stimuli. Following MS2-tagged *camKIIα* mRNA in cultured neurons, Bassell and coworkers were able to show an increase in kinesin-dependent dendritic targeting of *camKIIα* RNA and its associated RBP FMRP upon mGluR activation, and a concomitant increase in the association between FMRP and Kif5 Kinesin [127].

3.3.2 Visualizing Translation in Space and Time

Until recently, detection of proteins synthesized locally, in specific subcellular compartment was challenging. With the advent of novel tagging strategies, it is now possible to map mRNA translation with a high spatiotemporal resolution in living cells or organisms. In the TRICK method, for example, PP7 and MS2 tags recognized by distinct fluorescent peptides are inserted respectively in the coding region and 3′UTR of reporter RNAs, such that dually labeled RNAs lose their PCP signal upon ribosomal elongation [129]. By enabling the discrimination of translated from untranslated mRNAs, and the monitoring of the first round of translation, the TRICK method has been particularly useful in proving that *oskar* mRNA is not translated until it reaches the posterior pole of *Drosophila* oocyte. The use of the alternative SunTag approach to image translation of single mRNA molecules revealed for the first time cell-compartment specific heterogeneities of translation [118]. Live imaging of local translation in dendrites of primary hippocampal neurons, for example, provided evidence for a variability in mRNA translation rates, with translation rate higher in proximal than in the distal region of dendrites [118, 130]. Interestingly, and opposite to the previous assumption that mRNAs are transported in a silent translational state, those studies also demonstrated that active transport of mRNAs can occur after mRNAs have already initiated translation.

Together, RNA tagging has provided new insights into the kinetics and mechanisms of sequential RNA regulatory steps in living cells or organisms. Although most of the studies performed so far have used exogenously introduced reporter RNAs, implementation of the CRISPR/Cas9 strategy now enables to efficiently tag endogenous RNAs and work in a more physiological context. A current challenge is now to develop multicolor imaging and multiplexing methods to simultaneously image various RNAs, in the context of tissues or organisms.

4 Characterization of Ribonucleoprotein Complexes

Regulation of RNA production, maturation, transport, and expression involves the recruitment of RNA binding proteins (RBPs), and the formation of ribonucleoprotein (RNP) complexes of defined composition and structure [131]. Thus, uncovering the full landscape of RNA–protein interactions is of capital importance to understand RNA regulatory processes. Complementary protein-centric and RNA-centric methods have been developed to purify RNP complexes and identify their RNA and protein content [132]. These approaches have provided unprecedented insight into the molecular bases of RNA–protein interactions, but have also revealed the importance of RBP-ncRNA interactions, as well as the extent and complexity of the mRNA interactome. In vivo, RNAs and proteins are frequently packaged into dynamic high-order assemblies that contain multiple RNA and protein molecules. Recent studies exploring the physical and molecular bases of these assemblies have revealed that they exhibit characteristics of liquid droplets [133].

4.1 Identifying the Composition of RNP Complexes

4.1.1 Identifying RNAs Bound to RBPs

Protein-centric methods largely rely on immunoprecipitation of RBPs followed by large scale sequencing to identify their associated RNAs. While native populations of coprecipitated RNAs are identified with RIP-seq, RNA fragments cross-linked to the RBP of interest are sequenced and analyzed in CLIP methods, thereby providing precise information on the binding sites of RBPs to target RNAs [134–137]. Notably, recent implementation has been made to isolate the intramolecular and intermolecular RNA duplexes bound by given RBPs, which revealed in the case of the Staufen protein the high prevalence of long-range intramolecular RNA duplexes in the 3′UTRs of target RNAs [138]. Although RIP and CLIP approaches have provided invaluable information about posttranscriptional regulatory networks, these methods require a large quantity of material and are not well-adapted to map RNA–protein interactions in vivo, in specific cell types. To circumvent these issues, complementary approaches have been recently developed in which fusions between a given RBP and the catalytic domain of an RNA modifying enzyme are expressed in specific tissues. Transcriptomes are then sequenced to identify the transcripts specifically modified (and therefore bound) by the chimeric proteins. In the TRIBE method, for example, the catalytic domain of the RNA-editing enzyme ADAR was fused to three RBPs (HRP48, FRM1, and NonA), allowing for the identification of RNA targets from a subset of 150 fly neurons [139]. In the RNA tagging method, the *C. elegans* poly(U) polymerase PUP-2 was used to covalently mark the RNA targets of the yeast Puf3 protein [140].

4.1.2 Identifying RBPs Bound to RNAs

RNA-centric approaches are based on affinity capture of selected RNAs and subsequent identification of associated molecules [132], and have been particularly helpful to uncover the regulatory partners of noncoding RNAs. In in vitro approaches, synthetic RNA baits tagged with aptamers are used to capture proteins from cellular extracts. S1m aptamers, for example, were combined with AU rich elements (ARE) to identify ARE-binding proteins and potential regulators of mRNA degradation [141]. In in vivo approaches, native RNA–protein complexes assembled in cellular contexts are purified. This can be achieved by expression of aptamer-tagged RNA variants in cells or tissues followed by RNA-based affinity chromatography, as first optimized in bacteria for the purification of complexes containing MS2-tagged small regulatory RNAs [142]. Alternatively, RNP complexes can be purified by stringent purification methods, in which biotinylated antisense probes are used to capture endogenous RNAs. Coupled to mass spectrometry, such purifications were for example used to identify proteins interacting with the long noncoding RNA *Xist*, providing new insight on the role of this RNA in chromatin-dependent gene silencing [143–145].

4.1.3 An Increasing Interactome

As described, most methods implemented to characterize RNA–protein interactions provide information about the interactome of one RNA (or one RBP) at a time. In order to have a more comprehensive view of posttranscriptional gene regulatory networks, two groups have developed RNA interactome capture methods to systematically identify the proteome bound to poly(A) transcripts, and to globally map the sites of protein–RNA interactions [146–148]. Strikingly, mRNA interactome studies uncovered hundreds of proteins that were previously unknown to bind RNA and did not contain recognizable RNA interaction domains. Cross-linked RNA binders belong to a broad spectrum of protein families including kinases, metabolic enzymes, or isomerases implicated in spliceosome and ribonucleoprotein dynamics, raising the intriguing possibility that RNAs might control many cellular processes by directly tuning protein activity. Furthermore, mapping studies have shown on one hand the widespread binding of proteins to 3′UTR regions, and on the other hand, the prevalence of RNA binding to intrinsically disordered regions [148]. Initially developed in cultured cells, oligo(dT)-based capture of RNA interactomes has been implemented in living organisms such as yeasts, flies or plants [149–152]. Ephrussi and coworkers, in particular, compared the repertoire of RBPs bound to poly(A) RNAs prepared from early and late embryos, thus revealing that the RNA interactome exhibits an important plasticity during development [151].

4.2 Assembly of RNAs and Associated Proteins into Higher-Order, Dynamic Granules

In cells, various RNP assemblies control RNA biogenesis, transport, or expression, and are visualized as particles or granules found in both nuclear and cytoplasmic compartments [131, 153]. As illustrated in neuronal cells, where endogenous RNP granules characterized by specific markers were purified using multistep biochemical purification, granules are heterogeneous in term of RNA and protein contents [154]. To date, the precise stoichiometry of RNA granules is still unclear, but smFISH methods have revealed that the number of RNA molecules contained in individual RNPs is not uniform, and appears to vary in function of both granule-type and cellular context [155–160]. In the *Drosophila* germ line, for example, *nanos* is transported as single copies to the posterior pole of the oocyte, while *oskar* mRNA assembles into multiple copies prior to transport [158]. Furthermore, *nanos* granules are remodeled when reaching the posterior pole, such that *nanos* mRNA molecules assemble into homotypic clusters that recruit the RNA binding protein Vasa, generating germ cell granules [158, 159]. In mammals, quantitative analysis of the distribution of endogenous *MS2-β-actin* mRNA revealed that single copies of *β-actin* mRNA were present in RNP granules at the leading edge of primary fibroblasts [124], whereas about 25% of RNPs contained more than one *β-actin* mRNA molecule in primary cultures of neurons [157]. Interestingly, this number decreases with distance from the soma, and is modulated by neuronal activity. Furthermore, neuronal stimulation was shown to trigger a transient increase in mRNA granule accessibility, likely reflecting complex disassembly and engagement in local translation [157].

Dynamic remodeling and turnover of RNA granules is not restricted to germ cells or neurons, but is observed in various cell types in response to developmental signals or environmental stresses, raising the question of how these large complexes are dynamically assembled and regulated. As revealed by recent work, RNA granules may form through phase separation, generating reversible assemblies with semiliquid behavior [133, 161, 162]. RBPs, including translational repressors and RNA helicases, play a critical role in this process by promoting the establishment of multivalent interactions. Highlighting the need for dynamic interactions within RNP assemblies, mutations in the disordered regions of various RBPs have been shown to alter phase separation by generating toxic, aggregation-prone proteins inducing the abnormal compaction of RNPs into pathological inclusions [133, 163].

Remarkably, RNA granules not only set the basis for efficient and flexible compartmentalization of the cell cytoplasm, but are themselves organized into subdomains that result from the differential clustering of RNAs and proteins [159, 164–167]. In *Drosophila* oocyte, for example, in situ hybridization combined with electron microscopy revealed that *gurken* and *bicoid* mRNAs occupy distinct positions within P bodies: while *gurken* mRNA was found enriched at the periphery, *bicoid* mRNA was found in the central domain [165]. Interestingly, such a differential distribution reflects mRNA translational state, as *gurken* mRNA associates with its translational activator Orb at the edge of P bodies, where it is translated. Furthermore, centrally localized and repressed *bicoid* mRNA is released from P-bodies upon egg activation to become actively translated.

Together, our understanding of the molecular bases of RNA–protein recognition and assembly into RNP complexes has dramatically improved over the last past years. Efforts have however to be done to study RNA–protein interactions with a high spatiotemporal resolution, in living cells or organisms [168]. As a first step toward this goal, Singer and coworkers have combined endogenous single RNA and protein detection with two-photon fluorescence fluctuation analysis to directly measure the association of the translational repressor ZBP1 with *β-actin* mRNA in living fibroblasts. This revealed a stronger association between ZBP1 and *β-actin* mRNA at the nuclear periphery than at the leading edge, consistent with the localized translation of *β-actin* at the front of migrating cells [130].

5 Perspectives

With the advent of transcriptomic methods and the concomitant implementation of functional single-molecule imaging, our view on the "central dogma of molecular biology" has changed dramatically. Although it is now clear that RNA regulation is much more complex that initially anticipated, and that RNA has a large range of functions, methodological challenges are still ahead to continue improving spatiotemporal detection of RNA molecules. Optimization of spatially resolved fourth-generation sequencing technologies, for example, is needed to improve sensitivity and data interpretation [169]. Furthermore, improvements have to be made to visualize RNA molecules and regulatory partners in their

3D environment with high sensitivity and temporal resolution. In this respect, recently developed super-resolution microscopy techniques reaching unprecedented resolution hold great promises, as they enable highly accurate codetection of transcripts and associated molecules or cellular structures [170, 171]. A major future challenge will be to bridge the gap between visualization and functional study of RNAs in living samples. Latest developments in genome engineering techniques [172], together with the implementation of tools to remotely control RNA activity [173], make it now possible to manipulate RNAs. Applying such methods at the systems-level should help comprehensively explore RNA functions, and in particular elucidate the role of newly discovered noncoding species or the impact of RNA binding to protein activity. Importantly, this will also provide an integrative view on posttranscriptional strategies that have been adopted along evolution.

Acknowledgments

Work in F.B.'s lab is supported by the ARC (grant #PJA 20141201623), the ANR (through the RNAGRIMP research grant and the 'Investments for the Future' LABEX SIGNALIFE program # ANR-11-LABX-0028-01), and the JPND (FlySMALS grant). The authors apologize to colleagues whose relevant studies could not be cited because of space limitations.

References

1. Hangauer MJ, Carpenter S, McManus MT (2014) Discovering the complexity of the metazoan transcriptome. Genome Biol 15 (4):112. doi:10.1186/gb4172
2. Crosetto N, Bienko M, van Oudenaarden A (2015) Spatially resolved transcriptomics and beyond. Nat Rev Genet 16(1):57–66. doi:10.1038/nrg3832
3. Elkon R, Ugalde AP, Agami R (2013) Alternative cleavage and polyadenylation: extent, regulation and function. Nat Rev Genet 14 (7):496–506. doi:10.1038/nrg3482
4. Tian B, Manley JL (2013) Alternative cleavage and polyadenylation: the long and short of it. Trends Biochem Sci 38(6):312–320. doi:10.1016/j.tibs.2013.03.005
5. Sandberg R, Neilson JR, Sarma A, Sharp PA, Burge CB (2008) Proliferating cells express mRNAs with shortened 3′ untranslated regions and fewer microRNA target sites. Science 320(5883):1643–1647. doi:10.1126/science.1155390
6. Lianoglou S, Garg V, Yang JL, Leslie CS, Mayr C (2013) Ubiquitously transcribed genes use alternative polyadenylation to achieve tissue-specific expression. Genes Dev 27(21):2380–2396. doi:10.1101/gad.229328.113
7. Flavell SW, Kim TK, Gray JM, Harmin DA, Hemberg M, Hong EJ, Markenscoff-Papadimitriou E, Bear DM, Greenberg ME (2008) Genome-wide analysis of MEF2 transcriptional program reveals synaptic target genes and neuronal activity-dependent polyadenylation site selection. Neuron 60 (6):1022–1038. doi:10.1016/j.neuron.2008.11.029
8. Gilbert WV, Bell TA, Schaening C (2016) Messenger RNA modifications: form, distribution, and function. Science 352 (6292):1408–1412. doi:10.1126/science.aad8711
9. Roundtree IA, He C (2016) RNA epigenetics—chemical messages for

posttranscriptional gene regulation. Curr Opin Chem Biol 30:46–51. doi:10.1016/j.cbpa.2015.10.024

10. Dominissini D, Nachtergaele S, Moshitch-Moshkovitz S, Peer E, Kol N, Ben-Haim MS, Dai Q, Di Segni A, Salmon-Divon M, Clark WC, Zheng G, Pan T, Solomon O, Eyal E, Hershkovitz V, Han D, Dore LC, Amariglio N, Rechavi G, He C (2016) The dynamic N(1)-methyladenosine methylome in eukaryotic messenger RNA. Nature 530 (7591):441–446. doi:10.1038/nature16998
11. Linder B, Grozhik AV, Olarerin-George AO, Meydan C, Mason CE, Jaffrey SR (2015) Single-nucleotide-resolution mapping of m6A and m6Am throughout the transcriptome. Nat Methods 12(8):767–772. doi:10.1038/nmeth.3453
12. Spitale RC, Flynn RA, Zhang QC, Crisalli P, Lee B, Jung JW, Kuchelmeister HY, Batista PJ, Torre EA, Kool ET, Chang HY (2015) Structural imprints in vivo decode RNA regulatory mechanisms. Nature 519 (7544):486–490. doi:10.1038/nature14263
13. Zubradt M, Gupta P, Persad S, Lambowitz AM, Weissman JS, Rouskin S (2017) DMS-MaPseq for genome-wide or targeted RNA structure probing in vivo. Nat Methods 14 (1):75–82. doi:10.1038/nmeth.4057
14. Aw JG, Shen Y, Wilm A, Sun M, Lim XN, Boon KL, Tapsin S, Chan YS, Tan CP, Sim AY, Zhang T, Susanto TT, Fu Z, Nagarajan N, Wan Y (2016) In vivo mapping of eukaryotic RNA Interactomes Reveals Principles of higher-order organization and regulation. Mol Cell 62(4):603–617. doi:10.1016/j.molcel.2016.04.028
15. Lu Z, Zhang QC, Lee B, Flynn RA, Smith MA, Robinson JT, Davidovich C, Gooding AR, Goodrich KJ, Mattick JS, Mesirov JP, Cech TR, Chang HY (2016) RNA duplex map in living cells reveals higher-order transcriptome structure. Cell 165(5):1267–1279. doi:10.1016/j.cell.2016.04.028
16. Sharma E, Sterne-Weiler T, O'Hanlon D, Blencowe BJ (2016) Global mapping of human RNA-RNA interactions. Mol Cell 62 (4):618–626. doi:10.1016/j.molcel.2016.04.030
17. Vogel C, Marcotte EM (2012) Insights into the regulation of protein abundance from proteomic and transcriptomic analyses. Nat Rev Genet 13(4):227–232. doi:10.1038/nrg3185
18. Ingolia NT, Ghaemmaghami S, Newman JR, Weissman JS (2009) Genome-wide analysis in vivo of translation with nucleotide resolution using ribosome profiling. Science 324(5924):218–223. doi:10.1126/science.1168978
19. Brar GA, Weissman JS (2015) Ribosome profiling reveals the what, when, where and how of protein synthesis. Nat Rev Mol Cell Biol 16(11):651–664. doi:10.1038/nrm4069
20. Raj A, Wang SH, Shim H, Harpak A, Li YI, Engelmann B, Stephens M, Gilad Y, Pritchard JK (2016) Thousands of novel translated open reading frames in humans inferred by ribosome footprint profiling. Elife 5:e13328. doi:10.7554/eLife.13328
21. Wethmar K (2014) The regulatory potential of upstream open reading frames in eukaryotic gene expression. Wiley Interdiscip Rev RNA 5 (6):765–778. doi:10.1002/wrna.1245
22. Jan CH, Williams CC, Weissman JS (2014) Principles of ER cotranslational translocation revealed by proximity-specific ribosome profiling. Science 346(6210):1257521. doi:10.1126/science.1257521
23. Williams CC, Jan CH, Weissman JS (2014) Targeting and plasticity of mitochondrial proteins revealed by proximity-specific ribosome profiling. Science 346(6210):748–751. doi:10.1126/science.1257522
24. Heiman M, Schaefer A, Gong S, Peterson JD, Day M, Ramsey KE, Suarez-Farinas M, Schwarz C, Stephan DA, Surmeier DJ, Greengard P, Heintz N (2008) A translational profiling approach for the molecular characterization of CNS cell types. Cell 135(4):738–748. doi:10.1016/j.cell.2008.10.028
25. Shigeoka T, Jung H, Jung J, Turner-Bridger B, Ohk J, Lin JQ, Amieux PS, Holt CE (2016) Dynamic axonal translation in developing and mature visual circuits. Cell 166 (1):181–192. doi:10.1016/j.cell.2016.05.029
26. Liu S, Trapnell C (2016) Single-cell transcriptome sequencing: recent advances and remaining challenges. F1000Res 5. doi:10.12688/f1000research.7223.1
27. Grun D, Lyubimova A, Kester L, Wiebrands K, Basak O, Sasaki N, Clevers H, van Oudenaarden A (2015) Single-cell messenger RNA sequencing reveals rare intestinal cell types. Nature 525(7568):251–255. doi:10.1038/nature14966
28. Macosko EZ, Basu A, Satija R, Nemesh J, Shekhar K, Goldman M, Tirosh I, Bialas AR, Kamitaki N, Martersteck EM, Trombetta JJ, Weitz DA, Sanes JR, Shalek AK, Regev A, McCarroll SA (2015) Highly parallel genome-wide expression profiling of individual cells using nanoliter droplets. Cell

161(5):1202–1214. doi:10.1016/j.cell. 2015.05.002

29. Jaitin DA, Kenigsberg E, Keren-Shaul H, Elefant N, Paul F, Zaretsky I, Mildner A, Cohen N, Jung S, Tanay A, Amit I (2014) Massively parallel single-cell RNA-seq for marker-free decomposition of tissues into cell types. Science 343(6172):776–779. doi:10.1126/science.1247651
30. Zeisel A, Munoz-Manchado AB, Codeluppi S, Lonnerberg P, La Manno G, Jureus A, Marques S, Munguba H, He L, Betsholtz C, Rolny C, Castelo-Branco G, Hjerling-Leffler J, Linnarsson S (2015) Brain structure. Cell types in the mouse cortex and hippocampus revealed by single-cell RNA-seq. Science 347 (6226):1138–1142. doi:10.1126/science.aaa1934
31. Usoskin D, Furlan A, Islam S, Abdo H, Lonnerberg P, Lou D, Hjerling-Leffler J, Haeggstrom J, Kharchenko O, Kharchenko PV, Linnarsson S, Ernfors P (2015) Unbiased classification of sensory neuron types by large-scale single-cell RNA sequencing. Nat Neurosci 18(1):145–153. doi:10.1038/nn.3881
32. Bendall SC, Davis KL, Amir el AD, Tadmor MD, Simonds EF, Chen TJ, Shenfeld DK, Nolan GP, Pe'er D (2014) Single-cell trajectory detection uncovers progression and regulatory coordination in human B cell development. Cell 157(3):714–725. doi:10.1016/j.cell.2014.04.005
33. Tang F, Barbacioru C, Bao S, Lee C, Nordman E, Wang X, Lao K, Surani MA (2010) Tracing the derivation of embryonic stem cells from the inner cell mass by single-cell RNA-Seq analysis. Cell Stem Cell 6(5):468–478. doi:10.1016/j.stem.2010.03.015
34. Luo Y, Coskun V, Liang A, Yu J, Cheng L, Ge W, Shi Z, Zhang K, Li C, Cui Y, Lin H, Luo D, Wang J, Lin C, Dai Z, Zhu H, Zhang J, Liu J, Liu H, deVellis J, Horvath S, Sun YE, Li S (2015) Single-cell transcriptome analyses reveal signals to activate dormant neural stem cells. Cell 161(5):1175–1186. doi:10.1016/j.cell.2015.04.001
35. Achim K, Pettit JB, Saraiva LR, Gavriouchkina D, Larsson T, Arendt D, Marioni JC (2015) High-throughput spatial mapping of single-cell RNA-seq data to tissue of origin. Nat Biotechnol 33(5):503–509. doi:10.1038/nbt.3209
36. Satija R, Farrell JA, Gennert D, Schier AF, Regev A (2015) Spatial reconstruction of single-cell gene expression data. Nat Biotechnol 33(5):495–502. doi:10.1038/nbt.3192
37. Combs PA, Eisen MB (2013) Sequencing mRNA from cryo-sliced *Drosophila* embryos to determine genome-wide spatial patterns of gene expression. PLoS One 8(8):e71820. doi:10.1371/journal.pone.0071820
38. Junker JP, Noel ES, Guryev V, Peterson KA, Shah G, Huisken J, McMahon AP, Berezikov E, Bakkers J, van Oudenaarden A (2014) Genome-wide RNA tomography in the zebrafish embryo. Cell 159(3):662–675. doi:10.1016/j.cell.2014.09.038
39. Stahl PL, Salmen F, Vickovic S, Lundmark A, Navarro JF, Magnusson J, Giacomello S, Asp M, Westholm JO, Huss M, Mollbrink A, Linnarsson S, Codeluppi S, Borg A, Ponten F, Costea PI, Sahlen P, Mulder J, Bergmann O, Lundeberg J, Frisen J (2016) Visualization and analysis of gene expression in tissue sections by spatial transcriptomics. Science 353 (6294):78–82. doi:10.1126/science.aaf2403
40. Patel AP, Tirosh I, Trombetta JJ, Shalek AK, Gillespie SM, Wakimoto H, Cahill DP, Nahed BV, Curry WT, Martuza RL, Louis DN, Rozenblatt-Rosen O, Suva ML, Regev A, Bernstein BE (2014) Single-cell RNA-seq highlights intratumoral heterogeneity in primary glioblastoma. Science 344 (6190):1396–1401. doi:10.1126/science.1254257
41. Wills QF, Mead AJ (2015) Application of single-cell genomics in cancer: promise and challenges. Hum Mol Genet 24(R1): R74–R84. doi:10.1093/hmg/ddv235
42. Gaspar I, Ephrussi A (2015) Strength in numbers: quantitative single-molecule RNA detection assays. Wiley Interdiscip Rev Dev Biol 4(2):135–150. doi:10.1002/wdev.170
43. Pitchiaya S, Heinicke LA, Custer TC, Walter NG (2014) Single molecule fluorescence approaches shed light on intracellular RNAs. Chem Rev 114(6):3224–3265. doi:10.1021/cr400496q
44. Femino AM, Fay FS, Fogarty K, Singer RH (1998) Visualization of single RNA transcripts in situ. Science 280(5363):585–590
45. Raj A, van den Bogaard P, Rifkin SA, van Oudenaarden A, Tyagi S (2008) Imaging individual mRNA molecules using multiple singly labeled probes. Nat Methods 5 (10):877–879. doi:10.1038/nmeth.1253
46. Choi HM, Beck VA, Pierce NA (2014) Multiplexed in situ hybridization using hybridization chain reaction. Zebrafish 11 (5):488–489. doi:10.1089/zeb.2014.1501
47. Shah S, Lubeck E, Schwarzkopf M, He TF, Greenbaum A, Sohn CH, Lignell A, Choi HM, Gradinaru V, Pierce NA, Cai L (2016)

Single-molecule RNA detection at depth by hybridization chain reaction and tissue hydrogel embedding and clearing. Development 143(15):2862–2867. doi:10.1242/dev.138560

48. Player AN, Shen LP, Kenny D, Antao VP, Kolberg JA (2001) Single-copy gene detection using branched DNA (bDNA) in situ hybridization. J Histochem Cytochem 49 (5):603–612
49. Battich N, Stoeger T, Pelkmans L (2013) Image-based transcriptomics in thousands of single human cells at single-molecule resolution. Nat Methods 10(11):1127–1133. doi:10.1038/nmeth.2657
50. Pare A, Lemons D, Kosman D, Beaver W, Freund Y, McGinnis W (2009) Visualization of individual Scr mRNAs during *Drosophila* embryogenesis yields evidence for transcriptional bursting. Curr Biol 19 (23):2037–2042. doi:10.1016/j.cub.2009.10.028
51. Raj A, Peskin CS, Tranchina D, Vargas DY, Tyagi S (2006) Stochastic mRNA synthesis in mammalian cells. PLoS Biol 4(10):e309. doi:10.1371/journal.pbio.0040309
52. Trcek T, Larson DR, Moldon A, Query CC, Singer RH (2011) Single-molecule mRNA decay measurements reveal promoter- regulated mRNA stability in yeast. Cell 147 (7):1484–1497. doi:10.1016/j.cell.2011.11.051
53. Little SC, Tkacik G, Kneeland TB, Wieschaus EF, Gregor T (2011) The formation of the Bicoid morphogen gradient requires protein movement from anteriorly localized mRNA. PLoS Biol 9(3):e1000596. doi:10.1371/journal.pbio.1000596
54. van Zon JS, Kienle S, Huelsz-Prince G, Barkoulas M, van Oudenaarden A (2015) Cells change their sensitivity to an EGF morphogen gradient to control EGF-induced gene expression. Nat Commun 6:7053. doi:10.1038/ncomms8053
55. Vargas DY, Shah K, Batish M, Levandoski M, Sinha S, Marras SA, Schedl P, Tyagi S (2011) Single-molecule imaging of transcriptionally coupled and uncoupled splicing. Cell 147 (5):1054–1065. doi:10.1016/j.cell.2011.10.024
56. Levesque MJ, Raj A (2013) Single-chromosome transcriptional profiling reveals chromosomal gene expression regulation. Nat Methods 10(3):246–248. doi:10.1038/nmeth.2372
57. Kalo A, Kanter I, Shraga A, Sheinberger J, Tzemach H, Kinor N, Singer RH, Lionnet T, Shav-Tal Y (2015) Cellular levels of signaling factors are sensed by beta-actin alleles to modulate transcriptional pulse intensity. Cell Rep 11(3):419–432. doi:10.1016/j.celrep.2015.03.039
58. Little SC, Tikhonov M, Gregor T (2013) Precise developmental gene expression arises from globally stochastic transcriptional activity. Cell 154(4):789–800. doi:10.1016/j.cell.2013.07.025
59. Larsson C, Grundberg I, Soderberg O, Nilsson M (2010) In situ detection and genotyping of individual mRNA molecules. Nat Methods 7(5):395–397. doi:10.1038/nmeth.1448
60. Zhang K, Li JB, Gao Y, Egli D, Xie B, Deng J, Li Z, Lee JH, Aach J, Leproust EM, Eggan K, Church GM (2009) Digital RNA allelotyping reveals tissue-specific and allele-specific gene expression in human. Nat Methods 6(8):613–618. doi:10.1038/nmeth.1357
61. Nilsson M, Barbany G, Antson DO, Gertow K, Landegren U (2000) Enhanced detection and distinction of RNA by enzymatic probe ligation. Nat Biotechnol 18(7):791–793. doi:10.1038/77367
62. Grundberg I, Kiflemariam S, Mignardi M, Imgenberg-Kreuz J, Edlund K, Micke P, Sundstrom M, Sjoblom T, Botling J, Nilsson M (2013) In situ mutation detection and visualization of intratumor heterogeneity for cancer research and diagnostics. Oncotarget 4 (12):2407–2418. doi:10.18632/oncotarget.1527
63. Mito M, Kawaguchi T, Hirose T, Nakagawa S (2016) Simultaneous multicolor detection of RNA and proteins using super-resolution microscopy. Methods 98:158–165. doi:10.1016/j.ymeth.2015.11.007
64. Chen KH, Boettiger AN, Moffitt JR, Wang S, Zhuang X (2015) RNA imaging. Spatially resolved, highly multiplexed RNA profiling in single cells. Science 348(6233):aaa6090. doi:10.1126/science.aaa6090
65. Lubeck E, Cai L (2012) Single-cell systems biology by super-resolution imaging and combinatorial labeling. Nat Methods 9 (7):743–748. doi:10.1038/nmeth.2069
66. Lubeck E, Coskun AF, Zhiyentayev T, Ahmad M, Cai L (2014) Single-cell in situ RNA profiling by sequential hybridization. Nat Methods 11(4):360–361. doi:10.1038/nmeth.2892
67. Ke R, Mignardi M, Pacureanu A, Svedlund J, Botling J, Wahlby C, Nilsson M (2013) In situ sequencing for RNA analysis in preserved

tissue and cells. Nat Methods 10(9):857–860. doi:10.1038/nmeth.2563

68. Lee JH, Daugharthy ER, Scheiman J, Kalhor R, Yang JL, Ferrante TC, Terry R, Jeanty SS, Li C, Amamoto R, Peters DT, Turczyk BM, Marblestone AH, Inverso SA, Bernard A, Mali P, Rios X, Aach J, Church GM (2014) Highly multiplexed subcellular RNA sequencing in situ. Science 343(6177):1360–1363. doi:10.1126/science.1250212
69. Janiszewska M, Liu L, Almendro V, Kuang Y, Paweletz C, Sakr RA, Weigelt B, Hanker AB, Chandarlapaty S, King TA, Reis-Filho JS, Arteaga CL, Park SY, Michor F, Polyak K (2015) In situ single-cell analysis identifies heterogeneity for PIK3CA mutation and HER2 amplification in HER2-positive breast cancer. Nat Genet 47(10):1212–1219. doi:10.1038/ng.3391
70. Choi HM, Chang JY, Trinh le A, Padilla JE, Fraser SE, Pierce NA (2010) Programmable in situ amplification for multiplexed imaging of mRNA expression. Nat Biotechnol 28(11):1208–1212. doi:10.1038/nbt.1692
71. Sylwestrak EL, Rajasethupathy P, Wright MA, Jaffe A, Deisseroth K (2016) Multiplexed intact-tissue transcriptional analysis at cellular resolution. Cell 164(4):792–804. doi:10.1016/j.cell.2016.01.038
72. Battich N, Stoeger T, Pelkmans L (2015) Control of transcript variability in single mammalian cells. Cell 163(7):1596–1610. doi:10.1016/j.cell.2015.11.018
73. Bahar Halpern K, Tanami S, Landen S, Chapal M, Szlak L, Hutzler A, Nizhberg A, Itzkovitz S (2015) Bursty gene expression in the intact mammalian liver. Mol Cell 58 (1):147–156. doi:10.1016/j.molcel.2015.01.027
74. Symmons O, Raj A (2016) What's luck got to do with it: single cells, multiple fates, and biological Nondeterminism. Mol Cell 62 (5):788–802. doi:10.1016/j.molcel.2016.05.023
75. Sanchez A, Golding I (2013) Genetic determinants and cellular constraints in noisy gene expression. Science 342(6163):1188–1193. doi:10.1126/science.1242975
76. Hocine S, Raymond P, Zenklusen D, Chao JA, Singer RH (2013) Single-molecule analysis of gene expression using two-color RNA labeling in live yeast. Nat Methods 10 (2):119–121. doi:10.1038/nmeth.2305
77. Bahar Halpern K, Caspi I, Lemze D, Levy M, Landen S, Elinav E, Ulitsky I, Itzkovitz S (2015) Nuclear retention of mRNA in mammalian tissues. Cell Rep 13 (12):2653–2662. doi:10.1016/j.celrep.2015.11.036
78. Lin Y, Sohn CH, Dalal CK, Cai L, Elowitz MB (2015) Combinatorial gene regulation by modulation of relative pulse timing. Nature 527(7576):54–58. doi:10.1038/nature15710
79. Jambor H, Surendranath V, Kalinka AT, Mejstrik P, Saalfeld S, Tomancak P (2015) Systematic imaging reveals features and changing localization of mRNAs in Drosophila development. eLife 4. doi:10.7554/eLife.05003
80. Lecuyer E, Yoshida H, Parthasarathy N, Alm C, Babak T, Cerovina T, Hughes TR, Tomancak P, Krause HM (2007) Global analysis of mRNA localization reveals a prominent role in organizing cellular architecture and function. Cell 131(1):174–187. doi:10.1016/j.cell.2007.08.003
81. Wilk R, Hu J, Blotsky D, Krause HM (2016) Diverse and pervasive subcellular distributions for both coding and long noncoding RNAs. Genes Dev 30(5):594–609. doi:10.1101/gad.276931.115
82. Buxbaum AR, Haimovich G, Singer RH (2015) In the right place at the right time: visualizing and understanding mRNA localization. Nat Rev Mol Cell Biol 16(2):95–109. doi:10.1038/nrm3918
83. Medioni C, Mowry K, Besse F (2012) Principles and roles of mRNA localization in animal development. Development 139 (18):3263–3276. doi:10.1242/dev.078626
84. Mardakheh FK, Paul A, Kumper S, Sadok A, Paterson H, McCarthy A, Yuan Y, Marshall CJ (2015) Global analysis of mRNA, translation, and protein localization: local translation is a key regulator of cell protrusions. Dev Cell 35(3):344–357. doi:10.1016/j.devcel.2015.10.005
85. Li L, Feng J, Liu H, Tong L, Tang B (2016) Two-color imaging of microRNA with enzyme-free signal amplification via hybridization chain reactions in living cells. Chem Sci 7:1940–1945
86. Alonas E, Lifland AW, Gudheti M, Vanover D, Jung J, Zurla C, Kirschman J, Fiore VF, Douglas A, Barker TH, Yi H, Wright ER, Crowe JE Jr, Santangelo PJ (2014) Combining single RNA sensitive probes with subdiffraction-limited and live-cell imaging enables the characterization of virus dynamics in cells. ACS Nano 8(1):302–315. doi:10.1021/nn405998v
87. Jung J, Lifland AW, Zurla C, Alonas EJ, Santangelo PJ (2013) Quantifying RNA-protein

interactions in situ using modified-MTRIPs and proximity ligation. Nucleic Acids Res 41 (1):e12. doi:10.1093/nar/gks837

88. Tyagi S, Kramer FR (1996) Molecular beacons: probes that fluoresce upon hybridization. Nat Biotechnol 14(3):303–308. doi:10.1038/nbt0396-303
89. Alami NH, Smith RB, Carrasco MA, Williams LA, Winborn CS, Han SS, Kiskinis E, Winborn B, Freibaum BD, Kanagaraj A, Clare AJ, Badders NM, Bilican B, Chaum E, Chandran S, Shaw CE, Eggan KC, Maniatis T, Taylor JP (2014) Axonal transport of TDP-43 mRNA granules is impaired by ALS-causing mutations. Neuron 81(3):536–543. doi:10.1016/j.neuron.2013.12.018
90. Chen AK, Davydenko O, Behlke MA, Tsourkas A (2010) Ratiometric bimolecular beacons for the sensitive detection of RNA in single living cells. Nucleic Acids Res 38(14): e148. doi:10.1093/nar/gkq436
91. Hovelmann F, Gaspar I, Ephrussi A, Seitz O (2013) Brightness enhanced DNA FIT-probes for wash-free RNA imaging in tissue. J Am Chem Soc 135(50):19025–19032. doi:10.1021/ja410674h
92. Hovelmann F, Gaspar I, Loibl S, Ermilov EA, Roder B, Wengel J, Ephrussi A, Seitz O (2014) Brightness through local constraint–LNA-enhanced FIT hybridization probes for in vivo ribonucleotide particle tracking. Angew Chem 53(42):11370–11375. doi:10.1002/anie.201406022
93. Oomoto I, Suzuki-Hirano A, Umeshima H, Han YW, Yanagisawa H, Carlton P, Harada Y, Kengaku M, Okamoto A, Shimogori T, Wang DO (2015) ECHO-liveFISH: in vivo RNA labeling reveals dynamic regulation of nuclear RNA foci in living tissues. Nucleic Acids Res 43(19):e126. doi:10.1093/nar/gkv614
94. Paige JS, KY W, Jaffrey SR (2011) RNA mimics of green fluorescent protein. Science 333(6042):642–646. doi:10.1126/science.1207339
95. Ouellet J (2016) RNA fluorescence with light-up Aptamers. Front Chem 4:29. doi:10.3389/fchem.2016.00029
96. Lu Z, Filonov GS, Noto JJ, Schmidt CA, Hatkevich TL, Wen Y, Jaffrey SR, Matera AG (2015) Metazoan tRNA introns generate stable circular RNAs in vivo. RNA 21(9):1554–1565. doi:10.1261/rna.052944.115
97. Arora A, Sunbul M, Jaschke A (2015) Dual-colour imaging of RNAs using quencher- and fluorophore-binding aptamers. Nucleic Acids Res 43(21):e144. doi:10.1093/nar/gkv718
98. Guet D, Burns LT, Maji S, Boulanger J, Hersen P, Wente SR, Salamero J, Dargemont C (2015) Combining Spinach-tagged RNA and gene localization to image gene expression in live yeast. Nat Commun 6:8882. doi:10.1038/ncomms9882
99. Dolgosheina EV, Jeng SC, Panchapakesan SS, Cojocaru R, Chen PS, Wilson PD, Hawkins N, Wiggins PA, Unrau PJ (2014) RNA mango aptamer-fluorophore: a bright, high-affinity complex for RNA labeling and tracking. ACS Chem Biol 9(10):2412–2420. doi:10.1021/cb500499x
100. Zhang J, Fei J, Leslie BJ, Han KY, Kuhlman TE, Ha T (2015) Tandem Spinach Array for mRNA imaging in living bacterial cells. Sci Rep 5:17295. doi:10.1038/srep17295
101. Bertrand E, Chartrand P, Schaefer M, Shenoy SM, Singer RH, Long RM (1998) Localization of ASH1 mRNA particles in living yeast. Mol Cell 2(4):437–445
102. Campbell PD, Chao JA, Singer RH, Marlow FL (2015) Dynamic visualization of transcription and RNA subcellular localization in zebrafish. Development 142(7):1368–1374. doi:10.1242/dev.118968
103. Forrest KM, Gavis ER (2003) Live imaging of endogenous RNA reveals a diffusion and entrapment mechanism for nanos mRNA localization in Drosophila. Curr Biol 13 (14):1159–1168
104. Gagnon JA, Kreiling JA, Powrie EA, Wood TR, Mowry KL (2013) Directional transport is mediated by a dynein-dependent step in an RNA localization pathway. PLoS Biol 11(4): e1001551. doi:10.1371/journal.pbio.1001551
105. Lionnet T, Czaplinski K, Darzacq X, Shav-Tal Y, Wells AL, Chao JA, Park HY, de Turris V, Lopez-Jones M, Singer RH (2011) A transgenic mouse for in vivo detection of endogenous labeled mRNA. Nat Methods 8 (2):165–170. doi:10.1038/nmeth.1551
106. Larson DR, Zenklusen D, Wu B, Chao JA, Singer RH (2011) Real-time observation of transcription initiation and elongation on an endogenous yeast gene. Science 332 (6028):475–478. doi:10.1126/science.1202142
107. Daigle N, Ellenberg J (2007) LambdaN-GFP: an RNA reporter system for live-cell imaging. Nat Methods 4(8):633–636. doi:10.1038/nmeth1065
108. Yamada T, Yoshimura H, Inaguma A, Ozawa T (2011) Visualization of nonengineered single mRNAs in living cells using genetically

encoded fluorescent probes. Anal Chem 83 (14):5708–5714. doi:10.1021/ac2009405

109. Yoshimura H, Inaguma A, Yamada T, Ozawa T (2012) Fluorescent probes for imaging endogenous beta-actin mRNA in living cells using fluorescent protein-tagged pumilio. ACS Chem Biol 7(6):999–1005. doi:10.1021/cb200474a
110. Nelles DA, Fang MY, O'Connell MR, Xu JL, Markmiller SJ, Doudna JA, Yeo GW (2016) Programmable RNA tracking in live cells with CRISPR/Cas9. Cell 165(2):488–496. doi:10.1016/j.cell.2016.02.054
111. Garcia HG, Tikhonov M, Lin A, Gregor T (2013) Quantitative imaging of transcription in living Drosophila embryos links polymerase activity to patterning. Curr Biol 23 (21):2140–2145. doi:10.1016/j.cub.2013.08.054
112. Lucas T, Ferraro T, Roelens B, De Las Heras Chanes J, Walczak AM, Coppey M, Dostatni N (2013) Live imaging of bicoid-dependent transcription in Drosophila embryos. Curr Biol 23(21):2135–2139. doi:10.1016/j.cub.2013.08.053
113. Lenstra TL, Coulon A, Chow CC, Larson DR (2015) Single-molecule imaging reveals a switch between spurious and functional ncRNA transcription. Mol Cell 60 (4):597–610. doi:10.1016/j.molcel.2015.09.028
114. Martin RM, Rino J, Carvalho C, Kirchhausen T, Carmo-Fonseca M (2013) Live-cell visualization of pre-mRNA splicing with single-molecule sensitivity. Cell Rep 4 (6):1144–1155. doi:10.1016/j.celrep.2013.08.013
115. Coulon A, Ferguson ML, de Turris V, Palangat M, Chow CC, Larson DR (2014) Kinetic competition during the transcription cycle results in stochastic RNA processing. eLife 3. doi:10.7554/eLife.03939
116. Grunwald D, Singer RH (2010) In vivo imaging of labelled endogenous beta-actin mRNA during nucleocytoplasmic transport. Nature 467(7315):604–607. doi:10.1038/nature09438
117. Mor A, Suliman S, Ben-Yishay R, Yunger S, Brody Y, Shav-Tal Y (2010) Dynamics of single mRNP nucleocytoplasmic transport and export through the nuclear pore in living cells. Nat Cell Biol 12(6):543–552. doi:10.1038/ncb2056
118. Wang C, Han B, Zhou R, Zhuang X (2016) Real-time imaging of translation on single mRNA transcripts in live cells. Cell 165 (4):990–1001. doi:10.1016/j.cell.2016.04.040
119. Wu B, Eliscovich C, Yoon YJ, Singer RH (2016) Translation dynamics of single mRNAs in live cells and neurons. Science 352(6292):1430–1435. doi:10.1126/science.aaf1084
120. Yan X, Hoek TA, Vale RD, Tanenbaum ME (2016) Dynamics of translation of single mRNA molecules in vivo. Cell 165 (4):976–989. doi:10.1016/j.cell.2016.04.034
121. Morisaki T, Lyon K, DeLuca KF, DeLuca JG, English BP, Zhang Z, Lavis LD, Grimm JB, Viswanathan S, Looger LL, Lionnet T, Stasevich TJ (2016) Real-time quantification of single RNA translation dynamics in living cells. Science 352(6292):1425–1429. doi:10.1126/science.aaf0899
122. King ML, Messitt TJ, Mowry KL (2005) Putting RNAs in the right place at the right time: RNA localization in the frog oocyte. Biol Cell 97(1):19–33. doi:10.1042/BC20040067
123. St Johnston D (2005) Moving messages: the intracellular localization of mRNAs. Nat Rev Mol Cell Biol 6(5):363–375. doi:10.1038/nrm1643
124. Park HY, Lim H, Yoon YJ, Follenzi A, Nwokafor C, Lopez-Jones M, Meng X, Singer RH (2014) Visualization of dynamics of single endogenous mRNA labeled in live mouse. Science 343(6169):422–424. doi:10.1126/science.1239200
125. Zimyanin VL, Belaya K, Pecreaux J, Gilchrist MJ, Clark A, Davis I, St Johnston D (2008) In vivo imaging of oskar mRNA transport reveals the mechanism of posterior localization. Cell 134(5):843–853. doi:10.1016/j.cell.2008.06.053
126. Bullock SL, Nicol A, Gross SP, Zicha D (2006) Guidance of bidirectional motor complexes by mRNA cargoes through control of dynein number and activity. Curr Biol 16 (14):1447–1452. doi:10.1016/j.cub.2006.05.055
127. Dictenberg JB, Swanger SA, Antar LN, Singer RH, Bassell GJ (2008) A direct role for FMRP in activity-dependent dendritic mRNA transport links filopodial-spine morphogenesis to fragile X syndrome. Dev Cell 14(6):926–939. doi:10.1016/j.devcel.2008.04.003
128. Sinsimer KS, Lee JJ, Thiberge SY, Gavis ER (2013) Germ plasm anchoring is a dynamic state that requires persistent trafficking. Cell Rep 5(5):1169–1177. doi:10.1016/j.celrep.2013.10.045

129. Halstead JM, Lionnet T, Wilbertz JH, Wippich F, Ephrussi A, Singer RH, Chao JA (2015) Translation. An RNA biosensor for imaging the first round of translation from single cells to living animals. Science 347 (6228):1367–1671. doi:10.1126/science.aaa3380

130. Wu B, Buxbaum AR, Katz ZB, Yoon YJ, Singer RH (2015) Quantifying protein-mRNA interactions in single live cells. Cell 162(1):211–220. doi:10.1016/j.cell.2015.05.054

131. Mitchell SF, Parker R (2014) Principles and properties of eukaryotic mRNPs. Mol Cell 54 (4):547–558. doi:10.1016/j.molcel.2014.04.033

132. McHugh CA, Russell P, Guttman M (2014) Methods for comprehensive experimental identification of RNA-protein interactions. Genome Biol 15(1):203. doi:10.1186/gb4152

133. Courchaine EM, Lu A, Neugebauer KM (2016) Droplet organelles? EMBO J 35 (15):1603–1612. doi:10.15252/embj.201593517

134. Konig J, Zarnack K, Luscombe NM, Ule J (2011) Protein-RNA interactions: new genomic technologies and perspectives. Nat Rev Genet 13(2):77–83. doi:10.1038/nrg3141

135. Ule J, Jensen KB, Ruggiu M, Mele A, Ule A, Darnell RB (2003) CLIP identifies Nova-regulated RNA networks in the brain. Science 302(5648):1212–1215. doi:10.1126/science.1090095

136. Konig J, Zarnack K, Rot G, Curk T, Kayikci M, Zupan B, Turner DJ, Luscombe NM, Ule J (2010) iCLIP reveals the function of hnRNP particles in splicing at individual nucleotide resolution. Nat Struct Mol Biol 17(7):909–915. doi:10.1038/nsmb.1838

137. Hafner M, Landthaler M, Burger L, Khorshid M, Hausser J, Berninger P, Rothballer A, Ascano M, Jungkamp AC, Munschauer M, Ulrich A, Wardle GS, Dewell S, Zavolan M, Tuschl T (2010) PAR-CliP—a method to identify transcriptome-wide the binding sites of RNA binding proteins. J Vis Exp 41. doi:10.3791/2034

138. Sugimoto Y, Vigilante A, Darbo E, Zirra A, Militti C, D'Ambrogio A, Luscombe NM, Ule J (2015) hiCLIP reveals the in vivo atlas of mRNA secondary structures recognized by Staufen 1. Nature 519(7544):491–494. doi:10.1038/nature14280

139. McMahon AC, Rahman R, Jin H, Shen JL, Fieldsend A, Luo W, Rosbash M (2016) TRIBE: hijacking an RNA-editing enzyme to identify cell-specific targets of RNA-binding proteins. Cell 165(3):742–753. doi:10.1016/j.cell.2016.03.007

140. Lapointe CP, Wilinski D, Saunders HA, Wickens M (2015) Protein-RNA networks revealed through covalent RNA marks. Nat Methods 12(12):1163–1170. doi:10.1038/nmeth.3651

141. Leppek K, Stoecklin G (2014) An optimized streptavidin-binding RNA aptamer for purification of ribonucleoprotein complexes identifies novel ARE-binding proteins. Nucleic Acids Res 42(2):e13. doi:10.1093/nar/gkt956

142. Said N, Rieder R, Hurwitz R, Deckert J, Urlaub H, Vogel J (2009) In vivo expression and purification of aptamer-tagged small RNA regulators. Nucleic Acids Res 37(20):e133. doi:10.1093/nar/gkp719

143. Chu C, Zhang QC, da Rocha ST, Flynn RA, Bharadwaj M, Calabrese JM, Magnuson T, Heard E, Chang HY (2015) Systematic discovery of Xist RNA binding proteins. Cell 161(2):404–416. doi:10.1016/j.cell.2015.03.025

144. McHugh CA, Chen CK, Chow A, Surka CF, Tran C, McDonel P, Pandya-Jones A, Blanco M, Burghard C, Moradian A, Sweredoski MJ, Shishkin AA, Su J, Lander ES, Hess S, Plath K, Guttman M (2015) The Xist lncRNA interacts directly with SHARP to silence transcription through HDAC3. Nature 521 (7551):232–236. doi:10.1038/nature14443

145. Minajigi A, Froberg JE, Wei C, Sunwoo H, Kesner B, Colognori D, Lessing D, Payer B, Boukhali M, Haas W, Lee JT (2015) Chromosomes. A comprehensive Xist interactome reveals cohesin repulsion and an RNA-directed chromosome conformation. Science 349(6245). doi:10.1126/science.aab2276

146. Baltz AG, Munschauer M, Schwanhausser B, Vasile A, Murakawa Y, Schueler M, Youngs N, Penfold-Brown D, Drew K, Milek M, Wyler E, Bonneau R, Selbach M, Dieterich C, Landthaler M (2012) The mRNA-bound proteome and its global occupancy profile on protein-coding transcripts. Mol Cell 46 (5):674–690. doi:10.1016/j.molcel.2012.05.021

147. Castello A, Fischer B, Eichelbaum K, Horos R, Beckmann BM, Strein C, Davey NE, Humphreys DT, Preiss T, Steinmetz LM, Krijgsveld J, Hentze MW (2012) Insights into RNA biology from an atlas of mammalian mRNA-binding proteins. Cell 149 (6):1393–1406. doi:10.1016/j.cell.2012.04.031

148. Castello A, Fischer B, Frese CK, Horos R, Alleaume AM, Foehr S, Curk T, Krijgsveld J, Hentze MW (2016) Comprehensive identification of RNA-binding domains in human cells. Mol Cell 63(4):696–710. doi:10.1016/j.molcel.2016.06.029

149. Beckmann BM, Horos R, Fischer B, Castello A, Eichelbaum K, Alleaume AM, Schwarzl T, Curk T, Foehr S, Huber W, Krijgsveld J, Hentze MW (2015) The RNA-binding proteomes from yeast to man harbour conserved enigmRBPs. Nat Commun 6:10127. doi:10.1038/ncomms10127

150. Wessels HH, Imami K, Baltz AG, Kolinski M, Beldovskaya A, Selbach M, Small S, Ohler U, Landthaler M (2016) The mRNA-bound proteome of the early fly embryo. Genome Res 26(7):1000–1009. doi:10.1101/gr.200386.115

151. Sysoev VO, Fischer B, Frese CK, Gupta I, Krijgsveld J, Hentze MW, Castello A, Ephrussi A (2016) Global changes of the RNA-bound proteome during the maternal-to-zygotic transition in Drosophila. Nat Commun 7:12128. doi:10.1038/ncomms12128

152. Reichel M, Liao Y, Rettel M, Ragan C, Evers M, Alleaume AM, Horos R, Hentze MW, Preiss T, Millar AA (2016) In Planta determination of the mRNA-binding proteome of Arabidopsis etiolated seedlings. Plant Cell 28(10):2435–2452. doi:10.1105/tpc.16.00562

153. Buchan JR (2014) mRNP granules. Assembly, function, and connections with disease. RNA Biol 11(8):1019–1030. doi:10.4161/15476286.2014.972208

154. Fritzsche R, Karra D, Bennett KL, Ang FY, Heraud-Farlow JE, Tolino M, Doyle M, Bauer KE, Thomas S, Planyavsky M, Arn E, Bakosova A, Jungwirth K, Hormann A, Palfi Z, Sandholzer J, Schwarz M, Macchi P, Colinge J, Superti-Furga G, Kiebler MA (2013) Interactome of two diverse RNA granules links mRNA localization to translational repression in neurons. Cell Rep 5(6):1749–1762. doi:10.1016/j.celrep.2013.11.023

155. Amrute-Nayak M, Bullock SL (2012) Single-molecule assays reveal that RNA localization signals regulate dynein-dynactin copy number on individual transcript cargoes. Nat Cell Biol 14(4):416–423. doi:10.1038/ncb2446

156. Batish M, van den Bogaard P, Kramer FR, Tyagi S (2012) Neuronal mRNAs travel singly into dendrites. Proc Natl Acad Sci U S A 109(12):4645–4650. doi:10.1073/pnas.1111226109

157. Buxbaum AR, Wu B, Singer RH (2014) Single beta-actin mRNA detection in neurons reveals a mechanism for regulating its translatability. Science 343(6169):419–422. doi:10.1126/science.1242939

158. Little SC, Sinsimer KS, Lee JJ, Wieschaus EF, Gavis ER (2015) Independent and coordinate trafficking of single Drosophila germ plasm mRNAs. Nat Cell Biol 17(5):558–568. doi:10.1038/ncb3143

159. Trcek T, Grosch M, York A, Shroff H, Lionnet T, Lehmann R (2015) Drosophila germ granules are structured and contain homotypic mRNA clusters. Nat Commun 6:7962. doi:10.1038/ncomms8962

160. Tubing F, Vendra G, Mikl M, Macchi P, Thomas S, Kiebler MA (2010) Dendritically localized transcripts are sorted into distinct ribonucleoprotein particles that display fast directional motility along dendrites of hippocampal neurons. J Neurosci 30(11):4160–4170. doi:10.1523/JNEUROSCI.3537-09.2010

161. Weber SC, Brangwynne CP (2012) Getting RNA and protein in phase. Cell 149(6):1188–1191. doi:10.1016/j.cell.2012.05.022

162. Hyman AA, Weber CA, Julicher F (2014) Liquid-liquid phase separation in biology. Annu Rev Cell Dev Biol 30:39–58. doi:10.1146/annurev-cellbio-100913-013325

163. King OD, Gitler AD, Shorter J (2012) The tip of the iceberg: RNA-binding proteins with prion-like domains in neurodegenerative disease. Brain Res 1462:61–80. doi:10.1016/j.brainres.2012.01.016

164. Hubstenberger A, Noble SL, Cameron C, Evans TC (2013) Translation repressors, an RNA helicase, and developmental cues control RNP phase transitions during early development. Dev Cell 27(2):161–173. doi:10.1016/j.devcel.2013.09.024

165. Weil TT, Parton RM, Herpers B, Soetaert J, Veenendaal T, Xanthakis D, Dobbie IM, Halstead JM, Hayashi R, Rabouille C, Davis I (2012) Drosophila patterning is established by differential association of mRNAs with P bodies. Nat Cell Biol 14(12):1305–1313. doi:10.1038/ncb2627

166. Wang JT, Smith J, Chen BC, Schmidt H, Rasoloson D, Paix A, Lambrus BG, Calidas D, Betzig E, Seydoux G (2014) Regulation of RNA granule dynamics by phosphorylation of serine-rich, intrinsically disordered proteins in *C. elegans*. Elife 3:e04591. doi:10.7554/eLife.04591

167. Wheeler JR, Matheny T, Jain S, Abrisch R, Parker R (2016) Distinct stages in stress granule assembly and disassembly. Elife 5:e18413. doi:10.7554/eLife.18413
168. Zurla C, Jung J, Santangelo PJ (2016) Can we observe changes in mRNA "state"? Overview of methods to study mRNA interactions with regulatory proteins relevant in cancer related processes. Analyst 141(2):548–562. doi:10.1039/c5an01959a
169. Ke R, Mignardi M, Hauling T, Nilsson M (2016) Fourth generation of next-generation sequencing technologies: promise and consequences. Hum Mutat 37(12):1363–1367. doi:10.1002/humu.23051
170. Chen BC, Legant WR, Wang K, Shao L, Milkie DE, Davidson MW, Janetopoulos C, Wu XS, Hammer JA 3rd, Liu Z, English BP, Mimori-Kiyosue Y, Romero DP, Ritter AT, Lippincott-Schwartz J, Fritz-Laylin L, Mullins RD, Mitchell DM, Bembenek JN, Reymann AC, Bohme R, Grill SW, Wang JT, Seydoux G, Tulu US, Kiehart DP, Betzig E (2014) Lattice light-sheet microscopy: imaging molecules to embryos at high spatiotemporal resolution. Science 346 (6208):1257998. doi:10.1126/science.1257998
171. Ma J, Liu Z, Michelotti N, Pitchiaya S, Veerapaneni R, Androsavich JR, Walter NG, Yang W (2013) High-resolution three-dimensional mapping of mRNA export through the nuclear pore. Nat Commun 4:2414. doi:10.1038/ncomms3414
172. Doudna JA, Charpentier E (2014) Genome editing. The new frontier of genome engineering with CRISPR-Cas9. Science 346 (6213):1258096. doi:10.1126/science.1258096
173. You M, Jaffrey SR (2015) Designing optogenetically controlled RNA for regulating biological systems. Ann N Y Acad Sci 1352:13–19. doi:10.1111/nyas.12660

Chapter 2

Quantification of 2′-*O*-Me Residues in RNA Using Next-Generation Sequencing (Illumina RiboMethSeq Protocol)

Lilia Ayadi, Yuri Motorin, and Virginie Marchand

Abstract

RNA 2′-*O*-methylation is one of the ubiquitous nucleotide modifications found in many RNA types from bacteria, archaea, and eukarya. We and others have recently published accurate and sensitive detection of these modifications on native RNA at a single base resolution by high-throughput sequencing technologies. Relative quantification of these modifications is still under progress and would probably reduce the number of false positives due to 3D RNA structure. Therefore, here, we describe a reliable and optimized protocol for quantification of 2′-*O*-Methylations based on alkaline fragmentation of RNA coupled to a commonly used ligation approach followed by Illumina sequencing. For this purpose, we describe how to prepare in vitro transcribed yeast 18S and 25S rRNA used as a reference for unmodified rRNAs and to compare them to purified 18S and 25S rRNA from yeast total RNA preparation. These reconstructed rRNA mixes were combined at different ratios and processed for RiboMethseq protocol.

This technique will be applicable for routine parallel treatment of biological and clinical samples to decipher the functions of 2′-*O*-methylations in normal and pathologic processes or during development.

Key words 2′-O-methylation, High-throughput sequencing, RNA modification, Ribose methylation, Alkaline fragmentation

1 Introduction

Modulation of RNA properties by posttranscriptional mechanisms of RNA modification is a newly discovered layer for regulation of gene expression. At the level of epitranscriptome, these dynamic and regulated RNA modification events contribute to RNA-RNA and RNA-protein interactions, modulate alternative splicing, mRNA translation, RNA transport and localization [1, 2]. The current major challenge in the field is a careful mapping of different RNA modifications in coding and noncoding RNAs, as well as precise quantification of the modification rate for every given site. Taking into account that at least thousands of RNA modified nucleotides expected to be present in higher eukaryotic

Imre Gaspar (ed.), *RNA Detection: Methods and Protocols*, Methods in Molecular Biology, vol. 1649, DOI 10.1007/978-1-4939-7213-5_2, © Springer Science+Business Media LLC 2018

transcriptome, this task is unimaginable without appropriate high-throughput analysis techniques. Next-generation sequencing approaches invented for mapping of certain RNA modifications generally provide extremely valuable information on the location of RNA modification sites, but these methods are rarely able to give an estimation of the modification rate. This applies to popular antibody-based enrichment protocols, developed for m^6A and m^1A [3, 4], and also to variants of CMCT-based PseudoU-Seq [5–7]. Only bisulfite conversion/nonconversion of m^5C residues can deliver some quantitative information [8, 9], as well as both recently reported variants of RiboMethSeq developed for 2′-*O*-methylation analysis [10, 11].

Standard RiboMethSeq protocol relies on the protection of the 3′-adjacent phosphodiester bond in RNA from alkaline hydrolysis, when the ribose moiety is methylated at the 2′-OH position. All other phosphodiester bonds remain sensitive to alkaline cleavage, creating a more or less regular cleavage profile. When a given phosphodiester bond remains protected, this is an indication for the presence of 2′-*O*-methylation. In principle, such protection from cleavage can be used as a quantitative measure of methylation rate at a given nucleotide, assuming zero protection for unmodified nucleotide and complete protection for its modified counterpart. Evaluation of the modification rate can be done using calculated ScoreC (MethScore) which takes into account the variability in coverage for neighboring nucleotides and the protection at a given position [10, 11].

In this work we describe the method of 2′-*O*-Me quantification in RNA, demonstrating a linear dependence between the proportion of unmodified yeast rRNA in the mixture and the calculated values of MethScore. Individual quantification of methylated sites was performed using calibration mixtures composed of purified modified rRNA fractions and corresponding unmodified synthetic rRNA transcripts. MethScore values demonstrate linear dependence from the level of modification, providing a way for precise quantification.

2 Materials

Prepare all solutions using RNase-free water. Wear gloves to prevent degradation of RNA samples by RNases.

2.1 Yeast rDNA PCR Amplification

1. Specific forward and reverse primers (10 μM) for yeast 18S and 25S rDNA amplification.

   Two pairs of DNA oligonucleotides are used to amplify yeast full-length 18S (1800 pb) and 25S (3396 pb) rDNA fragments, respectively. Forward primers are designed with an

upstream T7 RNA polymerase promoter (underlined in the sequence) to perform T7 transcription.
18S forward primer:
5′-TAATACGACTCACTATAGGTATCTGGTTGATCCTG-CCAGTAG-3′.
18S reverse primer: 5′-TAATGATCCTTCCGCAGGTTC-3′.
25S forward primer:
5′-TAATACGACTCACTATAGGGTTTGACCTCAAATCA-GGTAGG-3′.
25S reverse primer: 5′-ACAAATCAGACAACAAAGGC-3′

2. 25 ng/μL plasmid DNA template pHW18 (*see* **Note 1**).
3. 2.5 U/μL Pfu DNA polymerase.
4. 10× Pfu DNA polymerase buffer.
5. dNTP mix: 1.25 mM each.
6. RNase-free water.
7. Individual RNase-free 0.2 mL PCR tubes.
8. PCR thermal cycler (e.g., Agilent SureCycler 8000).
9. RNAse-free 1.5 mL microcentrifuge tubes.
10. Phenol–chloroform mix (1:1, v/v).
11. 3 M Na-Acetate (NaOAc) in water, pH 5.2.
12. 96% ethanol.
13. 75% ethanol.
14. Tabletop centrifuge.

2.2 In Vitro Yeast rDNA Transcription and RNA Purification

2.2.1 In Vitro Yeast rDNA Transcription

1. 5× Transcription buffer: 400 mM HEPES–KOH, pH 7.5, 120 mM $MgCl_2$, 10 mM spermidine, 200 mM DTT.
2. 40 U/μL RiboLock RNase Inhibitor.
3. rNTP mix: 12.5 mM each.
4. 400 nM 18S and 450 nM 25S PCR templates.
5. 20 U/μL T7 RNA polymerase.
6. RNAse-free 1.5 mL microcentrifuge tubes.
7. 37 °C incubator.
8. 1 U/μL RQ1 RNase-free DNase.

2.2.2 Purification of In Vitro Transcripts

1. RNase-free water.
2. Phenol–chloroform mix (1:1, v/v).
3. Chloroform.
4. 5 M ammonium acetate (NH_4OAc).
5. 10 mg/mL glycogen coprecipitant.
6. Refrigerated tabletop centrifuge.
7. 96% ethanol.

8. 80% ethanol.
9. Mini Quick Spin RNA column (e.g., Roche).
10. UV spectrophotometer (e.g., NanoDrop 2000c).

2.3 Yeast Total RNA Extraction

1. Yeast cell culture (5 mL of yeast culture grown to an OD_{600} of 5–9).
2. AE buffer: 50 mM NaOAc in water, pH 5.2, 10 mM EDTA.
3. 10 w/v % SDS.
4. RNase-free water.
5. 1.5 mL RNase-free microcentrifuge tubes.
6. Acid phenol, pH 4.5.
7. Phenol–chloroform–isoamyl alcohol mix (25:24:1, v/v).
8. Chloroform.
9. 3 M NaOAc in water, pH 5.2.
10. 96% ethanol.
11. 80% ethanol.
12. Dry ice.
13. Refrigerated tabletop centrifuge.
14. Water bath or heating block, set to 65 °C.

2.4 Purification of 18S and 25S rRNA from Yeast Total RNA

1. Low Melting agarose (e.g., Lonza Nu Sieve® GTG® Agarose).
2. 10× TBE buffer: 108 g Tris, 55 g boric acid, 9.3 g EDTA, pH 8, H_2O qsp 1 L.
3. RNase-free water.
4. 1× TBE buffer: 100 mL of 10× TBE and 900 mL of RNase-free water.
5. Microwave oven.
6. Agarose gel electrophoresis chamber.
7. Electrophoresis power supply.
8. 3 μg/μL yeast total RNA preparation.
9. 1.5 mL RNase-free microcentrifuge tubes.
10. 6× DNA loading dye: 1.9 mM xylene cyanol, 1.5 mM bromophenol blue, and 25% glycerol.
11. SYBR® Gold Nucleic Acid Gel Stain 10,000× concentrate in DMSO (Invitrogen).
12. UV 365 nm transilluminator.
13. Scalpel or razor blade.
14. Heat block preheated at 65 °C.
15. Ultrapure phenol buffered with Tris–HCl pH 8.0.

16. Chloroform.
17. 3 M NaOAc in water, pH 5.2.
18. 96% ethanol.
19. 80% ethanol.
20. Refrigerated tabletop centrifuge.

2.5 Yeast rRNA Reconstruction

1. 18S and 25S in vitro transcripts.
2. 18S and 25S purified RNAs.
3. 3 μg/μL yeast total RNA preparation.
4. 1.5 mL RNase-free microcentrifuge tubes.
5. RNase-free water.
6. Agilent 2100 Bioanalyzer.
7. Agilent RNA 6000 Pico kit (50–5000 pg/μL).
8. Chip priming station (Agilent).

2.6 Alkaline Hydrolysis and RNA Fragmentation Quality Control

2.6.1 Alkaline Hydrolysis

1. Sodium bicarbonate buffer: 100 mM $NaHCO_3$ pH 9.2.
2. RNase-free water.
3. Individual RNase-free 0.2 mL PCR tubes.
4. PCR Thermal cycler (e.g., Agilent SureCycler 8000).
5. RNase-free 1.5 mL microcentrifuge tubes.
6. Refrigerate tabletop centrifuge.
7. 96% ethanol.
8. 80% ethanol.
9. 15 mg/mL GlycoBlue coprecipitant (e.g., Ambion).
10. 3 M NaOAc in water, pH 5.2.
11. Dewar containing liquid nitrogen.

2.6.2 RNA Fragmentation Quality Control

1. Agilent 2100 Bioanalyzer (Agilent Technologies).
2. Agilent RNA 6000 Pico kit (quantitative range 50–5000 pg/μL).
3. Chip priming station (Agilent Technologies).
4. RNase-free 1.5 mL microcentrifuge tubes.

2.7 End-Repair

1. RNase-free water.
2. 5 U/μL Antarctic Phosphatase.
3. Antarctic Phosphatase Buffer 10×.
4. 10 U/μL T4 PNK.
5. T4 PNK Buffer 10×.
6. 10 mM ATP.
7. 40 U/μL RiboLock RNase Inhibitor.

8. 0.2 mL RNase-free PCR tubes, strips of 8.
9. Flat PCR Caps, strips of 8.
10. PCR thermal cycler.

2.8 RNA Purification

1. RNeasy MinElute Cleanup kit (Qiagen).
2. 96% ethanol.
3. 80% ethanol.

2.9 Library Preparation Using NEBNext® Multiplex Small RNA Library Prep Set for Illumina®

1. NEBNext® Multiplex Small RNA Library Prep Set for Illumina® (set 1 or 2, New England Biolabs) (*see* **Note 2**).
2. 0.2 mL PCR tubes, strips of 8.
3. Flat PCR Caps, strips of 8.
4. PCR thermal cycler.

2.10 Library Purification Using GeneJET PCR Purification Kit

1. GeneJET® PCR Purification kit or equivalent.
2. RNase-free 1.5 mL microcentrifuge tubes.
3. Tabletop centrifuge.
4. 1.5 mL DNA low-binding tubes.

2.11 Library Quantification and Quality Assessment

2.11.1 Library Quantification

1. Fluorometer able to quantify DNA library with high sensitivity (e.g., Qubit® 2.0 fluorometer).
2. Qubit® dsDNA HS Assay kit (0.2–100 ng).
3. Thin-walled polypropylene tubes of 500 μL compatible with the fluorometer (e.g., Qubit® Assay Tube or Axygen® PCR-05-C tubes).

2.11.2 Library Quality Assessment

1. Agilent 2100 Bioanalyzer (Agilent Technologies).
2. Agilent HS DNA kit (quantitative range 5–500 pg/μL).
3. Chip priming station (Agilent Technologies).
4. RNase-free 1.5 mL microcentrifuge tubes.

2.12 Library Sequencing

1. An Illumina sequencer (starting from MiSeq to different HiSeq models).
2. Any appropriate sequencing kit for a single read length of 35–50 nt.

3 Methods

3.1 Yeast rDNA PCR Amplification

1. Mix in a PCR tube 1 μL of pHW18 template, 0.5 μL of each corresponding forward and reverse primers, 5 μL Pfu DNA polymerase buffer, 8 μL dNTP mix, and 1 μL Pfu DNA polymerase in a total volume of 50 μL.

2. Perform PCR on a thermal cycler using the following program parameters: 95 °C for 5 min; 25 cycles [denaturation 95 °C for 30 s, hybridization 55 °C for 30 s, elongation 72 °C for 4 min (18S) or 7 min (25S)]; 72 °C for 7 min. Cool down to 20 °C.
3. Optional: check PCR amplification using agarose gel (*see* **Note 3**).
4. Transfer your PCR reaction in a 1.5 mL RNase-free microcentrifuge tube.
5. Adjust the volume of the reaction to 200 μL with RNase-free water. Add 200 μL of phenol–chloroform mix and proceed to extraction. After centrifugation for 10 min at 16,000 × *g*, transfer the upper (aqueous) phase to a new 1.5 mL microcentrifuge tube.
6. Precipitate PCR template by the addition of 1/10th volume 3 M sodium acetate and three volumes of 96% ethanol. Mix by inverting the tube several times. Incubate for 15 min at −80 °C and centrifuge for 15 min at 12,000–16,000 × *g* at 4 °C.
7. Remove the supernatant and wash the pellet with 75% ethanol.
8. Dry and dissolve pellet in 20 μL of sterile water.

3.2 In Vitro Yeast rRNA Transcription and RNA Purification

1. In a 1.5 mL RNase-free microcentrifuge tube, settle the transcription reaction in a final volume of 50 μL by mixing at *room temperature* the components in the following order: 10 μL of transcription buffer, 1 μL of RNase inhibitor, 16 μL of rNTP mix, 12.5 μL (18S) or 11.1 μL (25S) of DNA template (100 nM final), and 3 μL of T7 RNA polymerase.
2. Incubate at 37 °C for 2 h (*see* **Note 4**).
3. Digest the DNA template by addition of 1 μL of DNase RQ1 and incubation for 20 min at 37 °C.
4. Adjust the volume of the reaction to 200 μL with sterile water and proceed to phenol–chloroform mix extraction followed by a chloroform extraction.
5. Precipitate the RNA transcript by addition of three volumes of 96% ethanol in the presence of 20 μL of ammonium acetate and 1 μL of glycogen. Mix by inverting the tube several times.
6. Incubate for 15 min at −80 °C and centrifuge for 15 min at 12,000–16,000 × g at 4 °C.
7. After centrifugation, the RNA pellet is washed with 80% ethanol, dried, and dissolved in RNase-free water.
8. Purify transcripts from unincorporated nucleotides by gel filtration on a Quick Spin RNA column according to the manufacturer's instructions.
9. Quantify RNA using the UV spectrophotometer.

10. Check the quality of the transcripts by using the Agilent 2100 Bioanalyzer (*see* **Note 5**) (*see* Subheading 3.4) and proceed to yeast rRNA reconstruction.

3.3 Yeast Total RNA Extraction

The protocol for total yeast RNA isolation using hot acid phenol is adapted from [12].

1. Transfer yeast cells in 1.5 mL microcentrifuge tubes and pellet cells by centrifugation at 2500 rpm (600 × *g*) for 5 min at room temperature. Discard the supernatant.
2. Resuspend the cells in 1 mL of RNase-free water. Centrifuge for 1 min at full speed at room temperature. Discard the supernatant.
3. Resuspend the cell pellet in 400 μL of AE buffer.
4. Add 40 μL of 10% SDS and vortex until the pellet is completely resuspended.
5. Add 440 μL of acid phenol. Vortex.
6. Incubate for 4 min at 65 °C and then cool rapidly the mixture on dry ice for 2–3 min.
7. Centrifuge the samples for 10 min at maximum speed at room temperature. Transfer carefully the aqueous (upper) phase to a new 1.5 mL microcentrifuge tube.
8. Add 420 μL of phenol–chloroform–IAA, vortex and centrifuge for 10 min at full speed at room temperature.
9. Transfer the aqueous phase to a new 1.5 mL centrifuge tube. Add 400 μL of chloroform. Vortex and centrifuge at full speed at room temperature for 10 min.
10. Transfer the aqueous phase to a new 1.5 mL centrifuge tube. Add 40 μL of 3 M NaOAc and 1 mL of 96% ethanol. Place at −80 °C for at least 30 min.
11. Centrifuge for 30 min at 4 °C at full speed.
12. Discard the supernatant and wash pellet with 500 μL of 80% ethanol.
13. Centrifuge for 10 min at 4 °C at full speed.
14. Discard the supernatant, centrifuge again your samples for a short spin.
15. Remove any liquid left.
16. Incubate your samples with open lid for 2 min at 37 °C and 5 min at room temperature.
17. Resuspend the pellet with 50 μL of RNase-free water.
18. Quantify yeast total RNA samples by measuring A_{260nm} using a UV-spectrophotometer (*see* **Note 6**). Check the quality of your samples by using the Agilent 2100 Bioanalyzer (*see* Subheading 3.4).

3.4 Purification of 18S and 25S rRNA from Yeast Total RNA

The protocol for RNA isolation is derived from [13]. The aim of this technique is to avoid any UV irradiation together with ribonuclease contamination that causes severe fragmentation of RNA entrapped in agarose gels.

1. Prepare 1% low-melting agarose by mixing 1 g of agarose with 100 mL of 1× TBE and heating the preparation in a microwave oven at intermediate power. This percentage of agarose gel is appropriate for separation of 18S from 25S rRNA (Fig. 1).

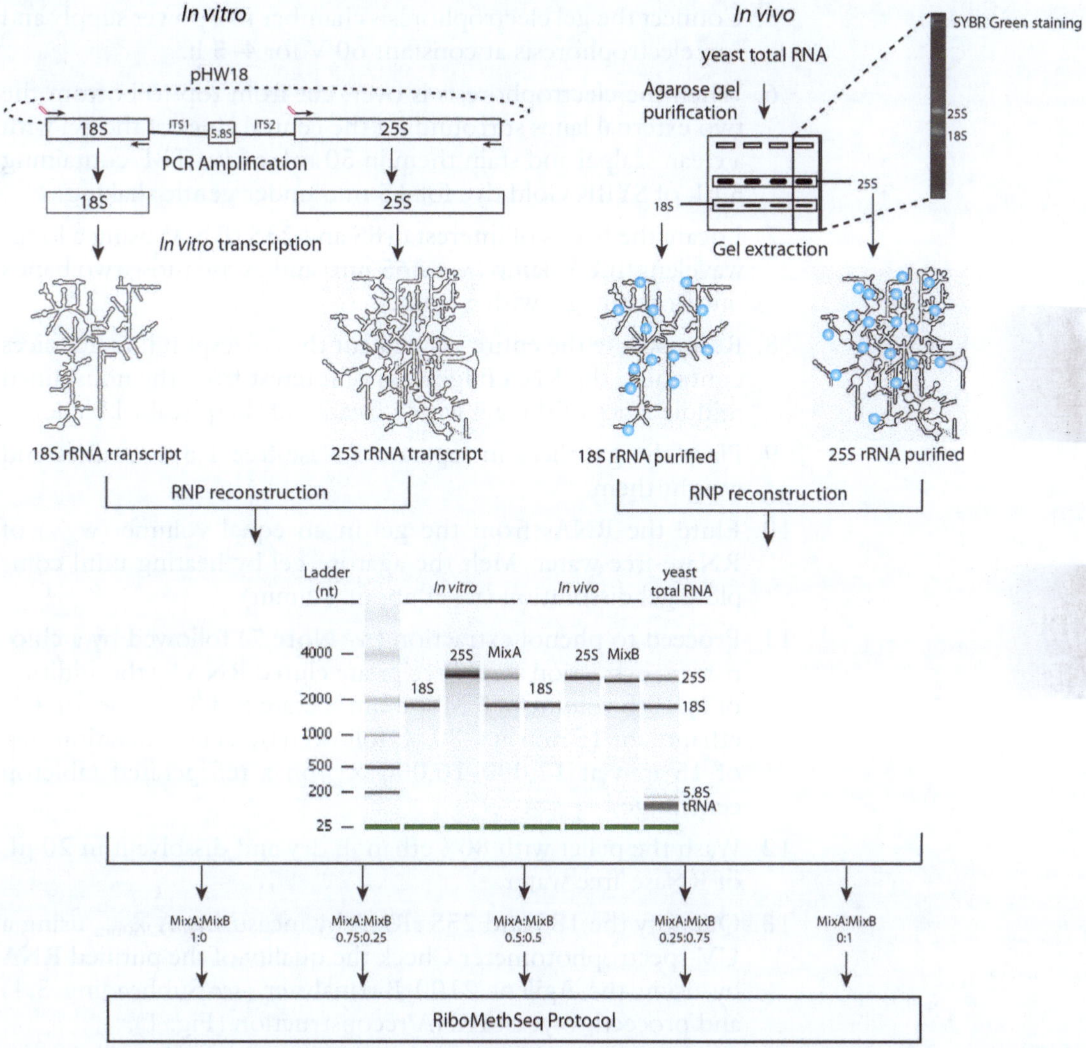

Fig. 1 General overview of 2′-O-methylation quantification using RNP reconstruction for 18S and 25S rRNA. Yeast 18S and 25S full-length rDNA were PCR amplified from plasmid pHW18 using specific forward (containing a T7 RNA polymerase promoter) and reverse primers. The PCR templates were then used for in vitro transcription and RNP reconstruction. In vivo 18S and 25S rRNA were gel-purified using a yeast total RNA preparation and used for RNP reconstruction. In vitro rRNA and in vivo rRNA were equimolar mixed and compared to yeast total RNA preparation and further used in different ratios for RiboMethSeq Protocol

Cool down the gel at room temperature and poor it to a gel chamber with appropriate combs until complete solidification.

2. When the gel is ready, place it in an agarose gel electrophoresis chamber, and then fill the chamber with 1× TBE until the gel surface is covered.
3. In a 1.5 mL RNase-free microtube, prepare a mix containing 1.3 μL of yeast total RNA with 5 μL of gel loading dye and 33.7 μL of RNase-free water.
4. Load 10 μL of the mix into 4 adjacent wells on the gel.
5. Connect the gel electrophoresis chamber to a power supply and run electrophoresis at constant 60 V for 4–5 h.
6. When the electrophoresis is over, cut from top to bottom the two external lanes surrounding the central lanes of the gel with a clean scalpel and stain them in 50 mL of 1× TBE containing 5 μL of SYBR Gold dye for 15 min under gentle shaking.
7. Locate the bands of interest (18S and 25S rRNA) using a long-wavelength UV-lamp ($\lambda = 365$ nm) and excise those two bands out from the gel with a scalpel.
8. Reconstitute the entire gel and cut the corresponding gel slices containing the RNA fragments of interest from the nonstained middle lanes of the gel with a clean and sharp scalpel blade.
9. Place the gel slices in separate RNase-free 1.5 mL tubes and weight them.
10. Elute the RNAs from the gel in an equal volume (w/v) of RNase-free water. Melt the agarose gel by heating until complete gel dissolution (takes about 30 min).
11. Proceed to phenol extraction (*see* **Note 7**) followed by a chloroform extraction step. Precipitate eluted RNA by the addition of 1/10th volume of 3 M sodium acetate and 3 volumes of 96% ethanol for 15 min at −80 °C followed by a centrifugation step of 15 min at 12,000–16,000 × *g* in a refrigerated tabletop centrifuge.
12. Wash the pellet with 80% ethanol, dry and dissolve it in 20 μL of RNase-free water.
13. Quantify the 18S and 25S rRNA by measuring A_{260nm} using a UV-spectrophotometer Check the quality of the purified RNA by using the Agilent 2100 Bioanalyzer (*see* Subheading 3.4) and proceed to yeast rRNA reconstruction (Fig. 1).

3.5 Yeast rRNA Reconstruction

1. Mix equimolar amounts of 18S and 25S in vitro RNA transcripts (*see* **Note 8**) in a 1.5 mL Eppendorf tube to get 100 ng at a final concentration of 10 ng/μL.
2. Follow the same procedure with purified 18S and 25S RNA.

3. Dilute your RNA mixes to 3–5 ng/μL with RNase-free water to be within the optimal range concentration of the assay (Pico RNA chip).
4. Dilute the yeast total RNA preparation used as a control to 3–5 ng/μL with RNase-free water.
5. Add 1–1 μL of your diluted RNA samples to 11 different 1.5 mL tubes already containing 5 μL of RNA marker (green cap vial) (*see* **Note 9**). Mix by pipetting up and down.
6. Mix 1 μL of the ladder (*see* **Note 10**) with 5 μL of RNA marker. Mix by pipetting up and down.
7. Prepare the chip priming station. Adjust the syringe clip to the highest top position.
8. Load 9 μL of the gel–dye mix in the well marked with a "G" surrounded by a black circle.
9. Close the chip priming station properly and press the plunger of the syringe until it is held by the clip.
10. Wait for 30 s and then release the clip.
11. Wait for 5 s until the plunger stops and pull it slowly back to the 1 mL position of the syringe.
12. Open the chip priming station and load 9 μL of the gel–dye mix in the 2 other wells marked "G."
13. Load 9 μL of the conditioning solution (white cap vial) in the well marked "CS."
14. Load 6 μL of the diluted ladder in the well marked with a ladder.
15. Load 6 μL of the diluted RNA samples in the wells marked 1–11.
16. Inspect the chip and make sure that no liquid spill is present on the edges of the wells.
17. Insert the chip in the Agilent 2100 Bioanalyzer and close the lid.
18. Select the following assay "Eukaryote Total RNA Pico series II" in the 2100 expert software screen.
19. Press "Start" to begin the chip to run (*see* **Note 11**).
20. After the run, immediately remove the chip and clean the electrodes with the electrode cleaner filled with 350 mL of RNase-free water (*see* **Note 12**).
21. Analyze the results of the chip (Fig. 1).
22. Inspect visually if the reconstructed mixes are comparable with the control (*see* **Note 13**).

23. Report each area for 18S and 25S rRNA indicated for each lane under the electropherogram and compare them between the lanes.
24. On this basis, adjust the quantity for each in vitro and purified rRNA compared to the control lane.
25. When these in vitro and purified rRNA reconstituted mixes are ready to use (*see* **Note 14**), combine them in different stoichiometric proportions to obtain 1:0; 0.75:0.25; 0.5:0.5; 0.25:0.75, and 0:1 mixes.
26. Proceed with alkaline hydrolysis.

3.6 Alkaline Hydrolysis and RNA Fragmentation Quality Control

3.6.1 Alkaline Hydrolysis

1. Prepare one 1.5 mL tube per sample to be analyzed containing 10 μL of 3 M NaOAc, 1 μL of GlycoBlue, and 1 mL of 96% ethanol for subsequent precipitation of the sample ("precipitation tubes," store at −20 °C until further use).
2. Dilute your RNA samples to a concentration of 10 ng/μL with RNase-free water.
3. To individual PCR tubes, add 5 μL of each of your diluted RNA samples, keep on ice until further use.
4. Add 5 μL of sodium bicarbonate buffer and mix by pipetting up and down.
5. Incubate in a thermal cycler preheated at 95 °C. Start a timer and incubate for 10 min (*see* **Note 15**).
6. Proceed with the next sample every 30 s.
7. Stop each reaction after 10 min at 95 °C by spinning down the PCR microcentrifuge tube and add the whole sample into the corresponding 1.5 mL precipitation tube from **step 1**.
8. Mix by inverting the tubes several times and throw them into liquid nitrogen.
9. Recover your samples from the liquid nitrogen and centrifuge them for 30 min at 4 °C at full speed in a microcentrifuge.
10. Remove the supernatant and make sure not to lose the pellet.
11. Wash with 600 μL of 80% ethanol.
12. Centrifuge your samples for 10 min at 4 °C at full speed.
13. Remove the supernatant.
14. Centrifuge your samples for a short spin.
15. Remove any liquid left.
16. Incubate your samples with open lid for 2 min at 37 °C and 5 min at room temperature.
17. Resuspend the pellet with 20 μL of RNase-free water.

3.6.2 RNA Fragmentation Quality Control

1. Prepare your samples by mixing 5 µL of RNA marker (green cap vial) with 1 µL of your fragmented RNA samples.
2. With the rest of gel–dye mix, load 9 µL in the well marked "G" surrounded with a black circle and proceed as described in Subheading 3.5 from **steps 9** to **19**.
3. Analyze the results obtained, the overall mean size of the RNA fragments should be around 50–100 nts.

3.7 End-Repair

1. Combine 16 µL of your treated RNA samples in a PCR tubes with 2 µL of phosphatase buffer, 1 µL of RiboLock RNase Inhibitor, and 1 µL of Antarctic Phosphatase.
2. Mix by pipetting up and down.
3. Incubate the PCR tubes for 30 min at 37 °C then for 5 min at 70 °C to inactivate the phosphatase and store for indefinite hold at 4 °C in a thermal cycler.
4. Add the following components to the previous mix: 17 µL of RNase-free water, 5 µL of PNK buffer, 5 µL of ATP, 1 µL of RiboLock RNase Inhibitor, 2 µL of PNK enzyme.
5. Incubate in a thermal cycler for 1 h at 37 °C and immediately proceed to the next step.

3.8 RNA Purification

All the reagents except ethanol used for RNA purification are part of RNeasy MinElute Cleanup kit.

1. Transfer the sample to a new 1.5 mL tube, and to adjust the final volume to 100 µL, add 50 µL of RNase-free water.
2. Add 350 µL of RLT buffer, mix by vortexing.
3. Add 675 µL of 96% ethanol and mix by inverting the tube up and down (*see* **Note 16**).
4. Transfer 700 µL of the sample to an RNeasy MinElute spin column (stored at 4 °C until use). Centrifuge for 30 s at 8000 × *g*.
5. Repeat the **step 4** with the rest of the sample. Then, add 500 µL of RPE buffer to the column. Centrifuge for 30 s at 8000 × *g*.
6. Discard the flow-through. Add 750 µL of 80% ethanol. Centrifuge for 2 min at 8000 × *g*.
7. Transfer the column to a new collection tube and centrifuge at full speed for 5 min with the lid open.
8. Transfer the column to a new 1.5 mL tube (provided with the kit). Add 10 µL of RNase-free water in the center of the column filter. Wait for 1 min.
9. Centrifuge at full speed for 1 min to elute. The recovered volume is about 9 µL.

3.9 Library Preparation Using NEBNext® Multiplex Small RNA Library Prep Set for Illumina®

1. Mix 6 μL of RNA sample with 1 μL of 3′SR adaptor (green cap vial) in a PCR tube.
2. Incubate for 2 min at 70 °C in a preheated thermal cycler. Transfer immediately to ice.
3. Add 10 μL of 3′ Ligation Buffer (green cap vial) and 3 μL of 3′ Ligation Enzyme (green cap vial).
4. Incubate for 1 h at 25 °C in a thermal cycler.
5. Add 4.5 μL of RNase-free water and 1 μL of SR RT primer (pink cap vial).
6. Incubate for 5 min at 75 °C, 15 min at 37 °C and 15 min at 25 °C.
7. Within the last 15 min of incubation, add 1.1 × n (n = number of samples) μL of the 5′ SR adaptor (yellow cap vial, previously resuspended in 120 μL of RNase-free water and stored at −80 °C) into an individual PCR tube.
8. Denature the 5′SR adaptor in a thermal cycler for 2 min at 70 °C and immediately place the tube on ice (*see* **Note 17**).
9. Add 1 μL of previously denatured 5′SR adaptor, 1 μL of 5′Ligation Reaction Buffer (yellow cap vial), and 2.5 μL of Ligase Enzyme Mix (yellow cap vial).
10. Incubate for 1 h at 25 °C in a thermal cycler.
11. Add the following components to the adaptor ligated RNA mix from the previous step: 8 μL of First strand synthesis reaction buffer (red cap vial), 1 μL of Murine RNase inhibitor (red cap vial), 1 μL of ProtoScript II reverse transcriptase (red cap vial) and mix well by pipetting up and down.
12. Incubate for 1 h at 50 °C.
13. Immediately proceed to PCR amplification (*see* **Note 18**). Add the following components to the RT reaction mix from the previous step: 50 μL of LongAmp Taq Master Mix (blue cap vial), 2.5 μL of SR primer (blue cap vial), 2.5 μL of index primer (*see* **Note 19**), and 5 μL of RNase-free water. Mix well.
14. Perform the following PCR cycling conditions: 1 cycle of initial denaturation for 30 s at 94 °C, 12–15 cycles of denaturation 15 s at 94 °C, annealing 30 s at 62 °C, extension 15 sec at 70 °C, 1 cycle of final extension for 5 min at 70 °C and store at 20 °C for indefinite hold.

3.10 Purification of the Library Using GeneJET PCR Purification Kit

1. Transfer the PCR mix to a 1.5 mL tube, and add 100 μL of binding buffer. Mix thoroughly.
2. Transfer the solution to the purification column. Centrifuge at full speed for 30 s. Discard the flow-through.

3. Add 700 μL of wash buffer to the column and centrifuge at full speed for 30 s. Discard the flow-through.
4. Centrifuge the empty column for 1 additional min.
5. Transfer the column to a clean 1.5 mL DNA low binding tube. Add 30 μL of Elution buffer to the center of the column membrane and centrifuge at full speed for 1 min.
6. Store the purified library at −20 °C until further use.

3.11 Library Quantification and Quality Assessment

3.11.1 Library Quantification

Before starting the experiments, incubate all solutions of the Qubit dsDNA HS assay kit at room temperature for at least 30 min. The kit provides the concentrated assay reagent, dilution buffer, and prediluted standards.

1. Prepare the dye working solution by diluting the concentrated assay reagent 1:200 in dilution buffer. Prepare 200 μL of working solution for each sample and two additional standards.
2. Prepare the two standards annotated "C" and "D" by mixing 10 μL of standard with 190 μL of working solution.
3. Add working solution to 1 μL of RNA sample to obtain 200 μL in total.
4. Vortex the tubes for 2 s and incubate them for 2 min at room temperature.
5. Insert the tubes into the Qubit® 2.0 Fluorometer and proceed with measurements: on the home screen of the Qubit® 2.0 Fluorometer, choose the type of assay (e.g., "HS DNA") for which you want to perform a new calibration.
6. Press "Yes" to read new standards.
7. When indicated, insert the standard tube and press "Read." Standard #1 and #2 correspond to standards "C" and "D," respectively.
8. Once the calibration is done, insert each sample and press "Read" to make the measurements. Check that the value of your samples is within the assay's range, and press "Calculate Stock Conc" (*see* **Note 20**).

3.11.2 Quality Assessment

Before starting the experiments, incubate all solutions of the Agilent High Sensitivity DNA kit at room temperature for at least 30 min in the dark. Vortex them and spin them down before use.

1. Add 15 μL of High sensitivity DNA dye concentrate (blue cap vial) into a High Sensitivity DNA gel matrix vial (red cap vial).
2. Vortex for 10 s and transfer the gel–dye mix to the center of the spin filter.
3. Centrifuge for 10 min at 2240 × *g* (*see* **Note 21**).

4. Add 1 µL of each of your library to 11 different tubes of 1.5 mL already containing 5 µL of RNA marker (green cap vial). Mix by pipetting up and down.
5. Mix 1 µL of the ladder (yellow cap vial) with 5 µL of High sensitivity DNA marker (green cap vial). Mix by pipetting up and down.
6. Prepare the chip priming station. Adjust the syringe clip to the lowest top position.
7. Load 9 µL of the gel–dye mix in the well marked with a "G" surrounded by a black circle.
8. Close the chip priming station properly and press the plunger of the syringe until it is held by the clip.
9. Wait for 1 min and then release the clip.
10. Wait for 5 s. Until the plunger stops and pull it slowly back to the 1 mL position of the syringe.
11. Open the chip priming station and load 9 µL of the gel–dye mix in the 3 other wells marked "G."
12. Load 6 µL of the diluted ladder in the well marked with a ladder.
13. Load 6 µL of the diluted library samples in the wells labeled 1–11.
14. Insert the chip in the Agilent 2100 Bioanalyzer, close the lid and select the following assay "High Sensitivity DNA" in the 2100 expert software screen.
15. Press "Start" to begin the chip to run.
16. After the run, immediately remove the chip and clean the electrodes with the electrode cleaner filled with 350 µL of RNase-free water.
17. Analyze the results of the chip. The range size of each library should be between 150 and 300 bp.

3.12 Library Sequencing

1. For sequencing, libraries are multiplexed and diluted to 8–12 pM final concentration.
2. Sequencing depth or coverage depends on the length of target RNA and can roughly be estimated at 1000 reads per RNA nt.
3. Sequencing length may vary from 35 to 50 nt in a single read mode.
4. Analyze the results (Fig. 2).

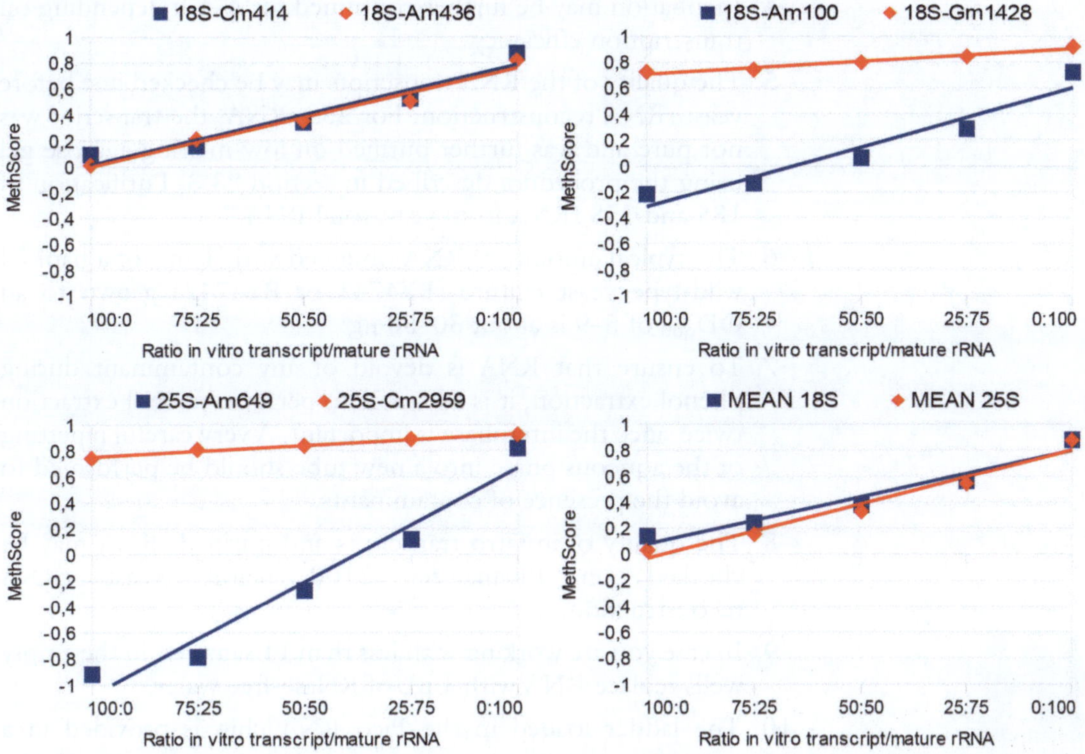

Fig. 2 MethScore values for 2′-*O*-methylation sites in 18S and 25S rRNA for various proportions of unmodified T7 rRNA transcript and naturally modified rRNA from yeast. Representative 2′-*O*-methylation sites were selected and their calculated MethScores are traced according to proportion of modified and unmodified rRNA sequence in the mix. Five points were measured: 100:0 (100% of in vitro rRNA transcript) to 0:100 (100% of mature rRNA), with intermediate 75:25, 50:50, and 25:75 ratios. The right bottom panel shows the mean MethScore values for all sites in 18S and 25S rRNA. Linear regression curves are also plotted. Correlation of observed MethScore and methylation rate is >0.95 for all tested positions

4 Notes

1. Plasmid pHW18 was a generous gift from Dr. D. Tollervey. It was used as a template to amplify yeast 18S and 25S rDNA.
2. The kit NEBNext® Multiplex Small RNA Library Prep Set for Illumina® (set 1) (NEB, E7300S) includes a set of 12 barcoding primers (numbered 1–12) that will be used for multiplexing reactions during PCR amplification. There is also a version set 2 with primers (numbers 13–24). If you do not need these barcoding primers, you may order a similar kit without the primers and use any other source of barcoding primers (Illumina, Epicentre, NEB).
3. After PCR, amplification may be checked by loading 1/10 of the PCR reaction on a 1.5% agarose gel.

4. Incubation may be further continued for 4–6 h depending on transcription efficiency.
5. The quality of the RNA transcripts may be checked just before yeast rRNA reconstruction. For 25S rRNA, the transcript was not pure and was further purified on low-melting agarose gel using the procedure described in section "3.3. Purification of 18S and 25S rRNA from yeast total RNA."
6. The typical amount of RNA obtained with 1 mL of a haploid wild-type yeast culture (BY4741 or BY4742) grown to an OD_{600} of 5–9 is about 30–50 μg.
7. To ensure that RNA is devoid of any contaminant during phenol extraction, it is advisable to perform phenol extraction twice since the interphase is important. A very careful pipetting of the aqueous phase into a new tube should be performed to avoid the presence of contaminants.
8. The quality of in vitro transcripts and purified rRNA may be checked on Bioanalyzer 2100 before yeast rRNA reconstruction.
9. In case you are working with less than 11 samples, in the empty wells replace RNA with 1 μL of RNase-free water.
10. The ladder loaded in the Pico RNA chip is provided in a separate package and may be prepared before the experiment: spin down the tube and transfer 10 μL to a RNase-free tube. Heat for 2 min at 70 °C. Cool down on ice and add 90 μL of RNase-free water. Prepare 5 μL aliquots using the Safe-Lock PCR tubes provided in the kit and store them at −70 °C. Before use, thaw one tube and keep it on ice. The ladder is quite stable at −70 °C and may be used at least 4 months.
11. The Agilent 2100 Bioanalyzer is very sensitive to vibrations and this may affect your results. Therefore make sure that no vibrations will occur during the run.
12. RNase contamination problems of the Bioanalyzer electrodes are very frequent and will affect the RNA integrity number of your samples. Therefore, if the Agilent 2100 Bioanalyzer is also frequently used to run DNA chips, it is strongly recommended to use a dedicated electrode cartridge only for RNA assays. In addition, we recommend for each chip to load an internal RNA control (total RNA preparation with a known RIN > 9). If you encounter contamination problems, soak the electrode cartridge into an RNaseZap® decontamination solution (Ambion) for at least 10 min, then rinse the electrodes with RNase-free water and let them dry out for at least one night.
13. The RNA quantity may be decreased to a minimal starting amount of 5–10 ng without considerably affecting coverage and calculation of the MethScore.

14. The rRNA ratio 25S/18S calculated for yeast total RNA preparation used as a control should be equivalent to 25S and 18S rRNA in vitro and in vivo mixes.
15. If the in vitro transcripts or total RNA samples (from yeast, human, and bacteria) are of good quality (RIN > 8), fragmentation time is around 10 min. However this time may be decreased with RNA of poor quality (4–6 min) or may be increased with RNA species of higher stability (i.e., tRNAs) (12–14 min). We recommend testing 3–4 different times of fragmentation to define the appropriate conditions for hydrolysis. In general for long RNAs the optimal size distribution is around 50–100 nt, while for short RNAs (<200 nts), 20–50 nt is the appropriate size distribution.
16. Ethanol quantity is increased compared to the manufacturer's recommendations in order not to lose the small RNA fragments during the RNA binding to the silica membrane.
17. Do not leave the heated adapter on ice for more than 5–10 min before proceeding to the next step; this may impact your library preparation.
18. We recommend proceeding immediately with PCR amplification. However, if it is not possible, inactivate the RT by heating for 15 min at 70 °C and cool down the reaction at 20 °C for 1–3 h or safely store the reactions at −20 °C for overnight.
19. Make sure to use only combinations of compatible primers for barcoding. Most Illumina sequencers use a green laser (or LED) to read G and T nucleotides and a red laser (or LED) to read A and C nucleotides. Within each sequencing cycle, at least one nucleotide for each color channel must be read in the index to ensure proper reading of the barcode sequence. Use as a reference the following guide (ScriptSeq™ Index PCR primers, Illumina) for verification of barcode compatibility or check compatibility with Illumina Experimental Manager software.
20. This quantification step is crucial. Make sure to quantify all your libraries properly since an underestimated or overestimated quantification will interfere with subsequent sequencing reads proportion and quality.
21. The High Sensitivity DNA gel–dye mix is stable for 1 month at 4 °C protected from light.

Acknowledgments

This work was supported by joint ANR-DFG grant HTRNAMod (ANR-13-ISV8-0001/HE 3397/8-1) and AO Lorraine University-Lorraine Region "Aberrant RNA methylation in cancer" funding to YM.

References

1. Liu N, Pan T (2016) N6-methyladenosine-encoded epitranscriptomics. Nat Struct Mol Biol 23(2):98–102. doi:10.1038/nsmb.3162
2. Fu Y, Dominissini D, Rechavi G, He C (2014) Gene expression regulation mediated through reversible m(6)A RNA methylation. Nat Rev Genet 15(5):293–306. doi:10.1038/nrg3724
3. Dominissini D, Moshitch-Moshkovitz S, Schwartz S, Salmon-Divon M, Ungar L, Osenberg S, Cesarkas K, Jacob-Hirsch J, Amariglio N, Kupiec M, Sorek R, Rechavi G (2012) Topology of the human and mouse m6A RNA methylomes revealed by m6A-seq. Nature 485(7397):201–206. doi:10.1038/nature11112
4. Dominissini D, Nachtergaele S, Moshitch-Moshkovitz S, Peer E, Kol N, Ben-Haim MS, Dai Q, Di Segni A, Salmon-Divon M, Clark WC, Zheng G, Pan T, Solomon O, Eyal E, Hershkovitz V, Han D, Dore LC, Amariglio N, Rechavi G, He C (2016) The dynamic N(1)-methyladenosine methylome in eukaryotic messenger RNA. Nature 530(7591):441–446. doi:10.1038/nature16998
5. Carlile TM, Rojas-Duran MF, Zinshteyn B, Shin H, Bartoli KM, Gilbert WV (2014) Pseudouridine profiling reveals regulated mRNA pseudouridylation in yeast and human cells. Nature 515(7525):143–146. doi:10.1038/nature13802
6. Schwartz S, Bernstein DA, Mumbach MR, Jovanovic M, Herbst RH, Leon-Ricardo BX, Engreitz JM, Guttman M, Satija R, Lander ES, Fink G, Regev A (2014) Transcriptome-wide mapping reveals widespread dynamic-regulated pseudouridylation of ncRNA and mRNA. Cell 159(1):148–162. doi:10.1016/j.cell.2014.08.028
7. Lovejoy AF, Riordan DP, Brown PO (2014) Transcriptome-wide mapping of pseudouridines: pseudouridine synthases modify specific mRNAs in *S. cerevisiae*. PLoS One 9(10): e110799. doi:10.1371/journal.pone.0110799
8. Squires JE, Patel HR, Nousch M, Sibbritt T, Humphreys DT, Parker BJ, Suter CM, Preiss T (2012) Widespread occurrence of 5-methylcytosine in human coding and non-coding RNA. Nucleic Acids res 40(11):5023–5033. doi:10.1093/nar/gks144
9. Schaefer M, Pollex T, Hanna K, Lyko F (2009) RNA cytosine methylation analysis by bisulfite sequencing. Nucleic Acids res 37(2):e12. doi:10.1093/nar/gkn954
10. Birkedal U, Christensen-Dalsgaard M, Krogh N, Sabarinathan R, Gorodkin J, Nielsen H (2015) Profiling of ribose methylations in RNA by high-throughput sequencing. Angew Chem Int 54(2):451–455. doi:10.1002/anie.201408362
11. Marchand V, Blanloeil-Oillo F, Helm M, Motorin Y (2016) Illumina-based RiboMeth-Seq approach for mapping of 2′-*O*-Me residues in RNA. Nucleic Acids res 44(16):e135. doi:10.1093/nar/gkw547
12. Schmitt ME, Brown TA, Trumpower BL (1990) A rapid and simple method for preparation of RNA from *Saccharomyces cerevisiae*. Nucleic Acids Res 18(10):3091–3092. PMCID: PMC330876
13. Ryohei U (2008) A rapid protocol for isolation of intact RNA fraction of specific molecular weight for synthesis of large-sized cDNAs. Ann Microbiol 58:765–769

Chapter 3

Identifying the m^{6}A Methylome by Affinity Purification and Sequencing

Phillip J. Hsu and Chuan He

Abstract

N^6-methyladenosine (m^{6}A) is the most abundant internal modification in eukaryotic mRNA, and is newly emerging as a key posttranscriptional mRNA regulator. Recent research has uncovered insight into the location and function of m^{6}A sites on a large scale, in part due to the transcriptome-wide identification of m^{6}A sites by high-throughput sequencing (m^{6}A-seq). Here, we present a protocol for m^{6}A-seq, which maps the m^{6}A methylome by affinity purification and sequencing.

Key words N^6-methyladenosine, Transcriptome, Methylome, Affinity purification, Sequencing

1 Introduction

Posttranscriptional modifications in eukaryotic messenger RNAs (mRNAs) play a crucial role in fine-tuning gene expression in response to changing conditions. Newly emerging as a key posttranscriptional mRNA regulator is N^6-methyladenosine (m^{6}A), the most abundant internal modification in mRNA [1]. m^{6}A regulates mRNA stability, localization, and translation; and is essential for life [2–4]. Research on m^{6}A, discovered over 4 decades ago, initially advanced slowly due to lack of available methods. Previous methods to identify m^{6}A sites were time-consuming and low-throughput, relying on isolation of specific transcripts followed by fragmentation to single nucleotides and biochemical analysis. Such analytical methods had the capacity to analyze only several transcripts at a time—much less the entire methylome. However, recent advances in next-generation sequencing have brought about techniques to gain insight into the location and function of m^{6}A sites on a large scale. In particular, m^{6}A-seq, a high-throughput sequencing method introduced in 2012 to accurately identify m^{6}A sites throughout the transcriptome, has begun to reveal the landscape of the m^{6}A methylome [5, 6].

Imre Gaspar (ed.), *RNA Detection: Methods and Protocols*, Methods in Molecular Biology, vol. 1649, DOI 10.1007/978-1-4939-7213-5_3,

m^6A-seq consists of affinity purification of m^6A-marked RNA fragments using an m^6A-specific antibody, followed by high-throughput sequencing and identification of m^6A-enriched sites. The affinity purified fragments are around 100–150 nucleotides long, and may contain the m^6A at any location. Analysis of the overlap between the fragments and the genome, as well as a comparison of fragments from affinity purified ("IP") and unpurified ("input") samples, allows organization of the fragments into "peaks." Identification of the midpoint of the peak helps to localize the m6A. Because m^6A is typically present on a consensus motif of DRACH (D = A, G, U; R = A, G; H = A, C, U), identification of this motif may help further identify the exact location of the m^6A [5, 7].

m^6A-seq is nonetheless limited in its ability to pinpoint m^6A at single-nucleotide resolution. Most DRACH sites across the genome do not contain m^6A, and peaks may not have a clear midpoint due to uneven fragmentation or clusters of m^6A residues. Recent updates to m^6A-seq have improved on this limitation. A photocrosslinking-assisted m^6A-sequencing strategy, PA-m^6A-seq, provides a higher resolution transcriptome-wide map of m^6A, and miCLIP, a method based on UV-induced antibody-RNA crosslinking, provides single-nucleotide resolution localization of m^6A across the transcriptome [8, 9]. Furthermore, m^6A-seq does not provide a stoichiometric analysis of the fraction of modified transcripts. A recent improvement, m^6A-LAIC-seq, quantitatively identifies the proportion of a m^6A-containing transcripts of a given gene (9, reviewed in 10). Although these limitations of m^6A-seq have been met, a remaining challenge and area of active research is the development of a method to sequence m6A in low-abundance samples or in single cells.

Here we present our protocol for m^6A-seq. Similar protocols have been published elsewhere [6, 12–14]. A detailed bioinformatics analysis pipeline has been published by the author of the original protocol [12].

2 Materials

Prepare all solutions using DEPC-treated nuclease-free water and molecular grade reagents. Be sure to prepare all reagents supplemented with SUPERase inhibitor or BSA freshly, as integrity may be compromised if left at 4 °C or RT.

2.1 Reagents

1. Cultured cells or tissues as a source of RNA. Any cell line or tissue is suitable for this procedure. As a reference, one confluent 10 cm plate of HeLa cells provides ~100 μg of total RNA, 2–5% of which is mRNA. At least 1 μg of polyA-selected mRNA is required.

2. RNeasy Plus Mini Kit (Qiagen, *see* **Note 1**).
3. mRNA Miniprep Kit (e.g., Sigma-Aldrich GenElute).
4. Gel Extraction Kit (e.g., Qiagen MinElute)
5. 4–20%TBE Gels, 10 well.
6. 0.5× TBE buffer.
7. RNA Loading Dye, (2×).
8. Low Range ssRNA Ladder.
9. SYBR Gold Nucleic Acid Gel Stain (10,000× Concentrate in DMSO).
10. TruSeq Stranded mRNA Library Preparation Kit Set A (Illumina).
11. β-mercaptoethanol.
12. 100% ethanol.
13. 70% ethanol, freshly prepared in DEPC-treated nuclease-free water.
14. 3 M sodium acetate pH 5.2.
15. 2.5% low melting point agarose gel with 0.5 μg/mL ethidium bromide.
16. 1× TAE buffer.
17. 50 mg/mL UltraPure BSA.
18. PBS, sterile.
19. Protein A beads for Immunoprecipitation (e.g., Thermo Scientific Dynabeads).
20. SUPERase In RNase Inhibitor (20 U/μL) (Ambion).
21. Agencourt AMPure XP Beads (Beckman Coulter).
22. FastStart Essential DNA Green Master (Roche).
23. 5× IP buffer: 50 mM Tris–HCl, pH 7.4, 750 mM NaCl, and 0.5% NP-40.
24. Wash buffer (kept on ice, prepared fresh): 5× IP buffer diluted to 1× in nuclease-free water, supplemented with 0.1% SUPERase inhibitor.
25. 20 mM m^6A solution: dissolve 10 mg of m^6A monophosphate sodium salt in 1.3 mL of DEPC-treated nuclease-free water. Aliquot and store at −20 °C; use within 12 months.
26. m^6A-specific antibody solution, 0.5 mg/mL: reconstitute 50 μg of affinity purified anti-m^6A rabbit polyclonal antibody (Synaptic Systems, cat. no. 202003) in 100 μL of DEPC-treated nuclease-free water. Aliquot and store at −20 °C; avoid multiple freeze–thaw cycles; use within 12 months.
27. Blocking buffer: wash buffer supplemented with 0.5 mg/mL UltraPure BSA.

28. Elution buffer (100 μL per reaction, kept on ice, prepared fresh): 20 μL of 5× IP buffer, 1 μL of SUPERase inhibitor, 33 μL of 20 mM m^6A solution, 46 μL of DEPC-treated nuclease-free water. Final concentrations: 1× IP buffer; 6.7 mM m^6A.

2.2 Equipment

1. Low-adhesion microcentrifuge tubes (1.5–1.75 mL).
2. Thin-walled PCR tubes with flat cap.
3. 0.65 mL Bioruptor Pico Microtubes (Diagenode).
4. Heating block.
5. Refrigerated benchtop microcentrifuge (capable of >16,000 × *g*).
6. Cell scrapers.
7. Magnetic rack for 1.6 mL tubes.
8. Head-over-tail rotator.
9. Thermal cycler.
10. Vortex mixer.
11. Spectrophotometer (e.g., NanoDrop Technologies ND-1000 or equivalent).
12. Sonication device (e.g., Diagenode Bioruptor Pico or equivalent).
13. Pipettes.
14. Pipette tips with filters.
15. Gel electrophoresis system.
16. Weigh boats.
17. Weighing scale.
18. Transilluminator.
19. Gel imager.
20. Cell lifters.

3 Methods

3.1 RNA Isolation

Timing: 3 h

1. Remove media from the cells by pouring or pipetting, and wash the cells gently with 10 mL of ice-cold PBS (*see* **Note 2**). Add 2 mL of ice-cold PBS to the cells, and scrape the cells from the plate using a cell lifter. Pipette the suspended cells into a 15 mL tube, and centrifuge at 4 °C for 5 min at 300 × *g*. Carefully remove the supernatant, and proceed immediately to **step 2**.

2. Isolate the RNA using the RNeasy kit following the manufacturer's protocol, being sure to use the gDNA Eliminator Columns to remove genomic DNA. Elute using 100 μL of DEPC-treated nuclease-free water.
3. (Recommended): determine RNA integrity using an Agilent 2100 Bioanalyzer or by agarose gel electrophoresis.
4. If desired, isolate the mRNA using the GenElute mRNA miniprep kit (*see* **Note 3**). Perform both the first and second elution using 50 μL of DEPC-treated nuclease-free water rather than the supplied elution buffer, as the components of the elution buffer may interfere with downstream steps. Measure the concentration of the mRNA via spectrophotometer.

3.2 RNA Fragmentation, Antibody Binding, and Affinity Purification

Timing: Day 1—8 to 11 h + overnight incubation; Day 2—1 h

1. Adjust the volume of 1 μg of polyA-enriched mRNA to 100 μL in a 0.65 mL Bioruptor Pico Microtube (*see* **Note 4**). Insert the tubes into the Bioruptor Pico sonication device prechilled to 4 °C. Fragment the RNA for 30 cycles of 30 s on/30 s off, which will produce fragments of 100–150 nucleotides.
2. (Recommended): verify RNA fragmentation using an Agilent 2100 Bioanalyzer or by TBE gel electrophoresis (*see* **Note 5**).
3. Save 5% (5 ng) of the fragmented mRNA as input at −80 °C.
4. Make IP mixture (500 μL per reaction): 950 ng (95 μL) of fragmented mRNA, 100 μL of 5× IP buffer, 12.5 μL of m^6A-specific antibody (0.5 mg/mL), 5 μL of SUPERase inhibitor, and 287.5 μL of DEPC-treated nuclease free water.
5. Rotate the IP mixture on a head-over-tail rotor at 4 °C for 2 h (*see* **Note 6**).
6. Gently resuspend the Protein A beads using a vortex mixer. Wash 60 μL of Protein A beads per reaction 3 times in ice-cold wash buffer using a magnetic rack. Resuspend the Protein A beads in 500 μL of blocking buffer and rotate for 1 h (*see* **Note** 7).
7. Remove the blocking buffer from the Protein A beads, and wash twice with 500 μL of wash buffer. Remove the wash buffer, and add the IP mixture to the washed Protein A beads, and rotate the mixture on a head-over-tail rotor at 4 °C for 2 h.
8. Wash the beads–RNA mixture in wash buffer 3 times using a magnetic rack.
9. Add 50 μL of elution buffer to the beads, and incubate the mixture for 1 h at 4 °C with continuous shaking (*see* **Note 8**).
10. Remove and save the supernatant as eluent in a 1.5 mL low adhesion tube, keeping it on ice.
11. Repeat **steps 9** and **10** and combine the two eluents.

12. Add 0.1 volumes (10 μL) of 3 M sodium acetate to the combined eluents, followed by 2.5 volumes (275 μL) of 100% ethanol. Mix well, and incubate the sample at −80 °C overnight, or longer if desired (the RNA is stable at −80 °C at this point). Do not add a glycogen carrier to the mixture, as it may interfere with downstream steps by precipitating free m^6A (*see* **Note 9**).
13. Centrifuge the tube at 16,000 × *g* at 4 °C for 25 min. Remove the supernatant, being careful to not disturb the pellet at the bottom of the tube. The pellet will not be visible.
14. Wash the pellet with 1 mL of 70% ethanol in DEPC-treated nuclease free water, and centrifuge the tube at 16,000 × *g* at 4 °C for 15 min. Remove all of the supernatant, being careful to not disturb the pellet at the bottom of the tube. Let the pellet air-dry on ice for 10 min, and resuspend the pellet with 13 μL of DEPC-treated nuclease free water. Do not measure the RNA concentration, as it is too low to measure at this point.

3.3 Library Preparation, Quality Control, and Sequencing

Timing: 1–2 days for library preparation and quality control; 5 days for sequencing

1. Use 5 μL of the affinity purified m^6A-enriched RNA ("IP") and 2 μL of the input saved in Subheading 3.2 as starting material for mRNA library preparation. Transfer the sample to an 8-tube PCR strip. Add 3 μL of DEPC-treated nuclease-free water to the input sample. Save the remaining mRNA as backup in case library preparation fails.
2. Add 12 μL of *Fragment, Prime, Finish Mix* from the Illumina TruSeq Stranded mRNA Library Preparation Kit to each sample. Place the PCR strip containing the sample onto the thermal cycler, and run at 94 °C for 20 s, 4 °C hold, to prepare the RNA for first strand synthesis. Immediately proceed to the next step.
3. Perform first strand cDNA synthesis, second strand cDNA synthesis, adenylation, and adapter ligation according to the manufacturer's instructions.
4. Measure the concentration of the cDNA library using qPCR: use 10 μL of 2× Roche FastStart Essential DNA Green Master mix, 2 μL of the Illumina PCR Primer Cocktail, 7 μL of DEPC-treated nuclease-free water, and 1 μL of the cDNA library. Determine the C_t at the midpoint of the qPCR amplification curve.
5. Enrich DNA fragments using the PCR kit provided in the TruSeq Stranded mRNA Library Preparation Kit. Use three fewer cycles than the C_t at the midpoint of the qPCR

amplification curve calculated in the previous step. Adjust the amount of Input and IP cDNA template so that the same number of PCR cycles can be applied (*see* **Note 10**).

6. Purify the cDNA from the PCR mixture using Agencourt AMPure XP Beads following the manufacturer's instructions.
7. Analyze RNA size on an Agilent 2100 Bioanalyzer. Ensure that fragments are 200–250 bp in length, and that few or no dimers are present (dimers are seen at ~127 bp).
8. If the Bioanalyzer analysis indicates that primer dimers are present, perform size selection using gel electrophoresis:
9. Run the PCR product on a 2.5% low melting point agarose gel with 0.5 μg/mL ethidium bromide in 1× TAE buffer at 90 V for 40 min along with the DNA marker provided in the Illumina kit.
10. Use a transilluminator to image the gel. Use a clean, sharp scalpel to isolate the higher band around 200–250 bp, which contains the cDNA library, being careful to avoid the lower band at 127 bp, which consists of primer dimers.
11. Weigh the gel fragment in a colorless tube, and add 3 volumes of Buffer QG from the MinElute Gel Extraction Kit to the gel slice.
12. Let the gel slice completely dissolve in the Buffer QG at room temperature (~10 min).
13. Isolate the cDNA library by following the remaining steps of the MinElute Gel Extraction Kit manufacturer's protocol.
14. Elute with 10 μL of DEPC-treated nuclease-free water (*see* **Note 11**).
15. Deep sequence the cDNA library using an Illumina HiSeq platform (or similar).

4 Notes

1. Although the RNeasy Plus Mini Kit is an efficient method to purify total RNA from limited samples of animal cells or tissues, it isolates only RNA molecules longer than 200 nt. This procedure enriches mRNA species, since RNAs below the <200 nucleotide cutoff, such as miRNAs, tRNAs, and smaller rRNAs, are excluded. If the user wishes to analyze RNA species shorter than 200 nt, an alternative is to use the Direct-zol™ RNA Miniprep Plus kit (Zymo Research)with on-column DNAse I digestion, following the manufacturer's protocol.
2. Do not detach the cells from the plate using trypsin, as this may compromise RNA integrity. If using easily detached cells, such

as HEK293T, pipette the ice-cold PBS gently onto the edge of the cell culture dish when washing to minimize loss of cells. Work quickly when scraping cells; scrape the cells in a 4 °C cold room if possible to minimize RNA degradation. Be sure to keep all samples on ice throughout the entirety of the procedure.

3. It is generally not necessary to remove any trace amounts of ribosomal RNA left over from the mRNA purification step, as the majority of the reads will be of mRNA. However, if the sequencing results show that there is too much ribosomal RNA, it is possible to perform a second round of polyA selection using the GenElute kit, or to further deplete ribosomal RNA using RiboMinus Eukaryote System v2 (Ambion). If you wish to analyze pre-mRNA as well as mRNA, it is possible to extract total RNA using the RNeasy kit, which will remove small RNAs, and then use the RiboMinus Eukaryote System v2 directly to remove ribosomal RNAs.
4. Be sure to use at least 1 μg of mRNA as starting material, or it may not be possible to construct the library. If library construction fails, consider starting with more mRNA, i.e., 2–10 μg.
5. If using TBE gel electrophoresis: wash the wells of a 4–20% TBE gel, thoroughly with 0.5× TBE buffer, and prerun for 10 min at 180 V at 4 °C in 0.5× TBE buffer. Dilute 15 ng (1.5 μL) of fragmented RNA to 10 μL, and add 10 μL of RNA Loading Dye, (2×). Denature the RNA at 70 °C for 3 min. Carefully load the RNA onto the gel along with 15 ng of NEB Low Range ssRNA ladder in an adjacent lane, and run at 180 V for 50 min at 4 °C. Add 3 μL of SYBR Gold Nucleic Acid Gel Stain to 30 mL fresh 0.5× TBE buffer, and mix well. Carefully remove the gel and allow it to soak in the SYBR Gold Nucleic Acid Gel Stain diluted in 0.5× TBE buffer for 10 min at room temperature. Image using a gel imager. The nucleic acid size should center around 100–150 nucleotides.
6. Be sure to use Low Adhesion tubes to allow more thorough washing and mixing, and to avoid sample loss. Be sure to wrap the caps of the tubes in Parafilm to prevent them from accidentally popping open.
7. Minimize bead loss by using low retention pipette tips. For each wash, remove the tube from the magnetic stand, and make sure the beads are well resuspended. Place the tube back on the magnetic rack, and make sure that all beads have cleared from solution before removing the supernatant by waiting 30–60 s.
8. Competitive elution produces less background compared to solvent extraction of the bead-antibody solution.
9. Do not mix the sodium acetate and isopropanol beforehand, as the mixture may precipitate.

10. It is recommended to perform the PCR reaction at half scale, saving half of the template in case more or fewer PCR cycles are needed.
11. DNA libraries may be stored at −20 °C until sequencing.

References

1. Desrosiers R, Friderici K, Rottman F (1974) Identification of methylated nucleosides in messenger RNA from Novikoff hepatoma cells. Proc Natl Acad Sci U S A 71:3971–3975. doi:10.1073/pnas.71.10.3971
2. Wang X, Lu Z, Gomez A et al (2014) N6-methyladenosine-dependent regulation of messenger RNA stability. Nature 505:117–120. doi:10.1038/nature12730
3. Wang X, Zhao BS, Roundtree IA et al (2015) N^6-methyladenosine modulates messenger RNA translation efficiency. Cell 161: 1388–1399. doi:10.1016/j.cell.2015.05.014
4. Bokar JA (2005) The biosynthesis and functional roles of methylated nucleosides in eukaryotic mRNA. In: Grosjean H (ed) Fine-tuning of RNA functions by modification and editing, vol 12. Springer, Berlin, pp 141–177. doi:10.1007/b106365
5. Dominissini D, Moshitch-Moshkovitz S, Schwartz S et al (2012) Topology of the human and mouse m6A RNA methylomes revealed by m6A-seq. Nature 485:201–206. doi:10.1038/nature11112
6. Meyer KD, Saletore Y, Zumbo P et al (2012) Comprehensive analysis of mRNA methylation reveals enrichment in 3′ UTRs and near stop codons. Cell 149:1635–1646. doi:10.1016/j.cell.2012.05.003
7. Wei C, Moss B (1977) Nucleotide sequences at the N6-methyladenosine sites of HeLa cell messenger ribonucleic acid. Biochemistry 16:1672–1676. doi:10.1021/bi00627a023
8. Chen K, Lu Z, Wang X et al (2014) High-resolution N(6)-methyladenosine (m(6) A) map using photo-crosslinking-assisted m(6) A sequencing. Angew Chem Int Ed Engl:1–5. doi:10.1002/anie.201410647
9. Linder B, Grozhik A V, Olarerin-George AO et al (2015) Single-nucleotide-resolution mapping of m6A and m6Am throughout the transcriptome. Nat Methods 12:767–772. doi:10.1038/nmeth.3453
10. Molinie B, Wang J, Lim KS, et al (2016) m 6 A-LAIC-seq reveals the census and complexity of the m 6 A epitranscriptome. doi: 10.1038/nmeth.3898
11. Shi H, He C (2016) A glance at N6-methyladenosine in transcript isoforms. Nat Methods 13:624–625. doi:10.1038/nmeth.3928
12. Dominissini D, Moshitch-Moshkovitz S, Salmon-Divon M et al (2013) Transcriptome-wide mapping of N(6)-methyladenosine by m(6)A-seq based on immunocapturing and massively parallel sequencing. Nat Protoc 8:176–189. doi:10.1038/nprot.2012.148
13. Dominissini D, Moshitch-Moshkovitz S, Amariglio N, Rechavi G (2015) Transcriptome-wide mapping of N6-methyladenosine by m6A-seq, 1st ed. RNA Modif. doi: 10.1016/bs.mie.2015.03.001
14. Schwartz S, Agarwala SD, Mumbach MR et al (2013) High-resolution mapping reveals a conserved, widespread, dynamic mRNA methylation program in yeast meiosis. Cell 155: 1409–1421. doi:10.1016/j.cell.2013.10.047

Chapter 4

PARIS: Psoralen Analysis of RNA Interactions and Structures with High Throughput and Resolution

Zhipeng Lu, Jing Gong, and Qiangfeng Cliff Zhang

Abstract

RNA has the intrinsic propensity to form base pairs, leading to complex intramolecular and intermolecular helices. Direct measurement of base pairing interactions in living cells is critical to solving transcriptome structure and interactions, and investigating their functions (Lu and Chang, Curr Opin Struct Biol 36:142–148, 2016). Toward this goal, we developed an experimental method, PARIS (Psoralen Analysis of RNA Interactions and Structures), to directly determine transcriptome-wide base pairing interactions (Lu et al., Cell 165(5):1267–1279, 2016). PARIS combines four critical steps, in vivo cross-linking, 2D gel purification, proximity ligation, and high-throughput sequencing to achieve high-throughput and near-base pair resolution determination of the RNA structurome and interactome in living cells. In this chapter, we aim to provide a comprehensive discussion on the principles behind the experimental and computational strategies, and a step-by-step description of the experiment and analysis.

Key words Psoralen, Cross-linking, AMT (4′-aminomethyltrioxsalen hydrochloride), RNA structure, RNA–RNA interaction, 2D gel electrophoresis, Proximity ligation, High-throughput sequencing

1 Introduction

RNA structures and interactions are the building blocks of many macromolecular machines that drive basic cellular processes. Notable examples include the rRNAs and tRNAs in protein synthesis, spliceosomal snRNAs in intron removal, microRNAs in mRNA repression, small nucleolar and small cajal body RNAs (snoRNAs/scaRNAs) in guiding RNA modifications, and ribozymes in catalyzing a variety of chemical reactions [1, 2]. The recent advent of high-throughput sequencing has led to the discovery of large numbers of long ncRNAs whose functions are under active investigations [3]. A picture is emerging for the lncRNAs where a multitude of sequence and structure motifs act as scaffolds in guiding the formation of functional protein–RNA complexes [4]. Even the previously well-studied smaller RNAs have been suggested to play more regulatory roles due to the variations in

Imre Gaspar (ed.), *RNA Detection: Methods and Protocols*, Methods in Molecular Biology, vol. 1649, DOI 10.1007/978-1-4939-7213-5_4, © Springer Science+Business Media LLC 2018

sequence, structure, expression pattern, RNA–protein complex compositions, and functions [5–8].

A detailed description of the physical basis of RNA molecules is an important step toward understanding their functions in physiology and pathology; however, structural analysis of RNA has been a daunting task for the entire RNA field. Since the first crystal structure was solved for the phenylalanine tRNA in the 1970s [9, 10], great progress has been made in the high-resolution analysis of many purified and "well-behaving" RNA molecules. Nevertheless, the vast majority of RNAs in living cells are large, highly dynamic and complex, and thus their structures are very challenging to solve by using most conventional methods, such as crystallography, nuclear magnetic resonance (NMR) and cryo-electron microscopy (cryo-EM). Recently developed chemical approaches, such as DMS-seq and icSHAPE, can measure RNA flexibility in cells [11–13], yet the measurements are one dimensional. In other words, interacting regions are not directly determined, but inferred based on the nucleotide reactivity. Modeling of structure with the flexibility measurement is often inaccurate, especially for long range, dynamic and complex structures [14].

On the molecular level, RNA structures and interactions are similar, both primarily made of stacking base pairs, or helices. To gain a global view of RNA helices in living cells, the key is to determine base pairing relationships, and the most commonly used chemical for such purpose is psoralen, a family of photo-cross-linkers that reversibly react with staggered pyrimidines on opposite strands [15]. Originally discovered in the early 1970s, psoralens were widely used for the analysis of the structures and interactions of abundant RNAs, such as rRNAs and snRNAs [16]. In conjunction with two-dimensional (2D) gel electrophoresis or northern blots using multiple probes, prominent psoralen cross-linking events can be used to determine base pairing partners, either within one RNA, or between two RNAs [17–20]. This classical method, however, is low throughput and laborious.

To directly determine double stranded RNA with high resolution on a global scale, we invented a new method, PARIS, which combines four critical techniques, psoralen cross-linking, 2D gel purification, proximity ligation, and high-throughput sequencing. The PARIS method employs a cell-permeable and reversible photo-cross-linker AMT (4′-aminomethyltrioxsalen) to covalently link RNA duplexes in living cells. The cross-linked RNA are partially digested with RNase and run through two-dimensional gels to selectively purify cross-linked RNA fragments, which make up a small fraction of all RNA fragments. Then proximity ligation is used to join the trimmed duplexes and the resulting chimeras can be sequenced and used to determine RNA structure and interaction. In this protocol, we provide a detailed step-by-step description, with explanations for the principle behind some critical steps. We also discussed some alternative solutions to tailor the analytic method for specific usage. Several of the library preparation steps have been optimized since the original protocol to make it more efficient [21].

2 Materials

All the materials used for RNA work should be RNase free. Common reagents and equipment are not listed here.

2.1 Basic Reagents

1. PBS.
2. S1 nuclease (e.g., ThermoFisher).
3. 10 w/v % SDS (sodium dodecyl sulfate) in water.
4. Proteinase K (e.g., Ambion).
5. ShortCut RNase III (e.g., NEB).
6. Gel loading dye, orange, 6×.
7. TRIzol LS reagent (e.g., Life Technologies).
8. Chloroform.
9. 100% ethanol.
10. 80% ethanol.
11. Isopropanol.
12. 100% DMSO.
13. 3 M Na-Acetate pH 5.5.
14. Glycogen (e.g., GlycoBlue, 15 mg/mL).
15. RNase inhibitor (e.g., Thermo Scientific SuperaseIn and RiboLock).
16. T4 RNA ligase 1 (ssRNA ligase), high concentration (e.g., NEB).
17. RecJf exonuclease (e.g., NEB, cat. no. M0264).
18. 5′ Deadenylase (e.g., NEB, cat. no. M0331).
19. Ultrapure TBE buffer, 10× (e.g., Life Technologies).
20. SequaGel UreaGel System, (National Diagnostics).
21. Ultrapure TEMED (e.g., Invitrogen).
22. Ammonium persulfate (APS).
23. Reverse transcriptase (e.g., Invitrogen SuperScript III).
24. 10 mM Deoxynucleotide solution mix (dNTPs).
25. Magnetic streptavidin beads (e.g., Life Technologies Dynabeads MyOne streptavidin C1).
26. Magnetic SPRI beads (e.g., Beckman-Coulter AMPure XP).

27. 20% N-Lauroylsarcosine sodium salt solution (e.g., Sigma-Aldrich).
28. Deoxycholic acid sodium salt (e.g., Thermo Fisher Scientific).
29. D-Biotin.
30. RNase cocktail enzyme mix (Ambion, cat. no. AM2286).
31. RNase H (e.g., Enzymatics).
32. DNA ladder, 25 bp.
33. Denaturing PAGE loading buffer (e.g., Ambion Gel loading buffer II).
34. SYBR Gold nucleic acid gel stain, 10,000× (e.g., Life Technologies).
35. 40 w/v % acrylamide–bis solution, 29:1.
36. CircLigase II ssDNA ligase (e.g., Epicentre, cat. no. CL9025K).
37. Phusion high-fidelity (HF) PCR master mix with HF buffer (e.g., NEB).
38. SYBR Green I nucleic acid gel stain, 10,000× (e.g., Life Technologies).

2.2 Oligos Used for Library Preparation

1. Preadenylated and 3′-biotin blocked RNA adapter (PAGE purified): /5rApp/AGATC GGAAG AGCGG TTCAG/3Biotin/. This designed adapter is important for efficient ligation on the 3′ end of RNA.
2. RT-primer-1 (DNA, standard desalting purification). /5phos/ WWW NNN ATCACG NNNNN TACCC TTCGC TTCAC ACACA AG/iSp18/GGATCC /iSp18/TACTG AACCGC, W = A/T and N = A/T/G/C are used to discriminate PCR duplicates, "ATCACG" is the specific experimental barcode, /iSp18/ is a spacer to prevent PCR from forming concatemers). The following is a list of 24 barcodes: 1. ATCACG, 2. CGATGT, 3. TTAGGC, 4. TGACCA, 5. ACAGTG, 6. GCCAAT, 7. CAGATC, 8. ACTTGA, 9. GATCAG, 10. TAGCTT, 11. GGCTAC, 12. CTTGTA, 13. AGTCAA, 14. AGTTCC, 15. ATGTCA, 16. CCGTCC, 17. GTCCGC, 18. GTGAAA, 19. GTGGCC, 20. GTTTCG, 21. CGTACG, 22. GAGTGG, 23. ACTGAT, 24. ATTCCT. The quality of the oligonucleotide synthesis should be verified by polyacrylamide gel electrophoresis; however, bulk PAGE purification before use in RT reactions is not necessary.
3. P3Tall v4 primer (PAGE purified): GGCAT TCCTG CTGAA CCGCT CTTCC GATCT. P6Tall v4 primer (PAGE purified): CTCTT TCCCC TTGTG TGTGA AGCGA AGGGT.

4. P3-Solexa PCR primer (PAGE purified): 5′-CAAGC AGAAG ACGGC ATACG AGATC GGTCT CGGCA TTCCT GCTGA ACCGC TCTTC CGATC T-3′.
5. P6-Solexa PCR primer (PAGE purified): 5′-AATGA TACGG CGACC ACCGA GATCT ACACT CTTTC CCTAC ACGAC GCTCT TCCGA TCT-3′.
6. P6_custom_SeqPrimer (PAGE purified): CACTC TTTCC CCTTG TGTGT GAAGC GAAGG GTA.

2.3 Regent Setup

1. AMT: dissolve 4′-aminomethyltrioxsalen in water at a final concentration of 1 mg/mL. Store it at −20 °C.
2. Urea/SDS solution containing 4 M urea and 0.1% SDS in water. Store it at room temperature.
3. 10 w/v % APS: Dissolve 5 g of APS in a final volume of 50 mL of nuclease-free water. Store it at 4 °C for up to 6 months.
4. 6 w/v % UreaGel denaturing PAGE solution. Prepare 1 L of UreaGel denaturing polyacrylamide gel stock solution by mixing 240 mL of UreaGel System concentrate, 660 mL of UreaGel System diluent, and 100 mL of UreaGel System buffer. Store it at 4 °C.
5. 6 w/v % Native PAGE solution: Prepare gel stock solution by mixing 100 mL of 10× TBE solution and 150 mL of 40% acrylamide–bis solution in a final volume of 1 L of nuclease-free water. Store it at 4 °C.
6. Native PAGE solution for first dimension gel. To make 20 mL gel solution, use 6 mL from 40% stock (Bio-Rad 29:1), 2 mL 10× TBE, 12 mL water, 20 μL TEMED, and 200 μL 10% APS. Make it fresh for use.
7. Gel Crushing Buffer: 500 mM NaCl, 1 mM EDTA, pH 8.0, and 0.1% SDS.
8. Bead Binding Buffer. The final component concentrations are 100 mM Tris–HCl pH 7.0, 1 M NaCl, 10 mM EDTA, and 0.2 v/v % Tween 20. Prepare the buffer in nuclease-free water. Store it at 4 °C.
9. Streptavidin beads. Prepare 20 μL of magnetic streptavidin beads for each PARIS sample by washing the beads twice with 1 mL of bead binding buffer. After washing, resuspend the beads in 10 μL of bead binding buffer per PARIS sample, and store them on ice until needed.
10. Bead Wash Buffer. Final component concentrations are 100 mM Tris–HCl pH 7.0, 4 M NaCl, 10 mM EDTA, and 0.2 v/v % Tween 20. Prepare the buffer in nuclease-free water. Store it at 4 °C.

11. 10 w/v % Sodium deoxycholic acid solution (NaDOC): Mix 5 g of deoxycholic acid sodium salt in 50 mL of nuclease-free water. Store it at room temperature.
12. RNase H buffer, 10×. Final component concentrations are 0.5 M HEPES pH 7.5, 0.75 M NaCl, 30 mM $MgCl_2$, 1.25 w/v % N-lauroylsarcosine sodium salt, 0.25 w/v % NaDOC, and 50 mM DTT. Prepare the buffer in nuclease-free water. If a precipitate forms, briefly warm it at 37 °C so that it dissolves into solution. Freshly prepare the solution.
13. SYBR Green I, 25×. Make a 1:400 dilution from SYBR Green I nucleic acid gel stain (10,000×) in DMSO. Store it at −20 °C.
14. Proximity ligation mixture (10 μL/sample): 2 μL 10× T4 RNA ligase buffer, 5 μL T4 RNA ligase (high concentration), 1 μL SuperaseIn, 1 μL 10 mM ATP, and 1 μL water.
15. Adapter ligation mixture (10 μL/sample): 2 μL 10× T4 RNA ligase buffer, 1 μL 0.1 M DTT, 5 μL 50 v/v % PEG8000, 1 μL 10 μM preadenylated and 3′-biotinylated RNA adapter, and 1 μL High Concentration T4 RNA ligase 1.
16. Reverse transcriptase enzyme mix (8 μL/sample): 4 μL 5× First Strand Buffer, 0.75 μL 40 U/μL RiboLock, 1 μL 100 mM DTT, 1 μL dNTPs (10 mM each), and 1.25 μL SuperScript III reverse transcriptase.
17. RNaseA/T1/H elution mix (50 μL/sample): 5 μL 10× RNase H Buffer, 12.5 μL 50 mM D-biotin, 1 μL P3-Tallv2 primer, 1 μL RNase cocktail enzyme mix, 1 μL RNase H, and 30.5 μL RNase-free water.
18. Circularization reaction mix (4 μL/sample): 2 μL 10× CircLigase II buffer, 1 μL 50 mM $MnCl_2$, and 1 μL CircLigase II.
19. qPCR reaction mixture (10.5 μL/sample): 10 μL 2× Phusion HF PCR master mix, 0.2 μL 25× SYBR Green I, and 0.25 μL P3/P6 TalI v4 PCR primer mix (20 μM forward and 20 μM reverse).

2.4 Equipment

1. 10–15 well gel cassette (e.g., BioRad mini-Protean 3).
2. PAGE apparatus (e.g., CSB Scientific).
3. High-voltage electrophoresis power supply (e.g., Bio-Rad).
4. Freeze dryer/lyophilizer (e.g., Labconco).
5. Spectrophotometer (e.g., Thermo Scientific NanoDrop).
6. Agilent Bioanalyzer.
7. Thermomixer.
8. UV cross-linker (e.g., Stratagene Stratalinker 2400 model).

9. 254 nm and 365 nm emission UV bulbs (e.g., Agilent Technologies).
10. Gel imager (e.g., Bio-Rad Gel Doc XR+).
11. 300 nm blue light transilluminator (e.g., Clare Chemical Research Dark Reader).
12. qPCR machine capable of removal of individual samples amid reaction (e.g., Stratagene Mx3005p).
13. 1.5 mL siliconized microcentrifuge tubes (e.g., MidSci).
14. 0.5 mL 10 kDa cutoff Centrifugal Filter Unit (e.g., Millipore Amicon Ultra-0.5 10 k).
15. Centrifuge tube filters, 0.45 μm (e.g., Sigma-Aldrich Corning Costar Spin-X).
16. DNA purification columns (e.g., Zymo Research, DNA Clean & Concentrator-5 columns).
17. RNA purification columns (e.g., Zymo Research, RNA Clean & Concentrator-5 columns).
18. Magnetic Eppendorf-tube stand (e.g., Thermo Fisher DynaMag2).

2.5 Software

1. Java, Perl, and Python interpreters.
2. Trimmomatic [22].
3. icSHAPE pipeline (https://github.com/qczhang/icSHAPE).
4. fastQC (http://www.bioinformatics.babraham.ac.uk/projects/fastqc/).
5. STAR [23].
6. SAMtools [24].
7. samPairingCalling.pl (https://github.com/qczhang/paris).
8. mfold [25].
9. sam2ngmin.py (https://github.com/zhipenglu/).
10. IGV genome browser [26].
11. maf_extract_ranges_indexed.py (bxpython package, https://github.com/bxlab/bx-python).
12. RNAalifold [27].
13. SISSIz [28].
14. Alternativestructure.py (https://github.com/zhipenglu/duplex).
15. RNAcofold (from the Vienna RNA package).

3 Methods

The following is a detailed description of the PARIS method (Fig. 1). Indicators of successful implementation are included for the critical steps. Standard experimental procedures like TRIzol extraction and ethanol precipitation of RNA are not described in details here and readers should refer to manufacturers' manuals.

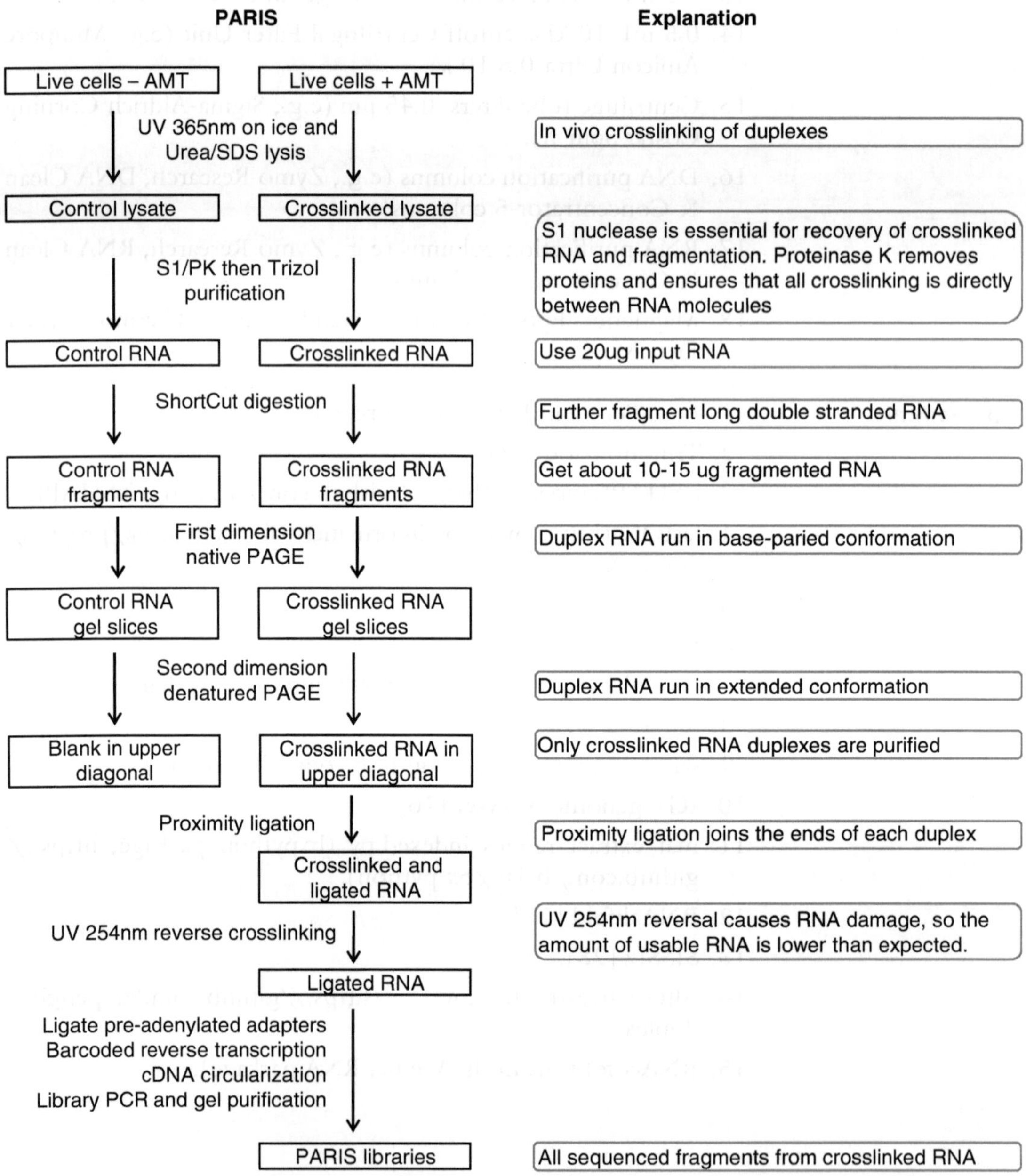

Fig. 1 Outline of the PARIS experimental strategy. Major steps are explained on the *right side*

The library preparation part (Subheadings 3.5 and 3.6) was adapted from a similar protocol for icSHAPE and modified [29]. The principles of data analysis have been described in the supplemental methods in the PARIS paper [21], and a step-by-step protocol is provided here (Fig. 4). Custom scripts for the analysis of PARIS data were published in Github: https://github.com/qczhang/paris, https://github.com/zhipenglu/duplex.

3.1 Psoralen Cross-Linking

1. Before performing psoralen cross-linking, design the experiment with the cross-linker capacity in mind. The Stratalinker 2400 model cross-linker can hold five 15 cm or ten 10 cm plates.
2. On the day of cross-linking, prepare the AMT cross-linking solution containing 0.5 mg/mL AMT and 1× PBS. Use PBS as control.
3. To cross-link RNA in vivo, adherent cells are cultured to ~70% confluent as usual. Remove media from tissue culture plates, wash with PBS and add 0.4 mL of PBS or AMT cross-linking solution to each 10 cm plate. Use 1 mL solution per 15 cm plate.
4. Incubate cells for 30 min at the normal cell culture conditions (e.g., 37 °C, 5% CO_2 in incubator for mammalian cells). At the same time make sure the 365 nm light bulbs are installed in the cross-linker. Cross-linking in the presence of AMT loosens cells from the plate for certain cell lines, such as HEK 293T, and, therefore, care should be taken in adding and removing liquid from the plate.
5. Place ice trays in the cross-linker and put cell culture plates on ice. Irradiate cells with 365 nm UV, 15 cm away from the light bulbs. Swirl the plates every 10 min and make sure that they are horizontal and the liquid covers all the cells (*see* **Note 1**).
6. Remove cross-linking solution after cross-linking. Scrape off cells in 1 mL chilled PBS, transfer to 1.5 mL tubes and centrifuge at 4 °C, 400 × *g* for 5 min.
7. Remove PBS from the tubes, snap-freeze cell pellets on dry ice and store pellets at −80 °C. Control pellets are white and AMT cross-linked pellets are yellowish. The difference in color is an indicator for successful cross-linking (Fig. 2a, *see* **Note 2**) The pellet from a 10 cm plate should be around 30 μL, optimal for a single S1/PK reaction.

3.2 S1/PK Extraction and ShortCut Digestion

1. Resuspend cell pellet from a 10 cm plate in 200 μL urea/SDS solution. Control samples can be very viscous and hard to dissolve. Use a pipet to dislodge pellet and shake vigorously to suspend the pellet. AMT cross-linked cell pellets are easier to

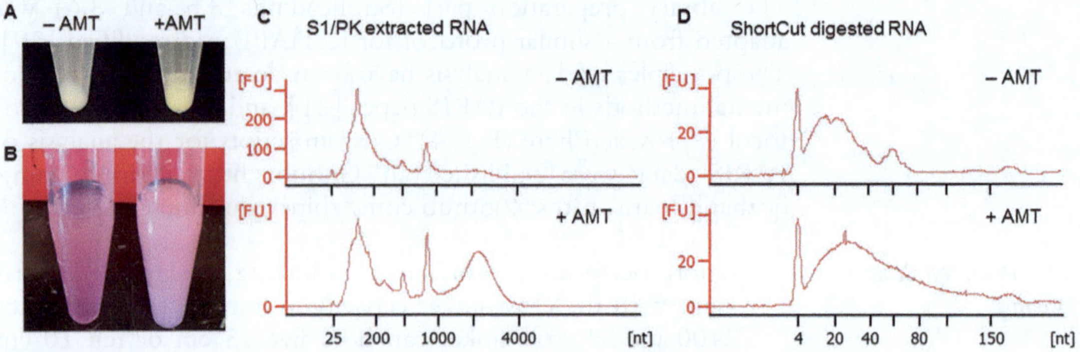

Fig. 2 Example results of AMT cross-linking and RNA fragmentation. (**a**) AMT cross-linked cell pellets (*right*) have a darker color than the non-cross-linked ones (*left*). (**b**) After S1/PK digestion, adding TRIzol and chloroform, phase separation is faster for the non-cross-linked cells (*left*) than the cross-linked cells (*right*), so the cross-linked samples have a milky appearance. (**c**) S1/PK extraction produces a characteristic broad peak between 2000 and 4000 nt in the Bioanalyzer electrophoretic trace of cross-linked cells. The height of the broad peak is variable among cell types and different batches of experiments. (**d**) ShortCut RNase III digestion reduces RNA to a smaller size (usually below 150 nt) to facilitate 2D gel purification and library preparation

break up and resuspend in urea/SDS solution as the DNA is fragmented by UV 365 nm in the presence of AMT.

2. To the lysed cells add 60 μL 5× S1 nuclease buffer, 2 μL S1 nuclease and mix well. Expect ~300 μL final volume. Perform the S1 digestion at room temperature for 10 min with frequent pipetting or vortexing to break viscous material.
3. To the S1 digested lysate add 33 μL 10% SDS (final concentration 1%) and 2 μL proteinase K (PK, 20 mg/mL, final 0.1 μg/μL). Samples should become clear soon after adding SDS. Perform the PK digestion at 50 °C for 30 min in a thermomixer, at 100 × *g* to mix the suspension (*see* **Note 3**).
4. After PK digestion, add 0.9 mL TRIzol LS reagent and mix vigorously. Then add 180 μL chloroform and mix vigorously. Phase separation is faster in control samples than +AMT samples, another visible difference between control and cross-linked samples (Fig. 2b). Complete the RNA extraction following the standard TRIzol protocol.
5. Quantify the purified RNA using a spectrophotometer and analyze the quality using Bioanalyzer. An example Bioanalyzer trace file is shown in Fig. 2c. Also *see* the previous PARIS publication for another example [21]. The AMT cross-linked sample should have a broad peak between 2000 and 4000 nt for human, mouse, and *Drosophila* cells, an important indicator of successful cross-linking and RNA extraction (*see* **Note 4**).
6. Use 20 μg S1/PK purified RNA for ShortCut RNase III digestion, in order to reduce the RNA fragments to smaller size (*see*

Note 5). Each ShortCut reaction in 50 μL volume includes 20 μg RNA, 4 μL ShortCut RNase III, 5 μL 200 mM $MnCl_2$, and 5 μL ShortCut buffer. The reaction is incubated at 37 °C for 20 min.

7. After ShortCut digestion, purify RNA using the standard TRIzol method and resuspend RNA in 15 μL water. Determine concentration of the samples by spectrophotometer and analyze size distribution using Bioanalyzer (Fig. 2d). Typically, the AMT cross-linked samples have a stronger tail above 100 nt than the control samples.

3.3 Purification of Cross-Linked RNA by 2D Gel

1. Prepare the 12% 1.5 mm thick native first dimension gel using the BioRad Mini-Protean 3 gel cassette (Fig. 3a). The 1.5 mm thick gel is used here to facilitate the removal of air bubbles from the second dimension. Use 15-well combs so that each lane is narrower and the second dimension has a higher resolution. After the gel solidifies, pull the comb very slowly to avoid the deforming well dividers.
2. To each 15 μL sample add 5 μL 6× Orange G loading dye. Load 3 μL dsRNA ladder as molecular weight marker. Run the first dimension gel at 100 V for 70 min in 0.5× TBE. Orange G should be 4/5 way to bottom. Usually we have a starting current of 15 mA and a starting power of 1.5 W.
3. After electrophoresis finishes, stain the gel with 2 μL SYBR Gold in 20 mL 0.5× TBE, incubate for 5 min. Image the gel using 300 nm transillumination (not the 254 nm epi-illumination, which reverses the psoralen cross-linking). Excise

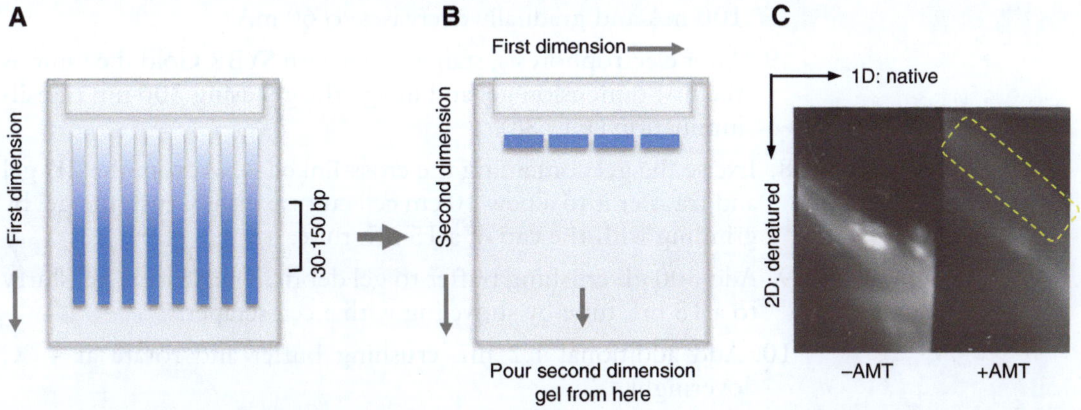

Fig. 3 Diagram and example result for 2D gel purification of cross-linked and digested RNA. (**a**) First dimension native gel. Gel slices containing RNA around 30–150 bp are usually cut out from the first dimension for the second dimension. (**b**) Second dimension denatured gel. Gel slices are aligned between the glass plates before pouring the urea denatured gel solution at the bottom of the gel plates (as indicated by the *arrow* in the *middle*). (**c**) An example second dimension gel. The *yellow-boxed area* indicate the cross-linked RNA to be extracted for library preparation

each lane between 30 and 150 bp from the first dimension gel (Fig. 3a). The second dimension gel can usually accommodate three gel splices.

4. Prepare the 20% 1.5 mm thick urea denatured second dimension gel using the UreaGel system. For 20 mL gel solution, use 16 mL UreaGel concentrate, 2 mL UreaGel diluent, 2 mL UreaGel buffer, 8 μL TEMED, and 160 μL 10% APS. Add TEMED and APS right before pouring the gel.
5. To make the second dimension gel, put the square plate horizontally and arrange gel slices in a "head-to-toe" manner with 2–5 mm gap between them (Fig. 3b). Leave 1 cm space at the top of the notched plate so that the second dimension gel would completely encapsulate the first dimension gel slices.
6. Apply 20–50 μL 0.5× TBE buffer on each gel slice to avoid air bubbles when placing the notched plate on top of the gel slices. Remove the excess TBE buffer after the cassette is assembled, and leave 2 mm space at the bottom of the notched plate to facilitate pouring the second dimension gel.
7. Pour and gel solution from the bottom of the plates, while slightly tilting the plates to one side to avoid air bubbles building up between the plates. If there are air bubbles, use the thin loading tips to draw them out.
8. Use ~60 °C prewarmed 0.5× TBE buffer to fill the electrophoresis chamber to facilitate denaturation of the cross-linked RNA. Run the second dimension at 30 W for 40 min to maintain high temperature and promote denaturation. Run the gel for 50 min. The voltage starts around 300 V and gradually increases to 500 V, while the current starts around 100 mA and gradually decreases to 60 mA.
9. After electrophoresis, stain the gel with SYBR Gold the same as the first dimension gel and image the gel using 300 nm transillumination (Fig. 3c).

8. Excise the gel containing the cross-linked RNA from the 2D gel and transfer it to a new 10 cm cell culture dish. Crush the gel by grinding with the cap of a 15 mL tube.
9. Add 300 μL crushing buffer to gel debris. Transfer the gel slurry to a 15 mL tube by shoveling with a cell scraper.
10. Add additional 1.2 mL crushing buffer and rotate at 4 °C overnight.
11. Transfer ~0.5 mL gel slurry to Spin-X 0.45 μm column. Spin at room temperature, 3400 × *g* for 1 min. Continue until all gel slurry is filtered.
12. Aliquot 500 μL of the filtered RNA sample to an Amicon 10 k 0.5 mL column. Spin at 4 °C, 12,000 × *g* for 5 min. Repeat

until all of the filtered RNA sample flowed through the column.

13. Wash the column with 300 μL water and spin the column at 4 °C, 12,000 × *g* for 5 min.
14. Invert and place the column in a new collection tube, and spin at 4 °C, 6000 × *g* for 5 min. Recover ~85 μL RNA from each column (~170 μL total from two columns).
15. Precipitate the RNA using the standard ethanol precipitation method, with glycogen as a carrier. Alternatively, the RNA can be purified using the Zymo RNA clean and concentrator-5 columns.
16. Reconstitute RNA in 11 μL water and dilute 1 μL RNA sample for Bioanalyzer analysis. The RNA sample should have a broad size distribution between 30 and 150 nt in the Bioanalyzer trace. The yield is typically 0.1–0.5% from 20 μg S1/PK extracted input RNA.

3.4 Proximity Ligation and Photoreversal of Cross-Linking

1. Add 10 μL proximity ligation mixture to 10 μL of RNA, mix well and incubate at room temperature overnight.
2. Boil the ligation mixture at 95 °C for 2 min. After heat denaturation, the samples may turn turbid. Spin down the insoluble material at 6000 × *g* for 5 min and retain the supernatant.
3. To the 20 μL clarified ligation product, add 40 μL water, 1 μL GlycoBlue and 6 μL 3 M sodium acetate, and mix well.
4. Then add 150 μL pure ethanol and precipitate by centrifugation at 16,000 × *g* for 20 min.
5. Wash the pellet with 80% ethanol and resuspend in 12 μL water.
6. To reverse the AMT cross-linking, put the samples on a clean surface with ice beneath it. Irradiate with 254 nm UV for 15 min and recover about 10 μL RNA.

3.5 Adapter Ligation and Reverse Transcription

1. Add 10 μL adapter ligation mixture to 10 μL RNA and perform the adapter ligation reaction for 3 h at room temperature.
2. After adapter ligation add the following reagents to remove free adapters: 3 μL 10× RecJf buffer, 2 μL RecJf, 1 μL 5′ deadenylase, 1 μL SuperaseIn, and 3 μL water. Incubate at 37 °C for 1 h.
3. Then purify RNA with Zymo RNA clean and Concentrator-5 or ethanol precipitation. Reconstitute RNA in 11 μL water.
4. To the purified RNA add 1 μL of custom RT primer 1 (with barcode).
5. Heat the samples to 70 °C for 5 min in a PCR block, cool the samples to 25 °C by stepping down 1 °C every 1 s (50 steps); hold at 25 °C.

6. Add 8 μL reverse transcriptase mix to the RNA and heat the samples at 25 °C for 3 min, 42 °C for 5 min, 52 °C for 30 min; hold at 4 °C.
7. Wash 10 μL of C-1 beads per sample with 100 μL Bead Binding Buffer three times.
8. Resuspend beads in 5 μL Bead Binding Buffer per sample. Add 5 μL bead suspension to the completed reverse transcription reaction.
9. Rotate at room temperature for 30 min.
10. Transfer bead suspension to new 1.5 mL tube. Let it settle on magnet and remove supernatant.
11. Add 500 μL Bead Wash Buffer to each sample and invert the tubes four times to mix.
12. Insert the samples into a magnetic stand for 1 min, and then remove the supernatant.
13. Wash 4 more times with 500 μL Bead Wash Buffer and twice with 500 μL 1× PBS.
14. Resuspend beads in 50 μL RNaseA/T1/H elution mix and incubate them at 37 °C for 30 min at 100 × *g* in a thermomixer.
15. Add 1 μL 100% DMSO and heat at 95 °C for 4 min. Briefly spin down samples and transfer eluted cDNA to a new tube.
16. Purify the cDNA using Zymo concentrator-5 columns.
17. Elute twice with 6 μL water to recover ~10 μL cDNA.

3.6 cDNA Circularization, Library PCR, and Sequencing

1. Add 4 μL circularization reaction mix to the cDNA sample and incubate them at 60 °C for 100 min, followed by 80 °C for 10 min. At the same time, warm up the lamps in the qPCR machine by starting the program (the warm-up takes about 20 min).
2. Add 10.5 μL qPCR reaction mix to each circularized cDNA sample. Transfer cDNA to optical PCR tubes (each tube should be separate so that individual tubes can be taken out of the qPCR machine when the fluorescence signal reaches a defined point).
3. Set up the following qPCR program. Choose SYBR, initial 95 °C, 45 s, 10 cycles of: 95 °C, 15 s; 65 °C, 30 s; 72 °C, 45 s, detect fluorescence at extension step (a set of nine cycles). Take sample out once amplification reaches exponential phase.
4. Transfer PCR product to 1.5 mL tube. Add 30 μL Ampure XP beads and 75 μL isopropanol.
5. Incubate for 5 min at RT.
6. Let the beads settle on the magnet for 5 min.

7. Remove the supernatant and wash the beads twice with 80% ethanol at RT.
8. Elute DNA with 10 μL water twice each tube (~20 μL eluate per sample). Pool elute and add 20 μL 2× Phusion HF mastermix and 2 μL 20 μM P3/P5 Solexa PCR primer (final 1 μM).
9. Run PCR reaction (95 °C, 45 s; 3 cycles of 95 °C, 15 s; 65 °C, 20 s; 72 °C, 45 s; and 4 °C on hold).
10. Purify reaction by standard Zymo concentrator-5 column protocol.
11. Elute with 8 μL water and add 2 μL Orange G loading dye.
12. Run a 6% native TBE gel at 100 V for 75 min, until the dye just ran off the gel.
13. Stain gel in SYBR Gold for 3 min. Image gel at 0.5, 1, and 2 s exposure times. Cut out the DNA from 175 bp and above (corresponding to >40 bp insert).
14. Use a syringe needle to punch a hole in the bottom of a 0.65 mL tube.
15. Transfer the gel slice to 0.65 mL tube and insert into a 2 mL collection tube. Spin at room temperature, 16,000 × *g* for 5 min. The gel slice gets sheared into slurry by passing through the hole.
16. Remove the 0.65 mL tube and add 300 μL water to the slurry. Shake at 55 °C, 100 × *g* overnight in a thermomixer.
17. Pass the gel slurry through a Spin-X 0.45 μm column to recover the DNA library.
18. Add 5× volume Zymo DNA binding buffer and flow-through Zymo concentrator-5 column.
19. Wash with 200 μL Washing buffer once and elute twice with 8 μL water (recover ~15 μL library). Quantify library by a high-sensitivity Bioanalyzer assay.
20. Barcoded libraries can be pooled together for sequencing if necessary.
21. Sequence the libraries on an Illumina sequencer using standard conditions and the P6_Custom_seqPrimer. Usually, a 70 nt single end sequencing reaction is enough for PARIS. The multiplexing and random barcodes are sequenced together with the insert.

3.7 Basic Analysis: Trimming, Mapping, and Filtering

1. Sequencing data are usually provided in zipped fastq files. First, the reads are trimmed to remove the adapter sequences. If the data are multiplexed, split the library. *See* Fig. S1H of the PARIS paper for an overview of the analysis strategy [21]. The analysis outline is summarized in Fig. 4. Assume that the

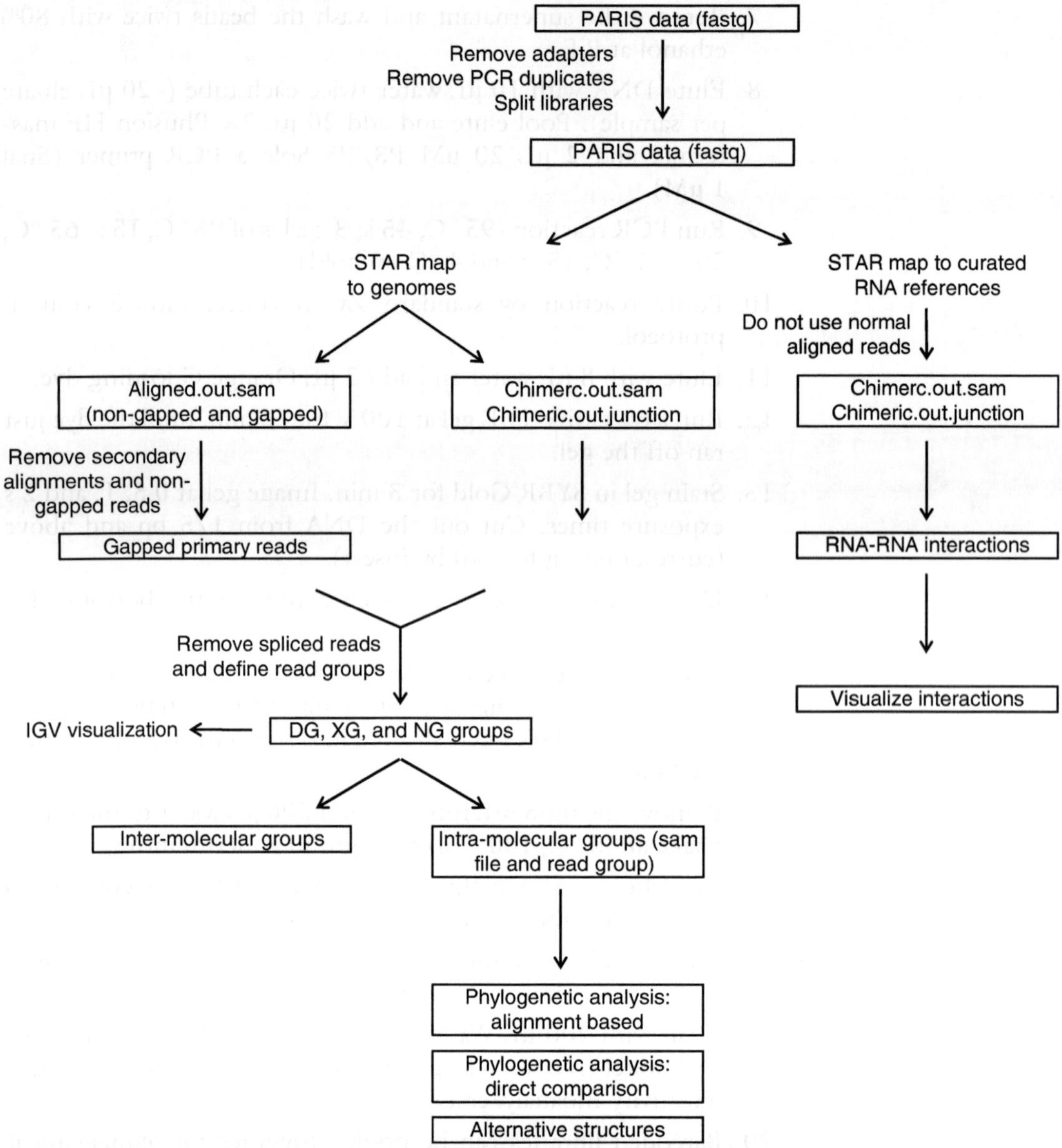

Fig. 4 The PARIS analysis pipeline, modified from the PARIS paper [21]. Major analyses outlined here include DG and NG assembly, visualization of RNA structure data and models in IGV, two approaches of phylogenetic analysis, analysis of alternative structures, identification and visualization of RNA–RNA interactions

file name prefix is x. The following are a preferred set of file name conventions. Having standard file names makes the management of files easier.

x.fastq.gz: zipped raw data x_trim3.fastq: after removing the 3′ end adapter sequencesx_trim3_nodup.fastq: after removing PCR duplicatesx_trim3_nodup_bcnn.fastq: split libraries with barcode nn, such as 01, 02 …x_trim_nodup_bcnn.fastq: after removing 5′

end adapter sequences x_trim_nodup_bcnn_starSSAligned.out. sam: reads mapped to STAR index for species SS, normally aligned. x_trim_nodup_bcnn_starSSChimeric.out.sam: same as above, for the chimeric reads

2. Trim the 3′ end adapter sequences using Trimmomatic [22].

```
java -jar trimmomatic-0.32.jar SE -threads 16
-phred33 x.fastq.gz x_trim3.fastq ILLUMINACLIP:
P6SolexaRC35.fa:3:20:10 SLIDINGWINDOW:10:30 MIN-
LEN:28
```

3. Remove the duplicates using the built-in random barcodes and readCollapse (from the icSHAPE pipeline).

```
readCollapse -U x_trim3.fastq -O x_trim3_nodup.
fastq
```

4. Split the trimmed and collapsed libraries using splitFastqLibrary (from the icSHAPE pipeline). L is a list of barcode and file name strings. D is the directory to store the split libraries

```
splitFastqLibrary -U x_trim3_nodup.fastq -l L -b 6:6
-s -d D
```

5. Trim the 5′ adapter sequences. HEADCROP:17 option is used to trim off the designed random and multiplexing barcodes.

```
java -jar trimmomatic-0.32.jar SE -threads 16
-phred33 x_trim3_nodup_bc01.fastq x_trim_no-
dup_bc01.fastq HEADCROP:17 MINLEN:20
```

6. Quality of the x_trim_nodup.fastq file was visualized using fastQC.

7. Map reads to a genome index of choice using STAR [23] (*see* **Notes 6** and **7**). The parameters chosen here are to reduce the penalty for gapped reads and allow mapping of chiastic reads. Some of the options can be adjusted for specific purposes, such as –outFilterMultimapNmax and –alignSJoverhangMin.

```
STAR -runMode alignReads -genomeDir /seq/STAR/in-
dex/starSS/ -readFilesIn x_trim_nodup_bc01.fastq
-outSAMtype SAM -outFileNamePrefix x_trim_no-
dup_bc01_starSS -outReadsUnmapped Fastq -outSAMat-
tributes All -alignIntronMin 1 -scoreGapNoncan -4
-scoreGapATAC -4 -chimSegmentMin 15 -chimJunctionO-
verhangMin 15 -runThreadN 8
```

8. Convert SAM to BAM format, sort the BAM file and extract primary mapped reads using SAMtools [24].

```
samtools view -b -F 0x900 -o x_trim_nodup_bc01_
starSSAligned_prim.out_sorted.bam x_trim_no-
dup_bc01_starSSAligned.out_sorted.bam
```

9. Extract gapped reads from the Aligned.out_sorted.bam file. The non-gapped reads do not contain the structure information and therefore discarded (*see* **Note 8**).

```
samtools view x_trim_nodup_bc01_starSSAligned_prim.
out_sorted.bam | awk '$6 ~ /N/' x_trim_nodup_bc01_
starSSAligned_prim_N.out_sorted.sam
```

10. To load SAM files into IGV for visualization, the SAMtools package is commonly used in the following three steps. These three steps can be used for any of the SAM files generated throughout the analysis.

```
samtools view -bS -o x.bam x.samsamtools sort x.bam
x_sorted
samtools index x_sorted.bam
```

3.8 Advanced Analysis and Visualization

1. Assemble duplex groups (DGs) using the `samPairingCalling.pl` script (https://github.com/qczhang/paris). Given the volume and complexity the RNA structures and RNA–RNA interactions. Here are two options for the assembly of DGs. First, the DGs can be assembled for all mapped reads. Second, DGs can be assembled for individual RNA transcripts. The second approach is preferred for detailed analysis since it is much faster than the first one. The processing first removed gapped reads that are gapped as a result of splicing and further removed PCR duplicates. Then the gapped reads are sorted by coordinates and then processed to obtain DGs with a two-step greedy algorithm. First, generate intermediate DGs. Each read is either added to an existing DG or used to establish a new DG based on the criteria: all reads in a DG must share at least 5 nt in both arms. Second, merge the intermediate DGs as long as the maximum gap of both arms is less than 10 nt and the maximum length of both arms of the final DG less than 30 nt. To guarantee the validity of the identified DG, we filter low quality DGs by two criteria: first, each DG must have two unique gapped reads that have different termini. Second, DG connection score(connection_A_B/sqrt(coverage_A*coverage_B), A and B representing the two arms) should be great than 0.01 (*see* **Note 9**).

```
perl samPairingCalling.pl -i x_Aligned_prim_N.sam
-j x_Chimeric.out.junction -s x_Chimeric.out.sam
-o x_geometric -g genome.fa -z chrom.sizes -a gen-
ome.gtf -t genome.fa -l 15 -p 2 -c geometric
1>x_geometric.stdout 2>x_geometric.log
```

Annotation of the command:x_Aligned_prim_N.sam: normal gapped reads x_Chimeric.out.sam and x_Chimeric.out.junction: chiastic reads and junctionsgenome.fa: genome referencechrom.sizes: two columns, chromosome name and sizegenome.gtf: genome annotation file from UCSC Genome Browsergeometric: using geometric mean of the coverage on the two arms for normalization.

This assembly step produces two main output files: *geometricsam and *geometric (while using the geometric option). The sam file can be used directly to assemble NGs in the next step, while the other file contains all connections that can be used for RNA–RNA interaction analysis (**step 6**).

2. Assemble nonoverlapping groups (NGs). To efficiently pack the DGs in the IGV genome browser, the DGs are further assembled into NGs using sam2ngmin.py (from the "duplex" scripts, https://github.com/zhipenglu/). NG is a new custom-defined tag in the SAM file format. Then convert the NG-assembled SAM file to indexed BAM file for visualization on IGV as above (**step 10**). The *geometricsam file is produced from the DG assembly step.

```
python sam2ngmin.py x_trim_nodup_norm_starhg38_geo-
metricsam x_trim_nodup_norm_starhg38_geometric_NG-
min.sam
```

3. The secondary structure model of an RNA can be prepared in a BED format, which is quite similar to the commonly used 'connect' format described in the mfold program [25]. This structure model can be uploaded to the IGV genome browser [26], in parallel with the PARIS DGs (Fig. 5). The file must include a track line specifying "track graphType=arc". Each record line must contain the first three columns of a bed file: chrom, start and end, where the start and end represent the base pair. Note that the start position follows standard BED file convention and is zero-based (first base on a sequence is position 0). Note that one can import a known secondary RNA structure after converting to the arc type bed (e.g., https://github.com/zhipenglu/duplex/ct2bed.py to convert the connection format to the arc bed type file).The following small

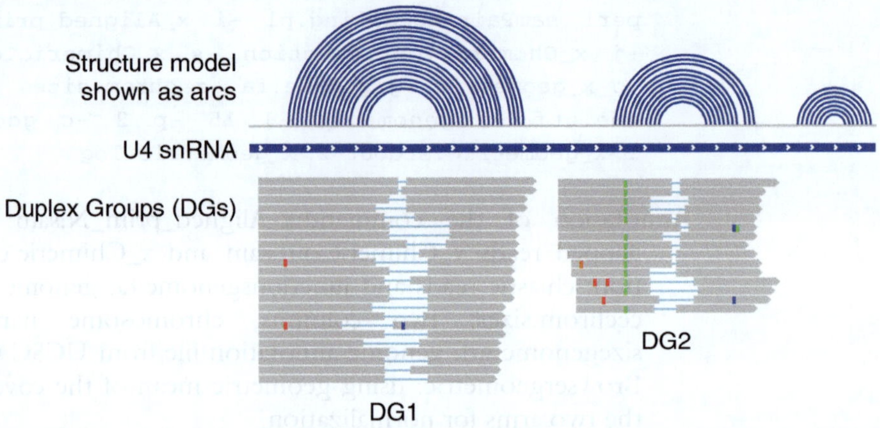

Fig. 5 An example visualization of RNA structures in IGV. The structure model is in the linear format (quasi-concentric arcs for an RNA duplex). A subset of gapped reads is assembled into two DGs. The two DGs can be assembled into one NG, since they do not overlap with each other. For more examples, see the PARIS paper [21]

example represents a hypothetical stem loop (also described here: https://www.broadinstitute.org/igv/node/284).

```
track graphType=arc
chr1 10 25 stemloop1
chr1 11 24 stemloop1
chr1 12 23 stemloop1
chr1 13 22 stemloop1
chr1 14 21 stemloop1
chr1 15 20 stemloop1
```

4. Visualization of RNA structure models and PARIS data on IGV genome browser. Upon loading files with DG, XG and NG tags to IGV (*see* **Note 10**), specify the following options by right clicking in the alignments panel. Color alignments by > tag > DG or XGGroup alignments by > tag > NG
5. For high-level visualization of RNA architecture, each DG can be represented by one arc (*see* Fig. 2c in the PARIS paper for an example [21]). To visualize all DGs in one transcript, extract the DGs from the *geometric file (from **step 1**). The start and end of the DG are used as the anchor points of the arc. The visualization is similar to the **step 3** (*see* **Note 11**).
6. Analysis of RNA–RNA interactions. PARIS directly identifies all RNA–RNA interactions. The comprehensiveness of PARIS makes the analysis and visualization challenging. To identify RNA–RNA interactions from PARIS data with high confidence, reads are mapped to indices of specific subset of RNAs, each one as a small "chromosome". For example, one can use all nonredundant human RNAs from Rfam, miRNAs

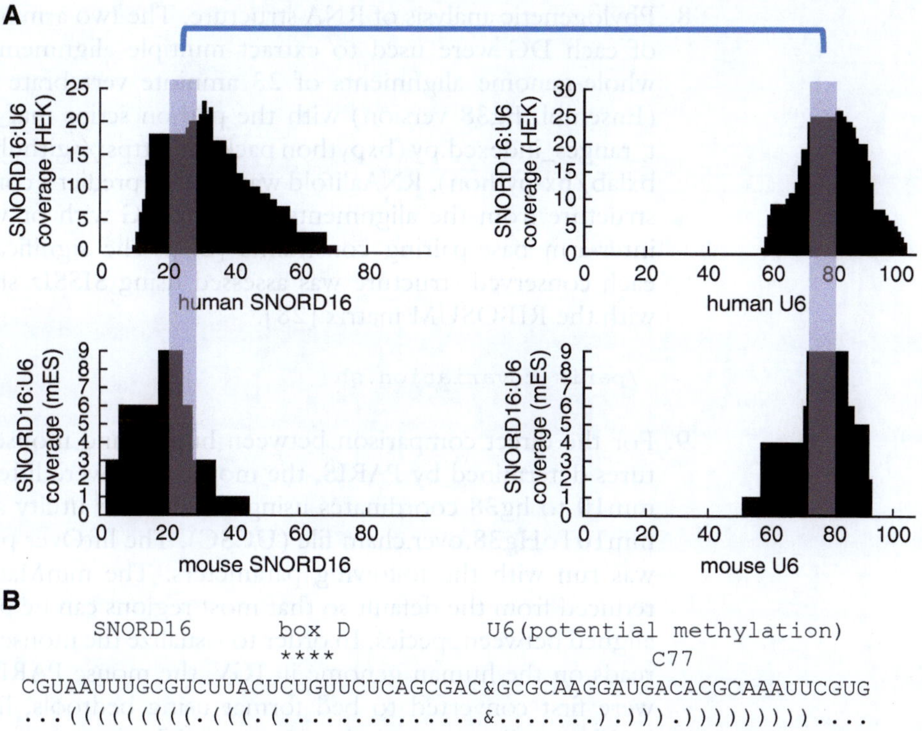

Fig. 6 Analysis and visualization of a new interaction between SNORD16 (or U16) and U6 in human HEK 293T cells and mouse ES cells. (**a**) The *blue* highlighted regions are the sequences involved in base pairing. (**b**) The interaction model. Box D is a characteristic sequence motif in the C/D box snoRNAs. SNORD16 is known to guide modification of 18S rRNA (snoRNABase) [30]. Here, we show that it may also guide the modification of U6 snRNA at C77, the same site as the SNORD10 target (snoRNABase) [31], but the base pairing is less perfect compared to the SNORD10–U6 interaction. For more examples, *see* Fig. 5 and Fig. S6 the PARIS paper [21]

from miRBase, and/or nonredundant RNAs from the human transcriptome annotations as references. After mapping reads to these references, chiastic reads were assembled into duplex groups and filtered to remove interactions within RNAs and ones with identical break points. Normal gapped reads are not used since these are all for RNA structures, not RNA–RNA interactions. The remaining chiastic reads are used to identify RNA–RNA interactions.

7. To visualize the interaction between a pair of RNAs, use the following script. This analysis produces a pair of tracks for the interacting RNAs, highlighting the regions involved in interaction (*see* Fig. 6 for two examples). The SAM files from the previous step are used as the input. Parameters are defined within the following script.

```
python intrxn_specificity.py
```

8. Phylogenetic analysis of RNA structure. The two arm intervals of each DG were used to extract multiple alignments from whole-genome alignments of 23 amniote vertebrate species (Ensembl, hg38 version) with the python script maf_extract_ranges_indexed.py (bxpython package, https://github.com/bxlab/bx-python). RNAalifold was used to predict a consensus structure from the alignments for each DG with or without inter-arm base-pairing constraints [27]. The significance of each conserved structure was assessed using SISSIz shuffling with the RIBOSUM matrix [28].

```
./paris_covariation.sh
```

9. For the direct comparison between human and mouse structures determined by PARIS, the mouse DGs were lifted from mm10 to hg38 coordinates using the liftOver utility and the mm10ToHg38.over.chain file (UCSC). The liftOver program was run with the following parameters. The minMatch was reduced from the default so that most regions can be properly aligned between species. In order to visualize the mouse PARIS reads on the human genome in IGV, the mouse PARIS reads were first converted to bed format using bedtools, lifted to hg38 coordinates, and then converted back to bam format using bedtools. It is noted that this strategy is limited by the quality of the available genome alignments, and improvement of these alignments is beyond the scope of the current study.

```
liftOver -minMatch=0.2 -minBlocks=0.2 -fudgeThick
```

10. Analysis of alternative structures using the alternativestructure.py script (https://github.com/zhipenglu/duplex). Alternative structures are defined as helices that overlap on one arm by more than 50%. In practice, DGs were intersected with each other to identify pairs of DGs that have one pair of overlapped arms (left-left, left-right or right-right), but not two pairs at the same time. Inter-arm structures were predicted using RNAcofold and significant overlapping of base pairs were used as another filter for alternative structures (at least 50% overlap). This script requires RNAcofold (from the Vienna RNA package) and python intervaltree module in proper paths. The x.bed file contains all the DGs in a BED format while the reference.fa contains the reference sequence. The x.alt is the output.

```
python alternativestructure.py x.bed reference.fa
x.alt
```

4 Notes

1. The 30 min irradiation time was chosen for strong cross-linking; however, shorter times may be useful for certain applications where rapid changes for RNA structures are expected. The cross-linking step performed on ice bed may cause cold shock and should be noted here. This condition, even though still likely to affect cell physiology, is very close to normal cell culture condition.
2. The quality and concentration of the solution can be measured by NanoDrop, which gives a characteristic spectrum with two major peaks at 250 and 300 nm [32]. The absorbance of AMT at 300 nm is 25,000/M cm [33].
3. Strong psoralen cross-linking turns RNA into extensive networks of interconnected molecules that are insoluble even under strong denatured and chaotropic conditions like TRIzol. Direct lysis of AMT cross-linked cells produces significant amount of insoluble material. S1 nuclease digestion is necessary to recover cross-linked RNA from the lysate. Psoralen cross-links RNA to protein to some extent and proteinase K digestion is also necessary to recover RNA from the lysate [34, 35]. S1 nuclease is resistant to certain denaturation conditions [36]. For example, S1 is active in 9 M urea. SDS can inactivate S1 activity at a concentration above 0.1% but proteins can complex with SDS to maintain S1 activity. When supplemented with manganese buffer, ShortCut converts double stranded RNA to 18–25 bp fragments with 5′-phosphate and 3′-hydroxyl groups that are can be directly used for proximity ligation.
4. The S1 and PK digestion of cell lysate is hard to control since the reaction is performed in suspension not solution. In general, the S1 digestion should greatly reduce the viscosity of the suspension. After S1/PK digestion, the addition of TRIzol should yield a clear solution. The RNA size distribution is variable especially regarding the height of the broad peak between 2000 and 4000 nt. Therefore, it is necessary to perform the cross-linking and digestion experiments at the same for a set of conditions that need to be compared.
5. Both S1 nuclease and ShortCut RNase III are chosen for the fragmentation because they produce 5′-P and 3′-OH that can be ligated directly in subsequent steps. In addition, the combination of S1 nuclease and ShortCut RNase III helps produce the maximal amount of cross-linked dsRNA fragments of appropriate size. RNase A and T1 do not produce clonable ends and therefore not used in these experiments.
6. The preprocessed PARIS sequencing reads can be mapped to the whole genome or a selected subset of gene, such as 45S

rRNAs, mitochondrial rRNAs, snRNAs, and XIST. A STAR index is needed for each reference. To make these mini-reference STAR indices, it is important to note that a custom value is needed for the `--genomeSAindexNbases` option, calculated as min(14, log2(GenomeLength)/2 − 1) (*see* the STAR manual for details).

7. For a smaller subset of RNA interaction analysis, the STAR mapping parameter may need to be adjusted for better performance. For instance, while looking the interaction landscape of rRNAs, mapping parameter -outFilterMultimapNxax is set to 10 to reduce ambiguously mapped reads.
8. The STAR mapping produces normally mapped reads (Aligned.out.sam) and chiastically mapped reads (Chimeric.out.junction and Chimeric.out.sam), all of which will be used for building RNA structures. Normal: LLLLLL-RRRRRR, chiastic: RRRRRR-LLLLLL, L for bases from the left arm, while R for the right arm. The Chimeric.out.sam is a naming convention from STAR, and it should not be confused with other definitions of "chimeric"; therefore, we use chiastic to refer to the RRRRRR-LLLLLL type alignments.
9. The coverage of the two arms in DG score calculation is different from the gapped reads connecting them, because each region could be covered by multiple DGs, and some of them are likely to be alternative structures, which are pervasive in the transcriptome.
10. Output files from the samPairingCalling.pl program contain both DG and XG (Chiastic Group) tags. The XG tag is used to differentiate normal gapped reads and chiastic reads, where the two arms are swapped in relative position. Chiastic reads also include additional reads that are mapped to different strands or different chromosomes. XG:i:0 for normal gapped, XG:i:1 for chiastic on the same strand of the same chromosome, and XG: i:2 for all others.
11. Visualization of entire bam files is not recommended since long-range and especially intermolecular interactions would extremely compress the DGs in arcs or read alignments. Instead, individual RNAs should be extracted from the bam files (aligned gapped reads) and bed files (arcs representing DGs) using the SAMtools and BEDtools programs. Examples and instructions can be downloaded from the following link https://www.dropbox.com/s/1oqkcfzlfafdahq/PARIS_visdata.tgz?dl=0.

Acknowledgment

The authors wish to thank Dr. Howard Chang at Stanford University for his generous support and encouragement. Z.L. is a Layton Family Fellow of the Damon Runyon-Sohn Foundation Pediatric Cancer Fellowship Award (DRSG-14-15). This work was supported by the National Natural Science Foundation of China (31671355) and the National Thousand Young Talents Program of China to Q.C.Z., and by the Jump Start Award for Excellence in Research to Z.L. (Stanford University, award No. 106).

References

1. Cech TR, Steitz JA (2014) The noncoding RNA revolution-trashing old rules to forge new ones. Cell 157(1):77–94. doi:10.1016/j.cell.2014.03.008
2. Mattick JS, Makunin IV (2006) Non-coding RNA. Human molecular genetics 15 Spec No 1:R17-29. doi:10.1093/hmg/ddl046
3. Rinn JL, Chang HY (2012) Genome regulation by long noncoding RNAs. Annu Rev Biochem 81:145–166. doi:10.1146/annurev-biochem-051410-092902
4. Wang KC, Chang HY (2011) Molecular mechanisms of long noncoding RNAs. Mol Cell 43(6):904–914. doi:10.1016/j.molcel.2011.08.018
5. Lu Z, Matera AG (2014) Developmental analysis of spliceosomal snRNA isoform expression. G3 5(1):103–110. doi:10.1534/g3.114.015735
6. O'Reilly D, Dienstbier M, Cowley SA, Vazquez P, Drozdz M, Taylor S, James WS, Murphy S (2013) Differentially expressed, variant U1 snRNAs regulate gene expression in human cells. Genome Res 23(2):281–291. doi:10.1101/gr.142968.112
7. Jia Y, JC M, Ackerman SL (2012) Mutation of a U2 snRNA gene causes global disruption of alternative splicing and neurodegeneration. Cell 148(1–2):296–308. doi:10.1016/j.cell.2011.11.057
8. Lu Z, Guan X, Schmidt CA, Matera AG (2014) RIP-seq analysis of eukaryotic Sm proteins identifies three major categories of Sm-containing ribonucleoproteins. Genome Biol 15(1):R7. doi:10.1186/gb-2014-15-1-r7
9. Kim SH, Quigley GJ, Suddath FL, McPherson A, Sneden D, Kim JJ, Weinzierl J, Rich A (1973) Three-dimensional structure of yeast phenylalanine transfer RNA: folding of the polynucleotide chain. Science 179 (4070):285–288
10. Robertus JD, Ladner JE, Finch JT, Rhodes D, Brown RS, Clark BF, Klug A (1974) Structure of yeast phenylalanine tRNA at 3 A resolution. Nature 250(467):546–551
11. Rouskin S, Zubradt M, Washietl S, Kellis M, Weissman JS (2014) Genome-wide probing of RNA structure reveals active unfolding of mRNA structures in vivo. Nature 505 (7485):701–705. doi:10.1038/nature12894
12. Ding Y, Tang Y, Kwok CK, Zhang Y, Bevilacqua PC, Assmann SM (2014) In vivo genome-wide profiling of RNA secondary structure reveals novel regulatory features. Nature 505 (7485):696–700. doi:10.1038/nature12756
13. Spitale RC, Flynn RA, Zhang QC, Crisalli P, Lee B, Jung JW, Kuchelmeister HY, Batista PJ, Torre EA, Kool ET, Chang HY (2015) Structural imprints in vivo decode RNA regulatory mechanisms. Nature 519(7544):486–490. doi:10.1038/nature14263
14. Lu Z, Chang HY (2016) Decoding the RNA structurome. Curr Opin Struct Biol 36:142–148. doi:10.1016/j.sbi.2016.01.007
15. Cimino GD, Gamper HB, Isaacs ST, Hearst JE (1985) Psoralens as photoactive probes of nucleic acid structure and function: organic chemistry, photochemistry, and biochemistry. Annu Rev Biochem 54:1151–1193. doi:10.1146/annurev.bi.54.070185.005443
16. Thompson JF, Hearst JE (1983) Structure of E. coli 16S RNA elucidated by psoralen crosslinking. Cell 32(4):1355–1365
17. Ross A, Brimacombe R (1979) Experimental determination of interacting sequences in ribosomal RNA. Nature 281(5729):271–276
18. Calvet JP, Pederson T (1979) Heterogeneous nuclear RNA double-stranded regions probed in living HeLa cells by crosslinking with the psoralen derivative aminomethyltrioxsalen. Proc Natl Acad Sci U S A 76(2):755–759

19. Calvet JP, Pederson T (1981) Base-pairing interactions between small nuclear RNAs and nuclear RNA precursors as revealed by psoralen cross-linking in vivo. Cell 26(3 Pt 1):363–370
20. Hausner TP, Giglio LM, Weiner AM (1990) Evidence for base-pairing between mammalian U2 and U6 small nuclear ribonucleoprotein particles. Genes Dev 4(12A):2146–2156
21. Lu Z, Zhang QC, Lee B, Flynn RA, Smith MA, Robinson JT, Davidovich C, Gooding AR, Goodrich KJ, Mattick JS, Mesirov JP, Cech TR, Chang HY (2016) RNA duplex map in living cells reveals higher-order transcriptome structure. Cell 165(5):1267–1279. doi:10.1016/j.cell.2016.04.028
22. Bolger AM, Lohse M, Usadel B (2014) Trimmomatic: a flexible trimmer for Illumina sequence data. Bioinformatics 30 (15):2114–2120. doi:10.1093/bioinformatics/btu170
23. Dobin A, Davis CA, Schlesinger F, Drenkow J, Zaleski C, Jha S, Batut P, Chaisson M, Gingeras TR (2013) STAR: ultrafast universal RNA-seq aligner. Bioinformatics 29(1):15–21. doi:10.1093/bioinformatics/bts635
24. Li H, Handsaker B, Wysoker A, Fennell T, Ruan J, Homer N, Marth G, Abecasis G, Durbin R, Genome Project Data Processing S (2009) The Sequence Alignment/Map format and SAMtools. Bioinformatics 25 (16):2078–2079. doi:10.1093/bioinformatics/btp352
25. Zuker M (2003) Mfold web server for nucleic acid folding and hybridization prediction. Nucleic Acids Res 31(13):3406–3415
26. Robinson JT, Thorvaldsdottir H, Winckler W, Guttman M, Lander ES, Getz G, Mesirov JP (2011) Integrative genomics viewer. Nat Biotechnol 29(1):24–26. doi:10.1038/nbt.1754
27. Lorenz R, Bernhart SH, Honer Z, Siederdissen C, Tafer H, Flamm C, Stadler PF, Hofacker IL (2011) ViennaRNA Package 2.0. Algorithms Mol Biol 6:26. doi:10.1186/1748-7188-6-26
28. Gesell T, von Haeseler A (2006) In silico sequence evolution with site-specific interactions along phylogenetic trees. Bioinformatics 22(6):716–722. doi:10.1093/bioinformatics/bti812
29. Flynn RA, Zhang QC, Spitale RC, Lee B, Mumbach MR, Chang HY (2016) Transcriptome-wide interrogation of RNA secondary structure in living cells with icSHAPE. Nat Protoc 11(2):273–290. doi:10.1038/nprot.2016.011
30. Kiss-Laszlo Z, Henry Y, Bachellerie JP, Caizergues-Ferrer M, Kiss T (1996) Site-specific ribose methylation of preribosomal RNA: a novel function for small nucleolar RNAs. Cell 85(7):1077–1088
31. Tycowski KT, You ZH, Graham PJ, Steitz JA (1998) Modification of U6 spliceosomal RNA is guided by other small RNAs. Mol Cell 2 (5):629–638
32. Hearst JE (1981) Psoralen photochemistry. Annu Rev Biophys Bioeng 10:69–86. doi:10.1146/annurev.bb.10.060181.000441
33. Grass JA, Hei DJ, Metchette K, Cimino GD, Wiesehahn GP, Corash L, Lin L (1998) Inactivation of leukocytes in platelet concentrates by photochemical treatment with psoralen plus UVA. Blood 91(6):2180–2188
34. Sastry SS, Ross BM, P'Arraga A (1997) Cross-linking of DNA-binding proteins to DNA with psoralen and psoralen furan-side monoadducts. Comparison of action spectra with DNA-DNA cross-linking. J Biol Chem 272(6):3715–3723
35. Sastry SS, Spielmann HP, Hoang QS, Phillips AM, Sancar A, Hearst JE (1993) Laser-induced protein-DNA cross-links via psoralen furanside monoadducts. Biochemistry 32 (21):5526–5538
36. Zechel K, Weber K (1977) Degradation of nucleic acids in cell lysates by S1 nuclease in the presence of 9 M urea and sodium dodecylsulfate. Eur J Biochem 77(1):133–139

Chapter 5

Axon-TRAP-RiboTag: Affinity Purification of Translated mRNAs from Neuronal Axons in Mouse In Vivo

Toshiaki Shigeoka, Jane Jung, Christine E. Holt, and Hosung Jung

Abstract

Translating ribosome affinity purification (TRAP) is a widely used technique to analyze ribosome-bound mRNAs in particular target cells that express a tagged ribosomal protein. We developed axon-TRAP-RiboTag, a TRAP-based method that allows purification and identification of translated mRNAs from distal neuronal axons in mouse, and identified more than 2000 of translated mRNAs in retinal ganglion cell (RGC) axons in vivo. The use of Cre-negative littermate control to filter out false-positive signals allows unbiased detection, and combining TRAP with in vitro ribosome run-off enables identification of actively translated mRNAs. Here, we describe a detailed protocol to identify translated mRNAs in RGC axons in mouse in vivo. This method can be applied to any neurons whose cell bodies and distal axons are anatomically separated.

Key words Ribosome, Translation, mRNAs, Immunoprecipitation, Axon, Neuron

1 Introduction

Translation is a key step that controls the abundance of proteins in the cell [1], and therefore it is crucial to profile translated mRNAs in a specific cell population in vivo to understand how cells regulate their gene expression within an organism. Translating ribosome affinity purification (TRAP), affinity purification of ribosomes carrying an epitope-tagged ribosomal protein (RP) from a specific cell population, is a useful approach to obtain cell type-specific profiles of ribosome-bound mRNAs [2–8]. This technique utilizes a genetically engineered animal that expresses an epitope-tagged RP, such as EGFP (enhanced green fluorescence protein)-Rpl10a and Rpl22-HA (hemagglutinin), in a genetically defined population of cells. Using a cell type-specific promoter that drives the expression of the tagged RP, ribosome–mRNA complexes can be specifically purified from the cells where the promoter is active in a tissue, an organ, or an organism. Compared to RNase protection-based ribosome profiling, TRAP has the advantage that it generates sequence

Imre Gaspar (ed.), *RNA Detection: Methods and Protocols*, Methods in Molecular Biology,
vol. 1649, DOI 10.1007/978-1-4939-7213-5_5, © Springer Science+Business Media LLC 2018

information on the untranslated regions (UTRs) of isoforms, which would be lost by RNase treatment, although the ribosome profiling can reveal the number of ribosomes bound to a single mRNA species which would accurately reflect the translational rate.

Previously, we developed axon-TRAP, a subcellular TRAP method that allows specific isolation of ribosome-bound mRNAs from the distal RGC axons in *Xenopus* tadpoles in vivo [4]. We recently extended this technique to mouse, using RiboTag [6], a knockin mouse line in which Cre-mediated recombination labels the 60S subunit ribosomal protein L22 (Rpl22) with hemagglutinin (HA) tags (Rpl22-HA). We crossed this mouse with a Pax6-alpha-Cre [9], in which distal neural retinal progenitors transiently express Cre, leading to permanent HA-labeling of ribosomes in RGCs (Fig. 1a, red area in the eye). This new technique, termed axon-TRAP-RiboTag, involves affinity purification of HA-tagged

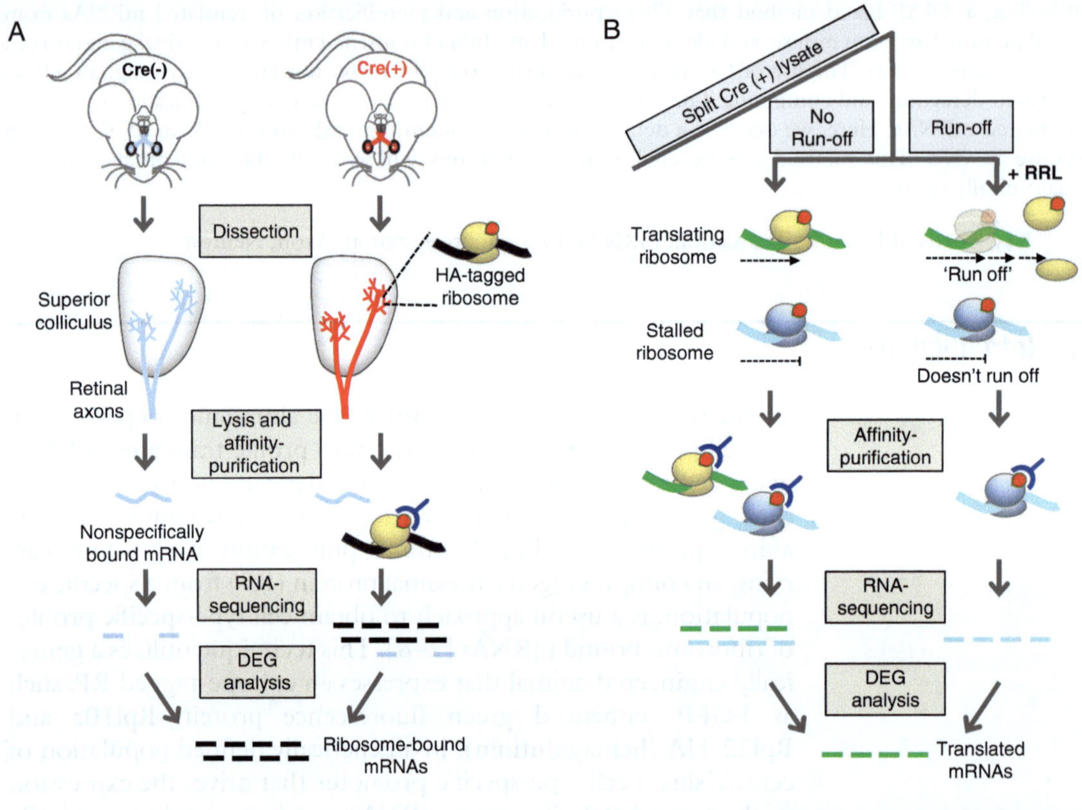

Fig. 1 Timeline of the experiments. (**a**) Axon-TRAP for RGCs. We crossed the RiboTag mouse with the Pax6-alpha-Cre to express HA-tagged ribosomes in RGCs. The affinity purification of HA-tagged ribosome–mRNA complexes from the superior colliculus (SCs), where RGC axons terminate, enables translational profiling of distal RGC axons in vivo by deep sequencing. To exclude false-positive signals, we compared the RNA-seq data from TRAPed mRNAs between the Cre-positive and -negative littermates. (**b**) In vitro ribosome run-off assay. To distinguish translated mRNAs from translation-stalled mRNAs, we carried out in vitro ribosome run-off assay, which allows translational elongation in vitro in the presence of rabbit reticulocyte lysate (RRL)

ribosome–mRNA complexes from the superior colliculus (SCs) in the midbrain, where RGC axons terminate, and enables translational profiling of distal RGC axons at different developmental stages in vivo by deep sequencing [5] (Fig. 1a).

Major technical problems of TRAP-based approaches stem from nonspecific binding of mRNAs expressed from cells that do not express the tagged ribosome during the affinity-purification step, which inevitably generates false-positive signals, particularly when combined with sensitive detection methods such as deep sequencing. This becomes particularly problematic when the mRNAs bound to tagged ribosomes represent only a small fraction of the total mRNA in the dissected tissue. For example, the mRNAs translated in distal RGC axons represent a very small fraction of the total mRNAs present in the superior colliculus, where many neurons and glia contain their own mRNAs. We addressed this problem by "differential expression analysis" on biological replicates of Cre-positive and -negative groups, which controls for all potential causes of nonspecific mRNA binding [5] (Fig. 1a).

Another key problem of TRAP-based approaches is that they purify all ribosome-bound mRNAs, not specifically translated mRNAs. As ribosome-bound mRNAs include not only translated mRNAs but also translation-stalled mRNAs, such as those bound to the translational repressor FMRP [10], TRAPed mRNAs do not necessarily represent the translatome (i.e., the entire set of translated mRNAs). To address this point, we carried out in vitro ribosome run-off assay to distinguish translated mRNAs from translation-stalled mRNAs. When allowed to continue translational elongation in vitro in the presence of rabbit reticulocyte lysate and translation initiation inhibitors, only the mRNAs that were being translated at the time of tissue preparation resume translation and finally "run-off" the tagged ribosome (Fig. 1b). Quantitative analysis of TRAPed mRNAs with and without run-off unbiasedly identifies translated mRNAs [5].

Here, we describe a detailed protocol for the axon-TRAP-RiboTag of RGC axons in vivo and a method for in vitro ribosome run-off to validate whether axon-TRAPed mRNAs were being translated. Although the protocol is designed for mouse RGC axons, this approach might be applied to other axons, subcellular compartments and cells.

2 Materials

2.1 Tissue Preparation

1. Mouse embryos or adult mice generated through crosses between a homozygotic RiboTag mouse and a Cre-expressing mouse (*see* **Note 1**). If a hemizygotic Cre-expressing mouse is used for the crossing, Cre-negative littermates can be used as the negative control for TRAP.

2. Liquid nitrogen in a container (*see* **Note 2**).
3. Dissection tools: forceps, fine scissors, and springbow dissecting scissors.

2.2 Lysis and Immunoprecipitation

Prepare all solutions using RNase-free water, buffers, and labware.

1. A Dounce homogenizer.
2. CHX-Lysis buffer: 20 mM HEPES–KOH, 5 mM $MgCl_2$, 150 mM KCl, 1 mM DTT, 1 v/v % NP40, 200 U/mL SUPERase In, 100 μg/mL cycloheximide, and Complete EDTA-free Protease Inhibitor Cocktail (*see* **Note 3**).
3. Wash buffer: 20 mM HEPES–KOH, 5 mM $MgCl_2$, 350 mM KCl, 1 mM DTT, 1 v/v % NP40, and 100 μg/mL cycloheximide.
4. Polyclonal anti-HA antibody (Abcam, ab9110).
5. Magnetic Protein G beads (e.g., Thermo Fisher Scientific Dynabeads).
6. A rotator for Eppendorf tubes in a cold room (4 °C).
7. Magnetic Eppendorf stand (e.g., Thermo Fisher Scientific DynaMag).
8. RNeasy mini kit (Qiagen).
9. 2-mercaptoethanol.

2.3 Run-Off Translation

1. A Dounce homogenizer.
2. RO-lysis buffer: 20 mM HEPES-KOH, 5 mM $MgCl_2$, 150 mM KCl, 1 mM DTT, 200 U/mL SUPERase In, Complete EDTA-free Protease Inhibitor Cocktail (Roche, cOmplete), 1% volume of amino acid mix (−leucine), and 1% volume of amino acid mix (−methionine) (amino acids mixes are included in Flexi Rabbit Reticulocyte Lysate System (Promega, L4540).
3. Flexi Rabbit Reticulocyte Lysate System (Promega, L4540).
4. 100 μg/mL Harringtonine stock solution.
5. 5 mM 4E1RCat stock solution.
6. 5 mg/mL cycloheximide.
7. Wash buffer: 20 mM HEPES-KOH, 5 mM $MgCl_2$, 350 mM KCl, 1 mM DTT, 1% NP40, and 100 μg/mL cycloheximide (*see* **Note 4**).

3 Methods

3.1 Mouse Breeding

1. A Pax6-alpha-Cre hemizygotic male mouse is mated with a RiboTag homozygotic female mouse. Each litter is expected to have roughly equal numbers of Cre-positive (the experimental group) and Cre-negative (the negative control) animals.

3.2 Tissue Preparation

In order to minimize the RNA degradation, prepare the snap-frozen tissue samples as quickly as possible.

1. Dissect the tissues that contain the target cells or subcellular compartments. Dissect another tissue for genotyping (*see* **Note 4**). In case of RGC axons, the rostral half of the superior colliculi is quickly dissected.
2. Place the tissue into a prelabeled Eppendorf tube.
3. Secure the lid, and then place the tube directly into the liquid nitrogen.
4. The frozen tissues can be stored in a liquid nitrogen storage tank or −80 °C freezer.
5. Genotype for the Cre transgene for each tissue sample prepared.

3.3 Lysis and Preclearing

1. Prepare the CHX-lysis buffer (lysis buffer with cycloheximide) immediately before the use.
2. Place the frozen tissues into a Dounce homogenizer containing 500 μL ice-cold CHX-lysis buffer (5–10 w/v % tissues in CHX-lysis buffer). If necessary, pull two to four tissue samples for the same genotype (i.e., pull Cre-positive tissues in one tube, and Cre-negative tissues in another tube). For RGC axons, we used three or four pairs of superior colliculi.
3. On ice, homogenize the tissues gently but thoroughly until lysate becomes clear. If using the same Dounce homogenizer for multiple samples, homogenize Cre-negative tissues first. Wash the homogenizer multiple times with RNase-free water, and once with CHX-lysis buffer.
4. Centrifuge the lysate at 16,000 × *g* at 4 °C for 10 min. Carefully take the supernatant and transfer to a prechilled tube. The supernatant contains tagged ribosome–mRNA complexes. If necessary, 5 μL of supernatant can be stored as the "input" for TRAP. During this step, prepare the magnetic Protein G beads as in **steps 5**–7.
5. Resuspend the magnetic Protein G beads in the vial, and transfer 40 μL to a new tube.
6. Place the tube on the magnet to separate the beads from the solution, and remove the supernatant.
7. Resuspend the beads in 40 μL of CHX-lysis buffer and repeat the **step 6**.
8. Transfer the supernatant after centrifugation (**step 4**) to the washed beads (**step** 7).
9. Rotate the tube (head-to-toe rotation) for 1 h in cold room (preclearing step to minimize nonspecific binding of mRNAs to the Protein G beads).

10. Place the tube containing lysate and the beads on the magnet and transfer the supernatant to a prechilled new tube on ice.

3.4 Immunoprecipitation and RNA Purification

1. Add 2.5 μL Rabbit anti-HA antibody to the precleared lysate (**step 10** of Subheading 3.3) and rotate the tube (head-to-toe rotation) in cold room overnight.
2. Wash 40 μL Dynabeads with CHX-lysis buffer as **steps 5**–7 of the Subheading 3.3, and add it to the lysate containing the antibody (**step 1**). Do not let the beads dry out.
3. After 4 h of head-to-toe rotation in cold room, place the tube on a prechilled magnet on ice and remove the supernatant. Do not let the beads dry out. The beads contain tagged ribosome–mRNA complexes. The supernatant at this step can be saved as the "unbound fraction."
4. Add 500 μL of wash buffer to the beads and rotate it for 5 min in cold room.
5. Repeat the washing step (**steps 3** and **4**) at least three times more (total four washing steps are recommended).
6. Add 500 μL of wash buffer to the beads, and transfer the buffer and beads to a new prechilled tube (*see* **Note 5**).
7. Remove wash buffer from the beads using a DynaMag magnet, and resuspend the beads in 100 μL of CHX-lysis buffer.
8. Add 350 μL of RLT buffer (RNeasy mini kit) containing 2-Mercaptoethanol (*see* **Note 6**). Vortex vigorously and incubate for 5 min at room temperature. During this step, mRNAs are dissociated from the tagged ribosome.
9. Centrifuge for 30 s using a benchtop centrifuge and collect the supernatant using a DynaMag magnet at room temperature. The supernatant contains axon-TRAPed mRNAs.
10. Add 250 μL of 100% ethanol to the supernatant (**step 9**).
11. Purify RNA using RNeasy mini kit (Qiagen). Perform on-column DNase digestion to remove potential DNA contamination.
12. Elute RNA from the column in 14 μL of RNase-free water (*see* **Note** 7).
13. Proceed to cDNA synthesis (*see* **Note 8**).

3.5 In Vitro Ribosome Run-Off

For the run-off experiments, we use run-off (RO)-lysis buffer that does not contain any detergent and cycloheximide (*see* **Note 9**).

1. Prepare the RO-lysis buffer immediately before the experiment.

2. Place the frozen tissues into the Dounce homogenizers containing 400 μL of ice-cold RO-lysis buffer (10 w/v % tissues in RO-lysis buffer).
3. Centrifuge the lysate at 16,000 × *g* at 4 °C for 10 min. During this step, prepare the rabbit reticulocyte lysate (RRL) as following **steps 4–6**.
4. Add the 8 μL harringtonine (100 μg/mL stock solution) and 4 μL 4E1RCat (5 mM stock solution) to 188 μL RRL, and mix gently. For the "no run-off" control, add 8 μL cycloheximide to the mix (*see* **Note 10**)
5. Preincubate RRL at 30 °C for 5 min
6. Transfer the supernatant (**step 3**) to the tube and mix it gently.
7. Split the lysate (**step 6**) into two tubes. One of two tubes can be used for the "no run-off" control.
8. Add 200 μL of RRL mix (**step 4**) to 200 μL of the lysate (**step** 7), and incubate at 30 °C for 30 min.
9. Add 800 μL ice-cold CHX-lysis buffer to the run-off reaction and incubate the tube on ice for 5 min to stop the translation elongation.
10. Proceed to immunoprecipitation, following the **steps 5–10** of the Subheading 3.3 and then Subheading 3.4.

4 Notes

1. Make sure that no cells in the target tissue (which contains axon terminals of interest and in this case is the rostral half of the superior colliculus) express tagged ribosomes. Histological assays using any Cre reporter mice (e.g., Rosa26-Stop-LoxP-EGFP) is a good start. Confirm this in the RiboTag × Cre mouse to be studied using HA immunohistochemical detection. PCR-based detection of chromosomal DNA that underwent Cre-mediated recombination is a more sensitive method. The "normal" RiboTag allele and "recombined" RiboTag allele can be distinguished by PCR (Fig. 2) [5]. The target tissue should not contain a detectible level of the recombined *RiboTag* allele. Use the following primers to detect normal and recombined *RiboTag* alleles: RiboTag fwd, 5′-GGGAGGCTTGCTGGATATG-3′, and HA rev, 5′-ACATCG-TATGGGTATAGATCC-3′ (Fig. 2).
2. Wear safety glasses or a face shield when using liquid nitrogen.
3. Prepare all solutions just before the experiment.

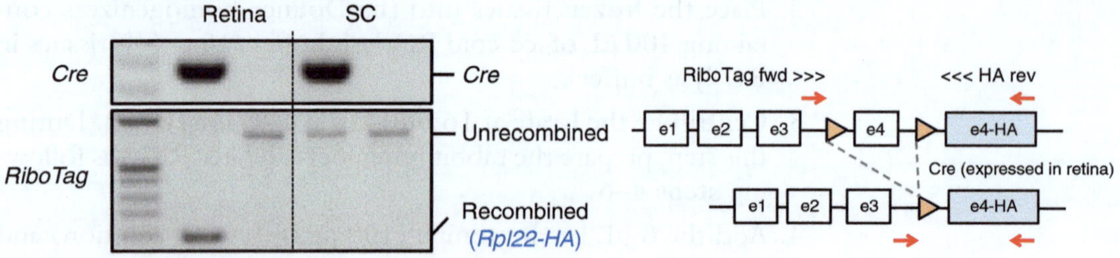

Fig. 2 PCR-based detection of Cre-mediated recombination. Cre-mediated recombination converts the normal RiboTag allele to the HA-tagged, recombined allele (*see* the gene model on the *right*). These two alleles can be distinguished by the primers depicted in the diagram. Genomic DNAs were extracted from the retina (containing the cell bodies of RGCs) and the superior colliculus (SC, containing the axon terminals of RGCs) of RiboTag mice with or without Pax6-alpha-Cre allele. Cre-mediated "RiboTag" is detectable only in the retina but not in the superior colliculus

4. If genotyping is needed, prepare tail or toe biopsies for genomic DNA PCR experiments. Clean the equipment to avoid carry-over/cross contamination between samples.
5. This step can reduce the false-positive signals due to nonspecific binding of mRNAs to the tube.
6. Alternatively, you can carry out a competitive elution using excessive HA peptides (Sigma, I2149). After **step** 7 of Subheading 3.4, resuspend the beads containing the tagged ribosome–mRNA complexes in 100 μL of 100 μg/mL HA peptide in CHX-lysis buffer. Vortex vigorously and incubate for 5 min at room temperature. Take the supernatant on a DynaMag magnet and transfer the supernatant to a prechilled tube on ice. The supernatant contains axon-TRAPed mRNAs. Proceed to **step 8**. Omit **step 9** (as the bead is removed). Although the competitive elution reduces the total amount of purified mRNAs, it can improve the signal-to-noise ratio by increasing the specificity.
7. We estimated that approximately 40% of HA-tagged translating ribosomes could be purified using this protocol.
8. If abundant mRNAs are isolated, they can be sequenced by typical RNA sequencing protocols. Alternatively, TRAPed mRNAs can be amplified by a PCR-based method before sequencing. We amplified cDNA by the method developed by Azim Surani and colleagues for single cell transcriptomics, which utilizes PCR-based amplification of polyadenylated RNAs [11]. We followed the detailed protocol up to Step 41 described in this paper [11] to make double-strand DNAs for deep sequencing. Successful axon-TRAP can be validated by agarose gel electrophoresis of cDNAs amplified from Cre-positive and Cre-negative samples. Cre-positive samples should generate stronger bands (Fig. 3).

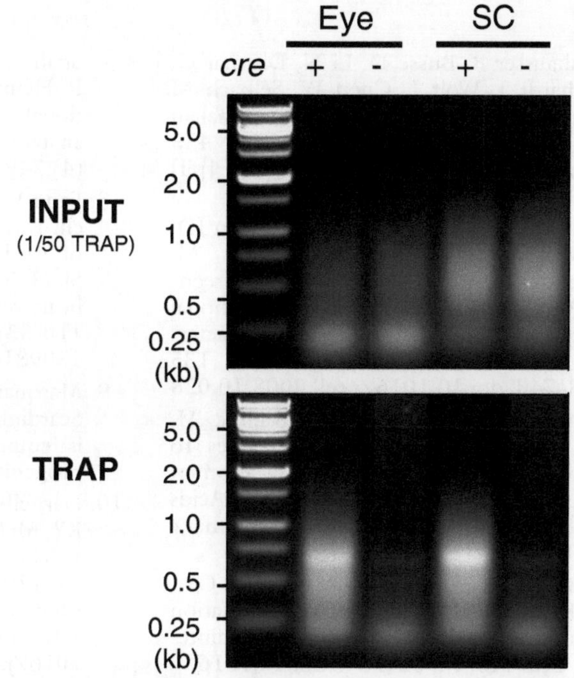

Fig. 3 Expected results. Double strand cDNAs made from TRAPed RNAs from neonatal mice (postnatal day 0.5). Three eyes or three pairs of superior colliculi were lysed in ~500 μL CHX-lysis buffer. Total 10 μL of this lysate was used as input (*upper*), and 500 μL lysate was TRAPed (*lower*). Polyadenylated RNA was made into double strand DNA using the method described by Tang and colleagues [11]

9. Because the efficiency of the lysis in this condition is significantly lower than in the condition using detergents and cycloheximide, a larger amount of starting materials is needed.
10. Harringtonine and 4E1RCat are added to inhibit the translation initiation [12].

Acknowledgments

This work was supported by Samsung Science and Technology Foundation under Project Number SSTF-BA1602-13 (to H.J.) and by Wellcome Trust Programme Grant (085314/Z/08/Z) (to C.E.H.).

References

1. Schwanhausser B, Busse D, Li N, Dittmar G, Schuchhardt J, Wolf J, Chen W, Selbach M (2011) Global quantification of mammalian gene expression control. Nature 473 (7347):337–342. doi: nature10098 [pii] 10.1038/nature10098
2. Heiman M, Schaefer A, Gong S, Peterson JD, Day M, Ramsey KE, Suarez-Farinas M, Schwarz C, Stephan DA, Surmeier DJ, Greengard P, Heintz N (2008) A translational profiling approach for the molecular characterization of CNS cell types. Cell 135 (4):738–748. doi:10.1016/j.cell.2008.10.028
3. Dougherty JD, Schmidt EF, Nakajima M, Heintz N (2010) Analytical approaches to RNA profiling data for the identification of genes enriched in specific cells. Nucleic Acids Res 38(13):4218–4230. doi:10.1093/nar/gkq130
4. Yoon BC, Jung H, Dwivedy A, O'Hare CM, Zivraj KH, Holt CE (2012) Local translation of extranuclear lamin B promotes axon maintenance. Cell 148(4):752–764. doi:10.1016/j.cell.2011.11.064
5. Shigeoka T, Jung H, Jung J, Turner-Bridger B, Ohk J, Lin JQ, Amieux PS, Holt CE (2016) Dynamic axonal translation in developing and mature visual circuits. Cell 166(1):181–192. doi:10.1016/j.cell.2016.05.029
6. Sanz E, Yang L, Su T, Morris DR, McKnight GS, Amieux PS (2009) Cell-type-specific isolation of ribosome-associated mRNA from complex tissues. Proc Natl Acad Sci U S A 106 (33):13939–13944. doi:10.1073/pnas.0907143106
7. Doyle JP, Dougherty JD, Heiman M, Schmidt EF, Stevens TR, Ma G, Bupp S, Shrestha P, Shah RD, Doughty ML, Gong S, Greengard P, Heintz N (2008) Application of a translational profiling approach for the comparative analysis of CNS cell types. Cell 135 (4):749–762. doi:10.1016/j.cell.2008.10.029
8. Fang Y, Gupta V, Karra R, Holdway JE, Kikuchi K, Poss KD (2013) Translational profiling of cardiomyocytes identifies an early Jak1/Stat3 injury response required for zebrafish heart regeneration. Proc Natl Acad Sci U S A 110(33):13416–13421. doi:10.1073/pnas.1309810110
9. Marquardt T, Ashery-Padan R, Andrejewski N, Scardigli R, Guillemot F, Gruss P (2001) Pax6 is required for the multipotent state of retinal progenitor cells. Cell 105(1):43–55
10. Darnell JC, Van Driesche SJ, Zhang C, Hung KY, Mele A, Fraser CE, Stone EF, Chen C, Fak JJ, Chi SW, Licatalosi DD, Richter JD, Darnell RB (2011) FMRP stalls ribosomal translocation on mRNAs linked to synaptic function and autism. Cell 146(2):247–261. doi:10.1016/j.cell.2011.06.013
11. Tang F, Barbacioru C, Nordman E, Li B, Xu N, Bashkirov VI, Lao K, Surani MA (2010) RNA-Seq analysis to capture the transcriptome landscape of a single cell. Nat Protoc 5 (3):516–535. doi:10.1038/nprot.2009.236
12. Cencic R, Hall DR, Robert F, Du Y, Min J, Li L, Qui M, Lewis I, Kurtkaya S, Dingledine R, Fu H, Kozakov D, Vajda S, Pelletier J (2011) Reversing chemoresistance by small molecule inhibition of the translation initiation complex eIF4F. Proc Natl Acad Sci U S A 108(3):1046–1051. doi:10.1073/pnas.1011477108

Chapter 6

LCM-Seq: A Method for Spatial Transcriptomic Profiling Using Laser Capture Microdissection Coupled with PolyA-Based RNA Sequencing

Susanne Nichterwitz, Julio Aguila Benitez, Rein Hoogstraaten, Qiaolin Deng, and Eva Hedlund

Abstract

LCM-seq couples laser capture microdissection of cells from frozen tissues with polyA-based RNA sequencing and is applicable to single neurons. The method utilizes off-the-shelf reagents and direct lysis of the cells without RNA purification, making it a simple and relatively cheap method with high reproducibility and sensitivity compared to previous methods. The advantage with LCM-seq is also that tissue sections are kept intact and thus the positional information of each cell is preserved.

Key words Motor neuron, Dopamine neuron, RNA sequencing, Laser capture microscopy, Smart-seq2, HistoGene, Cresyl violet, Rapid antibody staining

1 Introduction

An ability to decipher gene expression within individual cells and/or small, distinct neuronal populations will be fundamental to improve our understanding of normal biological processes and mechanisms of neuronal vulnerability to disease. Tissue samples are often in scarcity, in particular patient samples, and cellular heterogeneity can mask biological functions and/or disease pathways when tissues are analyzed as a whole or in bulks of hundreds to thousands of cells. It is therefore vital that gene expression can be studied in small tissue samples, minute cellular populations and even in individual cells. Toward this goal, we have developed a new method, LCM-seq, that combines laser capture microdissection/microscopy (LCM) with polyA-based Smart-seq2 RNA sequencing [1]. This method allows for efficient and robust

Susanne Nichterwitz and Julio Aguila Benitez contributed equally to this work.

Imre Gaspar (ed.), *RNA Detection: Methods and Protocols*, Methods in Molecular Biology, vol. 1649, DOI 10.1007/978-1-4939-7213-5_6, © Springer Science+Business Media LLC 2018

sequencing of neurons isolated from mouse and human postmortem tissues down to single cells. We utilized LCM as it enables precise isolation of individual cells from postmortem tissues. As tissue sections used for the procedure remain intact it preserves the positional information of each captured cell. LCM can successfully be applied to tissues from any embryonic or postnatal stage. It is particularly advantageous for adult animal tissues and human postmortem samples that do not easily allow for tissue dissociation. Smart-seq2 is a widely used method for single-cell RNA sequencing with increased sensitivity and reproducibility compared to other methods [2]. Furthermore, Smart-seq2 allows for a single tube reaction with direct lysis of live cells and subsequent cDNA library preparation, which we have adapted to cells isolated by LCM. Our newly developed method LCM-seq enables spatial transcriptomic profiling down to the single cell level, thus enabling the deduction of gene expression in intact tissues at a significantly improved resolution.

2 Materials

General note. Clean all work benches, instruments and equipment with RNaseZAP and wipe with wet tissue (H_2O or 70% ethanol). Use certified RNase/DNase-free materials and molecular biology grade reagents whenever available. Use filter tips. Wear gloves at all times when handling materials and change gloves regularly. Cleaning gloves regularly with RNaseZap wipes is recommended, particularly before touching LCM slides. Keep samples as cold as possible at all times and minimize handling to avoid degradation and contamination. Keep dedicated reagents, materials, and equipment for RNA work separately whenever possible.

2.1 Tissue Dissection

1. 5% avertin (2,2,2-tribromoethanol) in PBS.
2. 7 mL PE Pasteur pipette, cut to a spoon.
3. 2-methylbutane (Isopentane), approximately 10 mL.
4. Metal container for freezing reagent.
5. Styrofoam box with dry ice.
6. Dissection tools.
7. RNaseZap wipes.
8. Nuclease-free 1.5 mL tubes and aluminum foil for tissue storage, labeled and precooled on dry ice.
9. −80 °C freezer.

2.2 Tissue Cryosectioning

1. Cryostat, −20 °C.
2. RNaseZap wipes.

3. Embedding medium (OCT).
4. Embedding molds.
5. Block of dry ice.
6. Forceps, precooled on dry ice.
7. Razor blade or scalpel to trim tissue block.
8. Slide box.
9. LCM membrane slides (e.g., 1.0 PEN membrane glass slides, Zeiss).
10. Pencil to label slides, permanent marker to label embedding mold.
11. Brushes to handle tissue sections and cleaning of blade.

2.3 Tissue Staining and LCM

1. Leica LMD7000 (any LMD microscope can be used, but depending on the vendor it might change which microscope slides and collector tubes are used), preferably enclosed in a clean chamber equipped with a fan and a UV lamp.
2. Thermometer, hygrometer.
3. RNaseZap wipes.
4. One styrofoam box with dry ice and one with wet ice.
5. Staining containers (e.g., conventional 50 mL centrifuge tubes): 2× ddH_2O (for HistoGene staining only), 2× 75% ethanol, 1× 95% ethanol, and 1× 99.7% ethanol, 25 mL each.
6. Pipette boy, clean; 25 mL pipette tips, sterile.
7. 99.7% ethanol.
8. Nuclease-free ddH_2O.
9. HistoGene staining solution (Arcturus) or 1% cresyl violet acetate solution in 75% ethanol adjusted to pH 8.0 with 3 M Tris–HCl.
10. Plastic staining tray, 5 mL pipette tips, sterile, to place slides on.
11. Lint-free tissue wipes.
12. Slide forceps.
13. Slide box.
14. Pipettes and filter tips (1–10 μL and 200 μL).
15. Timer.
16. Lysis buffer: 0.2% Triton X-100, 2 U/μL RNase inhibitor, kept on ice, prepared freshly for every session.
17. 0.2 mL RNase/DNase-free PCR tubes (*see* **Note 11**).
18. Microcentrifuge (e.g., VWR Ministar silverline).
19. Parafilm.

2.4 Antibody Based Rapid Tissue Staining

1. Aluminum foil to cover light-sensitive staining solutions.
2. 15 mL centrifuge tubes.
3. Staining containers with 25 mL of ddH_2O and PBS (replaced with fresh liquids for every wash).
4. Staining containers for dehydration: 1× 50% ethanol, 1× 75% ethanol, 1× 95% ethanol and 1× 99.7% ethanol, 25 mL each.
5. Shaker.
6. Acetone, ice cold in a staining container (25 mL per staining).
7. Freshly made 0.25% Triton X-100 in PBS to dilute primary and secondary antibodies (500 μL per slide).
8. Primary antibody (volume is dependent on the antibody concentration needed), kept on ice until diluted for use.
9. Secondary antibody (the volume is dependent on the antibody concentration needed), kept on ice until diluted for use.
10. 2.5 mL of streptavidin–biotin solution (ABC solution, e.g., Vector Laboratories Vectastain ABC kit Elite) in a 15 mL tube (covered with foil and kept in motion for 30 min prior to use).
11. 2.5 mL of diaminobenzidine solution (DAB solution, e.g., Vector Laboratories DAB Kit) in a 15 mL tube (covered with foil and kept in motion for 2–3 min prior to use).
12. 1000 μL pipette and filter tips.

2.5 Library Preparation

1. Thermocycler, if possible dedicated to this procedure.
2. Iron thermal block stand.
3. Styrofoam box with ice.
4. Vortex.
5. Microcentrifuge.
6. Pipettes and filter tips for different volumes.
7. Timer.
8. Nuclease-free 1.5 mL tubes for preparation of master mixes.
9. PCR tubes.
10. Nuclease-free ddH_2O.
11. PEG (polyethylene glycol) 8000.

2.5.1 Reverse Transcription and PCR Amplification

1. Reverse transcription kit (e.g., Invitrogen SSRT II).
2. ERCC spike ins (2.5×10^5 dilution in ddH_2O, prepare aliquots in a sterile laminar flow hood and store at −80 °C, do not freeze-thaw).
3. dNTPs mix (10 mM each nucleotide).

4. oligo-dT: 5′AAGCAGTGGTATCAACGCAGAGTACTTTTTTTTTTT-TTTTTTTTTTTTTTTTTTTTVN3′ (HPLC purified, store at 10 μM concentration in ddH_2O at −20 °C for several months), can be freeze-thawed several times.
5. TSO-LNA oligo: 5′AAGCAGTGGTATCAACGCAGAGTACr GrG + G3′
   (HPLC purified, store at 100 μM concentration in ddH_2O at −80 °C, several freeze–thaw cycles are ok).
6. ISPCR primer: 5′AAGCAGTGGTATCAACGCAGAGT3′ (HPLC purified, store at 10 μM concentration in H_2O at −20 °C for several months), can be freeze-thawed several times.
7. High fidelity hot start PCR mix (e.g., Kapa HiFi HotStart Ready Mix).
8. Master mix 1 (Reverse transcription 1): 1 μL dNTPs mix, 1 μL Oligo dT, 0.1 μL ERCC spike-ins, 2.1 μL per sample.
9. Master mix 2 (Reverse transcription 2): 2 μL 5× SSRT II buffer, 0.5 μL 100 mM DTT, 2 μL 5 M Betaine, 0.1 μL 1 M $MgCl_2$, 0.25 μL RNase inhibitor, 0.5 μL SSRT II enzyme, and 0.1 μL TSO-LNA oligo; 5.45 μL sample.
10. Master mix 3 (PCR): 12.5 μL 2× KAPA HiFi HotStart Ready mix, 0.2 μL ISPCR primer, and 2.3 μL ddH_2O; 15 μL per sample.

2.5.2 Bead Purification

1. 80% ethanol, prepared fresh every week.
2. 96-well plates (V-bottom).
3. Magnetic stand for 96-well plates.
4. Vortex.
5. Magnetic beads, hydrophilic.
6. 19.5% and 24% PEG bead solution: magnetic beads in 19.5 or 24% PEG 8000, 1 M NaCl, 10 mM Tris–HCl pH = 8.0, 1 mM EDTA, 0.01% Igepal CA630, and 0.05% sodium azide.
7. Elution buffer (EB): 10 mM Tris–HCl, pH 8.5 (e.g., Qiagen).

2.5.3 Tagmentation and Sequencing Index Ligation

1. Nextera XT Index Kit (Illumina).
2. 5× TAPS-Mg buffer: 50 mM TAPS-NaOH pH 8.5 and 25 mM $MgCl_2$.
3. 0.2% SDS in ddH_2O.
4. High fidelity PCR enzyme (e.g., Kapa Hifi DNA polymerase kit with dNTPs).
5. Master mix 4 (Tagmentation): 5 μL 40% PEG 8000, 4 μL 5× TAPS-Mg, and 0.4 μL ~12.5 μM Tn5 enzyme (*see* **Note 20**) [3]; 9.4 μL per sample.

6. Master mix 5 (Index ligation and enrichment PCR): 10 μL 5× high fidelity PCR buffer, 1.5 μL dNTPs, 1 μL high fidelity PCR enzyme, and 2.5 μL $ddH_2 0$; 15 μL per sample.

3 Methods

3.1 Tissue Dissection

1. Clean work benches and tools and prepare all materials needed.
2. Anesthetize animals (using 5% avertin or other anaesthetic) and decapitate.
3. Dissect tissues as rapidly as possible, place the tissue into a small plastic spoon (cut from PE pasteur pipette) and snap-freeze by submerging the tissue in 2-methylbutane on dry ice (*see* **Note 1**).
4. Place tissues in RNase/DNase-free tubes or in clean aluminum foil (*see* **Note 1**) on dry ice for transportation and store at −80 °C until further processed.

3.2 Tissue Sectioning

1. Clean work benches and tools with 70% ethanol and RNaseZap.
2. If embedding of the tissue is required (*see* **Note 2**) use pre-chilled cryo embedding medium to avoid tissue thawing during embedding. Place a labeled mold filled with OCT on an even block of dry ice. Wait a few seconds until the OCT in the bottom of the mold is frozen (turns white), but the top layer is still liquid. Place the spinal tissues onto the bottom of the mold using cold forceps. Leave the tissue block on dry ice for 5–10 min until it is completely frozen. Place the mold in the cryostat to equilibrate to the sectioning temperature (wait at least 30 min before sectioning). Remove the plastic mold and trim away any excess OCT from the tissue block (using a razor blade). Brains (and other larger tissues) do not need embedding prior to sectioning. Mount the tissue or the tissue block (embedded tissue) on a chuck with OCT and wait until it is properly frozen.
3. Label LCM slides and place them into the cryostat in a plastic slide box (to cool down) that is kept in the cryostat at −20 °C during sectioning.
4. Section tissues at −20 °C at the required thickness (*see* **Note 3**). Place tissue sections on membrane glass LCM slides. To better attach sections, "melt" the tissue onto the LCM slide by placing a finger or a metal heat stick on the back of the slide just underneath the section (*see* **Note 4**).
5. Store slides with tissue sections at −80 °C until further processed (*see* **Note 5**).

3.3 Histological Tissue Staining for Laser Capture Microdissection

Transport slides (in slide boxes) on dry ice. Use forceps to move slides between staining containers, wear gloves, and clean these regularly with RNaseZap. For examples of histological stainings *see* Fig. 1.

1. Before initiating the staining, switch on the LCM system (microscope, laser, computer, start LMD software) and turn on the UV light and fan in the LCM chamber if available.

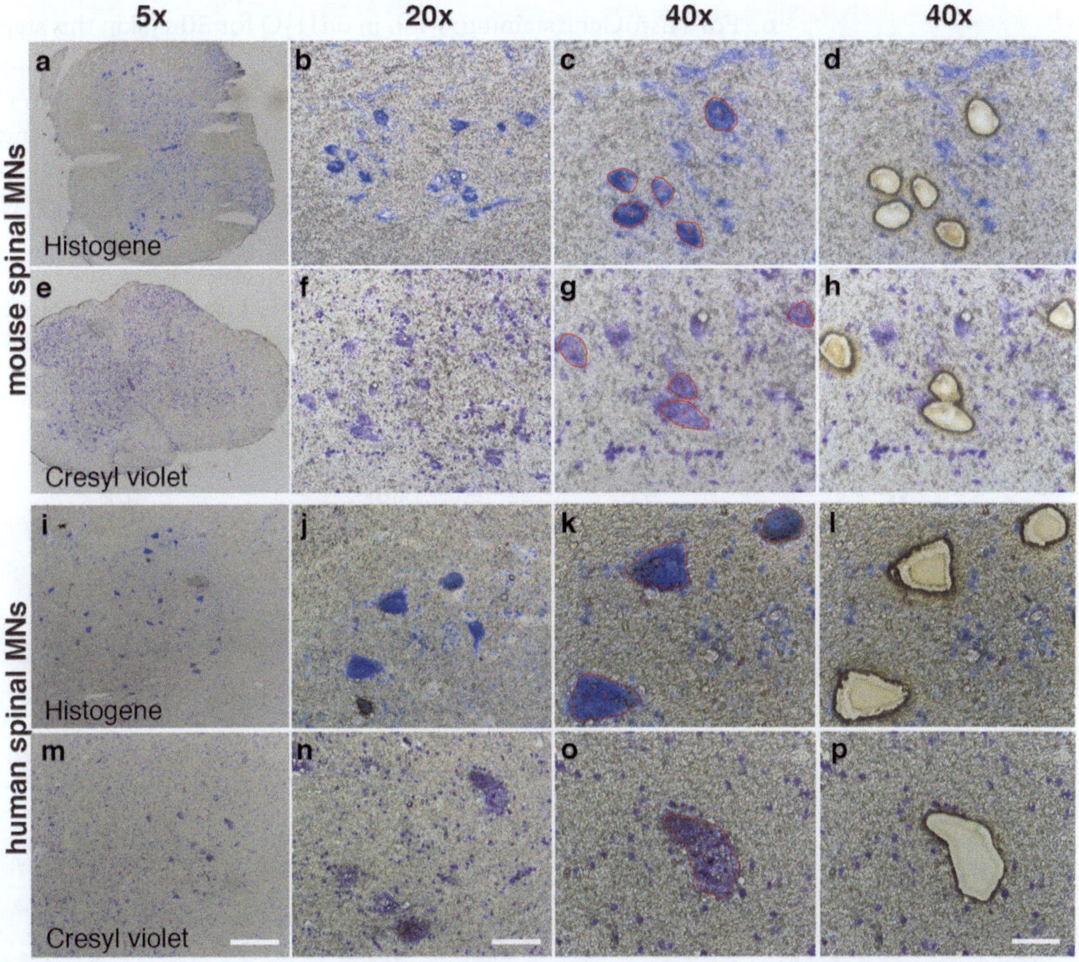

Fig. 1 Examples of HistoGene and cresyl violet stainings of mouse and human spinal cords at different magnifications. (**a–d, i–l**) Note that the HistoGene staining in general gives a better contrast. (**e–h, m–p**) Small nonneuronal cells are more visible after staining with cresyl violet. (**a, e, i, m**) Images taken at 5× magnification provide a good overview of the tissue morphology for reference. (**b, f, j, n**) Images taken prior to microdissection at 20× magnification are a helpful reference to identify cells of interest. (**c, g, k, o**) Note the slightly decreased contrast at 40× magnification with less distinct nuclei of the cells. To avoid contamination with surrounding cells, keep cutting outlines close to the cells. (**d, h, l, p**) Tissue after isolation of cells. Scale bar in **m** (applicable to **a, e, i, m**) = 400 μm, in **n** (applicable to **b, f, j, n**) = 100 μm and in **p** (applicable to **c, d, g, h, k, l, o, p**) = 50 μm)

2. Thaw slides for 30 s at room temperature by placing horizontally on a clean surface (e.g., 5 mL pipette tips on plastic tray).
3. Place into 75% ethanol for 30 s (*see* **Note 6**).
4. For HistoGene staining: incubate slides in ddH_2O for 30 s and subsequently dry by briefly tilting them onto lint-free tissue (skip this step for cresyl violet staining).
5. Place slides horizontally, cover with HistoGene or cresyl violet staining solution (~200 μL) for 20–60 s (*see* **Note 7**), then remove staining solution by tilting slides onto lint-free tissue.
6. For HistoGene staining: wash in ddH_2O for 30 s (skip this step for cresyl violet staining).
7. Dehydrate by moving slides through rising ethanol concentrations (75, 95 and 99.7% ethanol, 30 s each) followed by tilting the slides onto lint-free tissue.
8. Place slides into a dry, new slide box and initiate the LCM procedure.

3.4 Rapid Antibody Tissue Staining for Laser Capture Microdissection

Transport slides (in slide boxes) on dry ice. Use forceps to move slides between staining containers, wear gloves and clean these regularly with RNaseZap. For examples of antibody stainings *see* Fig. 2.

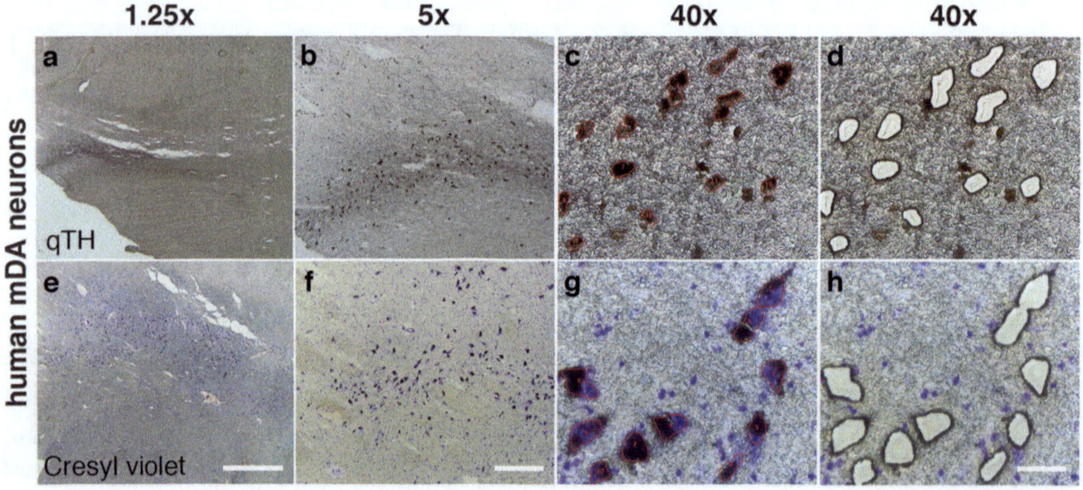

Fig. 2 Examples of tyrosine hydroxylase antibody staining and staining Nissl of human midbrain dopamine neurons. Midbrain dopamine (mDA) neurons in the ventral tegmental area visualized using a (**a–d**) quick TH antibody staining or (**e–h**) by using cresyl violet. (**a–b, e–f**) Low magnification images provide an important anatomical reference that later aid in interpreting the corresponding sequencing data. (**c, d**) While the TH antibody facilitates high specificity for laser dissection of dopaminergic cells, it lacks the ability to visualize surrounding nuclei of small cells that could be a source of contamination. (**g, h**) In contrast, when the cresyl violet is combined with the presence of neuromelanin, a feature of adult human midbrain dopamine neurons, the staining also reveals the nuclei of small cells and could aid in enriching the target population. Scale bars in **e** (applicable to **a, e**) = 2000 μm, in **f** (applicable to **b, f**) = 400 μm and in **h** (applicable to **c, d, g, h**) = 50 μm. Human tissues were kindly received from the Netherlands Brain Bank (NBB) and the National Disease Research Interchange (NDRI)

1. Before initiating the staining, switch on the LCM system (microscope, laser, computer, start LMD software) and turn on the UV light and fan in the LCM chamber if available. Thaw slides for 1 min at room temperature by placing them horizontally on a clean surface (e.g., 5 mL pipette tips on plastic tray).
2. Place into acetone for 5 min (*see* **Note 8**).
3. Incubate the slides three times in PBS (1 min each time) and subsequently dry the slides by briefly tilting these onto lint-free tissue. Note that after each step in the antibody staining procedure the solutions are removed and slides are dried briefly by simply tilting these onto lint-free tissue.
4. Place slides horizontally, cover with ~250 μL, of primary antibody staining solution for 4 min (*see* **Note 9**), then remove staining solution.
5. Wash the slides three times in PBS (2 min each time) and subsequently dry briefly.
6. Place slides horizontally and cover the tissues with ~250 μL secondary antibody staining solution for 4 min (*see* **Note 10**), then remove staining solution.
7. Wash the slides three times in PBS (2 min each time) and subsequently dry the slides briefly.
8. Place slides horizontally, cover with ~250 μL of ABC solution for 4 min, then remove staining solution.
9. Wash the slides three times in PBS (2 min each time) and subsequently dry the slides briefly.
10. Place slides horizontally, cover with ~250 μL of DAB solution for 1–2 min, then remove staining solution.
11. Wash the slides two times in PBS (2 min each time) and subsequently dry briefly.
12. Wash in ddH_2O for 30 s, then dehydrate the tissues by moving slides through rising ethanol concentrations (50, 75, 95, and 99.7% ethanol, 30 s each) followed by briefly drying the slides.
13. Place slides into a dry, new slide box, air-dry the slides completely and initiate the LCM procedure.

3.5 Isolation of Cells by Laser Capture Microdissection

Perform LCM (as exemplified by using a Leica LMD7000, cells captured by gravity, *see* **Note 11**):

1. Turn off the UV light in LCM chamber.
2. Place the open slide box (with slides in a horizontal position) into the LCM chamber with the fan still running for 3–5 min to dry slides properly. Subsequently turn off the fan.

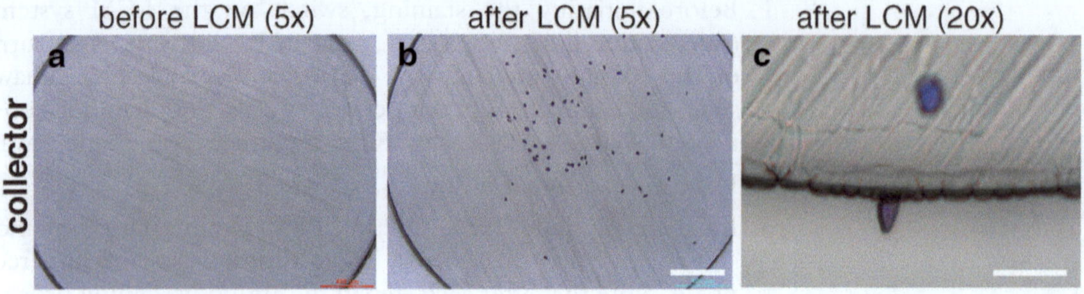

Fig. 3 PCR tube caps as seen in the tube holder of the LCM (Leica LMD7000). (**a**) Cap before and (**b**) after collection of individual cells by laser microdissection at 5× magnification. Note that the collected cells here have relatively large soma sizes of >200 μm^2 while smaller cells will be harder to *see*. (**c**) Collected cells can "hide" in uneven parts of the cap, especially when statics are high. If collection of a precise number of cells is required, it is advisable to check the collector carefully at a higher magnification (20×). Scale bar in **b** (applicable to **a, b**) = 400 μm and in **c** = 50 μm

3. Place 0.2 mL PCR tubes into the collector and check tubes for obvious contamination (change if contaminated) (*see* Fig. 3).
4. Place slides into the slide holder with the membrane facing down.
5. Create "fast overview" of slides at 1.25× magnification to facilitate navigation.
6. Search for cells of interest at 5 or 10× magnification (*see* **Note 12** and Figs. 1 and 2).
7. Microdissect at 40× magnification: draw cutting outlines closely to the cells to avoid contamination with other cell types (*see* **Notes 3** and **13** and Figs. 1 and 2).
8. Capture cells into RNase/DNase-free 0.2 mL PCR tubes (dry cap).
9. When sufficient numbers of cells have been collected, inspect the collector and unload the tube holder (*see* Fig. 3).
10. Add a small volume of lysis buffer and pipette up and down a few times to resuspend all cells (*see* **Note 14**).
11. Spin down samples in a microcentrifuge for 15 s.
12. Seal PCR tubes with Parafilm, label and snap-freeze on dry ice (*see* **Note 15**).
13. Store samples at −80 °C until further processed.

3.6 cDNA and Sequencing Library Preparation

All of the following steps are carried out on ice if not otherwise specified.

3.6.1 Reverse Transcription and PCR Amplification

1. Preheat thermocycler to 72 °C.
2. Prepare master mix 1 and 2 and place the metal tube rack on ice.
3. Retrieve LCM-isolated cell samples from −80 °C.
4. Spin down the samples in a microcentrifuge and transfer into PCR tube strips (optional, *see* **Note 16**).
5. Add 2.1 μL of master mix 1 (do not pipette up and down to avoid loss of material) and spin down.
6. Incubate for 3 min at 72 °C to open the secondary structures of the RNA molecules.
7. Spin down and place back on ice (oligo dTs anneal during snap cooling).
8. Add 5.45 μL of master mix 2 and spin down.
9. Initiate the RT reaction by heating samples at 42 °C for 90 min. Subsequently incubate samples at 50 °C for 2 min and at 42 °C for 2 min, and repeat this for 10 cycles. Finally, incubate samples at 70 °C for 15 min. The RT reaction takes approximately 2.5 h after which samples are spun down.
10. Prepare the PCR master mix 3, add 15 μL per sample and spin down.
11. Use the following program for PCR; 3 min at 98 °C, then 18 cycles of (20 s at 98 °C, 15 s at 67 °C, and 6 min at 72 °C) (*see* **Note 17**). End the PCR with a 5 min incubation at 72 °C. The PCR is approximately 2.5 h in length. Spin down the samples.

3.6.2 Bead Purification

1. Take the 19.5% PEG bead solution out of the fridge at least 20 min before starting the purification.
2. Transfer cDNA (25 μL) into a 96-well plate.
3. Mix the 19.5% PEG bead solution well by vortexing.
4. Add 25 μL of the 19.5% PEG bead mix to the cDNA, set the pipet to 50 μL and mix well by pipetting up and down ten times, while being careful not to produce bubbles.
5. Cover the plate and incubate at room temperature for 8 min.
6. Place the plate (covered) on a magnetic stand and leave for 5 min.
7. Take off the supernatant while being careful not to disturb the beads.
8. Wash by adding 200 μL of 80% ethanol toward the opposite side of the beads, shake gently, then remove ethanol, repeat once.

9. Leave the plate without cover to air-dry the beads. Dry bead pellets will have tiny cracks (*see* **Note 18**).
10. Add 17 μL EB; flush toward the beads.
11. Take the well plate off the magnetic stand.
12. Pipet up and down ten times and incubate for 2 min at room temperature (cover).
13. Place the plate back on the magnetic stand for 2 min (cover).
14. Collect a slightly smaller volume (16 μL), to not transfer any beads with the eluted DNA, and place this into low binding tubes.
15. Perform quality and quantity control of cDNA libraries using a Bioanalyzer instrument (*see* Fig. 4 for examples of cDNA profiles) or measure concentration in the next step using a Qubit instrument.

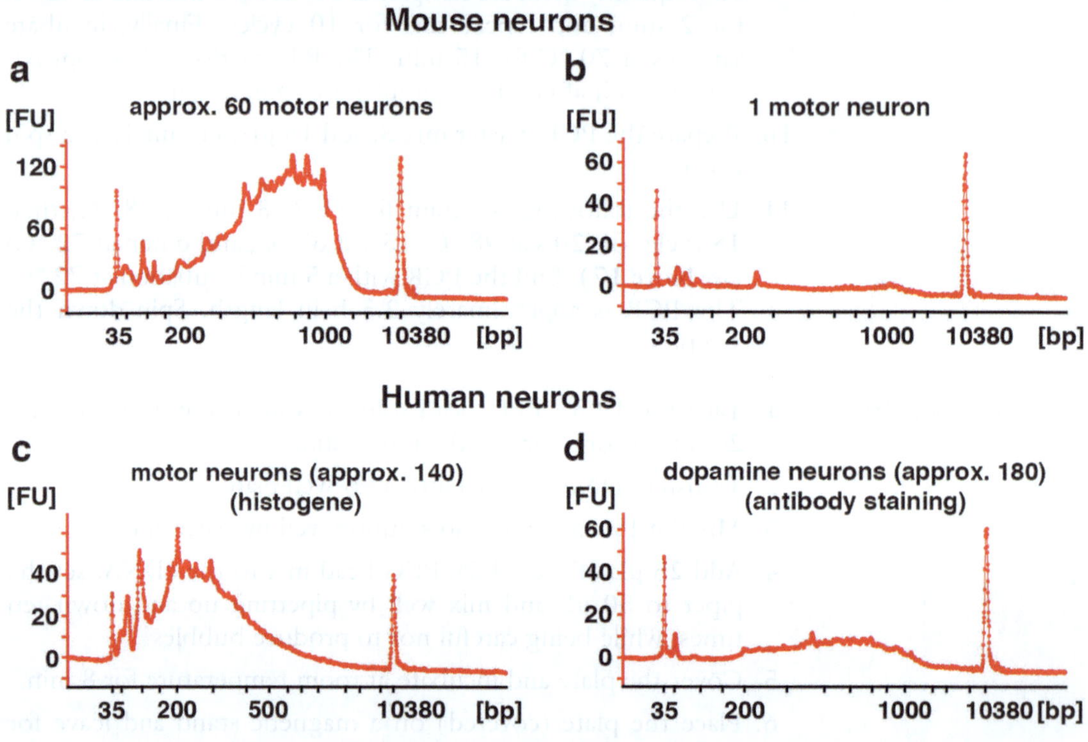

Fig. 4 cDNA library profiles generated from mouse and human LCM dissected cells. (**a, b**) Examples of Bioanalyzer profiles of cDNA libraries generated from mouse spinal motor neurons after HistoGene staining from 60 and 1 cells, respectively. Libraries from 10 to 150 mouse cells typically show a peak between 800 and 1200 bp. Depending on the degradation status/quality of the tissue, the curve can be flatter or show a more pronounced peak than in the 60 cell example depicted here. Libraries from singe cells typically have a barely recognizable peak at around 1200 bp and the concentration can be as low as 0.1 ng/μL (100–9000 bp range). (**c, d**) Human cDNA profiles typically indicate markedly more RNA degradation as reflected by a peak shift to smaller fragment sizes or (**c**), a flat curve without a distinct peak (**d**) and in general lower cDNA yield. FU = fluorescence units (amount of cDNA), bp = base pairs (fragment size)

3.6.3 Tagmentation and Sequencing Index Ligation

1. Preheat the PCR machine to 55 °C.

   Prepare the tagmentation master mix 4 on ice, mix well (*see* **Notes 19** and **20**).
2. Transfer 1 ng of cDNA per sample to a PCR tube and adjust the volume with ddH_2O to 10.6 μL.
3. Add 9.4 μL of master mix 4 (to a total volume of 20 μL), pipet up and down five times.
4. Incubate at 55 °C for 5 min.
5. Immediately add 5 μL 0.2% SDS (*see* **Note 21**) (pipet up and down five times) and incubate for 5 min at room temperature to strip enzymes from the DNA.
6. Prepare PCR master mix 5 on ice.
7. Place tubes back on ice.
8. Add 5 μL index primers (5 μL each, diluted 1:5 in ddH_2O). Choose a unique combination of two sequencing indexes indices (N7xx and S5xx) for each sample if they are to be sequenced on the same lane. Be careful not to confuse lids of different tubes to avoid cross contamination.
9. Add 15 μL of PCR master mix (master mix 5), pipet up and down (total volume 50 μL now).
10. Use the following program for the final enrichment PCR; 3 min at 72 °C, 30 s at 95 °C, then 10 cycles (*see* **Note 22**) of 10 s at 95 °C, 30 s at 55 °C, and 30 s at 72 °C. Finalize with 5 min incubation at 72 °C and set the PCR machine to hold at 4 °C thereafter.
11. Spin down samples and transfer to a 96-well plate for bead purification.

3.6.4 Bead Purification

1. Take the 24% PEG bead solution out of the fridge at least 20 min prior to initiating the purification.
2. For protocol *see* Subheading 3.6.2, use 24% PEG beads now and add 50 μL of bead mix (1:1) in **step 3**.

3.6.5 Pooling of Samples for Sequencing

1. Measure concentration of sequencing libraries on Qubit.
2. Pool equal amounts of DNA for sequencing in one lane (*see* **Note 23**).

4 Notes

1. The duration of the freezing process depends on the size of the tissue and can be judged by the color change. It takes approximately 7 s for an adult mouse brain to freeze and 3 s for a spinal cord. Do not submerge tissues too long as they may crack.

Spinal cord dissections take a maximum of 15 min from decapitating the animal and brains can be retrieved in less than 3 min. For storage of particularly brittle tissues, e.g., spinal cords from neonatal mice, use small tubes. Other tissues, e.g., brains, can be stored in aluminum foil.

2. Embed thin, brittle tissues, e.g., spinal cords, in cryo embedding medium (OCT) to avoid tissues from rupturing during sectioning. Avoid creating bubbles in the OCT (as these will render the sectioning uneven).
3. The thickness of the tissue sections depends on the size and shape of the cells of interest. To ensure that one cell is spanning the entire depth of the section, use a section thickness equal to the minimal **radius** of your cells as a guideline. Only capture cells that have a visible nucleus surrounded by cytoplasm. This will maximize the RNA content of your sample and minimize the risk of collecting several cell layers hidden underneath your cell.
4. When placing tissue sections on membrane LCM slides, avoid touching the membrane with sharp objects like forceps or cryoblades. The membrane is highly fragile and keeping it intact is crucial for good staining and drying of the slide for LCM.
5. Sectioned tissue can be stored for several weeks on slides prior to performing LCM. However, faster processing is recommended for better staining and LMD cutting outcomes.
6. When staining sections from OCT embedded tissue, carefully agitate the staining containers at every step to ensure that the OCT is properly removed.
7. The incubation time with HistoGene and cresyl violet staining solution is variable for different tissues and cell types. Determine the optimal staining time by starting with 20 s incubation. Prolong the incubation time if necessary, but keep it as short as possible to minimize RNA degradation. Additionally, in our hands cresyl violet seems to better visualize the nuclei of small glial cells that surround the neurons, enabling separate isolation also of these.
8. Acetone should be fresh for every staining and ice cold before use.
9. The primary antibody for the rapid staining is used at a high concentration. For example, we use the sheep-anti TH antibody at a 1/25 dilution for LCM while we normally utilize it at 1/1000 for immunohistochemistry and confocal imaging. In addition, the antibody needs to show a high degree of specificity to recognize only the protein and cells of interest. To increase the versatility of LCM-seq, we have tested and shown that an increased incubation time for the primary antibody for up to 1 h still results in sequencing data of high quality [1].

10. The secondary antibody should be of high quality and used at a high concentration to minimize incubation times and thus unnecessary degradation of tissues. Here, we use a 1/25 dilution of an antibody that is used at 1/500 in standard immunohistochemistry applications.
11. The relative humidity during LCM for downstream RNA processing should optimally be between 20 and 40%. RNA degradation increases with rising humidity whereas low humidity can increase statics and hamper the collection process if using a collection system based on gravity. However, we have successfully processed samples collected at a relative humidity between 10 and 60% and at temperatures between 20 and 29 °C. We recommend taking notes of temperature and relative humidity to enable a correlation of these parameters with sample quality. Furthermore, the use of anti-statics devices such as the Milty Zerostat gun may improve collection of cells. Note that also the choice of PCR tubes may have an influence on statics and in our hands, PCR soft tubes from Biozym Scientific work well.
12. We highly recommend saving microscope images of your stained sections pre-LCM at 10 or 20× magnification as well as post-microdissection as a reference. Pre-microdissection images can be helpful to identify cells with a clear nucleus and nucleolus, as the contrast at 40× magnification can be low.
13. Optimal laser settings depend on many factors including dryness of the slide, humidity in the chamber, thickness of the tissue, and tissue morphology, but also on personal preferences and may be adjusted throughout the LCM process. As a guideline, adjust the laser power and aperture from low to high at the beginning of every session. Move to the "no cap" position and start outlining an area outside of your region of interest, but with similar tissue morphology. Adjust laser settings until the cutting line is neat and thin and the membrane with the tissue detaches from the slide. Decreasing the speed of the laser will aid in keeping the cutting line thin as lower laser power and aperture are needed. When using a system based on gravity (such as Leica's), inspect the collector after isolating an easily countable number of cells to get an estimate of the actual number of cells that are collected into the cap. Keep in mind that statics can vary significantly throughout a session. Check the collector regularly if collecting large numbers of cells over several hours.
14. The volume of lysis buffer depends on the number of cells collected; for 50–150 cells, we use 5 μL. Do not pipette the lysis buffer up and down when collecting a very small number of cells (up to 10), to avoid losing cells in the pipette tip. Furthermore, we recommend to place the tube back into the

tube holder for inspection to ensure that no cells remain in the cap.

15. Keep time from thawing of slides for LCM to freezing the samples below 2 h.

16. For small numbers of collected cells, perform RT and PCR amplification in the collection tube to avoid loss of material. For a larger amount of cells collected, it is ok to transfer samples to PCR strips to facilitate handling.

17. The number of PCR cycles depends on the input material and should be determined for each project. 18 PCR cycles are sufficient for 1–150 LCM dissected motor neurons (>200 μm^2 area).

18. The time to air-dry the bead pellet is variable (5–15 min). Do not over-dry the magnetic beads. If some pellets are dry before others, already add the elution buffer to these while leaving the plate on the magnetic stand.

19. Pipette PEG slowly as it is very viscous and a final concentration in the 8–10% range is critical.

20. Preparation of sequencing libraries can be achieved using the Nextera XT DNA Library Preparation Kit (Illumina) using the manufacturer's recommendations. However, to reduce cost we have instead used a recombinant Tn5 enzyme produced in house as described in detail in (3). The volume of Tn5 needed for tagmentation can vary between batches of the enzyme and we have observed a loss of enzymatic activity over time. Test and adjust the volume for every new batch or when tagmentation is becoming inefficient.

21. Be precise when preparing and pipetting the 0.2% SDS solution. A slightly higher concentration of SDS may inhibit the PCR reaction.

22. The number of cycles can be increased to 12 if input DNA is much less than 1 ng.

23. The amount of samples that will be sequenced in one lane depends on the sequencing platform and the sequencing depth desired (i.e., a specific biological question). Only pool samples that have unique combinations of sequencing indices (N7xx and S5xx). For recommended combinations of indices see Illumina's guideline for sample pooling.

References

1. Nichterwitz S, Chen G et al (2016) Laser capture microscopy coupled with Smart-seq2 for precise spatial transcriptomic profiling. Nat Commun 7:12139. doi:10.1038/ncomms12139
2. Picelli S et al (2013) Smart-seq2 for sensitive full-length transcriptome profiling in single cells. Nat Methods 10:1096–1098. doi:10.1038/nmeth.2639
3. Picelli S et al (2014) Tn5 transposase and tagmentation procedures for massively scaled sequencing projects. Genome Res 24:2033–2040. doi:10.1101/gr.177881.114

Chapter 7

Spatial Transcriptomics: Constructing a Single-Cell Resolution Transcriptome-Wide Expression Atlas

Kaia Achim, Hernando Martínez Vergara, and Jean-Baptiste Pettit

Abstract

The method described here aims at the construction of a single-cell resolution gene expression atlas for an animal or tissue, combining in situ hybridization (ISH) and single-cell mRNA-sequencing (scRNAseq).

A high resolution and medium-coverage gene expression atlas of an animal or tissue of interest can be obtained by performing a series of ISH experiments, followed by a process of image registration and gene expression averaging. Using the overlapping fraction of the genes, concomitantly obtained scRNAseq data can be fitted into the spatial context of the gene expression atlas, complementing the coverage by genes.

Key words Spatial transcriptomics, Image registration, Single-cell mRNA-seq, Gene expression

1 Introduction

The molecular composition of cells defines their development, identity and function. mRNA expression is one of the components of this molecular composition. Recent advances in RNA expression studies provide excellent tools for the assessment of this part of cellular identity.

The method described here is an integrative approach that combines two cutting-edge high-throughput methods: universally applicable single-cell mRNA sequencing techniques, and the generation of ISH-based gene expression atlases. The integration of these two technologies combines their respective advantages, that is, transcriptomics and spatial profiling, allowing the generation of single-cell resolution transcriptome-wide full-body expression atlases.

For performing the ISH and scRNA-seq, researcher is free to use their favorite established protocol, such as ones described in this book. In this chapter, we summarize, first, the requirements for the acquisition and processing of the imaged ISH data to generate an expression atlas; second, the methods for single-cell dissociation and for obtaining an initial quality-control of the raw scRNAseq read count data; and third, the computational pipeline for the integration of these data (spatial mapping).

Imre Gaspar (ed.), *RNA Detection: Methods and Protocols*, Methods in Molecular Biology, vol. 1649, DOI 10.1007/978-1-4939-7213-5_7, © Springer Science+Business Media LLC 2018

2 Materials

2.1 In Situ Hybridization (ISH)

For ISH, researchers are free to use their favorite established protocol.

2.2 Imaging of the ISH Samples

1. Since large numbers of samples need to be imaged for the construction of the ISH atlas, access to a microscope with automatization possibility is advisable. We use Leica SP8 with 40× oil immersion objective, and use the mark-and-find function for automated serial acquisition of image stacks.

2.3 Generation of a Gene Expression Atlas

1. Fiji [1].
2. Python.
3. ITK [2].
4. R-Bioconductor [3] software.
5. Obtain the scripts published in Asadulina et al. 2012 [4] and Vergara et al. 2016 [5].

2.4 Single-Cell mRNA-Sequencing (scRNAseq) and Mapping the scRNAseq Data to the Reference Transcriptome

For the scRNAseq, the researcher is free to use any available approach (*see* **Note 1**) Here, we have used the C1 Single-cell Auto Prep platform by Fluidigm.

1. Obtain the C1 Single-cell Auto Prep IFC for mRNA-seq, designed to fit your cell size range (*see* **Notes 2** and **3**).
2. Obtain the following kits and reagents: SMARTer Ultra Low Input RNA Kit and ADVANTAGE-2 PCR kit (Clontech), ERCC spike-in RNA (Ambion,) Nextera XT DNA Sample Preparation kit (Illumina)
3. Install *bowtie2* [6] and *HTseq1* [7].

2.5 Spatial Mapping Tools

1. Install *R* [3].
2. The *R* scripts for spatial mapping are published in Achim et al. (2015) [8].

   Download these scripts via the following link: https://github.com/jbogp/nbt_spatial_backmapping.
3. The *BioWeb3D* [9] based interface for the visualization of the reference voxels and spatial mapping results can be accessed via the following link: http://www.ebi.ac.uk/~jbpettit/map_viewer/

2.6 Data Management Solutions

1. Since significant amounts of raw metadata will be produced and processed, a strategy for the storage, management, and streamlined data processing should be developed, preferably prior to starting with the experiments.

3 Methods

The methods described in this section cover, first, the principles and steps of generating gene expression atlases; second, the protocol for obtaining quality filtered and normalized scRNAseq count data; and third, the bioinformatic approach for the matching the scRNA-seq and ISH data together in its original spatial context.

3.1 Generation of a Gene Expression Atlas

The protocol described in this section makes use of high-throughput microscopy and different image analysis steps to find, for each gene, averaged expression patterns.

Gene expression atlases make use of corresponding landmarks between different images to align them using image registration. Alongside, the spatial profiles of different genes can be fitted to a common reference, allowing coexpression analysis in silico. This principle has been employed to build expression maps for highly stereotypic tissues achieving cellular resolution [10], extended to full-body atlases using universal markers [4] and more recently, combined with image analysis to achieve a general method for building single-cell resolution expression atlases for complex body plans. We have used this protocol to generate a whole-body atlas for the three-segmented worm *Platynereis dumerilii* [5].

3.1.1 High-Throughput Microscopy

In order to build an ISH-based gene expression atlas, we need raw microscopy data that consists of a series of stacks (virtual slices) of the samples (e.g., 3D tissue or animal), each containing at least two channels:

1. The first channel (DAPI) is common to all the stacks, and will provide landmark information to calculate the spatial transformation (registration). This channel is referred to as the target channel during the registration process.
2. The second channel (ISH signal) contains the specific gene-expression information that will be registered, guided by the target channel.
3. Image the samples including at least the target and gene expression channels. The acquisition resolution should be equal in x, y, and z axes to generate isotropic stacks that facilitate the subsequent 3D processing (*see* **Note 4**).
4. For each genetic marker, depending on the desired final resolution of the atlas, a minimum number of samples need to be imaged (*see* **Note 4**).

3.1.2 Generating a Reference

In this step, we build a reference template, which will be used to register all the individual samples to the same three-dimensional framework. This reference ideally constitutes an average representation of the target channel (DAPI) and can be built by iterative rounds of registration between different samples. The scripts necessary for these steps can be found in [4].

1. Select the target channels of a relatively large number of samples. This number depends on the stereotypicity of the animal and the quality of the imaging, but a rule of thumb is above a hundred.
2. Roughly orient each sample to the same position (e.g., anterior side up and ventral side forward).
3. Measure the size of all the samples, and select one of them (e.g., one that has a median size) as the preliminary reference.
4. Perform a round of rigid registration (e.g., using the plugin 'Rigid Registration' in Fiji) to the preliminary reference, and average over the aligned samples to create a "rigid reference."
5. Perform a round of affine registration over the oriented samples to the rigid reference, and average the resulting stacks to create an "affine reference."
6. Perform a round of affine registration, followed by deformable registration, over the oriented samples to the affine reference, and average the resulting stacks to create a "deformable reference."
7. At this point, it is advised to inspect the registered individuals and remove from the oriented samples those that do not register properly.
8. Repeat **step 6**, registering to the deformable reference, and average to create the "final reference." We have not found any significant improvements by repeating this step further, but it is advisable to check for each case, as further iteration rounds might improve the final reference.

3.1.3 Sample Registration and Averaging

The next step requires the alignment of the individual samples to the reference template. During the image registration process, the intensity patterns of the target channel are compared with those of the reference to calculate a geometrical transformation that can be based on rigid and/or nonrigid registration methods (scaling, translation, rotation, shearing and local deformations). The same geometrical transformation is then applied to the second channel, which overlays the expression signal onto the common reference.

1. For each sample for each gene, use the script "Affine_Deformable_Registration.py" in Vergara et al. (2016) [5] to register both channels.
2. Use "Average_registered.py" to average the signal channels and create a canonical representation of the gene expression.
3. Detemine, for each gene, a binary threshold for each expression pattern. This can be done manually (*see* [4]) or using a more systematic image analysis procedure that adjust the threshold in a gene and body region-dependent manner (*see* [5]).

3.1.4 Building the Gene Expression Atlas

The alignment of different expression patterns to the same reference can be used to generate a gene expression atlas, which consists of a matrix of determined three-dimensional points (voxels), each containing binary (e.g., expressed or not expressed) information for the expression of every included gene.

The script "ProSPr_6dpf_SuperVoxelPixCount.R" in [5] can be used to generate the matrix which will be referred to in **step 2** of Subheading 3.3.3.

3.2 Obtaining Single-Cell RNAseq Data

1. Prepare your single cell suspension (*see* **Note 5**).
2. Follow the Fluidigm company instructions for the chip priming, loading of the cells and on-chip lysis, reverse transcription and PCR. We recommend adding ERCC spike-in molecules to the Clontech lysis mix.
3. Harvest the cDNA from the Fluidigm IFC and follow the Illumina instructions for the DNA fragmentation, barcoding and paired-end sequencing on Illumina platform.
4. Map the raw sequence data against reference transcriptome using *bowtie2* [6].
5. Obtain the expression counts for each gene by running *HTseq1* [7].

3.3 Spatial Mapping to the Reference Transcriptome

In this section, we will match the scRNAseq data and the ISH atlas, to find the probable locations of the sequenced cells. We have used this method to identify the probable locations of cells sequenced from the 48 h post fertilization *Platynereis dumerilii* larvae, demonstrating that the spatial mapping successfully identified the locations of the original cells in 80% of cases [8]. Our spatial mapping approach has several advantages: (a) it does not require a priori labeling or known spatial landmarks, (b) using the specificity index (Subheading 3.3.4) rather than direct read count as a gene expression measure effectively removes technical noise problems and (c) a transcriptome-wide correspondence score (Subheading 3.3.5) provides an additional buffer for noise, as it means that the mapping cannot be driven by a single landmark gene with a spatial expression profile that differs from all other genes expressed in the cell of interest.

To start this protocol, we need, on the one hand, the spatially referenced gene expression data from the ISH-based atlas (Subheading 3.1), and, on the other hand, the gene expression matrix obtained by the scRNAseq method of choice (Subheading 3.2), and the spatial mapping *R* scripts (Subheading 2.2).

3.3.1 Quality Filtering of RNA-Seq Data

1. For quality filtering of the data, we removed the cells (a) where more than 10% of the total read counts accounted for the spike-in sequences, and (b) where less than 1200 unique transcripts were detected. Depending on the cell capture platform, additional quality filters may need to be developed in order to remove samples that contain multiple cells (**Note 6**).

3.3.2 Subsetting and Normalization of RNA-Seq Data

1. Select only the genes that are represented in the atlas, and create a ***RNA_seq.csv*** file, containing in rows the cells that you want to map against the atlas, and in columns, the genes for which expression data is present in the atlas.
2. For normalization, calculate tpm [11] for the final dataset.

3.3.3 Spatial Mapping: Formatting and Loading the Data

1. Load the downloaded analysis functions in R:

```
source("spatial_mapping.R")
```

2. The data should be organized in the following files:

   atlas: table containing binary expression data values for N voxels (rows) and M genes (columns), such as one created in Subheading 3.1.4.
   3D_coordinates_atlas: table containing the spatial coordinates for the voxels in the atlas. This table should contain no header.
   RNA_seq: table containing normalized RNA-seq counts values for the quality-filtered cells in rows and the genes in columns. This table should contain only the genes common between the RNA-seq profiles and the selected atlas, and the order of the genes must be identical in the ***RNA_seq.csv*** and ***atlas.csv*** files.
3. Load your data in R:

```
atlas <- read.table("atlas",header=TRUE,sep="\t")
coordinates <- read.table("3D_coordinates_atlas",header=FALSE,sep=",")
rna_seq <- read.table("RNA_seq",header=TRUE,sep="\t")
```

3.3.4 Spatial Mapping: Specificity Index

1. In order to filter out noise and better match the RNAseq and ISH data, we define a specificity ratio (*see* **Note 7**), $r_{c,m}$ for each cell-gene combination:

$$r_{c,m} = \frac{D_{c,m}}{\frac{1}{C}\sum_{a=1}^{C} D_{a,m}}$$

where $C \times M$ is the read count matrix D with C cells and M genes, so that $D_{c,m}$ describes the normalized number of reads mapped to cell c for gene m.

   The specificity index is calculated at the spatial mapping script step:

```
specificity_matrix <- specificity_scores(rna_seq)
```

Optionally, we can apply specificity scores also on the atlas (symmetrical penalization, *see* **Note 8**). For that, run the following script:

```
number_of_points_in_altas <- 1000
atlas_specificity_score <- 1/apply(atlas,2,function(x){length(x[x > 0])/number_-
of_points_in_altas})
```

2. If you decide to apply asymmetrical penalization, i.e., only penalize mismatches when the gene is expressed in the scRNA-seq and not in the atlas, set the `atlas_specificity_score` as a vector of 0's:

```
atlas_specificity_score <- as.numeric(vector(length = num-
ber_of_points_in_altas))
```

3.3.5 Spatial Mapping: Correspondence Score

1. To map back each sequenced single cell to its localization in the reference ISH dataset, we use a scoring system where we calculate the correspondence between every cell-voxel combination.

The correspondence scores are calculated at the spatial mapping script step:

```
results_scores = sapply(seq_along(rna_seq[,1]),function(cell_num) {
            print(paste("Mapping cell",cell_num))
            cell = rna_seq[cell_num,]
            specificity_score <- specificity_matrix[cell_num,]

            # Set the threshold for binarization of the RNA-seq counts # data
            count_threshold <- 10
            binary_expression_cell<-sapply(cell,function(x){ifelse
(x>count_threshold,1,0)})

            # Launch the mapping against the atlas for the cell
            mapping_result_cell <-spatial_map_scoring(specificity_score,binary_
expression_cell, atlas_specificity_score,atlas)

            # Create the folder for mapping results, if it doesn't exist:
            dir.create(file.path(".", "mapping_results"), showWarnings = FALSE)

            # Save the mapping results in a file in mapping_results folder:
           write.table(file=paste("mapping_results/mapping_result_cell_", cell_num,
sep=""), mapping_result_cell)
           mapping_result_cell
    })
```

This scoring system compares the binary vector of expression provided by the whole mount ISH data with a binarized version of the expression pattern for each sequenced cell. To binarize the expression vectors, we used a threshold of 10 reads above which a gene was considered expressed. This was configured by the variable *count_threshold* in the script above.

Mathematically, the score $S_{e_c, e_{ref}}$ between the binary expression vector e_c from single cell c and e_{ref} from voxel ref in the ISH dataset is defined as:

$$s_{c,\mathrm{ref}} = \sum_{m=1}^{M} f_{r_{c,m}}\left(e_{c,m}, e_{\mathrm{ref},m}\right)$$

with

$$f_{r_{c,m}}\left(e_{c,m}, e_{ref,m}\right) = \begin{cases} t\left(r_{c,m}\right), & e_{c,m} = e_{ref,m} = 1 \\ -t\left(r_{c,m}\right), & e_{c,m} = 1, \ e_{ref,m} = 0 \\ 0, & \text{Otherwise.} \end{cases}$$

and

$$t\left(r_{c,m}\right) = \frac{r_{c,m}}{1 + r_{c,m}}$$

This scoring scheme is designed to assess the correspondence between a single cell and each reference voxel with regard to the specificity ratio of each gene for the considered single cell. The specificity scores are transformed to fall in the interval [0,1] following an algebraic function, t, which avoids giving too much weight to exceptionally specific genes and quickly reduces the weight of nonspecific genes that may hinder the precision of the mapping.

3.3.6 Spatial Mapping: Selecting the Confidence Thresholds

1. For a single cell c, once the scores against every voxel in the reference dataset are computed and sorted, we need to define a score threshold above which we consider the voxels as the potential area where the single cell was located.

   To find this threshold, we will perform a simulation study by generating random "simulated single cells." We start with 100x coverage (100 simulated samples per sequenced cell). Each simulated single cell is created by randomly shuffling the specificity scores for all genes in each sequenced cell.

   The simulated dataset is generated at the spatial mapping script step:

```
generate_simulated_data(specificity_matrix,100,"simulated_data/")
```

2. The command above will create C datasets containing 100 simulated cells each. Each dataset has two files:

   n.data : table of specificity scores
   n_bin.data : table of binary expression inferred from the specificity scores

3. In subsequent steps, we apply the mapping method to this set of randomly generated samples and summarize the results for different thresholds.

   From this null distribution, we choose three different thresholds corresponding to different levels of confidence in the mapping:

   High confidence mapping: choose the number of voxels and the mapping score combination that is true for maximum of 10% of simulated samples, i.e., the probability of a simulated cell mapping back with the same criteria is less than 10%;

   Medium confidence mapping: the probability of a simulated cell mapping back with these criteria should be less than 30%;

   Low confidence mapping: the probability of a simulated cell mapping back with these criteria should be less than 50%.

4. Example of the mapping of 10 simulated datasets:

```
#Iterate over 10 simulated datasets:
cell_counter = 1
mapping_simulated_list = list()
for(dataset in 1:10) {
     randomCells = read.table(paste("simulated_data/",dataset,"_bin.data",
sep=""))
     randomRatios = read.table(paste("simulated_data/",dataset,".data",sep=""))

     # map each simulated cell
     for(cell_i in 1:100) {
            print(cell_i)
            mapping_simulated_list[[cell_counter]] <- spatial_map_scoring(randomRa-
tios[cell_i,],randomCells[cell_i,],atlas_specificity_score,atlas)
            cell_counter = cell_counter +1
     }
}

#convert to a data frame:
mapping_simulated = data.frame(mapping_simulated_list)
colnames(mapping_simulated)=paste("cell_",1:length(mapping_simulated_list),sep="")
```

5. In the next step, we set the range of threshold scores and the number of voxels above that score for the evaluation of the results.

   As the thresholds are arbitrary and dataset dependent, choose the threshold_score numbers based on the mean scores across the whole simulated dataset.

6. Set the threshold_number around the number of voxels that constitute 1–2 cell sizes. For example, if the mean score is 0 and one cell in the atlas is covered by 5–10 voxels:

```
threshold_score <- seq(-2,2,1)
threshold_number <- seq(5,15,5)

# Compute the number of simulated cells meeting the requirements # for the selected
combinations of scores and the number of voxels # above that score:
number_cells_above = matrix(nrow=length(threshold_score),ncol=length(threshold_-
number))
for(j in 1:length(threshold_number)){
    print(j)
    number_cells_above[,j] = unlist(lapply(seq_along(threshold_score),function(i){
        sup = apply(mapping_simulated,2,function(x){
length(x[x>threshold_score[i]])
       })
       length(sup[sup>threshold_number[j]])
       }))
   }
number_cells_above = data.frame(number_cells_above)
rownames(number_cells_above) = as.character(threshold_score)
colnames(number_cells_above) = as.character(threshold_number)

# Compute the proportions
proportion_cells_thresholds = number_cells_above/(dim(mapping_simulated)[2])
proportion_cells_thresholds
```

7. Running this script should generate a table just as follows:

Voxels **Score**	**5**	**10**	**15**
−2	0.69	0.51	*0.5*
−1	0.39	**0.31**	0.25
0	0.25	0.25	0.19
1	0.17	0.17	0.17
2	0.15	**0.07**	0.07

Based on this table, we can define the values for the thresholding of the mapping results for sequenced cells:

```
# high confidence: as seen from the table (bold italics), 7% of # simulated samples reach
threshold 2 for at least 10 voxels
# set the minimum score:
h_thres = 2
# and minimum number of mapped voxels:
h_thres_num = 10

# medium confidence values (marked in the table in bold):
m_thres = -1
m_thres_num = 10

# low confidence values (marked in the table in italics):
l_thres = -2
l_thres_num = 15
```

8. Using these values, we now summarize the results for sequenced cells:

```
results <- summary_results(results_scores,h_thres,m_thres,l_thres,h_thres_num,
m_thres_num,l_thres_num)
```

9. In this final step, summary output table will be generated, containing the number of voxels that passed the threshold for each sequenced cell, and the mapping confidence level.

3.4 Visualization

1. We can visualize the reference datapoints and the voxels therein that correspond to the single cells according to the mapping results, using the mapping visualization tool at: http://www.ebi.ac.uk/~jbpettit/map_viewer/ (based on *BioWeb3D* [9]).
2. For importing to *Bioweb3D*, the mapping results need to be scaled:

```
scaled_results <-scale_res(results_scores,h_thres,m_thres,l_thres,h_thres_num,
m_thres_num,l_thres_num,rownames(rna_seq))
write.table(file="scaled_results.csv",sep=",",scaled_results,row.names=FALSE,
quote=FALSE)
```

3. For visualization, import the `3D_coordinates_atlas` table as the main dataset and `scaled_results` table as cluster data.

4 Notes

1. **scRNAseq method**: An array of different scRNAseq approaches are available to date: C1 Single Cell AutoPrep System (Fluidigm), CEL-seq2 [12], Smart-seq2 [13], Drop-seq [14], and InDrops [15], to name just a few. In these methods, single cells are first captured using either droplet- or laminar flow based microfuidic devices or fluorescence-activated cell sorting (FACS) for the distribution of cells into reaction wells. The capture and amplification of single cell RNA relies on the recognition of the poly-A tail of mRNA molecules by reverse transcription primer. Reverse transcribed cDNA molecules are subsequently amplified by PCR (Smart-seq, Smart-seq2, Drop-seq) or in vitro transcription (CEL-seq2, InDrops).
2. **Variations in cell size and transcriptional activity**: As mentioned, several scRNAseq methods are available to date. Some of these methods are limited to a particular cell size range, which sets limits to the coverage of the whole cell population, as well as increases technical noise and complicates the normalization procedures. It is known that larger or transcriptionally more active cells with more mRNA molecules yield more complex RNAseq libraries, due to technical reasons.
3. **Complexity of the cell population**: The approach described here relies on the assumption that the data are a diverse collection of cells from several different cell types. Thus, while our approach is ideal for the identification of different cell types from a relatively balanced population, detection of rare and underrepresented cell types is more challenging.
4. **Critical parameters for building a gene expression atlas**: The two most critical parameters to consider when building a gene expression atlas are the imaging resolution (pixel size) and the minimum number of samples required to get a good averaged expression for each individual marker (e.g., gene), as both impact the final atlas resolution. A proper analysis of these and other technical parameters is advised before acquiring the data to build the atlas. For an example on one method to do this, *see* Vergara et al. 2016 [5].
5. **Cell preparation and cell viability**: Good quality of scRNAseq data is essential for reliable results. Low complexity scRNAseq samples map to spatial reference with poor confidence and broadly. By our experience, good cell viability (>90% viable cells) and short preprocessing time of single-cell preparations is key to good quality sequencing samples.

   We encourage testing different treatments to obtain single cell suspensions. The cell dissociation procedure should be optimized for speed and completeness, while minimizing the cell death events. In most cases, enzymatic treatment of the tissue is required. Some examples of enzymes suitable for gentle cell dissociation are Collagenase II (e.g., Roche Liberase TM) or Papain (E.G. Worthington Biochemical Corporation). Addition of calcium chelators such as EDTA can facilitate the dissociation process and reduce the probability of reaggregation.

Cell numbers and concentration can be quantified for example using FACS, MACS or hemocytometer. Single-cell mRNAseq methods typically require large numbers of input cells, of which only a fraction is eventually captured.

In our case, we performed no labelling or staining of cells prior to capture, allowing extremely short (<1 h) preparation times before cell lysis, and obtained extremely high quality scRNA data, while even DNA staining with Hoechst reduced this quality significantly.

6. **The possibility of multiple cells existing per sample:** Most of the available scRNAseq methods (e.g., droplet- and FACS-sorting based methods) capture cells blindly, i.e., it is not possible to determine the number of cells per sample by imaging of visual observation. Therefore, the possibility of multiple cells per sample should be considered. Naturally, in the cases of multiple cells in the sequencing sample the spatial mapping should perform with lower confidence. Therefore, the cases of multiple cells could be identified by lower mapping precision, or multiple mapped sites per sample, however, it is also possible that the cell type of same molecular profile occurs in multiple sites in the tissue or animal, complicating this interpretation. Thus, the identification of cell numbers per sample should include other parameters, such as number or transcripts detected, the amount of cDNA, or ERCC spike-in ratios.

7. **Considerations to the specificity score calculation**: To calculate specificity of each gene, we compare its level of expression in a cell with its average expression level across all cells. This dampens technical noise, and efficiently filters out housekeeping genes that are expressed highly, but ubiquitously across all cells. However, calculating the gene's specificity only against the background of sampled cells can lead to both false positive and false negative calls, if the sampled cell set is small. We recommend sampling at least 10x the number of cell types that are hypothetically present in the tissue of interest, and even higher numbers if some of the expexted cell types are under-represented. Transcriptional bursts or random fluctuations in expression can lead to an uncharacteristic, short-term high-level expression of a gene, also resulting in a false-positive expression calls. In the mapping step, using a scoring function that

incorporates information from all genes mitigates against this limitation.

8. **Scoring the mismatches between scRNAseq and spatial reference: symmetric vs asymmetric penalization scheme:** For the correspondence score calculation, the user can choose between two penalization schemes: symmetric and asymmetric. The asymmetric scheme only penalizes mismatches when the gene is expressed in the RNA-seq data and not in the reference atlas, while the symmetric scheme also penalizes similarly mismatches that arise when a gene is not expressed in the RNA-seq but expressed in the reference atlas.

   For symmetrical penalization, specificity scores need to be computed on the whole reference dataset. This can be done in the same way as described for the single cell data. Since the reference dataset is binarized, the resulting specificity score for each gene will fall in the interval [0,1] thus suppressing the need to transform these data. In this context the scoring system becomes:

$$s_{c,\mathrm{ref}} = \sum_{m=1}^{M} f_{r_{c,m}}\left(e_{c,m}, e_{\mathrm{ref},m}\right)$$

with

$$f_{r_{c,m}}\left(e_{c,m}, e_{\mathrm{ref},m}\right) = \begin{cases} t\left(r_{c,m}\right), & e_{c,m} = e_{\mathrm{ref},m} = 1 \\ -t\left(r_{c,m}\right), & e_{c,m} = 1,\ e_{\mathrm{ref},m} = 0 \\ -r'_{\mathrm{ref},m}, & e_{c,m} = 0,\ e_{\mathrm{ref},m} = 1 \\ 0, & \text{Otherwise} \end{cases}$$

and

$$\mathrm{t}\left(r_{c,m}\right) = \frac{r_{c,m}}{1 + r_{c,m}}$$

and

$$r'_{\mathrm{ref},m} = \frac{D'_{\mathrm{ref},m}}{\frac{1}{N}\sum_{a=1}^{N} D'_{a,m}}$$

With ||Ref|| x M read count matrix, D', where $D'_{\mathrm{ref,m}}$ is the binary expression in voxel ref for gene m. N is the number of voxels in the reference atlas.

References

1. Schindelin J, Arganda-Carreras I, Frise E, Kaynig V, Longair M, Pietzsch T, Preibisch S, Rueden C, Saalfeld S, Schmid B, Tinevez JY, White DJ, Hartenstein V, Eliceiri K, Tomancak P, Cardona A (2012) Fiji: an open-source platform for biological-image analysis. Nat Methods 9(7):676–682. doi:10.1038/nmeth.2019
2. Yoo TS, Ackerman MJ, Lorensen WE, Schroeder W, Chalana V, Aylward S, Metaxas D, Whitaker R (2002) Engineering and algorithm

design for an image processing Api: a technical report on ITK–the insight toolkit. Stud Health Technol Inform 85:586–592

3. R Core Team (2014) R: a language and environment for statistical computing. http://www.R-project.org/
4. Asadulina A, Panzera A, Veraszto C, Liebig C, Jekely G (2012) Whole-body gene expression pattern registration in Platynereis larvae. Evodevo 3(1):27. doi:10.1186/2041-9139-3-27
5. Vergara HM, Bertucci PY, Hantz P, Tosches MA, Achim K, Vopalensky P, Arendt D (2017) Whole-organism cellular gene-expression atlas reveals conserved cell types in the ventral nerve cord of Platynereis dumerilii. Proc Natl Acad Sci U S A. 114:5878–5885. doi:10.1073/pnas.1610602114
6. Langmead B, Salzberg SL (2012) Fast gapped-read alignment with bowtie 2. Nat Methods 9 (4):357–359. doi:10.1038/nmeth.1923
7. Anders S, Pyl PT, Huber W (2015) HTSeq–a python framework to work with high-throughput sequencing data. Bioinformatics 31(2):166–169. doi:10.1093/bioinformatics/btu638
8. Achim K, Pettit JB, Saraiva LR, Gavriouchkina D, Larsson T, Arendt D, Marioni JC (2015) High-throughput spatial mapping of single-cell RNA-seq data to tissue of origin. Nat Biotechnol 33(5):503–509. doi:10.1038/nbt.3209
9. Pettit JB, Marioni JC (2013) bioWeb3D: an online webGL 3D data visualisation tool. BMC Bioinformatics 14:185. doi:10.1186/1471-2105-14-185
10. Tomer R, Denes AS, Tessmar-Raible K, Arendt D (2010) Profiling by image registration reveals common origin of annelid mushroom bodies and vertebrate pallium. Cell 142 (5):800–809. doi:10.1016/j.cell.2010.07.043
11. Wagner GP, Kin K, Lynch VJ (2012) Measurement of mRNA abundance using RNA-seq data: RPKM measure is inconsistent among samples. Theory Biosci 131(4):281–285. doi:10.1007/s12064-012-0162-3
12. Hashimshony T, Senderovich N, Avital G, Klochendler A, de Leeuw Y, Anavy L, Gennert D, Li S, Livak KJ, Rozenblatt-Rosen O, Dor Y, Regev A, Yanai I (2016) CEL-Seq2: sensitive highly-multiplexed single-cell RNA-Seq. Genome Biol 17:77. doi:10.1186/s13059-016-0938-8
13. Picelli S, Bjorklund AK, Faridani OR, Sagasser S, Winberg G, Sandberg R (2013) Smart-seq2 for sensitive full-length transcriptome profiling in single cells. Nat Methods 10 (11):1096–1098. doi:10.1038/nmeth.2639
14. Macosko EZ, Basu A, Satija R, Nemesh J, Shekhar K, Goldman M, Tirosh I, Bialas AR, Kamitaki N, Martersteck EM, Trombetta JJ, Weitz DA, Sanes JR, Shalek AK, Regev A, McCarroll SA (2015) Highly parallel genome-wide expression profiling of individual cells using Nanoliter droplets. Cell 161(5):1202–1214. doi:10.1016/j.cell.2015.05.002
15. Klein AM, Mazutis L, Akartuna I, Tallapragada N, Veres A, Li V, Peshkin L, Weitz DA, Kirschner MW (2015) Droplet barcoding for single-cell transcriptomics applied to embryonic stem cells. Cell 161(5):1187–1201. doi:10.1016/j.cell.2015.04.044

Chapter 8

Single mRNA Molecule Detection in *Drosophila*

Shawn C. Little and Thomas Gregor

Abstract

Single molecule fluorescent in situ hybridization (smFISH) enables quantitative measurements of gene expression and mRNA localization. The technique is increasingly popular for analysis of cultured cells but is not widely applied to intact organisms. Here, we describe a method for labeling and detection of single mRNA molecules in whole embryos of the fruit fly Drosophila melanogaster. This method permits measurements of gene expression in absolute units, enabling new studies of transcriptional mechanisms underlying precision and reproducibility in cell specification.

Key words Single molecule, FISH, Quantitation, Fluorescence imaging, Drosophila

1 Introduction

Single molecule approaches to biology provide unprecedented access to direct observations of many key biological activities [1]. Single molecule mRNA counting has proven invaluable for examining gene expression heterogeneity among otherwise identical cells [2–5]. Quantifying gene expression through single molecule fluorescent in situ hybridization (smFISH) has become increasingly popular as a method for absolute quantification of gene expression in cultured cells [6–9]. Although the use of smFISH in intact organisms has not yet been as widely adopted, interest and demonstrated applications of the method are increasing [10–13].

Here, we present a detailed protocol for single molecule detection of mRNA in Drosophila embryos using multiple short oligonucleotides directly conjugated to fluorophore [4, 14–17]. We provide details regarding probe preparation, embryo fixation, and in situ hybridization conditions for smFISH. We detail three methods for mounting embryos to maximize signal, minimize noise, and maintain tissue morphology. We discuss confocal scanning conditions for optimal generation of high-quality images for quantitative analysis.

Imre Gaspar (ed.), *RNA Detection: Methods and Protocols*, Methods in Molecular Biology, vol. 1649, DOI 10.1007/978-1-4939-7213-5_8, © Springer Science+Business Media LLC 2018

2 Materials

2.1 Generating Fluorescent Probes with DNA–Fluorophore Coupling

Except where noted, use purified deionized water for preparing solutions.

1. Oligonucleotides: 48–96 DNA oligonucleotides, each 20 nucleotides in length, complementary to a transcript of interest and bearing a 3′ amine modification (*see* **Note 1**). Each oligo should be at concentration of 100 μM. Store at −20 °C.
2. Fluorophore: Dissolve 1 mg of NHS ester-bearing fluorescence dye in 200 μL of N,N-dimethylformamide (DMF, molecular biology grade). Prepare immediately prior to use (*see* **Note 2**).
3. 1 M sodium bicarbonate: Weigh 0.84 g $NaHCO_3$ and transfer to a 15 mL conical tube containing 9.2 mL water. Mix well by vortexing until dissolved. Add 5 N NaOH to obtain pH 9.0. Use immediately or aliquot and store at −20 °C (*see* **Note 3**).
4. 5 M potassium acetate: Weigh 49.1 g KCH_3COO. Add gradually to beaker containing 45 mL water with stirring to dissolve. Use a 100 mL graduated cylinder to bring volume to 100 mL with additional water. Store at room temperature.
5. DEPC-treated water (*see* **Note 4**).

2.2 HPLC Purification of Labeled Oligonucleotides

1. Buffer A: 0.1 M triethylammonium acetate, pH 6.5, filtered and degassed.
2. Buffer B: Acetonitrile, HPLC grade.
3. C18 reverse phase HPLC column (10 μm particle size, 300 Å pore size, 4.6 mm i.d. × 250 mm).
4. Spectrophotometer.

2.3 Embryo Collection and Cross-Linking

1. Apple juice agar plates, yeast paste, and collection cages (*see* **Note 5**).
2. Wire mesh collection baskets: Use wire-cutting scissors to cut steel gauze (e.g., Alfa Aesar Type 304 stainless steel mesh) into small squares approximately 1.5 × 1.5 cm. Fold the edges upward and inward to create roughly circular-shaped metal baskets.
3. Scintillation vials, 20 mL volume.
4. Pasteur pipettes.
5. Heptane, HPLC grade.
6. Bleach: 7% sodium hypochlorite solution.
7. Distilled or reverse osmosis water.
8. Phosphate-buffered saline (PBS, 10×): in 1 L beaker, add to 800 mL water 80 g NaCl, 2 g KCl, 14.4 g dibasic sodium

phosphate (Na_2HPO_4), and 2.4 g monobasic potassium phosphate (KH_2PO_4). Stir until dissolved.

9. 4% paraformaldehyde fixation buffer: Combine 1 mL 20% methanol-free paraformaldehyde solution (e.g., Electron Microscopy Sciences 15713) with 0.5 mL 10× PBS in 3.5 mL water in scintillation vial. Vortex briefly to mix.
10. Methanol, certified ACS reagent grade.

2.4 In Situ Hybridization

1. PTw: PBS (phosphate-buffered saline, 1×) + 0.1 v/v % Tween 20. Store at room temperature.
2. SSC, 20×: In 1 L beaker, add to 800 mL water 175.3 g NaCl and 88.2 g sodium citrate. Stir until dissolved. Adjust pH to 7.2 with HCl. Tranfer to 1 L graduated cylinder and adjust volume to 1 L with additional water. Stir again, then autoclave. Store at room temperature.
3. FISH wash buffer: Combine 7 mL deionized formamide (*see* **Note 6**) with 4 mL 20× SSC and 9 mL DEPC-treated water. Add Tween 20 to 0.1 v/v %. Make fresh on the day of use.
4. Hybridization buffer: 10 w/v % dextran sulfate, 0.01 w/v % salmon sperm ssDNA, 1 v/v % (2 mM) vanadyl ribonucleoside, 0.2 w/v % RNase-free BSA, 4× SSC, 35 v/v % deionized formamide, and 0.1 v/v % Tween 20 in DePC-treated water (*see* **Note 7**).
5. DAPI (4′,6-diamidino-2-phenylindole) solution: Make a stock solution by dissolving 10 mg of powder in 10 mL of methanol. The solution can be stored at 4 °C for an indefinite period. Make a working solution by diluting the stock by a factor of 4000 volume:volume in PTw and use immediately.
6. Nutator rotating platform shaker.

2.5 Mounting Embryos on Slides

1. Mounting media: Aqua-Poly/Mount (Polysciences), Prolong Gold Antifade Mountant (Life Technologies), or Vectashield Antifade (Vector Laboratories) mounting medium (*see* **Note 8**).
2. 23 gauge needle.
3. Nail polish.
4. Glass slides.
5. No. 1.5 cover glass.

3 Methods

1. Calculate the volume of each individual oligonucleotide to use for conjugation. 1 mg of fluorophore dissolved in 200 μL DMF provides adequate fluorophore to conjugate 266 μL of

3.1 Probe-Fluorophore Chemical Coupling

combined oligonucleotide at a concentration 100 μM. Therefore, the volume of each individual oligonucleotide will change depending on the number of oligonucleotides per probe set. For example, a set of 48 oligonucleotides uses 266/48 = 5.5 μL of each oligonucleotide. Smaller volumes of combined oligonucleotides may also be used as long as all of the following volumes are scaled appropriately.

2. Combine all oligos in a single 1.5 mL Eppendorf.
3. Add 0.25× volumes of fresh 1 M sodium bicarbonate and briefly vortex.
4. Add fluorophore in DMF at concentration of 5 μg/μL. The volume of fluorophore should be 0.6× the combined volume of oligonucleotides and sodium bicarbonate.
5. Apply gentle agitation at room temperature for 2 h protected from light on a rocking platform.
6. If necessary to contain the combined volume of the reagents added in **steps** 7 and **8**, divide solution into 2 tubes of equal volume.
7. Add 0.11× volumes of 5 M potassium acetate pH ~5.2 and vortex briefly.
8. Add 2.5× volumes of 100% ethanol to each tube and vortex briefly.
9. Immerse tubes in a bath of dry ice and ethanol for 2 h.
10. Centrifuge for 30 min at full speed in a microfuge at 4 °C. A colored pellet will form.
11. Completely remove ethanol without disturbing pellet. Allow to air dry briefly.
12. Add 50 μL DEPC-treated water, or 25 μL per tube if divided into two tubes, and resuspend pellet by vigorous tapping. The contents can then be combined into a single tube if divided.
13. The oligonucleotide mixture in water may be stored at 4 °C for >1 week.

3.2 HPLC Purification of Labeled Oligonucleotide Probes

After conjugation, a fraction of oligonucleotides will remain unlabeled and must be separated from labeled probes by HPLC.

1. Load material from the conjugation reaction on column equilibrated in 95% Buffer A/5% Buffer B.
2. Linearly increase percentage of Buffer B to 50% over the first 25 min. Unlabeled oligonucleotides will tend to elute from the column prior to minute 14. In our experience, oligonucleotides labeled with Atto 565 will elute between about minute 14 and minute 16. Oligonucleotides labeled with Atto 633 elute

between about minutes 16 and 21. Collect eluate from instrument during these times.

3. At minute 26, increase Buffer B to 95% and maintain through minute 40. Excess unconjugated fluorophore will elute from the column (*see* **Note 9**).
4. Buffer B can then be decreased to 5% at minutes 41 through 50 to reequilibrate the column. The labeled oligonucleotides may be kept in elution buffer at 4C for up to a week.
5. Remove solvent from probes using a vacuum concentrator "speed vac." Place no more than 500 μL of eluate in each tube. With a moderate amount of heating, solvent will evaporate in about 1.5–2 h. Do not overdry, otherwise it may be difficult to resuspend pellet. Make a stock solution by resuspending in 100 μL DePC treated water (*see* **Note 10**).
6. Measure the probe concentration of a 1:50 or 1:100 dilution of stock solution with a spectrophotometer or NanoDrop to determine absorbance at the appropriate wavelength. This value can then be used with the molar extinction coefficient of the fluorophore to calculate the concentration of probe. Typically, the stock concentration is between 0.5 and 5 μM of each individual labeled oligonucleotide, depending on the amount of starting material and the efficiency of labeling.
7. Prepare a working dilution of probe in hybridization buffer (*see* **Note 11**).

3.3 Embryo Fixation in Paraformaldehyde

1. Exchange agar plate on cup of adult flies for new plate with a dollop of yeast paste. Leave plates on for desired period of time at 25 °C. Use timed collections to increase fraction of embryos at a given age.
2. Collect agar plate containing embryos and replace with new plate on cup. Dissolve chorion by adding bleach for 1 min with gentle swirling.
3. Pour embryos into wire mesh basket. Wash extensively with DI water to remove remaining chorion and excess bleach.
4. In a scintillation vial, place 5 mL of heptane and 5 mL of fixation buffer. Vortex for 15 s, then allow phases to separate for 1–2 min (*see* **Note 12**).
5. Immerse mesh basket containing embryos. Be sure the basket drops below the interface between the aqueous and organic phases. This is where the embryos will accumulate.
6. Remove mesh basket from the vial. Check that no embryos remain on the basket. If some are present, the basket can be immersed again. After removing the basket, rinse it in water to ensure no embryos remain.

7. Cap vial and place onto an orbital shaker set to about 200 rpm. Allow to shake for 20–25 min.
8. Remove vial from shaker. Using a Pasteur pipette, remove all the paraformaldehyde solution (the lower layer) while minimizing the removal of embryos.
9. When only heptane and embryos remain in the vial, quickly add 5 mL methanol. Recap the vial and vortex vigorously for at least 30 s and up to 1 min. After vortexing, allow vial to rest for 5 min. This step extracts most embryos from the vitelline membrane. Extracted embryos will sink to the bottom of the vial.
10. With a fresh Pasteur pipette, remove sunken embryos from vial and place into a fresh Eppendorf tube. Allow embryos to settle to bottom of tube. With a fresh Pasteur pipette remove as much methanol from the tube as possible without exposing embryos to air.
11. Add 1–1.5 mL methanol to tube, ensuring that embryos are mixed and dislodged from the bottom of the tube. This ensures efficient washing. Allow embryos to settle. With a fresh Pasteur pipette, remove as much methanol as possible.
12. Repeat **step 11** at least two and up to five more times. This ensures that heptane is removed from embryos. The same Pasteur pipette may be used for these methanol washes.
13. Add 1 mL methanol. Store embryos at −20 °C until use (*see* **Note 13**).

3.4 In Situ Hybridization

1. Using a Pasteur pipette, transfer a volume containing approximately 25–50 μL of fixed embryos in a fresh Eppendorf tube. Allow embryos to settle and remove as much liquid as possible without exposing embryos to air.
2. Pipet 1 mL PTw, mix, allow embryos to settle, remove PTw using a Pasteur pipet, and repeat this step 3 more times.
3. Pipet 1 mL PTw and place embryos on Nutator for 20 min. Remove PTw.
4. Pipet 1 mL FISH wash buffer and mix by inverting tube slowly a few times. Allow embryos to settle. Remove wash buffer. Repeat this step one more time.
5. Pipet 1 mL FISH wash buffer and mix by inverting. Place tube on Nutator for 20 min.
6. While incubation is underway, preheat diluted probes to 37 °C.
7. Remove tube from Nutator and allow embryos to settle. Remove nearly all the wash buffer without removing embryos.
8. Slowly pipet 80–100 μL probe at working dilution. Embryos will rise and float in a layer of wash buffer on top of viscous hybridization buffer. To mix embryos and probe, tap the side of

the tube many times and continue until "Schlering lines" are no longer seen (*see* **Note 14**).

9. Place embryos at 37 °C for at least 1.5 h and up to 16 h (*see* **Note 15**). For short incubation times (less than 6 h), mix by vigorous tapping every 20–30 min. Protect samples from exposure to light.
10. Remove diluted probe with a pipetman. Used probe can be stored at −20 °C for future use and can be reused at least two more times.
11. Add 1 mL wash buffer preheated to 37 °C. Allow embryos to settle. Remove wash buffer with a Pasteur pipette.
12. If using long incubation times (>6 h), add 1 mL preheated wash buffer, place tube at 37 °C for 1 h, remove buffer, and repeat. For shorter incubation times, repeat **step 11** two more times.
13. Add 1 mL PTw. Allow embryos to settle, then remove PTw. Repeat this step once.
14. If desired, stain embryos with DAPI by diluting DAPI stock 1:4000 in PTw. Add 1 mL diluted DAPI to tube. Place on Nutator platform rocker for 30 s. Remove and allow embryos to settle (usually takes another 30 s). Remove DAPI solution. Add 1 mL PTw, allow embryos to settle, remove PTw, and repeat three more times (*see* **Note 16**).

3.5 Mounting Embryos for Imaging in Auqa-Poly/Mount

1. Mounting in Aqua-Poly/Mount: Aqua-Poly/Mount is a water-soluble gelling mounting medium. It provides the best option for maintaining structural integrity of embryos and for carefully orienting embryos on slides. However, it also yields the lowest signal-to-noise ratio. It is appropriate for experiments that utilize large numbers of probes and thereby generate a high signal.
2. Using a wide-mouth pipet tip rinsed several times in PTw, transfer up to 400 embryos to a glass slide that has been cleaned with 70% ethanol and dried.
3. Under stereo dissecting microscope, push embryos into a small pile in the middle of the slide. Use a small piece of paper towel to wick away most of the PTw without allowing embryos to dry.
4. Deposit directly on the pile of embryos a drop of Aqua-Poly/Mount. Using 23 gauge needle, gently mix embryos into medium until "Schlering lines" are no longer visible.
5. Use needle to arrange embryos in desired orientation on slide. If medium begins gelling, add a few microliters of water.

6. When embryos are in desired orientation, allow to dry for a few minutes at room temperature. This minimizes embryos shifting when cover glass is added.
7. Place a small droplet of Aqua-Poly/Mount on a cover glass and gently spread toward edges using a closed pair of forceps, avoiding introducing bubbles.
8. Carefully lay cover glass onto mounted embryos. It often works best to hold the cover glass at one edge an angle and slowly tilting the glass down onto the embryos. This helps reduce the introduction of air bubbles.
9. If desired, embryos can be flattened slightly by placing a small weight on the cover glass (*see* **Note 17**).
10. Allow slide to gel for >6 h before imaging (*see* **Note 18**). Protect slides from light.

3.6 Mounting Embryos in Prolong Gold

1. Prolong Gold is a hardening mounting medium that offers high signal-to-noise compared to Aqua-Poly/Mount. However, unlike Aqua-Poly/Mount, embryos have a stronger tendency to shrivel and become distorted if left for more than a short period before applying the cover glass. In addition, Prolong Gold is much less pliable than Aqua-Poly/Mount as it begins to cure. The curing process begins within minutes of air exposure. This allows for much less time to arrange or orient embryos on a slide. Nevertheless, if orientation of every embryo is not critical, Prolong Gold is preferred for generating high signal-to-noise data.
2. Using a wide-mouth pipet tip rinse several times in PTw, transfer up to 400 embryos to a glass slide that has been cleaned with 70% ethanol and dried. Remove the majority of the excess PTw with a piece of paper towel.
3. Under stereo dissecting microscope, use a 23 gauge needle to separate and arrange embryos on the glass slide. Do not allow embryos to become overly dry, as they can shrink dramatically and introduce artifacts. Add more PTw if too much buffer evaporates.
4. When embryos are in desired arrangement on the slide, place a droplet (about 25–30 μL) of Prolong Gold on a clean cover glass and set aside.
5. Use a small folded piece of paper towel to wick away most of the PTw surrounding the embryos. Perform this step in under 1 min to avoid the start of the curing process of the mountant on the cover glass.
6. Carefully lay cover glass onto mounted embryos. It often works best to hold the cover glass at one edge an angle and slowly

tilting the glass down onto the embryos. This helps reduce the introduction of air bubbles.

7. Use a paper towel to remove excess Prolong Gold from around the edges of the cover glass.
8. Allow medium to cure for 48 h at room temperature before imaging. Protect slides from light during curing process (*see* **Note 19**).

3.7 Mounting Embryos in Vectashield

1. Vectashield provides the highest signal-to-noise ratio of the mountants we have tested. However, it is not a hardening medium. It is not possible to orient embryos. It can be challenging to maintain separation between embryos during mounting. Nevertheless, Vectashield is useful for generating usable data from probes/mRNAs that generate low signal, such as in experiments with small numbers of probes. Although a hardening version of Vectashield is available, it is not appropriate for thick specimens such as embryos.
2. Using a wide-mouth pipet tip rinsed several times in PTw, transfer embryos to a glass slide that has been cleaned with 70% ethanol and dried. Remove the majority of the excess PTw with a piece of paper towel.
3. Under stereo dissecting microscope, use a 23 gauge needle to separate and arrange embryos on the glass slide. Do not allow embryos to become overly dry, as they can shrink dramatically and introduce artifacts. Add more PTw if too much buffer evaporates.
4. When embryos are in desired arrangement on the slide, place a droplet (about 25–30 μL) of Vectashield on a clean cover glass and set aside.
5. Use a small folded piece of paper towel to wick away most of the PTw surrounding the embryos.
6. Carefully lay cover glass onto mounted embryos. It works best to keep the droplet of Vectashield in the middle of the cover glass and drop the glass directly onto the embryos.
7. Use a paper towel to remove excess Vectashield from around the edges of the cover glass. This will also help to flatten the embryos. Do not overflatten.
8. Seal the cover glass to the slide with nail polish and allow to dry. Slide may be imaged immediately (*see* **Note 20**).

3.8 Confocal Microscopy

1. Acquire 3D stacks by laser scanning confocal microscopy using a high magnification, high NA objective (e.g., 63× NA 1.4). Example images are shown in Fig. 1. We have principally used a Leica SP5 equipped with GaAsP "HyD" avalanche photodiode detectors in photon counting mode (*see* **Note 21**). For thick

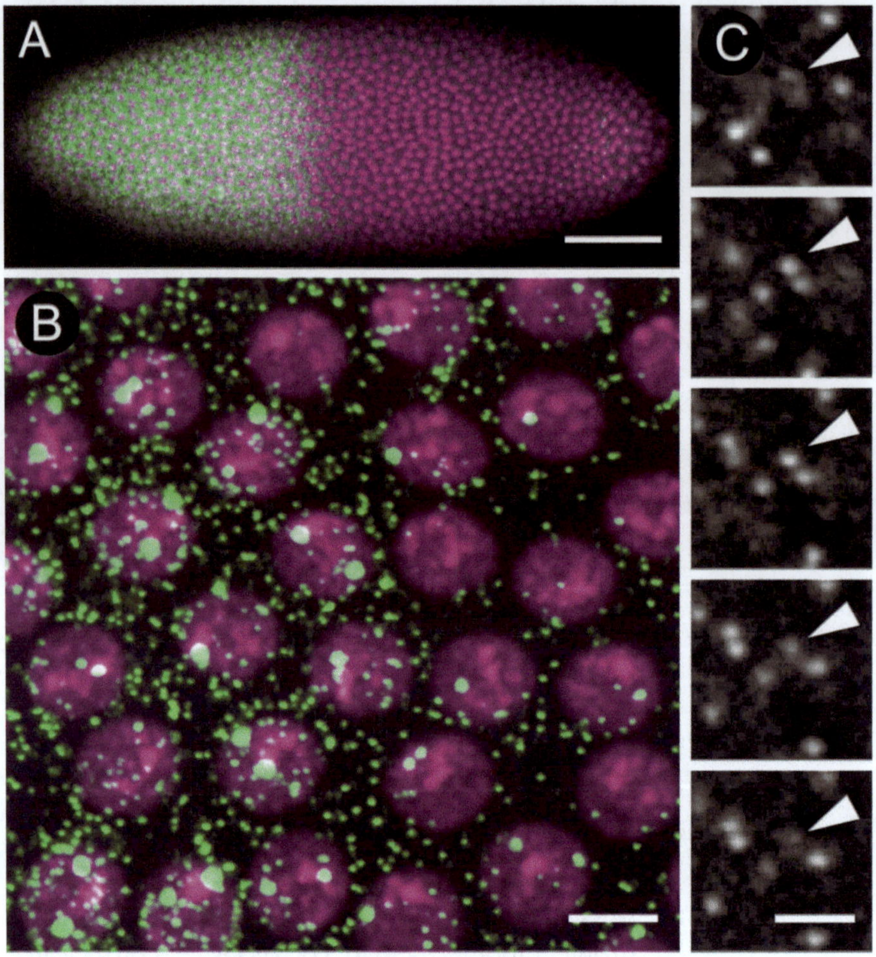

Fig. 1 Single molecule mRNA detection in Drosophila embryos. (**a**) Low magnification image of an embryo approximately 2 h after fertilization following the twelfth nuclear division cycle and labeled with probes complementary to the gene *hunchback* (*hb*). Embryo is oriented with anterior to the left. *Green*: *hb* expression; *magenta*: DAPI stain to highlight nuclei. (**b**) Single confocal slice through the nuclear layer illustrating nascent transcription sites and single mRNAs. (**c**) Series of adjacent confocal slices surrounding a single mRNA illustrating detection on multiple z planes. *Scale bars*: 50 μm (**a**), 5 μm (**b**), 2 μm (**c**)

specimens such as embryos, confocal microscopy is essential to visualize diffraction-limited objects. Wide-field fluorescence microscopy, used in many studies of cultured cells, is inappropriate because of the overwhelming signal from out-of-focus planes.

2. Voxel size: For most applications, diffraction-limited objects are most effectively detected by our custom MATLAB software in images obtained with pixels of 76 × 76 nm and a z increment of between 250 and 420 nm (*see* **Note 22**).

3. Total section thickness: We typically collect image stacks spanning 10–20 μm in z (*see* **Note 23**).

4. Scanner settings and image acquisition: To improve our ability to locate and measure the signal of diffraction-limited objects, we sum the fluorescence of multiple scans, usually between 4 and 8, of the same volume. This is usually best accomplished by setting "line accumulation" to between 4 and 8 iterations. For dual-color imaging, we use sequential scans, alternating channels with every line, in order to minimize cross talk between channels. Because of the small numbers of fluorophores, the scanning speed often must be 2–8 times slower than that typically used for most confocal imaging applications.
5. Laser power: As with all fluorescent imaging, an important goal is to obtain images of high quality while minimizing photobleaching. Therefore it is desirable to use as little laser power as necessary to provide sufficient excitation to lift signal above background or autofluorescence. The laser power necessary to achieve this goal is experiment-specific and depends on the number of probes employed, the efficiency of the hybridization, the amount of background and autofluorescence, and the sensitivity of the photodetectors. The necessary laser power must therefore be determined empirically (*see* **Note 24**).
6. As controls for single molecule counting, it is advisable when possible to perform the labeling and imaging on embryos that are heterozygous for mRNA-null mutations in a gene of interest. Counts and particle densities can be compared between similarly staged embryos of different genotypes. In the absence of compensation mechanisms, heterozygotes will produce mRNAs at half the rate of wild-type embryos. In addition, mRNA-null homozygous embryos can be used as controls for background fluorescence.

3.9 Analysis

Upon request, custom MATLAB analysis software is available which provides semiautomated threshold detection and which outputs spatial coordinates and estimates of fluorescence intensity from TIFF image stacks.

4 Notes

1. For a given mRNA target, we usually utilize at least 48 oligonucleotides designed using the oligo RNA FISH probe design tool provided online by Biosearch Technologies (http://biosearchtech.com/products/rna-fish/custom-stellaris-probe-sets). Using this tool, we design 20-mers complementary to a transcript of interest with a minimum spacing of two nucleotides between oligonucleotides. This ensures that presence of fluorophore does not hinder binding of neighboring probe. Larger numbers of probes increases signal of single transcripts. To

maximize utilization of unique sequences, implement repeat masking. CG content of each oligonucleotide should be between 40 and 60%. Avoid short repeats and continuous stretches of A/T longer than 5 nt. We obtain 3′-amine-modified oligonucleotides from Biosearch Technologies (Petaluma, CA) in 96-well format. Each oligo bears a single free amine at its 3′ end. We order oligos bearing the mdC(TEG-Amino) 3′ modification at the 10 nmol delivered scale. Other modifications that add additional free amines may also be used, although we have not tested them in our protocol. Oligos from Biosearch Technologies may be ordered in solution at a concentration of 100 μM. This is essentially a lifetime supply for most genes we have studied. Biosearch Technologies states that the amine group will remain reactive for 6 months when stored at −20 °C with minimal freeze–thaw cycles. We have successfully conjugated fluorophore to oligos over the course of more than 3 years even with several freeze–thaw cycles.

2. We have had excellent success with the "Atto" family of fluorophores. We typically use Atto 565 NHS ester and Atto 633 NHS ester. The Atto dyes exhibit photostability and minimal photobleaching even under prolonged excitation. We have also successfully used Alexa 514 NHS ester and Alexa 594 NHS ester. The NHS reactive group is unstable. For best results, use dyes in the coupling reaction immediately upon arrival. If necessary, store unopened package at −20 °C. Prolonged storage is not recommended. Condensation of water in the air onto the fluorophore will tend to inactivate reagent. Therefore, if stored at −20 °C, allow package to reach room temperature before opening. During the coupling reaction, protect the dye from light as much as possible. It is usually convenient to order 1 mg of fluorophore. Immediately before performing the coupling, open the packaging and dissolve the 1 mg of powder in 200 μL DMF.

3. Freshly made sodium bicarbonate solution is the most effective. Alternately, small aliquots can be made and stored at −20C. After thawing, discard any unused portion.

4. Add 0.1 mL diethylpyrocarbonate (DEPC) to 100 mL of water in 250 mL beaker. Stir for >12 h. Autoclave for >15 min to remove DEPC.

5. Reference 18 provides a concise description of a simple method for obtaining large numbers of embryos [18].

6. Formamide is often deionized prior to shipment from supplier. However, in our experience, superior results are obtained by deionizing formamide in-house, even when using material marked by the manufacturer as deionized. To deionize formamide, place a large volume (100–500 mL) of formamide in a

500 mL beaker with a stir bar and add 10 w/v % deionizing resin, e.g., mixed bed resin for deionizing (Sigma-Aldrich M8032). Stir for 1 h at low speed using the lowest stirring speed that does not allow the resin to settle at the bottom of the beaker. After 1 h, filter out the resin using a vacuum-driven filter sterilizer. Aliquot and store at −20 °C until needed. With our FISH wash buffer recipe with 35% formamide, it is convenient to make either 3.5 or 7.0 mL aliquots for 10 or 20 mL of wash buffer.

7. The solution of dextran sulfate, SSC, and water can be heated to ~60 °C and vortexed repeatedly until the dextran has dissolved. Allow the solution to cool to room temperature before adding the formamide and other components.
8. Different mountants are useful for different applications. If orientation and morphology of embryos is crucial, Aqua-Poly/Mount is preferred. For small numbers of probes where the signal-to-noise is too low in Aqua-Poly/Mount, Vecta-Shield is preferred, although morphology is often distorted. For a balance between orientation/morphology and signal-to-noise, Prolong Gold offers a reasonable compromise.
9. There can be unconjugated free fluorophore after the reaction. Do not collect the free fluorophore. Atto 633 elutes at around minute 24 and 25 and shows no absorption at 260 nm. Atto 565 elutes around minute 18 and shows weak absorption at 260 nm.
10. Probes may be stored for years at −20 °C until ready for use. We have found that repeated freeze–thaw cycles have no discernable impact on probe performance.
11. We have obtained best results with a working concentration of around 1–2 nM for each individual labeled oligonucleotide. Note that the hybridization buffer is viscous compared to the stock solution. Vigorous vortexing for at least 1 min is recommended for thorough mixing of probe stock in hybridization buffer. Diluted probe can be stored indefinitely at −20 °C. Diluted probe tolerates many freeze–thaw cycles without loss of performance. If using previously diluted probe stored at −20 °C, warm to 37 °C prior to adding to embryos.
12. The mixing and vortexing can be performed during the chorion removal (**step 2**).
13. Superior results tend to be obtained from embryos stored <1 week at −20 °C.
14. Mixing of embryos in all solutions is essential. When mixing embryos and probe, embryos that contact with a surface that has not been first coated with hybridization buffer will tend to stick to the tube. Thus, gentle tapping is advised at first. This

will begin mixing embryos and hybridization buffer while also splashing the sides of the tube in hybridization buffer. As mixing continues, the tapping can become more and more vigorous. Embryos will also tend to become more and more transparent as mixing continues into hybridization buffer.

15. For some well-behaved probe sets, a minimal incubation time will produce high quality images. Others probe sets or mRNAs require longer incubation times. Optimal incubation time is best determined empirically.
16. Stained embryos may be stored at 4 °C for at least a week, but best results are obtained by proceeding directly to mounting.
17. A small amount of weight generated by, for example, a stack of small coins, can be placed onto a piece of Kimwipes folded over onto the cover glass. This will slightly flatten the embryos, which can be advantageous for placing many nuclei within the same imaging plane. However, caution is warranted: too much weight or too long application will distort the tissue. These distortions will alter the apparent density of mRNA molecules and will lead to increased measurement error in assessing mRNA density.
18. Embryos mounted in Aqua-Poly/Mount may be stored at 4 °C for up to a month prior to imaging. However, in our experience, superior results are obtained when imaged within a few days of mounting.
19. Embryos mounted in Prolong Gold may be stored at room temperature for up to a month prior to imaging. In our experience, superior results are obtained when imaged within a few days of mounting.
20. Embryos mounted in Vectashield may be stored at 4 °C for up to a month after mounting. In our experience, superior results are obtained when imaged within 1 or 2 weeks of mounting.
21. Low noise, high sensitivity detectors are essential for imaging dim objects labeled with few fluorophores. The "hybrid detectors" found on most current Leica scanning confocal systems offer high performance. These detectors can be operated in either "standard" mode or photon counting mode. Standard mode offers the option to apply gain and offset settings to amplify signal and reduce putative background. This can be useful for counting objects, but can also be misleading for making quantitative measurements. For quantitative measurements of fluorescence, photon counting mode is strongly preferred.
22. Efficient detection of diffraction limited objects requires high-resolution images. A simple calculation of Nyquist sampling for most commercial confocal setups can be found at https://svi.

nl/NyquistCalculator. This provides an initial estimate for adequate sampling density, that is, the number of pixels per unit of distance. In practice, for the fluorophores we employ on a Leica SP5, we have had good success with pixels of dimension 76 × 76 nm in xy. In addition, we deliberately sample in the axial (z) direction more frequently than strictly required, in order to ensure that all true objects appear on multiple adjacent imaging planes. We use this feature as a powerful means of separating true objects from background signal. This procedure is described in detail in reference 4. We have used z sampling intervals as small as 250 nm [4] and as large as 420 nm [15].

23. Total section thickness depends on the particular experiment or application. For example, to count most or all zygotically expressed mRNAs typically requires a total imaging thickness of at least 15 μm. Alternatively, to estimate the total number of all transcripts in whole embryos, then it is required to image a very thick section spanning tens of microns. At the other extreme, if objects of interest are found in roughly the same imaging plane, this shortens data collection time. Because the quantification software requires multiple z planes, it is imperative to image at least 3 z slices, and probably more are advised in most cases.

24. Our experiments typically require us to image both relatively dim and relatively bright objects, for example, single mRNAs that produce a low level of signal and nascent transcription sites which can contain the equivalent of 50–100 mRNAs. To simultaneously image both single mRNAs and bright sites of transcription, we employ Leica's HyD detectors in photon counting mode. The low noise of these detectors allows us to easily discern single mRNAs from background and simultaneously to measure the fluorescence of transcription sites without saturating the detector. We can thus apply a single scan to capture all data simultaneously. We determine the magnitude of laser power empirically for every probe set with the goal of minimizing photobleaching during prolonged scanning while still providing adequate signal to separate true objects from noise. As a general rule, objects that are clearly discernable by eye in confocal stacks will be most readily separated from background noise during the analysis steps. We have found that traditional photomultiplier tubes (PMTs) do not offer sufficient dynamic range for both reliable detection of single mRNAs and nonsaturated transcription sites in the same scan. However, traditional PMTs may still be used by performing two scans of the same sample at two different laser intensities. A low-power scan is first taken for measuring the intensities of transcription sites, followed by a high power stack that

saturates the signal from transcription sites but provides high signal from individual mRNAs so that they may be separated from imaging noise. We have not extensively tested other microscopes or imaging configurations. Overall, to determine optimal parameter settings, a simple rule of thumb is that an image that looks good by eye will always generate more reliable results; the more distinct and the brighter objects appear compared to the surrounding background, the more easily and reliably those objects will be detected by the software.

References

1. Chen H, Larson DR (2016) What have single-molecule studies taught us about gene expression? Genes Dev 30(16):1796–1810. doi:10.1101/gad.281725.116
2. Chong S, Chen C, Ge H, Xie XS (2014) Mechanism of transcriptional bursting in bacteria. Cell 158(2):314–326. doi:10.1016/j.cell.2014.05.038
3. Gandhi SJ, Zenklusen D, Lionnet T, Singer RH (2011) Transcription of functionally related constitutive genes is not coordinated. Nat Struct Mol Biol 18(1):27–34. doi:10.1038/nsmb.1934
4. Little SC, Tikhonov M, Gregor T (2013) Precise developmental gene expression arises from globally stochastic transcriptional activity. Cell 154(4):789–800. doi:10.1016/j.cell.2013.07.025
5. Raj A, Peskin CS, Tranchina D, Vargas DY, Tyagi S (2006) Stochastic mRNA synthesis in mammalian cells. PLoS Biol 4(10):e309. doi:10.1371/journal.pbio.0040309
6. Chen KH, Boettiger AN, Moffitt JR, Wang S, Zhuang X (2015) RNA imaging. Spatially resolved, highly multiplexed RNA profiling in single cells. Science 348(6233):aaa6090. doi:10.1126/science.aaa6090
7. Schwabe A, Bruggeman FJ (2014) Single yeast cells vary in transcription activity not in delay time after a metabolic shift. Nat Commun 5:4798. doi:10.1038/ncomms5798
8. Skinner SO, Xu H, Nagarkar-Jaiswal S, Freire PR, Zwaka TP, Golding I (2016) Single-cell analysis of transcription kinetics across the cell cycle. Elife 5:e12175. doi:10.7554/eLife.12175
9. Taniguchi Y, Choi PJ, Li GW, Chen H, Babu M, Hearn J, Emili A, Xie XS (2010) Quantifying E. coli proteome and transcriptome with single-molecule sensitivity in single cells. Science 329(5991):533–538. doi:10.1126/science.1188308
10. Bahar Halpern K, Tanami S, Landen S, Chapal M, Szlak L, Hutzler A, Nizhberg A, Itzkovitz S (2015) Bursty gene expression in the intact mammalian liver. Mol Cell 58(1):147–156. doi:10.1016/j.molcel.2015.01.027
11. Bayer LV, Batish M, Formel SK, Bratu DP (2015) Single-molecule RNA in situ hybridization (smFISH) and immunofluorescence (IF) in the Drosophila egg chamber. Methods Mol Biol 1328:125–136. doi:10.1007/978-1-4939-2851-4_9
12. Bolkova J, Lanctot C (2016) Quantitative gene expression analysis in Caenorhabditis elegans using single molecule RNA FISH. Methods 98:42–49. doi:10.1016/j.ymeth.2015.11.008
13. Stapel LC, Lombardot B, Broaddus C, Kainmueller D, Jug F, Myers EW, Vastenhouw NL (2016) Automated detection and quantification of single RNAs at cellular resolution in zebrafish embryos. Development 143(3):540–546. doi:10.1242/dev.128918
14. Femino AM, Fay FS, Fogarty K, Singer RH (1998) Visualization of single RNA transcripts in situ. Science 280(5363):585–590
15. Little SC, Tkacik G, Kneeland TB, Wieschaus EF, Gregor T (2011) The formation of the Bicoid morphogen gradient requires protein movement from anteriorly localized mRNA. PLoS Biol 9(3):e1000596. doi:10.1371/journal.pbio.1000596
16. Petkova MD, Little SC, Liu F, Gregor T (2014) Maternal origins of developmental reproducibility. Curr Biol 24(11):1283–1288. doi:10.1016/j.cub.2014.04.028
17. Raj A, van den Bogaard P, Rifkin SA, van Oudenaarden A, Tyagi S (2008) Imaging individual mRNA molecules using multiple singly labeled probes. Nat Methods 5(10):877–879. doi:10.1038/nmeth.1253
18. Bachmann A, Knust E (2008) The use of P-element transposons to generate transgenic flies. Methods Mol Biol 420:61–77. doi:10.1007/978-1-59745-583-1_4

Chapter 9

Detection and Automated Analysis of Single Transcripts at Subcellular Resolution in Zebrafish Embryos

L. Carine Stapel, Coleman Broaddus, and Nadine L. Vastenhouw

Abstract

Single molecule fluorescence in situ hybridization (smFISH) is a method to visualize single mRNA molecules. When combined with cellular and nuclear segmentation, transcripts can be assigned to different cellular compartments resulting in quantitative information on transcript levels at subcellular resolution. The use of smFISH in zebrafish has been limited by the lack of protocols and an automated image analysis pipeline for samples of multicellular organisms. Here we present a protocol for smFISH on zebrafish cryosections. The protocol includes a method to obtain high-quality sections of zebrafish embryos, an smFISH protocol optimized for zebrafish cryosections, and a user-friendly, automated analysis pipeline for cell segmentation and transcript detection. The software is freely available and can be used to analyze sections of any multicellular organism.

Key words smFISH, Zebrafish, Cryosections, Automated cell segmentation, Transcript detection

1 Introduction

An important feature of multicellular organisms is their large variety of cell types. Each cell type is characterized by a specific gene expression profile which provides information about the function of the cell. Spatial information on gene expression is often obtained by in situ hybridization [1, 2]. However, this technique provides limited cellular resolution and is not quantitative due to nonlinear signal amplification [1, 2]. Quantitative information on gene expression is often obtained by qPCR or RNA-sequencing but these techniques only detect highly abundant transcripts when single cells are analyzed, and precise spatial information is lost [3–5].

Single molecule fluorescence in situ hybridization (smFISH) is a method to visualize single transcripts at subcellular resolution in their original tissue context [6–11]. Stellaris smFISH is a popular method because it is straightforward and affordable, and its quantitative nature has been thoroughly tested [7]. The method makes

Imre Gaspar (ed.), *RNA Detection: Methods and Protocols*, Methods in Molecular Biology, vol. 1649, DOI 10.1007/978-1-4939-7213-5_9, © Springer Science+Business Media LLC 2018

use of up to 48 individually labeled 20 nucleotide-long oligonucleotides that hybridize to their target RNA [7]. The accumulation of a large number of probes on a single transcript produces a diffraction limited signal that can easily be distinguished from background [7]. However, smFISH gives the best results when used on thin samples like single cells or tissue sections, and no protocols were available for zebrafish yet. Furthermore, available analysis pipelines for quantification of transcripts at cellular resolution in multicellular organisms relied on manual cell segmentation which is very labor-intensive and time-consuming [8, 12–14].

We previously developed a protocol for smFISH on zebrafish cryosections as well as an automated analysis pipeline for transcript detection and cell segmentation (Fig. 1) [15]. As input samples for our smFISH protocol, we use cryosections of zebrafish embryos embedded in OCT. Although sectioning at early zebrafish stages is difficult because cells are large and embryos fragile, our protocol generates high quality sections at a broad range of developmental stages. We then adapted a protocol for Stellaris smFISH on tissue sections [8] for use on these zebrafish sections (Fig. 1a). With the resulting protocol, even very low transcript levels can be detected at high specificity and sensitivity. To increase image analysis speed and throughput, we developed a freely available analysis pipeline for automated transcript detection and semiautomated cell segmentation. This pipeline can be applied to sections of zebrafish and other multicellular organisms (Fig. 1b) [15]. The transcript analysis pipeline enables quantification of transcript levels, as well as quantification of the number of active transcription sites (transcription foci), and the number of transcripts per focus (Fig. $1b_6$). Nuclear segmentation is integrated in the transcript analysis pipeline to be able to assign transcripts to nuclei or cytoplasm. The membrane segmentation pipeline we developed consists of three parts. First, a random forest pipeline in KNIME is used to predict for each pixel whether it is part of the membrane, a membrane intersection point (vertex), or background (Fig. $1b_2$). Then, the PathFinder plugin in Fiji uses these predictions to generate a cell mask (Fig. $1b_3$). Finally, the Fiji Cell annotation tool can be used to correct small errors in the segmentation, and to group cells according to cell type (Fig. $1b_4$). The resulting cell mask can be used in combination with the transcript analysis pipeline in Fiji to assign transcripts to individual cells and nuclei (Fig. $1b_6$).

Here, we describe (1) a method for high-quality cryosectioning of zebrafish embryos, (2) an optimized smFISH protocol for use on zebrafish cryosections, and (3) a pipeline for (semi)automated transcript detection and cell segmentation that can be applied to smFISH samples of any multicellular organism.

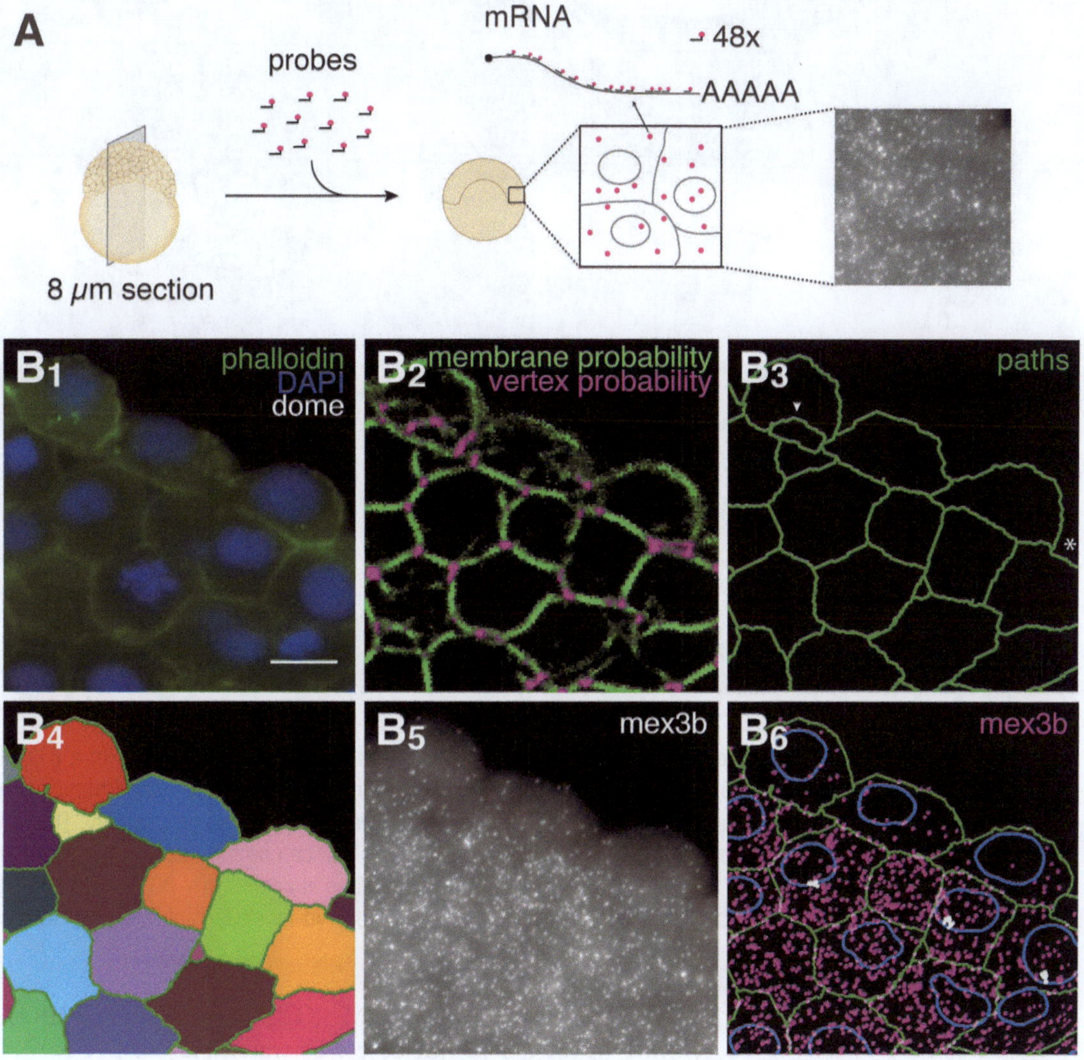

Fig. 1 Protocol and analysis pipeline for smFISH in zebrafish. (**a**) Overview of the smFISH method on sections of zebrafish embryos. (**b**) Membrane segmentation and transcript detection pipeline. *Scale bar*: 10 μm. Membrane staining (**b$_1$**) is used to calculate the probability that pixels belong to membrane (*green*), or membrane intersection points (vertices, *magenta*) (**b$_2$**). Paths are traced from vertices over the membranes to generate a cell segmentation (**b$_3$**). Small errors in the segmentation (*asterisk* and *arrowhead* in **b$_3$**) are manually corrected to finalize the cell mask (**b$_4$**). smFISH signal (**b$_5$**) is used to identify single transcripts (*magenta*) and transcription foci (*white*) and combined with the cell (*green*) and nuclear (*blue*) segmentations (**b$_6$**). Images are maximum projections of 17 z-slices spaced by 0.3 μm

2 Materials

2.1 Embryo Embedding

1. Zebrafish embryos.
2. Embryo medium: 5 mM NaCl, 0.17 mM KCl, 0.33 mM $CaCl_2$, 0.33 mM $MgSO_4$ in deionized water.

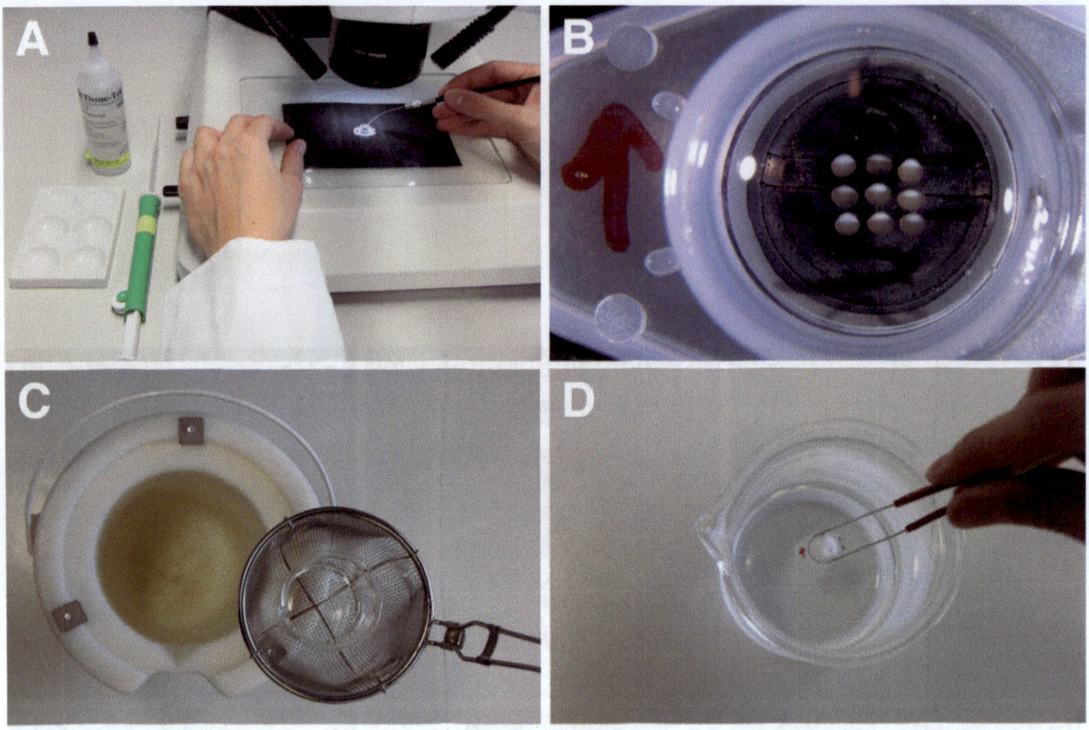

Fig. 2 Embedding of zebrafish embryos for cryosectioning. (**a**) Samples are equilibrated in OCT in a 6-well staining dish and embedded in the cap of an Eppendorf tube under a dissecting scope. (**b**) Embryos are oriented laterally with their animal caps facing in the same direction to ensure optimal section quality. The red arrow on the Eppendorf cap indicates their orientation so that the cryo-block can be mounted on the cryostat with the animal caps facing toward the blade. (**c**) A beaker with isopentane is cooled to −80 °C in a bucket with liquid nitrogen. A sieve may be used to hold the beaker. (**d**) The cryo-block is rapidly frozen in the precooled isopentane

3. PBS: 137 mM NaCl, 2.7 mM KCl, 4.3 mM Na_2HPO_4, 1.47 mM KH2PO4 in nuclease-free water. Adjust pH to 7.4 with HCl.
4. PBT: 0.1 v/v % Tween 20 in PBS.
5. Fixative: 4% formaldehyde in PBT.
6. Cryoprotection solution: 30% sucrose in PBS.
7. Tissue-Tek® OCT compound (optimal cutting temperature compound) (Fig. 2a).
8. Cryo incubation solution: prepare a solution of 30% sucrose in a 50/50 mixture of PBS with OCT.
9. Isopentane.
10. Petri dish, 100 mm diameter.
11. Caps of Eppendorf tubes (*see* **Note 1**).
12. Pipet pump for glass pipets to transfer dechorionated embryos (e.g., The Pipet Pump, Bel-Art Scienceware) (Fig. 2a).

13. Glass pipet.
14. 6-well staining plate (Fig. 2a).
15. Embryo manipulator (e.g., nickel-plated pin holder with curved pin, Fine Science Tools).
16. Large liquid nitrogen container (Fig. 2c).
17. Large sieve (Fig. 2c).
18. 250 mL beaker.
19. Sharp forceps.
20. Blunt-end forceps.
21. Low-temperature thermometer (to −100 °C).
22. Incubator set to 28 °C.
23. Stereomicroscope.

2.2 Cryosectioning

1. Glass cleaning detergent (e.g., Mucasol).
2. Milli-Q water.
3. 100% ethanol.
4. 0.1 w/v % poly-L-lysine (MW 150,000–300,000 Da) in Milli-Q water.
5. 22 × 22 mm selected #1.5 coverslips (0.17+/− 0.0005 mm) (*see* **Note 2**).
6. Coverslip holder (e.g., XL Wash-N-Dry coverslip rack, Diversified Biotech) (Fig. 3b).
7. Slide staining dish or beaker to fit coverslip holder (Fig. 3b).
8. Anti-roll plate for cryostat.
9. Cryostat blade.
10. Specimen stage for cryostat.
11. Thick and thin brush.
12. 6-well plate.
13. Parafilm.
14. Sonication bath.
15. Cryostat (e.g., Microm HM560).

2.3 smFISH (Incl. Imaging)

1. 70% ice-cold ethanol in Milli-Q water (store at −20 °C).
2. 2× SSC diluted from 20× SSC (commercial, RNase free).
3. 5 μg/mL proteinase K in 2× SSC.
4. smFISH wash buffer: 10% formamide and 2× SSC in nuclease-free water (*see* **Note 3**).
5. smFISH hybridization buffer: 10 w/v % dextran sulfate, 10 v/v % formamide, 1 mg/mL E. coli tRNA, 2× SSC, 0.02 w/v % BSA, and 2 mM Vanadyl-ribonucleoside complex in nuclease-free water (*see* **Note 3**).

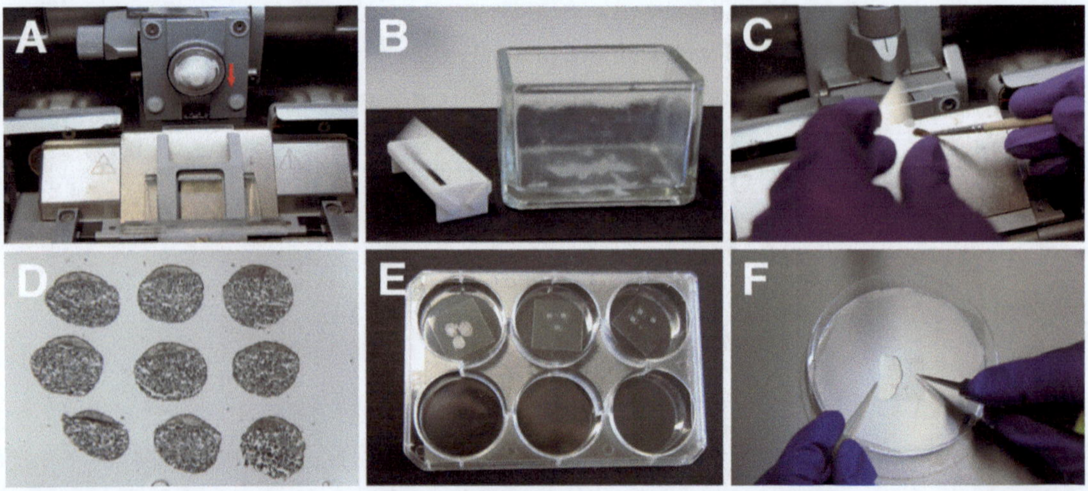

Fig. 3 Sample preparation for smFISH. (**a**) The cryo-block is mounted on the cryostat with the animal caps facing toward the blade of the cryostat (*red arrow*). (**b**) Coverslips are placed in a coverslip holder and are coated with poly-L-lysine in a slide staining dish to increase adhesion of sections to the coverslips. (**c**) Multiple sections can be placed on a single coverslip. (**d**) Cryosections on a coverslip. Note the orientation. While the yolk is damaged, the animal caps are intact. (**e**) Coverslips with sections are placed section-up in a 6-well plate for smFISH washes. (**f**) Coverslips are placed section-down on a drop of hybridization mix in a Parafilm-coated petri dish for probe hybridization overnight

6. Custom Stellaris smFISH probes, diluted to 25 μM in Milli-Q water (*see* **Note 4**).
7. 1 mg/mL DAPI.
8. Phalloidin–Alexa 488 (*see* **Note 5**).
9. GLOX buffer: 10 mM Tris, pH 7.5, 2× SSC, and 0.4 w/v % glucose (in nuclease-free water).
10. GLOX mounting medium: GLOX buffer, 47 μg/mL glucose oxidase, and 0.58 mg/mL catalase suspension.
11. Microscope slides.
12. Nail polish.
13. Epifluorescence microscope with suitable filter sets, a high NA objective, and a sensitive camera (*see* **Note 6**).

2.4 Data Analysis

For additional information on how to install the plugins, you may consult the documentation associated with Stapel et al. 2016 [15].

1. Fiji (http://fiji.sc/Fiji)
   (a) Activate update sites "MS-ECS-2D" and "3D ImageJ Suite".
2. KNIME

   Background information and more details on how to install KNIME for the applications described in this chapter can be found at *http://tinyurl.com/KNIME-MS-ECS*. Briefly:

(a) In the KNIME.ini file in your KNIME installation folder: set Xmx to –Xmx4g or –Xmx6g and –XX:MaxPermSize to –XX:MaxPermSize = 512 m.

(b) Ensure your KNIME update sites include "KNIME Analytics Platform Update Site" and "Stable Community Contributions".

(c) Add update site "MPI-CBG" with location 'https://community.knime.org/download/de.mpicbg.knime.ip.update'

(d) Install the following tools (*see* **Note 7**):
- KNIME External Tool Support (Labs).
- KNIME Image Processing.
- KNIME Image Processing—Supervised Image Segmentation.
- KNIME Quick Forms.
- KNIME Quick Forms (legacy).
- KNIME Virtual Nodes.

(e) Install the KNIME MS-ECS-2D workflow via "tinyurl.com/KNIME-MS-ECS". On this page you can also find more background information on how to install KNIME.

(f) The pipeline is available in versions for Mac or Linux.

3 Methods

Carry out all washes and rinses in 1 mL unless otherwise specified.

Use RNase-free solutions and wear gloves at all times to prevent RNase contamination.

3.1 Embryo Embedding

1. Collect embryos in the chorion in a petri dish and let them develop in embryo medium to the desired stage (*see* **Note 8**).
2. Fix embryos in a round bottom 2 mL tube in 4% formaldehyde in PBT overnight at 4 °C or for 4 h at RT.
3. Rinse embryos in PBT and manually dechorionate them in PBT under a dissecting scope using sharp forceps. Put embryos back in a 2 mL tube with PBT.
4. Equilibrate embryos in 30% sucrose in PBS until they sink (*see* **Note 9**).
5. Rinse embryos twice in fresh 30% sucrose in PBS (*see* **Note 10**).
6. Replace with fresh 30% sucrose in cryo incubation solution. Make sure that embryos are mixed well with the medium and leave for 5 days at 4 °C (*see* **Note 11**).
7. Equilibrate embryos in OCT by moving them through two consecutive baths of OCT (*see* **Notes 12** and **13**) (Fig. 2a).

8. Embed multiple embryos in the cap of an Eppendorf tube (*see* **Note 14**) (Fig. 2b). Fill the cap with OCT so that a dome of OCT extends from the cap.
9. Cool ~100 mL isopentane in a 250 mL beaker to −80 °C in a liquid nitrogen bath (*see* **Note 15**) (Fig. 2c). Use blunt-end tweezers to carefully immerse the sample in the precooled isopentane until it freezes and the OCT turns white (about 5 s) (Fig. 2d).
10. Use blunt-end tweezers to take the frozen block out of the isopentane. Drain off excess isopentane, wrap the block in plastic wrap and aluminum foil (mark stage and date on the foil) and store in a tightly sealed bag at −80 °C (*see* **Note 16**).

3.2 Coating Coverslips

1. Prepare a sonication bath with clean demi water.
2. Load selected #1.5 coverslips (*see* **Note 2**) into a coverslip holder and place them in a glass container (Fig. 3b).
3. Fill the container with 1:20 Mucasol in Milli-Q water until the coverslips are covered, and sonicate for 10 min.
4. Rinse the coverslips and the container with Milli-Q water until all traces of detergent are gone.
5. Place the coverslip holder back in the container and fill with 100% ethanol. Sonicate for 10 min.
6. Drain excess ethanol from the coverslip holder. Submerge coverslips in 1:10 poly-L-lysine in Milli-Q water for 30 min to coat (*see* **Note 17**).
7. Drain excess poly-L-lysine from the coverslip holder and let the coverslips air-dry overnight.

3.3 Cryosectioning

1. Set cryostat block and blade temperature to −17 °C (*see* **Note 18**).
2. Take a cryo-block out of the −80 °C freezer and mount it to a specimen stage using OCT (*see* **Note 19**). Let the block equilibrate to cryostat temperature for 30 min.
3. Align the anti-roll plate parallel to the cryostat blade with a small distance between plate and blade to guide sections between.
4. Mount the sample on the cryostat so that the blade will hit the yolk last (*see* **Note 20**) (Fig. 3a).
5. Make 8–10 μm sections and quickly mount the sections on the coated coverslips that you prepared under Subheading 3.2 (*see* **Notes 21–23**) (Fig. 3c).
6. Let the sections dry at room temperature for a couple of minutes (*see* **Note 23**) before storing them at −80 °C (*see* **Notes 24** and **25**) (Fig. 3d).

3.4 smFISH

1. Take coverslips with sections from −80 °C and place them in a 6-well plate (Fig. 3e). Post-fix the sections in 4% formaldehyde in PBS for 15 min (*see* **Notes 26** and **27**).
2. Rinse sections twice with PBS.
3. Rinse with 70% ice cold ethanol and replace with fresh 70% ethanol. Keep for 4–8 h at 4 °C to permeabilize the sample (*see* **Note 28**).
4. Rehydrate the sections in 2× SSC.
5. Treat with 5 μg/mL proteinase K in 2× SSC for 10 min while gently shaking (*see* **Note 29**).
6. Wash 2 × 5 min with 2 mL 2× SSC while gently shaking.
7. Rinse sections with wash buffer and replace with fresh wash buffer. Let equilibrate for 5 min.
8. Thaw 100 μL hybridization buffer per sample and add 0.7 μL smFISH probe (25 μM) (*see* **Notes 4, 30** and **31**).
9. Coat the bottom of a petri dish with Parafilm and place a 95 μL drop of hybridization buffer with probe on the Parafilm (Fig. 3f). Take a coverslip with sample and carefully remove as much wash buffer as possible with a piece of filter paper without touching the sections. Carefully place the coverslip section-down on the drop of hybridization buffer (Fig. 3f), close but do not seal the petri dish and incubate overnight for 14–16 h at 30 °C.
10. The next day, pipet 100 μL wash buffer on the corner of the coverslip and carefully peel the coverslip off the Parafilm, making sure to not dislodge the sample. Place the coverslip in a 6-well plate with the sections facing up.
11. Rinse once with wash buffer.
12. Wash twice with wash buffer for 30 min at 30 °C without shaking. For membrane staining, add 1:100 phalloidin–Alexa 488 (or another fluorophore) to the second wash step (*see* **Note 5**). For DNA staining, add 0.5 μg/mL DAPI to the second wash step.
13. Rinse once with GLOX buffer and leave in fresh GLOX buffer at 4 °C until mounting (*see* **Note 32**).
14. Mount in GLOX mounting medium. Place 25 μL GLOX mounting medium on a microscope slide and slowly lower the smFISH sample down on the drop to prevent bubbles. Remove excess mounting medium by carefully touching a piece of filter paper to the side of the coverslip.
15. Tightly seal the sample with nail polish to prevent evaporation of the mounting medium and proceed to imaging.

16. Image your samples on an epifluorescence microscope with a high NA objective and a sensitive camera (*see* **Notes 6** and **33**) (Fig. 1b_5). Use z-spacing of 0.3 μm or less to capture each transcript in multiple z-slices (*see* **Note 34**).

3.5 Image Analysis

Here, we describe all steps required for image analysis. For more background information on the image analysis tools, you may consult the documentation associated with Stapel et al. 2016 [15].

3.5.1 Membrane Segmentation

These steps can be skipped if one is not interested in assigning transcripts to individual cells.

1. Open your image in Fiji and duplicate the central membrane slice of the z-stack (e.g., slice 10 for a z-stack of 19 slices) (Fig. 1b_1) using the function "Duplicate" (*see* **Note 35**).
2. If you are analyzing multiple images at the same time, resize all images to the exact same size. This is a requirement for the membrane prediction pipeline in KNIME that we will use in the next steps. You can use Fiji function "Canvas size" for this. Save the image in '.tiff' format and collect all membrane images for which you want to generate segmentations in a single folder.
3. Start KNIME and open the workflow "MS-ECS-2D_2.0" (Fig. 4). Double click on the node "Image reader" at step 2.1 (Fig. 4) and select the membrane images that you generated in the previous steps (*see* **Note 36**).
4. Reset the node "Prediction" at step 2.2 (Fig. 4) by right-clicking on the node and selecting "Reset".
5. Double click on the node "Prediction" and then on "Run cascaded RF" to open these nodes. Now double click on the node "Image Resizer" to configure this node. Set X, Y, and Z to downsample your images and speed up the analysis (*see* **Note 37**).
6. Wait until the prediction is finished. The node will show a green check mark. Now use the node "Image Writer" at step 2.3 to write the result images to a folder on your computer (Fig. 4).
7. Sort the membrane probability and the vertex probability images (Fig. 1b_2) into separate folders.
8. Open Fiji and start the plugin PathFinder which is part of the MS-ECS-2D update site (*see* **Note 38**). Select the folder with your original membrane images (*see* **step 2**), the folder with the membrane probability images (*see* **step** 7), and the folder with vertex probability images (*see* **step** 7) as prompted (*see* **Note 39**). While the PathFinder plugin is running, several image windows will pop up and disappear again.
9. Once the run is finished, all pop-up windows will have disappeared and you will find a folder "results" inside the folder with

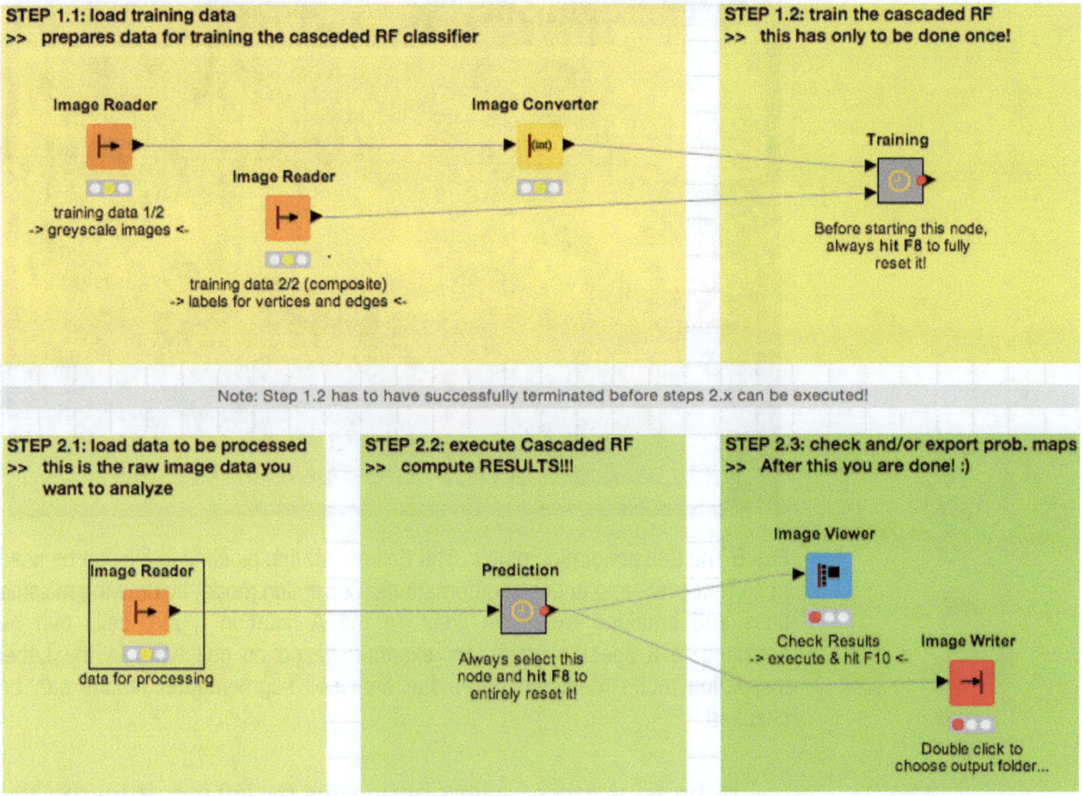

Fig. 4 KNIME cascaded random forest pipeline. The KNIME cascaded random forest pipeline consists of two steps. In *step 1* the pipeline is trained to recognize membrane, membrane intersection points (vertex points) and background on a small set of representative membrane images (*see* **Note 36**). Details on training can be found at "tinyurl.com/KNIME-MS-ECS". Once the pipeline has been trained it can be used repeatedly to predict membrane and vertex points in *step 2*. In *step 2.1*, membrane images are loaded. In *step 2.2* the cascaded random forest pipeline is run. Before running this step, the downsampling factor needs to be adjusted to the pixel size of the images (*see* **Note 38**). In *step 2.3* the results can be viewed and written to the computer for further processing

your original membrane images. Open the image that ends on "_scaled" in Fiji and duplicate the first channel using the "Duplicate" function. This channel contains a mask of the segmented cells ("Cell Masks", Fig. 1b$_4$). In addition, open the original membrane image that you used to run the segmentation (*see* **step 2**).

10. Start the Fiji plugin "Cell annotation" that is part of the MS-ECS-2D update site (Plugins > Cell transcript > Cell annotation) (*see* **Note 40**) (Fig. 5). Select the Original image and its matching "Cell Masks" image. You can ignore the Nuclei channel and Membrane channel fields.
11. In the Cell annotation tool, set the mode to "Correction". Correct any under-segmentations by drawing missing lines pressing the left mouse button. Correct any over-segmentations

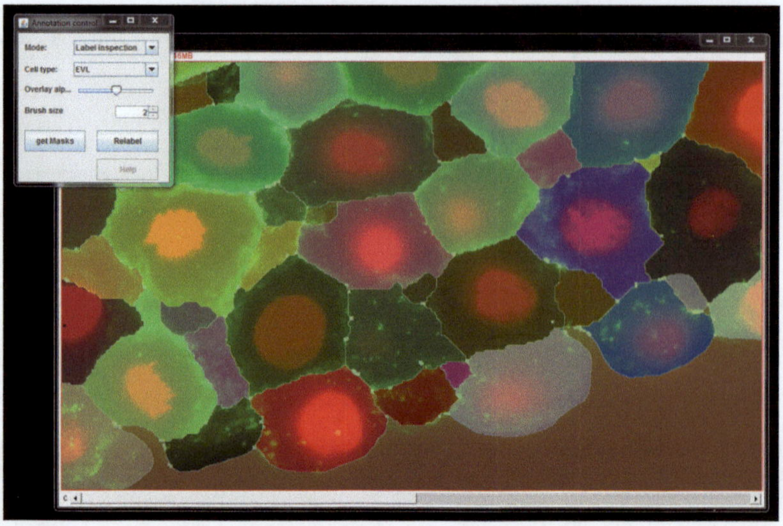

Fig. 5 The Cell annotation plugin. The Cell annotation pipeline in Fiji can be used to correct over- and under-segmentations (Correction mode) by drawing missing lines and breaking excessive lines. In the Annotation mode, cells can be assigned to a specific group, for example based on cell type. In the Label inspection mode (depicted here), the final cell segmentation results can be checked

by breaking excessive lines by pressing the left mouse button and the Alt key simultaneously. Go to "Label inspection" mode to check whether you have made all necessary corrections (Fig. 5).

12. Use "get Masks" or simply close the Annotation control window to finalize your masks image and save it (*see* **Note 41**) (Fig. 5).

3.5.2 Transcript Detection

1. Open the original z-stack image that you acquired on the microscope. In addition, open the Cell mask image that you generated in the previous steps in Fiji if you would like to assign transcripts to cells (*see* **Note 42**).
2. Start the "Cell transcript analysis" plugin which is part of the MS-ECS-2D update site (plugins > Cell transcript > Cell transcript analysis). Select the smFISH image and cell mask and adjust the settings to your sample (Fig. 6a).
3. First, run the Transcript analysis plugin once to generate a plot of maximum distributions ('_MaxDistrib_*.png') (*see* Fig. 6b). This plot will help you to determine the appropriate transcript detection threshold. The threshold should be set between background and transcript signal peaks (*see* Fig. 6b). You can zoom into the plot by dragging a rectangle around the area of interest (*see* Fig. $6b_2$). Identify the transcript detection threshold that is appropriate for your smFISH sample.

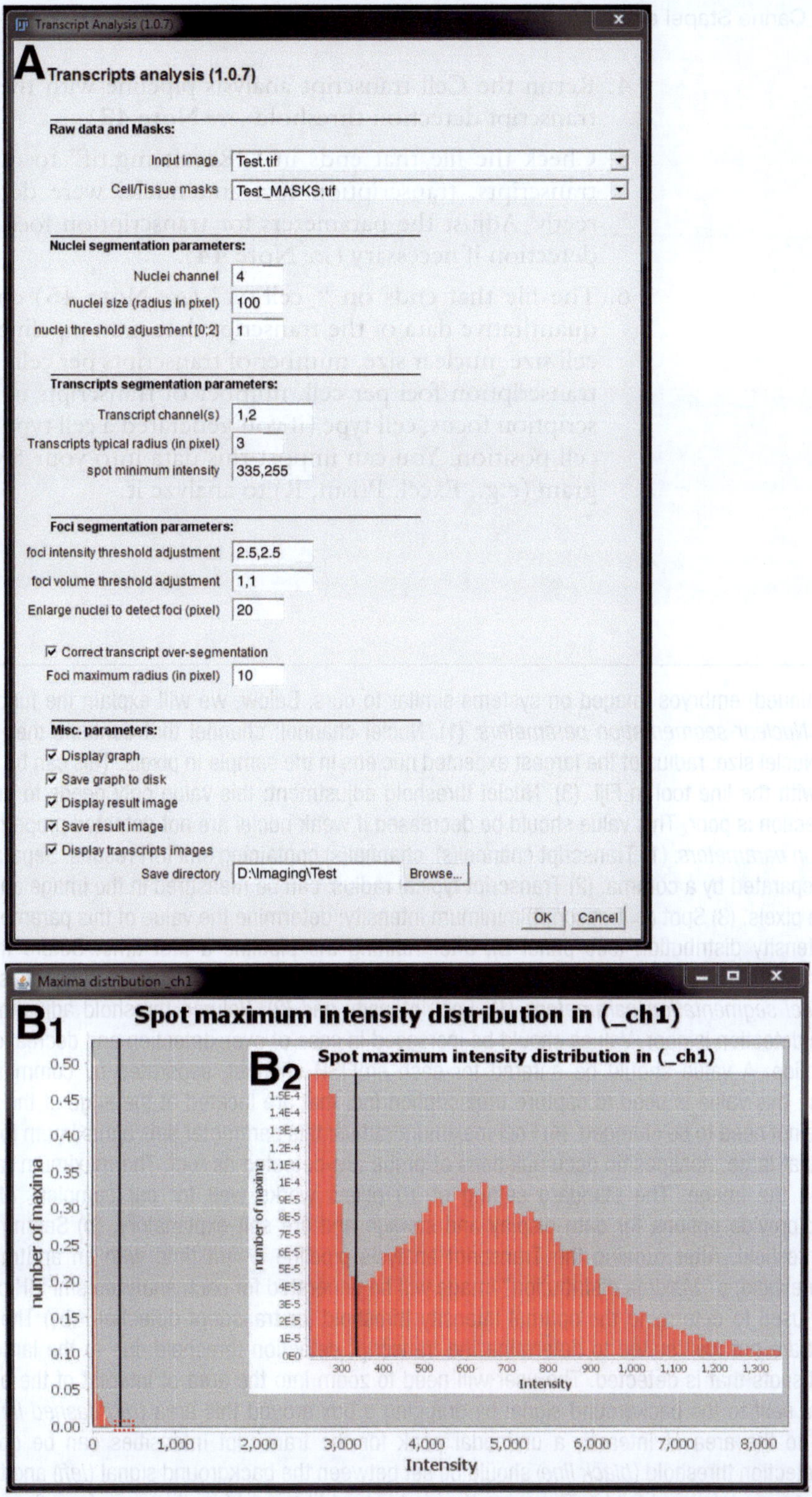

Fig. 6 The Transcript analysis plugin. (**a**) User interface for the Transcript analysis plugin. Before running the Transcript analysis plugin in Fiji, the user is prompted to select smFISH input and mask images, and to set parameters for nuclear segmentation, transcript detection, foci detection and data display in a user interface. The values depicted in this screenshot are good starting values for the analysis of pregastrulation stage

4. Rerun the Cell transcript analysis pipeline with the identified transcript detection threshold (*see* **Note 43**).
5. Check the file that ends in "_ResultImg.tif" to determine if transcripts, transcription foci and nuclei were detected correctly. Adjust the parameters for transcription foci and nuclei detection if necessary (*see* **Note 44**).
6. The file that ends on "_cell.txt" (*see* **Note 45**) contains the quantitative data of the transcript detection pipeline including cell size, nuclear size, number of transcripts per cell, number of transcription foci per cell, number of transcripts in each transcription focus, cell type (if you generated a cell type mask) and cell position. You can import this data into your favorite program (e.g., Excel, Prism, R) to analyze it.

Fig. 6 (continued) embryos imaged on systems similar to ours. Below, we will explain the function of each parameter. *Nuclear segmentation parameters*: (1). Nuclei channel: channel that contains the image of the nuclei. (2). Nuclei size: radius of the largest expected nucleus in the sample in pixels. This can be measured in the image with the line tool in Fiji. (3). Nuclei threshold adjustment: this value only needs to be modified if nuclear detection is poor. This value should be decreased if weak nuclei are not detected properly. *Transcript segmentation parameters*: (1) Transcript channel(s): channel(s) containing smFISH results. Separate channels should be separated by a comma. (2) Transcript typical radius: can be measured in the image and is typically two or three pixels. (3) Spot (= transcript) minimum intensity: determine the value of this parameter based on the spot intensity distribution (*see* panel B), after running the pipeline a first time. Before the transcript detection threshold has been determined, fill in arbitrary values for each smFISH channel, separated by commas. *Foci segmentation parameters*: (1). Foci intensity and (2). Volume threshold adjustments can be made if foci detection is poor. Values should be increased in case of over-detection and decreased in case of under-detection. A value should be entered for each smFISH channel, separated by commas. (3) Nuclei enlargement: this value is used to capture transcription foci that are located at the edge of the nucleus and usually does not need to be changed. (4) Foci maximum radius: this parameter sets a maximum to the foci size to prevent that large, nonspecific accumulations of probe are detected as foci. The maximum radius can be measured in the image. The standard setting of 10 pixels works well for our samples. *Miscellaneous parameters*: provide options for data display and storage and are self-explanatory. (**b**) Setting a transcript detection threshold. After running the Transcript analysis pipeline a first time with an arbitrary transcript detection threshold, a "Maxima distribution" image will be generated for each analyzed smFISH channel. This plot can be used to determine the optimal intensity threshold for transcript detection. (**b$_1$**) The axes of the original plot are not well suited to determine the transcript detection threshold due to the large number of background spots that is detected. The user will need to zoom into the area of interest of the plot, which is located right next to the background signal by dragging a box around this area (*red dashed line*). (**b$_2$**) After zooming in to the area of interest, a unimodal peak for the transcript intensities can be observed. The transcript detection threshold (*black line*) should be set between the background signal (*left*) and the unimodal peak for the transcript signal (*right*). The same threshold should be used for all images from the same sample (coverslip) that were acquired with the same microscope settings

4 Notes

1. We use Eppendorf caps to embed our samples. Alternatively, commercial molds can be used but we prefer the small size of Eppendorf caps. For early stages, coloring the bottom of the cap/embedding mold with a black marker can improve visibility of the embryos (Fig. 2b).
2. Mounting sections on coverslips instead of slides improves light transmission and image quality because it places the sample directly at the microscope objective without a barrier of mounting medium. We prefer using 22 × 22 mm coverslips. Although protocols from the company that sells Stellaris smFISH probes suggest to use #1 coverslips, we obtain optimal results with #1.5 coverslips. Selected #1.5 coverslips are optimized for the light path of most microscopes.
3. The concentration of formamide in the smFISH and hybridization buffers can be increased from 10% to up to 25% to reduce background signal. However, we find that this often results in a loss of signal and find that optimizing probe concentration is a more efficient way to increase the signal to noise ratio (*see* **Note 28**).
4. Probes can be designed using freely available software from Stellaris (*https://www.biosearchtech.com/support/tools/design-software/stellaris-probe-designer*). We recommend blasting the probe sequences that are suggested by this webtool against the zebrafish genome to make sure they exclusively hybridize to your transcript of interest. When choosing a fluorophore for probe labeling, select fluorophores in the red or far-red spectrum. Fluorophores with shorter wavelengths produce poor results due to auto fluorescence in zebrafish samples. Which exact fluorophore is optimal for your needs will depend on the filter sets on your microscope.
5. We use phalloidin to mark cell outlines and enable automated cell segmentation. Alternatively, membranes can be visualized using a transgenic line (e.g., a lyn::fluorescent protein line), or by injecting mRNA that codes for a membrane localized fluorescent protein at the 1-cell stage. For the choice of fluorescent protein, we have obtained good results with tdTomato, which was still visible after the smFISH protocol. GFP, on the other hand, loses its activity in the smFISH protocol and will need to be detected using an antibody (*see* **Note 29**).
6. We use a Delta Vision system equipped with an Olympus UPlan SApochromat 100× 1.4 oil objective, an EDGE/sCMOS camera and the following filter sets: 435/48 (DAPI), 525/36 (Alexa 488), 594/45 (CalFluor 610-labeled smFISH probes), 676/34 (Quasar 670-labeled smFISH probes. We

have also obtained good results with a 60× 1.3NA silicon objective. It is also possible to image your samples on a spinning disk microscope with comparable objectives and camera, but results might be more difficult to interpret for probe sets with low signal to noise ratios.

7. Uncheck "Group items by category" to be able to search through an alphabetically ordered list of all tools. This makes it easy to find the required tools.

8. While we optimized the protocol for embryonic stages up to gastrulation (shield stage), it also works well at later stages. We will indicate small changes in the embedding procedure that can be taken into account when working with post-gastrulation stages.

9. Embryos at pregastrulation stages will sink within 30 min; later stages might take longer to sink.

10. It is important to rinse the embryos several times in sucrose/PBS to remove any Tween remaining from the fixation step, as this decreases embedding quality.

11. While a 5-day incubation time seems excessive, in our hands it has led to greatly improved section quality at (pre)gastrulation stages without loss of RNA signal. Make sure to keep embryos in the dark if you are working with a fluorescent transgenic line (*see* **Note 5**). At post-gastrulation stages, 5-day incubation is not necessary and one can proceed with the next steps of the protocol as soon as embryos sink to the bottom of the tube.

12. We prefer using 6-well staining plates and an embryo manipulator (Fig. 2) to move embryos through OCT (*see* Subheading 2.1, **items 14** and **15**; Fig. 2a).

13. From this step onward keep embryos and liquids at 4 °C as much as possible to enhance freezing speed and improve embryo integrity.

14. Make sure to orient all embryos in the same direction and note the orientation on the embedding cap/mold with a marker. This will improve section quality later (*see* **Note 20**) (Fig. 2b).

15. Check the temperature with a low-temperature thermometer. After use, isopentane can be stored in a glass bottle and reused.

16. Section quality will be optimal if blocks are sectioned within 1 week. However, it is possible to obtain good sections when blocks have been stored for several months.

17. Poly-L-lysine for coverslip coating can be stored in a plastic container and reused up to three times.

18. Temperatures might need to be optimized depending on your cryostat. However, we recommend to use relatively high temperatures for the best results. As a reference, we use block and

blade temperatures of around −20 °C for many other sample types on the same cryostat.

19. After mounting the cryo-block to the specimen stage, remove the mold and check whether any cracks are present. If so, repair cracks with OCT (use the quick-freeze option on your cryostat to rapidly freeze the newly applied OCT). Cracks can occur due to rapid freezing of the OCT in a confined space. Although cracks can be prevented by freezing at lower temperatures, this also decreases embryo integrity and is therefore not recommended.
20. The yolk can fracture during sectioning. By sectioning through the yolk last, you ensure that this does not affect embryo integrity.
21. Sectioning results are best when the cryo-block is relatively warm. If the block is too cold and your sections contain cracks or roll up tightly it can help to briefly warm up the block before cutting a section by pressing your thumb to the block for a couple of seconds. Make sure to wear clean gloves to not contaminate the sample with RNases!
22. The anti-roll plate will keep the section flat. Lift the plate to mount your section to a coverslip. You can use a thin brush to prevent the section from rolling up (Fig. 3c).
23. We use 6–12 coverslips per embryo and put sections obtained from multiple positions in the embryo on each single coverslip (*see* Fig. 2e). Make sure to not leave the sections out at RT for more than 20 min before storing them at −80 °C as this can affect smFISH quality.
24. We store sections in 6-well plates, sealed with Parafilm. Sections can be stored at −80 °C for a long time. Up to 6 months, we have not observed a decrease in sample quality.
25. Sections should be kept at −80 °C for at least 1 h before continuing with the smFISH protocol. This will improve smFISH results.
26. Add the fixative to each section immediately after taking it out of the freezer, without letting it thaw. This will prevent RNA degradation.
27. Add solutions to the side of the well and not directly on the coverslip to avoid dislodging the sections.
28. In some protocols, sections are kept in ethanol for up to a month. However, in our hands this reduces smFISH quality.
29. This treatment is optional but improves signal/noise for most of our probe sets. Mild proteinase K treatment can digest proteins that cover the RNA of interest and reveal smFISH probe binding sites. Make sure to optimize treatment

conditions to your aliquot of proteinase K since excessive protein digestion will affect sample integrity.

30. If signal/noise is poor in your samples, the best way to optimize this is by changing probe concentrations in the hybridization buffer. We usually test ranges from 0.2–1.5 μL probe (25 μM stock solution) per 100 μL buffer.
31. smFISH can easily be combined with antibody staining to, for example, boost the signal of a transgenic line, or to detect the protein encoded by your mRNA of interest. For antibody staining, add the primary antibody to the overnight hybridization step (Subheading 3.4, **step 9**) and the secondary antibody to the first wash step (Subheading 3.4, **step 12**). Antibody staining can also be performed after smFISH staining. Be aware that this will lead to a reduction of smFISH signal.
32. GLOX mounting medium is a simple and quick mounting medium which can be used when imaging within 24 h of sample preparation (coverslips can be stored in GLOX buffer at 4 °C until mounting and imaging). For longer-term storage, other mounting mediums like prolong GOLD or Vectashield are recommended. When using mounting mediums like prolong GOLD or Vectashield, make sure that no liquid is left on the sample before mounting. A small amount of liquid can form a barrier between the sample and the mounting medium, preventing the mounting medium from penetrating the sample. This will lead to rapid bleaching of your sample.
33. Acquire images starting with the longest wavelength and work your way down to shorter wavelengths. This will minimize bleaching across fluorophores because shorter wavelengths carry higher energy levels and induce more bleaching.
34. Using z-spacing of maximally 0.3 μm ensures that individual transcripts are detected in more than one z-slice and that no transcripts are missed in the acquisition. This aids computational transcript detection.
35. Pressing the letter "l" in Fiji will open a command searching window. Although all commands are accessible through the menu, this is a quick way to find commands without having to browse through the menu structure.
36. The KNIME workflow consists of a training phase and a prediction phase (Fig. 4). After training the workflow once, it can be used to produce predictions for different samples. We have trained the pipeline on zebrafish membrane images, and have produced high-quality predictions for samples from zebrafish as well as other species using this pipeline. Therefore, we suggest that you use our pretrained pipeline on your samples. If the available pipeline produces poor results for your samples, you can consider retraining the pipeline with your own

samples. Details on how to do this can be found on the following webpage: "tinyurl.com/KNIME-MS-ECS".

37. The downsampling factor that you use in the prediction phase should correspond to the downsampling factor that was used to train the pipeline. To train the pipeline we used a factor of 0.33 for data with pixel size 0.1072 μm. If the pixel size in your samples is different, you should scale the downsampling factor for the prediction phase accordingly. For example, if your pixel size is 1.5× larger than ours, use a downsampling factor of 0.5.

38. We have found that the standard settings of the PathFinder plugin will give good results in many cases. If you observe strong over- or under-segmentation you may change the Segmentation parameters. Increasing the threshold for vertex, membrane path, or membrane pixel detection will reduce over-segmentation. Decreasing the small cell removal parameter will also reduce over-segmentation.

39. Image names do not need to match between folders, but make sure that the order of images is identical between folders as the PathFinder plugin will work its way down the file list.

40. Cell segmentation will not be fully accurate. You can use the Cell annotation tool (Fig. 5) to correct small segmentation errors that might have been made in the PathFinder tool. In addition to cell segmentation correction, the "Cell annotation" tool can be used to define different cell types. Options are preset for early embryonic cell types, namely EVL (enveloping layer, #1), YSL (yolk syncytial layer, #2), DEL (deep layer, #3), and Outside (#4) for regions outside of the embryo.

41. Before opening the Cell annotation tool again to start with your next image, you will need to Reset startup tools in Fiji (press ≫ in the Fiji menu bar and then Restore Startup Tools).

42. If you are not interested in assigning transcripts to single cells, you do not need to use the membrane segmentation pipeline to generate a cell mask. Instead, you can use the "Cell annotation" plugin to outline your areas of interest and use the resulting file as a mask or use no mask at all.

43. To ensure reproducibility, we recommend to use the same transcript detection thresholds for all images of the same sample (coverslip) that were acquired in the same imaging session.

44. To ensure reproducibility, we recommend to use the same parameters for nuclei segmentation and foci segmentation for all images of similar embryonic stages that were acquired at the same pixel size.

45. Additional data files that are generated are: a file with the analysis parameters ('_parameters.txt'), a file with transcript detection information per channel ('_spot_*.txt'), a log file

('.log'), and a scatter plot that indicates which spots have been identified as transcripts and which have been identified as foci, based on their intensity and size ('_ScatterPlot_*.png'). These files provide more background information on the results and the analysis.

Acknowledgments

This work was supported by MPI-CBG core funding, a Human Frontier Science Program Career Development Award [CDA-00060/2012-C] to NLV; and a Boehringer Ingelheim Fonds PhD fellowship to LCS. We thank Pavel Vopalensky for critically reading the manuscript, Julia Eichhorn for taking photos of the sectioning procedure, and Jan Philipp Junker and Alexander van Oudenaarden for initial advice on smFISH.

References

1. Thisse C, Thisse B (2008) High-resolution in situ hybridization to whole-mount zebrafish embryos. Nat Protoc 3:59–69. doi:10.1038/nprot.2007.514
2. Tomancak P, Berman BP, Beaton A et al (2007) Global analysis of patterns of gene expression during drosophila embryogenesis. Genome Biol 8:R145. doi:10.1186/gb-2007-8-7-r145
3. Grün D, Kester L, van Oudenaarden A (2014) Validation of noise models for single-cell transcriptomics. Nat Methods 11:637–640. doi:10.1038/nmeth.2930
4. Junker JP, Noël ES, Guryev V et al (2014) Genome-wide RNA tomography in the zebrafish embryo. Cell 159:662–675. doi:10.1016/j.cell.2014.09.038
5. Satija R, Farrell JA, Gennert D et al (2015) Spatial reconstruction of single-cell gene expression data. Nat Biotechnol 33:495–502. doi:10.1038/nbt.3192
6. Raj A, Peskin CS, Tranchina D et al (2006) Stochastic mRNA synthesis in mammalian cells. PLoS Biol 4:e309. doi:10.1371/journal.pbio.0040309
7. Raj A, van den Bogaard P, Rifkin SA et al (2008) Imaging individual mRNA molecules using multiple singly labeled probes. Nat Methods 5:877–879. doi:10.1038/nmeth.1253
8. Lyubimova A, Itzkovitz S, Junker JP et al (2013) Single-molecule mRNA detection and counting in mammalian tissue. Nat Protoc 8:1743–1758. doi:10.1038/nprot.2013.109
9. Wang F, Flanagan J, Su N et al (2012) Technical advance. J Mol Diagn 14:22–29. doi:10.1016/j.jmoldx.2011.08.002
10. Little SC, Tikhonov M, Gregor T (2013) Precise developmental gene expression arises from globally stochastic transcriptional activity. Cell 154:789–800. doi:10.1016/j.cell.2013.07.025
11. Boettiger AN, Levine M (2013) Rapid transcription fosters coordinate snail expression in the drosophila embryo. Cell Rep 3:8–15. doi:10.1016/j.celrep.2012.12.015
12. Bahar Halpern K, Tanami S, Landen S et al (2015) Bursty gene expression in the intact mammalian liver. Mol Cell 58:147–156. doi:10.1016/j.molcel.2015.01.027
13. Itzkovitz S, Lyubimova A, Blat IC et al (2011) Single-molecule transcript counting of stem-cell markers in the mouse intestine. Nat Cell Biol 14:106–114. doi:10.1038/ncb2384
14. Oka Y, Sato TN (2015) Whole-mount single molecule FISH method for zebrafish embryo. Sci Rep 5:8571. doi:10.1038/srep08571
15. Stapel LC, Lombardot B, Broaddus C et al (2016) Automated detection and quantification of single RNAs at cellular resolution in zebrafish embryos. Development 143:540–546. doi:10.1242/dev.128918

Check for updates

Chapter 10

Super-Resolution Single Molecule FISH at the *Drosophila* Neuromuscular Junction

Joshua S. Titlow, Lu Yang, Richard M. Parton, Ana Palanca, and Ilan Davis

Abstract

The lack of an effective, simple, and highly sensitive protocol for fluorescent in situ hybridization (FISH) at the *Drosophila* larval neuromuscular junction (NMJ) has hampered the study of mRNA biology. Here, we describe our modified single molecule FISH (smFISH) methods that work well in whole mount *Drosophila* NMJ preparations to quantify primary transcription and count individual cytoplasmic mRNA molecules in specimens while maintaining ultrastructural preservation. The smFISH method is suitable for high-throughput sample processing and 3D image acquisition using any conventional microscopy imaging modality and is compatible with the use of antibody colabeling and transgenic fluorescent protein tags in axons, glia, synapses, and muscle cells. These attributes make the method particularly amenable to super-resolution imaging. With 3D Structured Illumination Microscopy (3D-SIM), which increases spatial resolution by a factor of 2 in X, Y, and Z, we acquire super-resolution information about the distribution of single molecules of mRNA in relation to covisualized synaptic and cellular structures. Finally, we demonstrate the use of commercial and open source software for the quality control of single transcript expression analysis, 3D-SIM data acquisition and reconstruction as well as image archiving management and presentation. Our methods now allow the detailed mechanistic and functional analysis of sparse as well as abundant mRNAs at the NMJ in their appropriate cellular context.

Key words smFISH, Single molecule fluorescence in situ hybridization, Structured Illumination, Super-resolution imaging, 3D-SIM, *Drosophila melanogaster*, Larval neuromuscular junction, mRNA localization, Synapse

1 Introduction

In situ hybridization has been a mainstay of cell and developmental biology for determining where and when genes are expressed in wild-type or mutant cells and tissues. The recent development of single molecule fluorescent in situ hybridization (smFISH) methods have increased the sensitivity, ease of application of FISH methodology, and enabled multiplexing with antibodies against

The original version of this chapter was revised. An erratum to this chapter can be found at https://doi.org/10.1007/978-1-4939-7213-5_32

Imre Gaspar (ed.), *RNA Detection: Methods and Protocols*, Methods in Molecular Biology, vol. 1649, DOI 10.1007/978-1-4939-7213-5_10, © The Author(s) 2018

specific proteins [1–3]. This next generation FISH approach uses approximately 50 short, fluorochrome-labeled DNA oligonucleotide (oligos) probes, which are approximately 20 bp in length. Such tiled oligonucleotides sets are designed to bind to nonoverlapping regions of a transcript. The large number of probes means that the technique is sensitive enough to detect the majority of individual mRNA molecules in a tissue, achieving a very high signal–noise ratio. The detected individual transcripts appear as bright foci and any off-target labeling by individual oligonucleotides appears as dim, diffuse signal, or low-intensity punctae [1]. Using shorter probes also provides better tissue penetration and enables less harsh hybridization conditions, maintaining antigenicity for antibody staining and making the technique especially suitable for whole mounted tissues.

The study of RNA biology in neuroscience has been held back by the lack of suitable methods for high quality in situ hybridization in some key experimental models and tissues. *Drosophila* in particular is an excellent model system for elucidating molecular mechanisms of neuronal development and function in all parts of the nervous system [4–6]. One of the key models for studying synaptic plasticity and physiology is the larval neuromuscular junction (NMJ) preparation of the body wall musculature. This system also has tremendous potential for studying the role of RNA metabolism in plasticity and physiology [7, 8]. However, while smFISH has been used successfully in *Drosophila* oocytes and embryos [9, 10], only traditional RNA FISH methods have been used in the NMJ [11–13]. Such methods have not been widely adopted due to variability, poor signal–noise ratios, and limited sensitivity for sparse transcript expression. Here, we describe our modified smFISH protocol for visualizing single mRNA molecules in the larval NMJ together with endogenous fluorescent proteins and antibody markers. To complement the single transcript sensitivity of smFISH, we used 3D structured illumination microscopy (3D-SIM), a super resolution imaging technique that provides enhanced spatial information regarding the RNA's subcellular environment [14]. The increased optical resolution of methods like 3D-SIM [15] provide a more accurate representation of whether a transcript resides in or is adjacent to a particular RNP granule or subcellular compartment (*see* **Note 1**). Furthermore, the relatively mild hybridization and wash conditions required for smFISH allow tissue morphology to be well preserved for meaningful biological interpretations.

Resolving individual transcripts in an intact tissue is extremely powerful for investigating gene expression and mRNA localization. To fully realize the benefits of single transcript detection, an automated quantification workflow saves time and reduces variability. Various computer programs have been developed to automate segmentation and quantification of the number of foci in an image. We used FindFoci, an open source ImageJ (Fiji) plugin that is part of the GDSC suite [16]. We also used an open source

MatLab program called FISHQuant that allows automated segmentation and fluorescence intensity calculations [17], and a user-friendly commercial solution, namely the spot counting algorithm in Imaris. We found that all three programmes performed similarly with our in situ data in automated quantitation of transcript numbers. To quality control acquisition of raw 3D-SIM data and the 3D-SIM reconstructions we used the ImageJ (Fiji) plugin SIMcheck [18]. Finally, we managed the relatively large number and size of image files with OMERO and created summary figures with OMERO-Figure, a platform that enables public distribution of the raw image files.

2 Materials

2.1 smFISH Probes

DNA oligonucleotide probes (Stellaris® RNA FISH) were purchased from LGC BioSearch Technologies (California, USA) and sequences were selected using the company's online probe designer. Alternatively, one could design 30–50 3′ primary amine labeled 18-mer DNA oligonucleotides that cover a region of the chosen gene, order a plate of HPLC-purified oligonucleotides (available from most manufacturers who synthesize PCR primers) and conjugate fluorochromes to the probes oneself [3]. Probes described here were labeled by the manufacturer with either Quasar 570 or Quasar 670 dyes, as using orange and red emitters minimizes background from autofluorescence in the NMJ.

2.2 Larva Neuromuscular Junction Dissection

1. Dissecting microscope with light source.
2. 35 mm petri dish.
3. Sylgard or similar elastomer [19].
4. Insect pins.
5. Microdissection scissors and forceps.
6. Saline buffer: 70 mM NaCl,5 mM KCl, 20 mM $MgCl_2$, 10 mM $NaHCO_3$, 5 mM trehalose, sucrose 115 mM, and 5 mM HEPES, pH 7.2.

2.3 Fixation and Hybridization (See Note 2)

1. Fix solution: PBS, 0.3%Triton X, 4% formaldehyde from freshly thawed aliquots of 16% EM grade PFA.
2. PBTX: PBS, 0.3%Triton X.
3. Bovine serum albumin (nuclease-free).
4. 70% ethanol.
5. Wash buffer: 10% 20× SSC (3 M NaCl, 0.3 M sodium citrate, pH 7.0), 10% freshly thawed deionized formamide, 80% DEPC-treated water.
6. Hybridization buffer: 10 w/v % dextran, 250 nM smFISH probe in Wash buffer.

7. (Optional) for immunohistochemistry: appropriate primary and secondary antibodies diluted in Wash buffer.
8. (Optional) to counterstain axon terminals: dye conjugated anti-horseradish peroxidase antibody diluted 1:100 in Wash buffer.
9. (Optional) to counterstain nuclei: 1 μg/mL DAPI in Wash buffer.

2.4 Mounting

1. Glass slide and glass coverslips (High Precision No. 1.5 coverslips for 3D-SIM).
2. Double-sided adhesive tape.
3. Vectashield mounting medium.
4. 100 nm Tetraspek beads.

2.5 Image Acquisition

1. For conventional imaging: wide-field epifluorescence microscope, spinning disk or laser scanning confocal microscope with a 60× or 100× 1.3–1.4NA oil or silicone oil immersion objective. We used an Ultra- VIEW VoX from PerkinElmer mounted on an IX81 Olympus microscope with 60× 1.35 NA oil immersion objective and an electron-multiplying charge-coupled device camera (ImagEM; Hamamatsu Photonics).
2. For 3D-SIM microscopy: images are acquired on a DeltaVision OMX, V3-Blaze (GE) with 60× 1.3 NA silicone oil immersion objective from Olympus (*see* **Note 3**).

2.6 Image Processing and Analysis

1. ImageJ/FIJI with SIMcheck and FindSpot plug-ins.
2. fairSIM [24].
3. OMERO server (https://www.openmicroscopy.org/site/support/omero5.2/sysadmins/unix/server-installation.html).
4. (Optional) Matlab with the FISHQuant script.
5. (Optional) Imaris.

3 Methods

3.1 Larva Neuromuscular Junction Dissection

1. Video protocols for *Drosophila* larva dissection are available online [20, 21]. Pin the larva dorsal side up on a 35 mm Petri dish filled half way with Sylgard, by placing pins at the anterior and posterior ends.
2. Cover the larva with a few drops of saline buffer.
3. Use microdissection scissors to create a small incision at the centre of the dorsal midline.

4. Extend the incision along the dorsal midline toward the posterior end, then from the centre towards the anterior end of the larva, make the cuts as superficial as possible so as not to damage the underlying nervous system and muscle tissues.
5. Carefully remove gut tissue by holding the trachea with forceps and cutting the tracheal attachments at each abdominal segment. After cutting the trachea on either side the gut tissue and other organs can be carefully removed all at once, leaving the brain and nerves intact.
6. Place two pins into the outer "shoulders" of the anterior body wall and gently stretch the tissue away from the midline. Do the same for the posterior side.
7. At this point the brain can either be removed, by cutting the nerves just above the muscle tissue, or properly positioned for in situ imaging by gently stretching the head pin.

3.2 Fixation

1. Replace the dissection buffer with fix solution and incubate by gentle rocking at room temperature for 25 min.
2. Remove the fix buffer and rinse 3× with PBTX.
3. (Optional) If immunohistochemistry is to be performed, block the tissue by incubating for 60 min in PBTX with 1% RNAse free bovine serum albumin.
4. Carefully transfer the tissue to a 0.75 mL microcentrifuge tube filled with 0.2 mL 70% ice-cold ethanol and incubate for 4–24 h at 4 °C.

3.3 Hybridization

1. Replace the ethanol with 0.2 mL wash buffer and incubate for 10 min at 38 °C with gentle rocking.
2. Replace the wash buffer with 0.1 mL hybridization buffer and incubate for at least 4 h (ideally overnight) at 38 °C with gentle rocking.

3.4 Washing and Counterstain

1. Remove the hybridization buffer and rinse 3× with 0.2 mL wash buffer.
2. Incubate the tissue in 0.2 mL wash buffer for 45 min at 38 °C with gentle rocking.
3. (Optional) For counterstaining, add secondary antibodies (1:500 dilution) and/or DAPI. To label axon terminals in the NMJ use dye conjugated anti-horseradish peroxidase antibody (1:100 dilution) (*see* **Note 4**).
4. (Optional) If tissues are counterstained, remove excess dye by washing 3× in the wash solution and incubating at room temperature for 15 min with gentle rocking.

3.5 Mounting

1. Remove the wash buffer and incubate tissue in Vectashield for several minutes (*see* **Note 5**).
2. Place thin strips of double-sided tape across a glass microscope slide spaced about as wide as the coverslip.
3. Place a drop (~30 μL) of Vectashield at the centre of the slide, between the strips of tape.
4. Position the tissues dorsal side up in the Vectashield and carefully place a coverslip on the strips of tape (*see* **Note 6**).
5. Seal the coverslip with multiple layers of clear nail varnish, taking care not to let the varnish come in contact with the Vectashield.
6. For super-resolution imaging with silicone immersion lens (NA 1.3): dilute Vectashield to 70% with water and use High Precision No. 1.5 coverslips.

3.6 Image Acquisition on the Spinning Disk Confocal Microscope

1. Acquire optical sections of the region of interest using optimal imaging configuration for your system, i.e., choosing appropriate beam splitter, emission filter, laser power, and pixel size.
2. Exposure times of 600–800 ms are often required for camera-based imaging systems. The slowest scan speed and line averaging are often necessary on scanning confocal systems.
3. Single transcripts generally appear as discrete punctae with consistent intensities. An exception is in the nucleus where a high concentration of nascent transcripts form a much larger and brighter fluorescent focus at the gene locus (Fig. 1b).

3.7 Image Acquisition for 3D-SIM

1. Acquire 3D-SIM data according to manufacturer's guidelines and good imaging practices, balancing signal–noise and bleaching while correcting for spherical aberration [22, 23]. Check raw data with the open source ImageJ plugin SIMcheck (Fig. 2).
2. Perform image reconstruction using commercial software that accompanies the instrument (in our case, SoftWORX by GE for the OMX V3) or an open-source alternative, such as fairSIM [24]. For multichannel imaging you will need to register channels using Tetraspec beads data and an appropriate image registration software.
3. Check the quality of reconstruction using SIMcheck (Fig. 2).

3.8 Image Management Using the OMERO Database (Open Microscopy Environment)

1. Install an OMERO server and import your data to it with the OMERO.insight client software [25–27]. Use the "tagging" facilities to organize your imaging data. The initial installation and subsequent super-user management of the server requires some degree of system administration experience. OMERO software is open source and released by the OME Consortium at www.openmicroscopy.org. Look at the online video tutorials, such as http://help.openmicroscopy.org/importing-data-

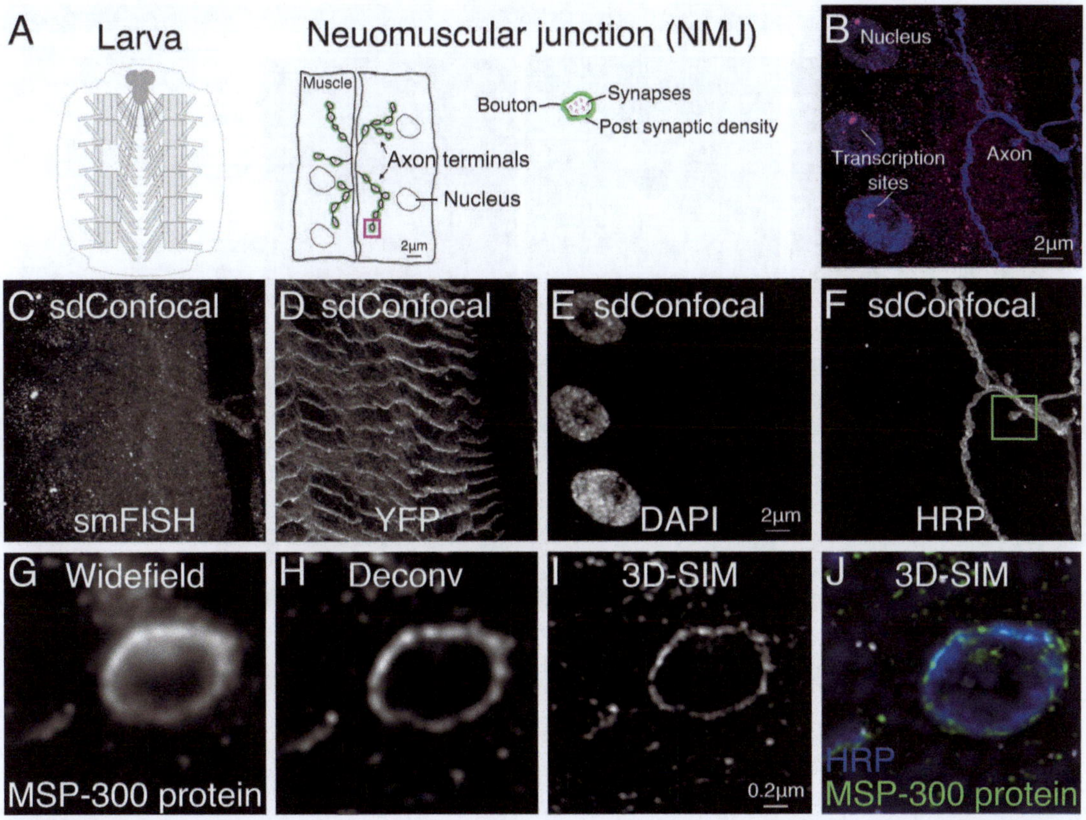

Fig. 1 Example of smFISH data acquired from the larva neuromuscular junction (NMJ) with spinning disk confocal (sdConfocal) or 3D structured illumination microscopy (3D-SIM). (**a**) Schematic of the larva fillet preparation indicating the location of an NMJ and the major subcellular compartments. (**b**) Merged 3D projection of a specimen labeled with smFISH probe (*magenta*) and Alexa 647-conjugated anti-HRP counterstain (*blue*). Image was acquired on a spinning disk confocal with 60× 1.35 NA oil objective. (**c**) sdConfocal image of MSP-300-smFISH: a coding region of MSP-300 mRNA was hybridized with a probe set containing 48 short oligos (18 nts) individually labeled with Quasar 570. (**d**) The MSP-300::YFP fusion protein is easily detected in the smFISH preparation. Nuclei and NMJ axons were labeled with DAPI and Alexa 647-conjugated anti-HRP respectively (**e, f**). Box in **f** shows the relative region of a bouton, (*see* **g–j**). Enhancement in resolution can be seen by comparing widefield (**g**) and deconvolved (**h**) images of the MSP-300 label to 3D-SIM reconstructions (**i, j**)

5.html and at the installation instructions. This client-server software integrates visualization, data mining, and image analysis of biological microscopy images. OMERO through its use of the Bio-Formats importer (http://www.openmicroscopy.org/site/products/bio-formats) and conversion to OME-TIF supports over 140 image file formats and the raw data can be managed from the web or exported from the online platform to a third party software like ImageJ (Fiji).

2. Use the OMERO web browser to view and organize the primary imaging data using searchable tags.
3. Use OMERO to share the data between collaborating scientists from any location with Internet access.

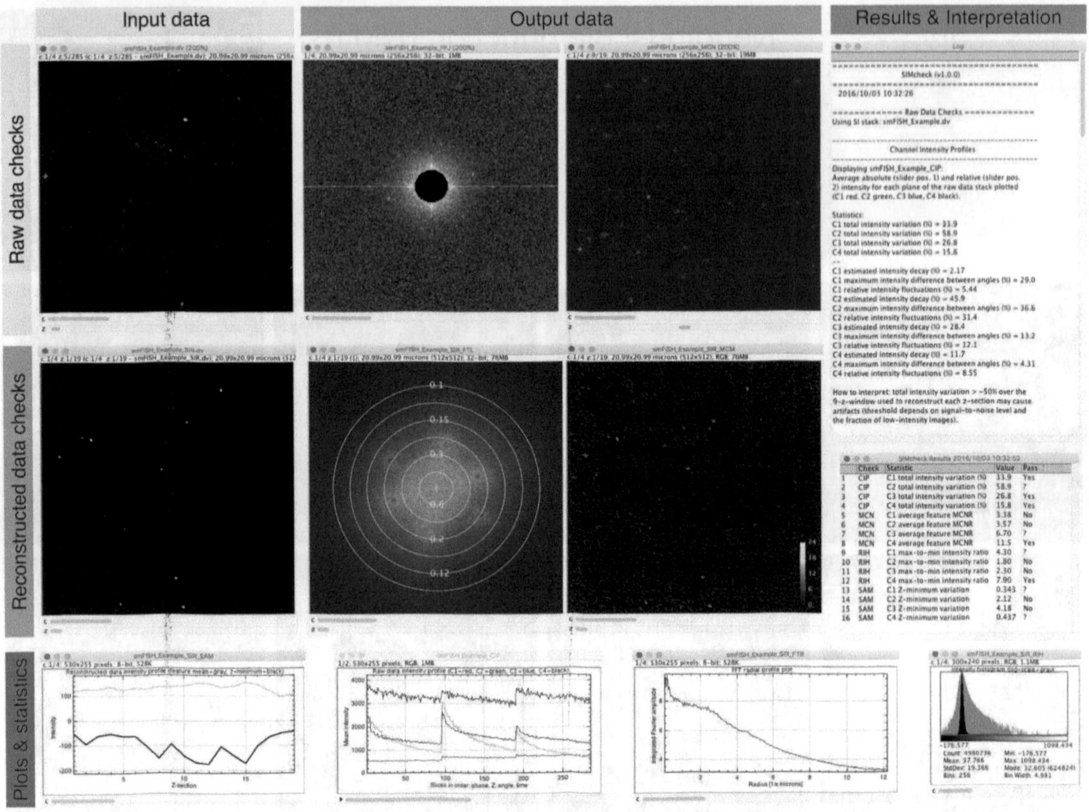

Fig. 2 Representative output from SIMcheck. For detailed explanation of these plots and statistics *see* [18]

4. Create and highlight figures from typical data sets using OMERO.figure (video http://figure.openmicroscopy.org/videos.html). OMERO.figure uses unique OMERO IDs for each image, from which the figure panels are made, to link to the original raw image data. Therefore, figures can be adjusted with great ease and other scientists in a team can easily view the original data. While OMERO-Figure can be used to add some annotations to the figure panels, we find that publication ready figures require the use of other image manipulation software.

3.9 Image Analysis (See Note 7)

3.9.1 Find Foci

1. Install and open the FindFoci GUI application in ImageJ; Plugins >GDSC > FindFoci > FindFocus GUI [16].
2. Open an image in ImageJ and split the channels; Image > Color > Split Channels.
3. Select the smFISH channel in the FindFoci GUI.
4. The GUI has a live preview mode that displays identification of points under various threshold settings. Using the configuration shown in Fig. 3, adjust the "Background param" slider until labels appear over each spot (Fig. 4b).
5. Number and features of the foci can be obtained from the measurement table or exported as a text file.

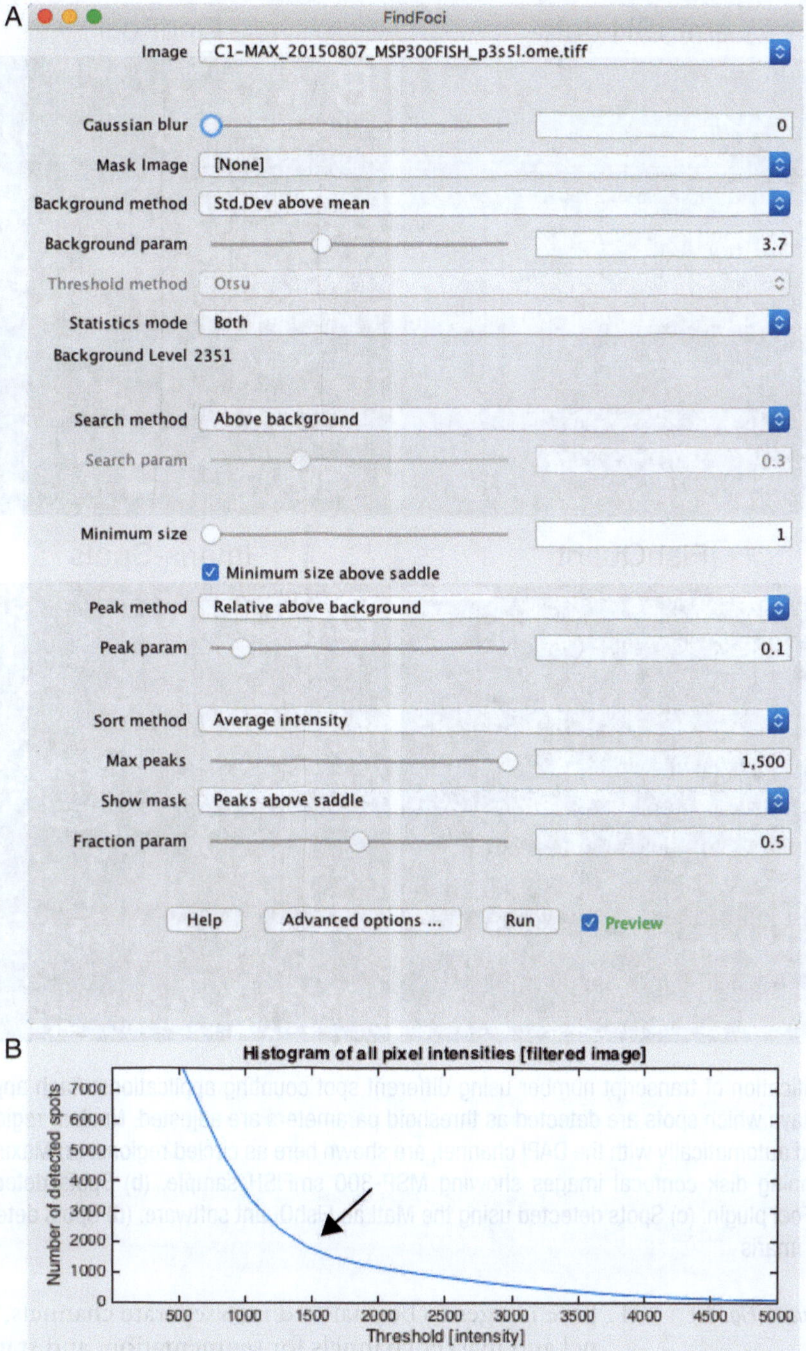

Fig. 3 Thresholding parameters for counting spots in the FindFoci ImageJ plugin. With these settings (**a**) the "Background param" slider is adjusted until all spots are identified in the image. The FISHQuant and Imaris Spots applications use intensity thresholding (**b**), which is used to provide an initial separation between background and high intensity spots (*arrow*), which are then refined using additional parameters

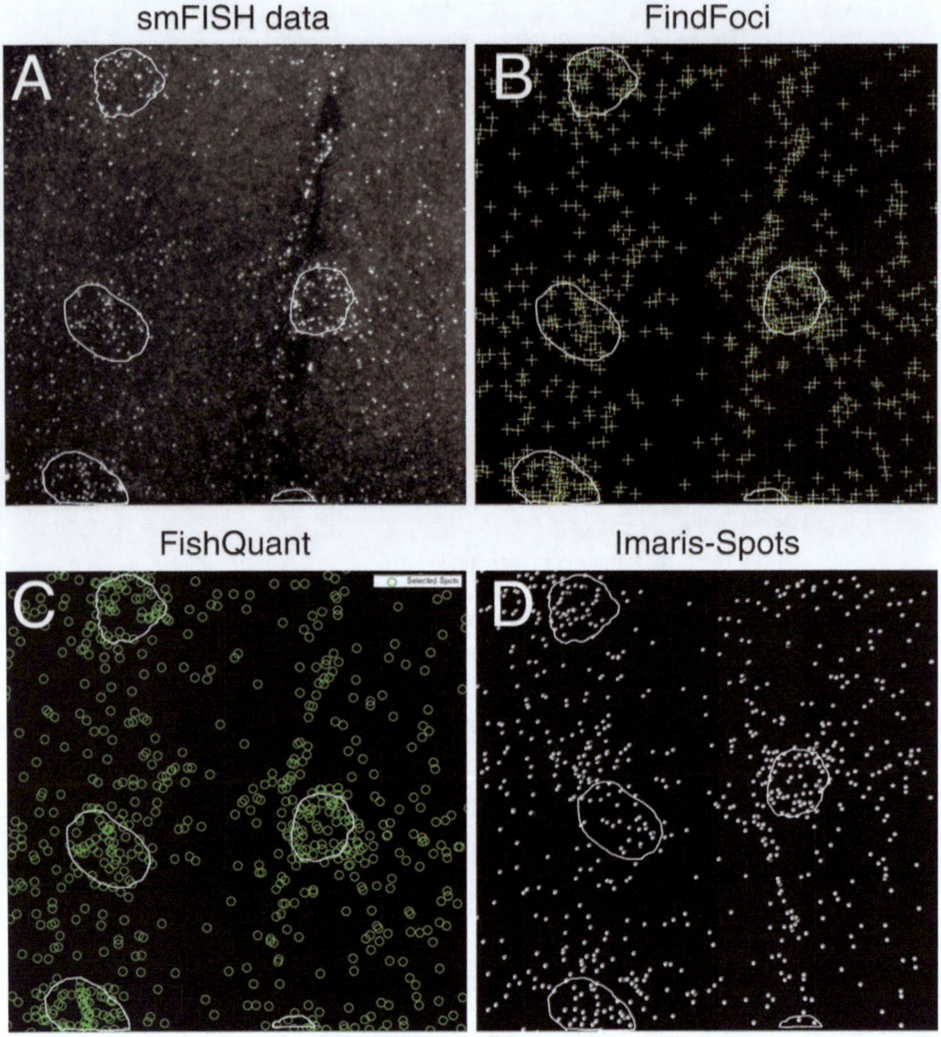

Fig. 4 Quantification of transcript number using different spot counting applications. Each application has a GUI that displays which spots are detected as threshold parameters are adjusted. Nuclear regions, which can be segmented automatically with the DAPI channel, are shown here as circled regions. (**a**) Maximum projected stack of spinning disk confocal images showing MSP-300 smFISH sample. (**b**) Spots detected using the ImageJ FindFoci plugin. (**c**) Spots detected using the MatLab FishQuant software. (**d**) Spots detected using the Spots tool in Imaris

3.9.2 FISHQuant (Fig. 4c)

1. Save images to be analyzed into separate channels, FISH channel and marker channels for segmentation, and start the FISH-Quant application in Matlab [17].
2. Follow the FISHQuant manual for loading data, filtering the image, and thresholding the spots.
3. Adjust the threshold parameters until each of the spots are marked in the GUI tool. The intensity profile will typically show an obvious separation between background and real spots (Fig. 3b).

4. Export the thresholded spots text file to determine the number of transcripts.

3.9.3 Imaris Spots (Fig. 4d)

1. Open the image with Imaris.
2. Choose the Spots tool.
3. Provide an estimated diameter for the spots (350 nm works well).
4. Slide the spot quality threshold tool until foci are accurately identified. The intensity profile will typically show an obvious separation between background signal and labeled mRNA (Fig. 3b).
5. Save the statistics text file to determine the number of transcripts.

4 Notes

1. Conventional confocal and widefield deconvolution microscopy techniques have a lateral resolution limit of ~250 nm and ~500 nm in the axial direction for green-emitting fluorochromes. 3D-SIM can enhance the lateral resolution to ~125 nm and axial resolution to ~250 nm. The effective measured diameter (full width half max) of a 1 kb folded mRNA molecule is ~150 nm in the red channel [28].
2. Fresh reagents (especially SSC and deionized formamide) are important for obtaining optimal signal-to-noise ratio. For best results, flash-freeze 1 mL aliquots of deionized formamide with liquid nitrogen and store at −80 °C. Reagents should be prepared with DEPC-treated water and autoclaved whenever possible.
3. It is important to correct for spherical aberration by matching the refractive index of the mountant with the immersion oil, or by adjusting the correction collar of the objective. 3D-SIM is very sensitive to artefacts caused by spherical aberration, particularly when imaging at depths greater than a few microns from the coverslip. While it is possible to correct spherical aberration to some extent when imaging deep with an oil immersion 1.42 NA objectives, we find the best results are obtained when imaging deep with a silicone immersion objective, such as the 60× /1.3 from Olympus. Adaptive optics approaches hold the most promise for correcting aberration and remote focusing [29], but have not yet been popularized in off-the-shelf instruments.
4. Endogenous fluorescent proteins are well-preserved for confocal imaging but often bleach too quickly to acquire high quality SIM images. To overcome this problem, label fluorescent proteins with antibodies, such as the Chromotek lama anti-GFP antibody, coupled to a highly photo-stable Alexa Fluor or Atto

dye of choice. The further the dye chosen emits into the red wavelengths, the better the signal–noise ratio because of reduced tissue autofluorescence, but the lower the resolution achievable, which is particularly important if 3D-SIM is used. The further the dye chosen emits into the red wavelengths the better the signal to background ratio because of reduced tissue autofluorescence, but the lower the resolution achievable, which IS particularly important if 3D-SIM is used.

5. To limit aberrations, the mounting medium must penetrate the tissue evenly so that the refractive index inside the cell matches the lens immersion oil, as much as possible.
6. After placing NMJ preparations on the glass slide, pipette 1 μL of 100 nm Tetraspek beads directly onto one of the preparations to use for aligning the different channels and for testing the quality of the point spread function (PSF).
7. All three software solutions support batch analysis. FindFoci performance is not as accurate as the others for data with low signal-to-noise. It is difficult to identify transcripts in Imaris Spots with segmented regions of interest, e.g., the nucleus.

Acknowledgments

We thank Talila Volk (Weizmann Institute of Science, Rehovot, Israel) for the Msp300 antibody; Flybase and the Bloomington Drosophila Stock Center for resources. We also thank David Ish-Horowicz and members of the Davis lab for discussions and comments on the manuscript. This work was supported by a Wellcome Trust Senior Basic Biomedical Research Fellowship (096144) to I. D., Wellcome Trust Strategic Awards (091911 and 107457/Z/15/Z) supporting advanced microscopy at Micron Oxford (http://micronoxford.com), and a Clarendon scholarship to LY.

References

1. Femino AM et al (1998) Visualization of single RNA transcripts in situ. Science 280 (5363):585–590
2. Raj A et al (2008) Imaging individual mRNA molecules using multiple singly labeled probes. Nat Methods 5(10):877–879
3. Batish M, Raj A, Tyagi S (2011) Single molecule imaging of RNA in situ. Methods Mol Biol 714:3–13
4. Dubnau J, Tully T (1998) Gene discovery in Drosophila: new insights for learning and memory. Annu Rev Neurosci 21:407–444
5. Skoulakis EM, Grammenoudi S (2006) Dunces and da Vincis: the genetics of learning and memory in Drosophila. Cell Mol Life Sci 63(9):975–988
6. Walkinshaw E et al (2015) Identification of genes that promote or inhibit olfactory memory formation in Drosophila. Genetics 199 (4):1173–1182
7. Pradhan SJ et al (2012) The conserved P body component HPat/Patl negatively regulates synaptic terminal growth at the larval Drosophila neuromuscular junction. J Cell Sci 125(Pt 24):6105–6116
8. Nesler KR et al (2013) The miRNA pathway controls rapid changes in activity-dependent synaptic structure at the Drosophila melanogaster neuromuscular junction. PLoS One 8(7):e68385
9. Abbaszadeh EK, Gavis ER (2016) Fixed and live visualization of RNAs in Drosophila oocytes and embryos. Methods 98:34–41

10. Trcek T et al (2015) Drosophila germ granules are structured and contain homotypic mRNA clusters. Nat Commun 6:7962
11. Karr J et al (2009) Regulation of glutamate receptor subunit availability by microRNAs. J Cell Biol 185(4):685–697
12. Gardiol A, St Johnston D (2014) Staufen targets coracle mRNA to Drosophila neuromuscular junctions and regulates GluRIIA synaptic accumulation and bouton number. Dev Biol 392(2):153–167
13. Packard M et al (2015) Nucleus to synapse Nesprin1 railroad tracks direct synapse maturation through RNA localization. Neuron 86(4):1015–1028
14. Gustafsson MG et al (2008) Three-dimensional resolution doubling in wide-field fluorescence microscopy by structured illumination. Biophys J 94(12):4957–4970
15. Schermelleh L, Heintzmann R, Leonhardt H (2010) A guide to super-resolution fluorescence microscopy. J Cell Biol 190(2):165–175
16. Herbert AD, Carr AM, Hoffmann E (2014) FindFoci: a focus detection algorithm with automated parameter training that closely matches human assignments, reduces human inconsistencies and increases speed of analysis. PLoS One 9(12):e114749
17. Mueller F et al (2013) FISH-quant: automatic counting of transcripts in 3D FISH images. Nat Methods 10(4):277–278
18. Ball G et al (2015) SIMcheck: a toolbox for successful super-resolution structured illumination microscopy. Sci rep 5:15915
19. Verstreken P, Ohyama T, Bellen HJ (2008) FM 1-43 labeling of synaptic vesicle pools at the Drosophila neuromuscular junction. Methods Mol Biol 440:349–369
20. Brent JR, Werner KM, McCabe BD (2009) Drosophila larval NMJ dissection. J Vis Exp (24):–1107
21. Smith, R. and J.P. Taylor, (2011). Dissection and imaging of active zones in the Drosophila neuromuscular junction. J Vis Exp, (50): 2676
22. Dobbie IM et al (2011) OMX: a new platform for multimodal, multichannel wide-field imaging. Cold Spring Harb Protoc 2011 (8):899–909
23. Demmerle J et al (2015) Assessing resolution in super-resolution imaging. Methods 88:3–10
24. Muller M et al (2016) Open-source image reconstruction of super-resolution structured illumination microscopy data in ImageJ. Nat Commun 7:10980
25. Allan C et al (2012) OMERO: flexible, model-driven data management for experimental biology. Nat Methods 9(3):245–253
26. Burel JM et al (2015) Publishing and sharing multi-dimensional image data with OMERO. Mamm Genome 26(9–10):441–447
27. Li S et al (2016) Metadata management for high content screening in OMERO. Methods 96:27–32
28. Gopal A et al (2012) Visualizing large RNA molecules in solution. RNA 18(2):284–299
29. Kner P et al (2010) High-resolution wide-field microscopy with adaptive optics for spherical aberration correction and motionless focusing. J Microsc 237(2):136–147

Chapter 11

Detection of mRNA and Associated Molecules by ISH-IEM on Frozen Sections

Catherine Rabouille

Abstract

The use of tagged RNA probes to directly hybridize frozen sections of chemically fixed tissues, followed by the tag detection with specific antibodies and gold conjugates form the core of the in situ hybridization (ISH)-immunoelectron microscopy (IEM) method that we have developed and successfully used to detect endogenous *gurken* and *bicoid* mRNAs in *Drosophila* oocytes.

Key words In situ hybridization, Electron microscopy, Frozen sections, Immunoelectron microscopy, ISH-IEM, Ultrastructure, mRNA detection

1 Introduction

Cell polarization is established by the differential localization of proteins. This can occur through their posttranslational targeting mediated by protein domains interacting with specific effectors. It can also occur through localized translation followed by the asymmetric cellular localization of their mRNAs [1]. In this regard, the best-studied examples of localized mRNAs are *gurken*, *bicoid*, and *oskar* in the developing *Drosophila* oocyte [2, 3].

Determining protein localization at the ultrastructural level at the electron microscopy (EM) level is classically achieved by immuno-EM on ~70 nm ultrathin frozen sections of chemically fixed tissues that are obtained by a process called cryo-sectioning. These preserve both the cell ultrastructure and protein antigenicity. Primary antibodies against specific antigens are then used that are detected by gold conjugates [4].

On the other hand, mRNA localization is achieved by RNA in situ hybridization (ISH) that is a method of choice for mRNA detection in cells or tissues. This uses tagged antisense RNA strands (RNA probes) that specifically hybridize to their complementary mRNAs. The tag (digoxigenin (DIG) or biotin) is then visualized

Imre Gaspar (ed.), *RNA Detection: Methods and Protocols*, Methods in Molecular Biology, vol. 1649, DOI 10.1007/978-1-4939-7213-5_11, © Springer Science+Business Media LLC 2018

with specific antibodies that are detected by enzymatic reactions (alkaline phosphatase and peroxidase). So far, this technique has been mostly developed for light microscopy methods that do not inform on the surrounding ultrastructure, such as the presence of cytoskeletal elements, organelles, or membrane-less assemblies. To allow for the ultrastructural analysis of mRNA localization at the EM level, we have developed a post embedding method combining ISH with EM, that we called ISH-IEM [5]. Of note, this is not the first EM method to be developed (*see* references in [5]), but the one presented in this article allows the simultaneous visualization of mRNAs and proteins on the same frozen section [6, 7].

Although we use frozen sections to preserve protein antigenicity, the cellular morphology is somewhat compromised and one important aspect of this ISH-IEM method is to postfix the frozen sections before their hybridization.

The fixed frozen sections are then hybridized (ISH) with DIG or biotin labeled antisense RNA probes overnight in a humid chamber. This allows an efficient hybridization (transfer and annealing of the antisense probe to the target mRNA). At least for *gurken* mRNA in *Drosophila* egg chambers, the optimal temperature is 55 °C, and addition of dextran sulfate to the hybridization buffer allows for a more efficient detection [5].

The tagged probes are then detected by immunoelectron microscopy (IEM) using anti-DIG or anti-biotin antibodies. Those are visualized by protein A conjugated to colloidal gold particles (PAG) according to the protocol developed by [4].

As protein antigenicity is retained in frozen sections, they can be detected using specific primary antibodies together with mRNAs, as performed for classical double immunolabeling [4].

We used this protocol on *gurken* mRNA and *bicoid* mRNA both localized at the (dorso)-anterior corner of the stage 9 Drosophila oocyte and successfully showed that there were localized to P-bodies (Fig. 1) [5–7]. Note that the procedure that is described below can easily be adapted to fluorescence detection on thick sections [5].

2 Materials

Many reagents are toxic and should be handled with care in the fume hood and using gloves.

2.1 Fixatives (See Notes 1 and 2)

1. 4% paraformaldehyde (PFA): prills 95 w/v % to be diluted in 0.1 M phosphate buffer (not PBS) at pH 7.4 to make a 16% stock that can be frozen. Prepare 4% solution fresh from frozen 16% stock.

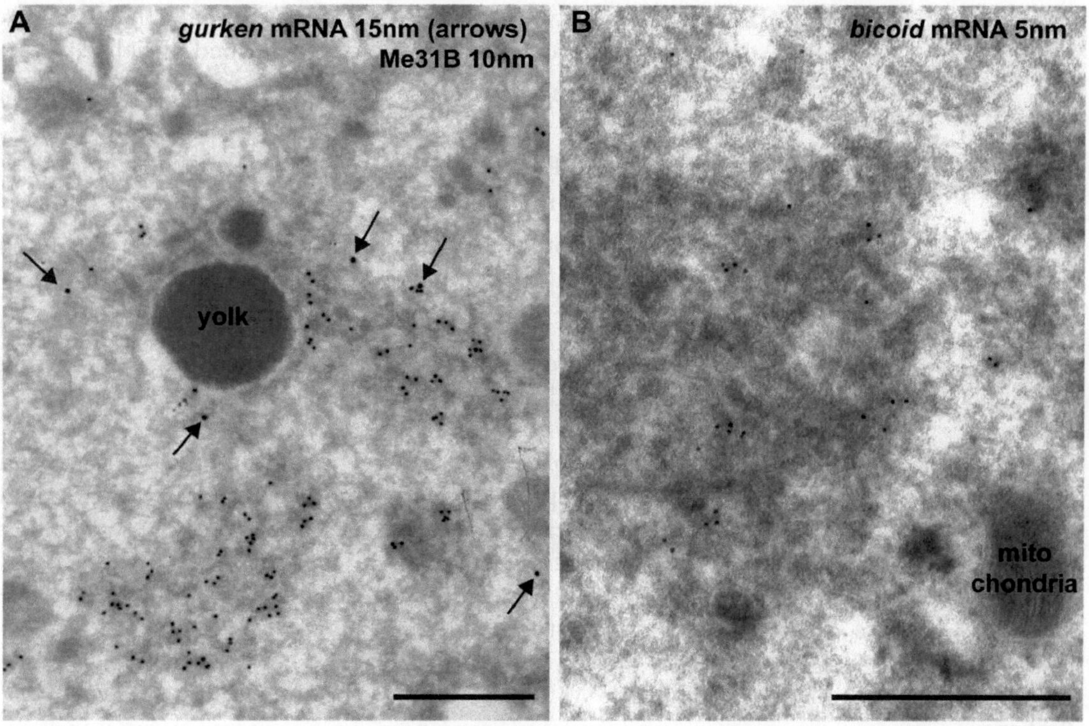

Fig. 1 ISH-IEM of localized mRNAs in Drosophila oocytes. (**a**) High magnification visualization of endogenous *gurken* mRNA in stage 9 Drosophila oocyte by ISH-IEM on ultrathin frozen sections of a 4% paraformaldehyde (PFA)-fixed stage 9 wild-type oocyte hybridized with a biotin-labeled *gurken* RNA probe followed by a rabbit anti-biotin antibody and protein A gold (PAG 15 nm) (*arrows*) and Me31B (PAG 10 nm) marking the P-bodies. Note that *gurken* mRNA is at the edge of the P-bodies as described in [7]. (**b**) High magnification visualization of endogenous *bicoid* mRNA in similar section but detected by a DIG labeled *bicoid* probe followed by sheep anti-DIG antibody, a rabbit anti-sheep antibody and PAG 5 nm. Note that *bicoid* mRNA is preferentially found in the interior of the P-bodies [7]. Scale bars: 500 nm

2. 2% PFA + 0.2% glutaraldehyde (GA) in 0.1 M phosphate buffer at pH 7.4. Prepare fresh from 16% PFA and 8% EM grade GA stocks.
3. 1% PFA for postfixation in 0.1 M PBS at pH 7.4 for sample storage. Prepare fresh from stock.
4. 1% GA for postfixation in 0.1 M PBS at pH 7.4. Prepare fresh from stock.

2.2 Tissue Preparation and Sectioning

1. 12% Gelatin: (Twee Torens). Prepare 12 w/v % gelatin as described in [4]. Briefly, resuspend 12 g gelatin in 100 mL 0.1 M phosphate buffer. Stir for 5 min at RT (~21 °C). Incubate for 6 h at 60 °C and stir every 30 min. After the gelatin is dissolved, lower the temperature to 37 °C and add 100 μL of 20 w/v % sodium azide. Filter the solution in 5 mL vials. Place in the refrigerator until use (*see* **Note 3**).
2. 2.3 M sucrose (D(+)-saccharose).

3. Methylcellulose (MC): 25 centipoises. 2 w/v% MC stock solution was prepared and stored as described in [4].
4. Methylcellulose–sucrose: Mix 2 w/v % methylcellulose with 2.3 M sucrose 1:1 ratio and stir for at least 15 min at 4 °C before use. Prepare fresh.
5. Roughened aluminum rivets.
6. Ultracryomicrotome (e.g., Leica Microsystems Ultracut FCS) to cut ultrathin frozen sections.
7. Diamond knife (e.g., Drukker International).
8. Nickel grids: hexagonal, 50 mesh (VECO-Stork) supporting a carbon-coated formvar film as described in [4] (*see* **Note 4**).

2.3 Probe Preparation

1. 1 μL of DNA plasmid (pGEM-T including promoter SP6 and T7 or Blue Script including promoter T7 and T3) can be used to generate sense and antisense RNA probe using either DIG-RNA labeling kit (e.g., Roche, cat. no. 11175025910) or Biotin-UTP (e.g., Roche, cat. no. 11388908910). The probes can be stored at −20 °C.

2.4 ISH on Frozen Sections

1. 20× SSC: 3 M NaCl, 300 mM sodium citrate.
2. Prehybridization buffer: 50% deionized formamide (from a minimum 99.5% stock), 2× SSC from a 20× stock in dH_2O, pH 6.5 treated with diethyl pyrocarbonate (DEPC). The buffer must be freshly prepared.
3. Hybridization buffer: 50% deionized formamide, 2× SSC (as above), 10 w/v % dextran sulfate (sodium salt from *Leuconostoc* spp.), 50 μg/mL heparin (sodium salt, grade I-A from porcine intestinal mucosa, and 100 μg/mL *E. coli* transfer RNA (from *Escherichia coli*Bacteria strain W) in DEPC-treated dH_2O, pH 6.5.
4. 2 μg/mL DIG or biotin labeled antisense probe diluted in hybridization buffer.
5. Hybridization chamber made of 6 and 9 cm glass petri dishes and filter paper soaked with 50% formamide in dH_2O (*see* **Note 5** and also [5]).
6. Hot plate at 55 °C.
7. Microfuge caps.

2.5 Antibodies and Detection

Antibodies and Protein-A Gold (PAG) conjugates are diluted in 1% BSA in PBS. All dilutions should be prepared fresh.

1. 0.15 w/v % glycine in PBS.
2. 1 and 0.1 w/v % BSA in PBS.

3. Rabbit anti-biotin (Rockland) to detect the biotinylated probe, diluted 1:10,000 in 1% BSA in PBS.
4. To detect the DIG-labeled probe: Sheep anti-DIG Fab fragments coupled to alkaline phosphatase (DIG-AP, 1:500, Roche), or sheep anti-DIG Fab fragments coupled to HRP (DIG-HRP, 1:1000, Roche), diluted in 1% BSA in PBS.
5. Rabbit anti-sheep IgG (1:750, Nordic) in 1% BSA in PBS.
6. Protein-A Gold (PAG) conjugates diluted in 1% BSA in PBS (*see* [4]). PAG can be purchased at the Department of Cell Biology, UMC-Utrecht by subscription. For contact and information: g.posthuma@umcutrecht.nl.

2.6 Contrasting

1. 4 w/v % uranyl acetate (UA) (SPI-CHEM) pH 4.0 prepared and stored as described in [4].
2. 2% UA-oxalate, pH 7.0. Mix 4% UA and 0.3 M oxalic acid in 1:1 ratio. Adjust pH to 7.0 with 25% NH_4OH.
3. Uranyl acetate–methyl cellulose (UA-MC) pH 4.0. Mix 4% UA with 2% MC in 1:9 ratio by stirring gently. The mixture can be stored in the dark at 4 °C for up to 3 months. This reagent is required for embedding and contrasting.
4. Wire loops with plastic handle. Loop diameter should slightly exceed the diameter of the EM grids.

2.7 Viewing

Transmission electron microscope (e.g., Jeol) to view sections. Note that any electron microscope with an 80 kV power is adequate.

3 Methods

3.1 Tissue Fixation

From anaesthetize fattened 2-day-old *Drosophila* female flies under CO_2, excise the whole ovaries in Ringer buffer using two pairs of forceps and transfer them immediately to one of the two fixatives (above) for 3–4 h at room temperature. When using 4% PFA, continue with an overnight incubation at 4 °C. After fixation, replace the fixative by 1% PFA in 0.1 M phosphate buffer at 4 °C for storage (*see* **Note 2**) [5, 8].

3.2 Embedding and Ultrathin Cryosectioning

1. Dissect the ovaries into ovarioles and staged egg chambers in gelatin and make blocks (as described in [5, 8]).
2. Incubate blocks in 2.3 M sucrose overnight at 4 °C in a rotator.
3. Freeze blocks on roughened aluminum rivets with flat head in liquid nitrogen.

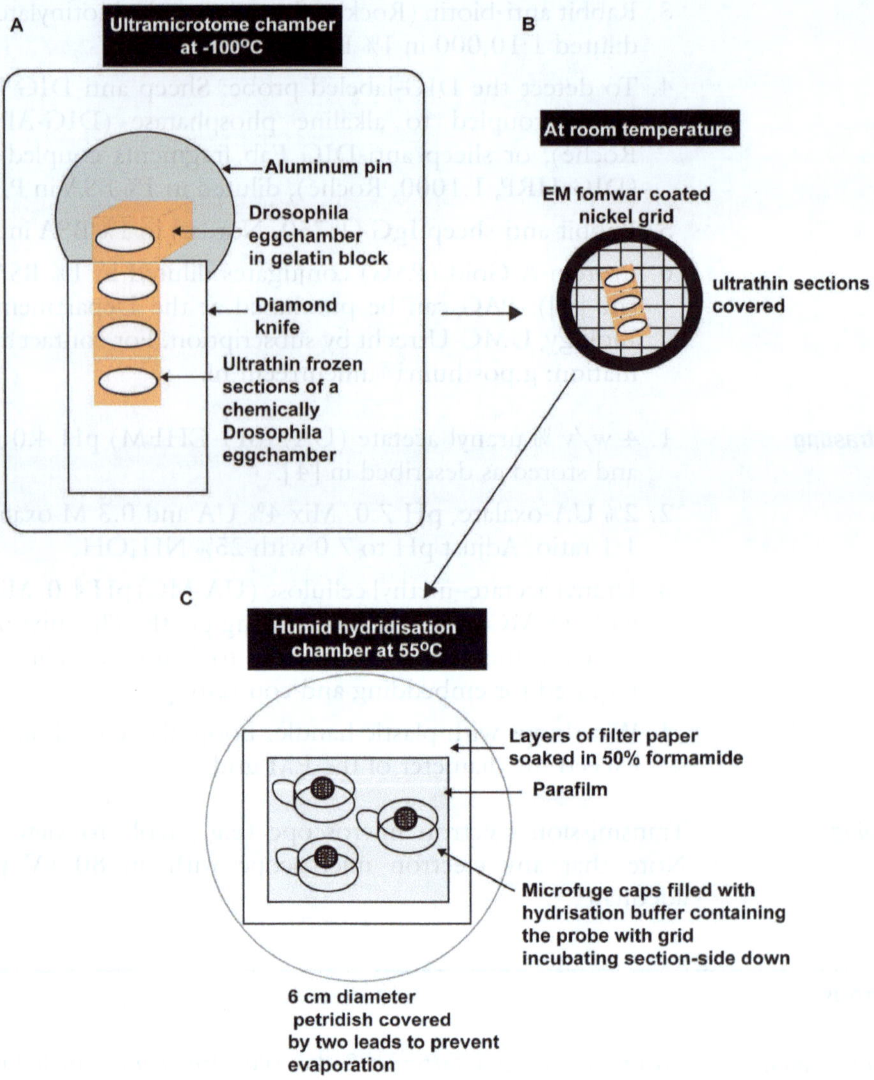

Fig. 2 Schematics representation of critical steps of the ISH-IEM method. (**a**) Ultrathin frozen sections of a Drosophila egg chamber embedded in a gelatin block are cut on a diamond knife in an ultramicrotome. (**b**) The sections are collected on a formvar coated EM nickel grid. (**c**) Grids are incubated in a humid hybridization chamber in microfuge caps

4. Cut ultrathin frozen sections (70–80 nm) at −120 °C in an ultracryomicrotome [4] and collect them with a drop of methylcellulose–sucrose on Nickel 50 mesh grids supporting a carbon-coated formvar film (Fig. 2).

3.3 ISH on Grids

1. After washing the excess of MC–sucrose by floating the grids (sections-side down) on drops of DEPC-treated PBS for 10 min at 37 °C, post-fix them in 1% GA in PBS for 5 min at RT or in 4% PFA in PBS for 15 min at 37 °C (*see* **Note 6**)

2. Prehybridization: After rinsing the fixative on three drops of PBS at 37 °C, incubate the grids (sections side down) in prehybridization buffer for 15 min at 37 °C, followed by 5 min on a hot plate at 55 °C (*see* **Note 7**).
3. Denature the probe at the final concentration of 2 μg/mL of hybridization buffer in a boiling water for 10 min followed by cooling down on ice for 2 min (*see* **Note 8**).
4. Hybridization: This is performed at at 55 °C (for *grk* mRNA, *see* **Note 7**) in a formamide-saturated humid chamber made a 6 cm petri dish covering a piece of Parafilm that is placed on top of filter papers soaked in 50% formamide.
5. Place cutoff caps of 0.5 mL microfuge tubes flat-side down on the Parafilm and fill them with 200 μL hybridization buffer containing the denatured probe.
6. Place the chamber on a hot plate at 55 °C and transfer the grids onto the probe-containing hybridization buffer in the microfuge caps. Cover the caps with 6 cm petri dish, cover again with a 9 cm petri dish (to avoid evaporation) and incubate for overnight. It is important to ensure that the chamber will be moist for 16 h. Excess filter papers and formamide can be used (*see* **Note 5**) (Fig. 2) [5].

3.4 Double Immunolabeling (IEM) of mRNA and Protein

1. After rinsing the grids in prehybridization buffer and cooling down at RT, float the grids section side on PBS, then on 0.15 w/v % glycine in PBS (glycine reduces the residual aldehyde groups from the fixative to enable antibody reactivity) and then on 1% BSA in PBS to reduce nonspecific binding of the antibodies during the next steps.
2. Detect the antisense-tagged probe using a primary antibody to the tag (DIG and biotin) and PAG conjugates as described [4, 8].
3. Incubate the grid with anti-biotin or anti-DIG antibody:
   (a) For biotinylated probes: Incubate grid for 1 h on a 5 μL droplet of polyclonal rabbit anti-biotin antibody diluted in 1% BSA in PBS.
   (b) For DIG labeled probes: Incubate grid for 20 min on a 5 μL droplet of sheep anti-DIG antibody diluted in 1% BSA in PBS. Then rinse in five drops of 1% BSA in PBS for 10 min. Incubate grid for 20 min on a 5 μL droplet of rabbit anti-sheep antibody diluted in 1% BSA in PBS.
4. Rinse in five drops of 1% BSA in PBS for 10 min.
5. Incubate grid for 20 min on a 5 μL droplet of PAG (5, 10, 15, or 20 nm size) diluted in 1% BSA in PBS (*see* **Note 9**).
6. Rinse in two drops of 0.1% BSA in PBS and then in PBS.
7. Stabilize the reaction by a 5 min incubation on 1% GA in PBS.

8. For protein detection after mRNA, proceed with double or triple labeling of the sections, by repeating **steps 1–4** using appropriate primary antibodies and PAG of different sizes [4]. Rabbit polyclonal antibodies are directly detected by PAG (**step 5**). Mouse and sheep monoclonal antibodies are decorated by a rabbit anti-mouse and anti-sheep IgGs, respectively, followed by PAG (**step 3**a).

3.5 Contrasting

1. Wash the grids in ten drops of dH_2O for 10 min.
2. Contrast the sections by floating the grids for 5–10 min on 2% UA oxalate.
3. Pass the grids over one or two drops of UA-MC [6] and leave them in the third drop for 5–10 min on ice.
4. Use a wire loop to scoop the grid from the UA-MC, remove excess UA-MC by dragging the loop with grid over a filter paper and let the grid dry in air for at least 10 min.
5. Pinch out the grid from the wire loop and store the grid in a grid box.
6. Visualize the results by using a transmission electron microscope at 80 kV.

4 Notes

1. Most of the reagents involved in the procedure are toxic. They need to be manipulated in the fume hood and with gloves.
2. For easier comparison, the two ovaries from the same female can be fixed differently. Fixation is the key step in this protocol. The resulting morphology and labeling depend on the fixative used. Furthermore, fixation should be performed at RT, not at 4 °C and the fixative diluted in phosphate buffer pH 7.4, not PBS. The ovaries have to be fixed immediately upon excision and should not be stored in the Ringer buffer. Other tissues can also be used but their fixation might need to be done by perfusion (not only by immersion) to keep an optimal ultrastructure, for instance for rat liver or brain [5]
3. Most commercial gelatins give a precipitate (probably calcium phosphate) when prepared in phosphate buffer, but they can be prepared in different buffers (e.g., Tris–HCl or Tris-buffered saline). The concentration should be adjusted to the desired stiffness of the resulting pellets [4].
4. Nickel grids: 50 mesh grids are optimal for viewing sections of large cells such as the Drosophila oocytes. For small cells, smaller meshes are adequate and provide a stronger support. However, only nickel grids are suitable for the ISH procedure

as copper dissolves in the hybridization buffer. Sections might fold or be lost from the grids during the procedure and this can be fixed by preparing many grids for each labeling [5]

5. Make sure the humid chamber remains moist for the entire procedure. This is critical. Grids should not dry out. If necessary, add filter papers to the chamber and an excess of 50% formamide [5]
6. Hybridization might lead to a loss of morphology because of harsh extraction during formamide incubation and to a low immune-reactivity. This can be solved by cutting thicker sections, and post-fixing them harder (with GA or acrolein), but be aware that it might, for some mRNAs, lead to near or complete loss of detection). The choice of fixatives depends on the antigenicity of the protein you want to visualize with the mRNA.
7. The melting temperature (T_m, at which 50% of the RNA molecules are single-stranded) increases linearly with the %G and the %C present in the RNA.

   The T_m is also dependent on the molar content of protons in the solution and the formamide percentage in the buffer. This is summarized in the following equation:

   $T_m = 81.5\ °C + 16.6 \times \log[Na+] + 0.41 \times (\%GC) - 0.63 \times (\%\ formamide)$ [9, 10]. The annealing temperature of nucleotides is considered to be 5 °C lower than the T_m.
8. The specificity of the antisense probes needs to be checked using different controls. The best are hybridizations with the sense probe or without probe, and preincubation with RNaseA (10 μg/mL – 1 in PBS at 37 °C for 15 min) before the hybridization procedure.

   In case of nonspecific labeling and/or high background, you can assess which steps create it by omitting the primary or secondary antibodies (if used) and skip the double labeling. Also use sections from other species where the mRNA is not expressed.
9. Of note, the anti-DIG antibody used in this protocol is raised in sheep, and cannot be directly recognized by PAG and necessitates the need of a bridging rabbit anti sheep antibody that can lead to the clustering of the PAG. This could be an advantage for less abundant transcripts.

Acknowledgments

I thank Dr. Bram Herpers (Leiden, NL) in collaboration with Ilan Davis and his group (Oxford, UK), and Tim Weil (Cambridge, UK) for developing and using this technique.

References

1. Lawrence JB, Singer RH (1986) Intracellular localization of messenger RNAs for cytoskeletal proteins. Cell 45(3):407–415
2. Grunert S, St Johnston D (1996) RNA localization and the development of asymmetry during Drosophila oogenesis. Curr Opin Genet Dev 6(4):395–402
3. Riechmann V, Ephrussi A (2001) Axis formation during Drosophila oogenesis. Curr Opin Genet Dev 11(4):374–383
4. Slot JW, Geuze HJ (2007) Cryosectioning and immunolabeling. Nat Protoc 2 (10):2480–2491. doi:10.1038/nprot.2007.365
5. Herpers B, Xanthakis D, Rabouille C (2010) ISH-IEM: a sensitive method to detect endogenous mRNAs at the ultrastructural level. Nat Protoc 5(4):678–687. doi:10.1038/nprot.2010.12
6. Delanoue R, Herpers B, Soetaert J, Davis I, Rabouille C (2007) Drosophila Squid/hnRNP helps Dynein switch from a gurken mRNA transport motor to an ultrastructural static anchor in sponge bodies. Dev Cell 13 (4):523–538. doi:10.1016/j.devcel.2007.08.022
7. Weil TT, Parton RM, Herpers B, Soetaert J, Veenendaal T, Xanthakis D, Dobbie IM, Halstead JM, Hayashi R, Rabouille C, Davis I (2012) Drosophila patterning is established by differential association of mRNAs with P bodies. Nat Cell Biol 14(12):1305–1313. doi:10.1038/ncb2627
8. Herpers B, Rabouille C (2004) mRNA localization and ER-based protein sorting mechanisms dictate the use of transitional endoplasmic reticulum-golgi units involved in gurken transport in Drosophila oocytes. Mol Biol Cell 15 (12):5306–5317. doi:10.1091/mbc.E04-05-0398
9. Howley PM, Israel MA, Law MF, Martin MA (1979) A rapid method for detecting and mapping homology between heterologous DNAs. Evaluation of polyomavirus genomes. J Biol Chem 254(11):4876–4883
10. Le Guellec D (1998) Ultrastructural in situ hybridization: a review of technical aspects. Biol Cell 90(4):297–306

Chapter 12

Hybridization Chain Reaction for Direct mRNA Detection Without Nucleic Acid Purification

Yao Xu and Zhi Zheng

Abstract

Hybridization chain reaction (HCR) provides a feasible solution for nucleic acid detection without target amplification. By highly specific sandwich hybridization, target RNA can be directly captured onto solid support and detected using HCR with fluorescent dyes. Here, we describe a novel method for malaria RNA detection based on sandwich hybridization and two-dimensional HCR, without involving nucleic acid purification or any enzymatic reaction, using ordinary oligonucleotides without labeling or modification.

Key words Hybridization chain reaction, mRNA detection, Sandwich hybridization

1 Introduction

Hybridization chain reaction is a new class of enzyme-free fluorescent signal amplification method for nucleic acid detection. In this procedure, the target DNA/RNA initiates a hybridization cascade between two hairpin sequences through toehold mediated strand displacement to realize signal amplification [1, 2]. The two species of hairpin monomers of the HCR are metastable and can coexist in the solution until triggered by the target DNA to form a nicked DNA double helix analogous to alternating copolymers (Fig. 1), which can be seen in agarose gel electrophoresis (Fig. 2). Although HCR-based signal amplifications have been described for nucleic acid biosensing [3, 4], there are several drawbacks significantly impeding the practical application of these methods: (1) as a highly sensitive signal amplification method, HCR inevitably amplifies backgrounds, decreasing specificity. These backgrounds could be produced by physical nonspecific adsorption of hairpin sets, or caused by leakiness in toehold-mediated HCR (i.e., hairpin polymerization in the absence of initiating target DNA), especially for real sample detection when the components are complicated. (2) The sensitivity of toehold-mediated HCR amplification is not

Imre Gaspar (ed.), *RNA Detection: Methods and Protocols*, Methods in Molecular Biology, vol. 1649, DOI 10.1007/978-1-4939-7213-5_12, © Springer Science+Business Media LLC 2018

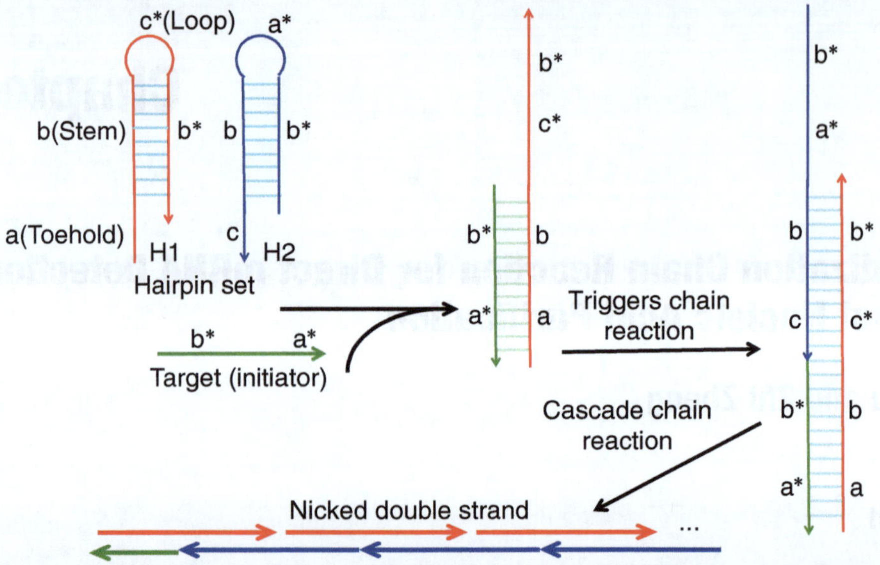

Fig. 1 Linear hybridization chain reaction in solution. All hairpin probes have a structure of toehold-stem-loop. *Letters marked with asterisk* are complementary to the corresponding *unmarked letter*. Target DNA hybridizes with hairpin in H1 via base pairing to single-stranded toehold "a", mediating a branch migration that opens the stem of the hairpin, exposing a "c*-b*" sticky end. This complex further opens hairpin H2 by base pairing to "c*" to form a new complex containing a "b*-a*" sticky end, which is the same to the target. Thus, a cascade chain reaction is generated to form a nicked double strand polymer. Figure adapted from [5] with permission from Elsevier

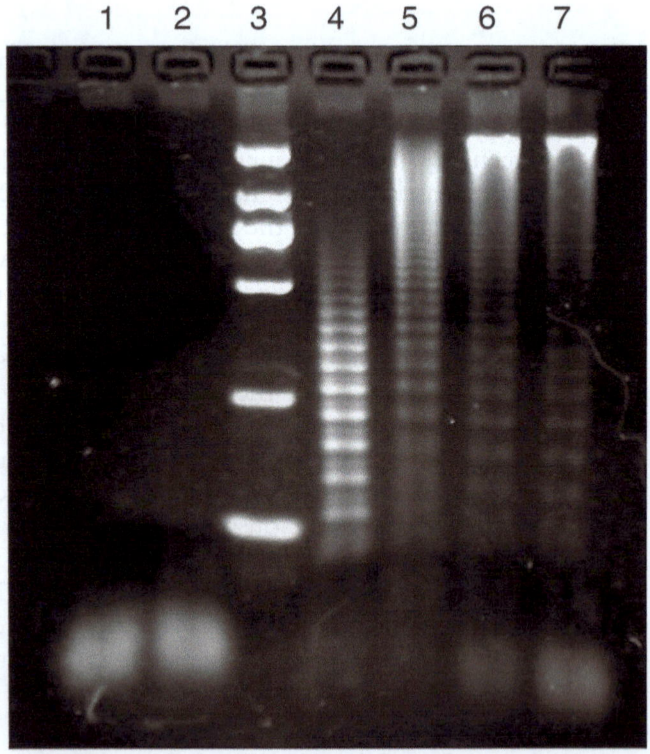

Fig. 2 Effect of initiator concentration on HCR amplification. *Lanes 1* and *2*: 1 μM A1 and 1 μM A2, respectively. *Lane 3*: DL2000 DNA marker. *Lanes 4–7*: four different concentrations of initiator (100, 50, 10, and 0 nM) in a 1 μM mixture of A1 and A2. Figure adapted from [5] with permission from Elsevier

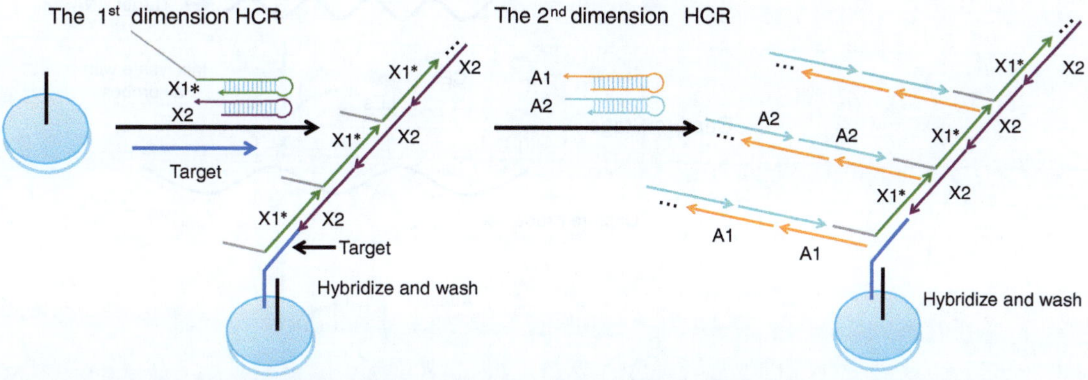

Fig. 3 Two-dimensional HCR on solid surface (*see* **Note 10**). In the 2D HCR, The target probe hybridizes on its 3′-half with the capture probe conjugated on the solid surface; the 5′-half of the target then opens the hairpin probe X1*, triggering the cascade chain reaction of the hairpin set X* to form a nicked double helix with extra single-strand hangout branches. After washing, hairpin set A is added; the overhang branches from X1* opened A1 and initiated the HCR of the hairpin set A, generating a 2D HCR product. SYBR Green I is employed in the detection phase to generate fluorescent signals. We also detected the fluorescent signal generated by hairpin sets with the same sequences but labeled with fluorescein isothiocyanate (FITC). It was observed that unlabeled hairpin probes generated higher signal than labeled probes [5]. Figure adapted from [5] with permission from Elsevier

sufficient for clinical applications. Linear, one-dimensional chain reactions are adopted in most current HCR amplification methods. Although branched HCR or other higher dimensional amplification may presumably give much higher sensitivity, the challenges of steric hindrance and the difficulties to form branched oligonucleotides in solution still make it hard to realize.

Here, we demonstrate an improved HCR-based RNA detection method (Figs. 3 and 4) [5]. By adopting a sandwich RNA capturing assay, the nucleic acid extraction procedure is bypassed, allowing high-throughput, ELISA-like sample processing in 96-well plates. Simultaneously, a novel, onsite two-dimensional branched HCR assembly is made possible by detecting target captured on solid support. The sensitivity and the specificity of RNA detection were improved compared with the traditional linear HCR.

2 Materials

All oligonucleotides are purchased commercially and are PAGE-purified. Ultrapure water and analytical grade reagents are used in all runs. 5× SSC buffer (750 mM NaCl, 75 mM sodium citrate, pH 7.4) is used for all hybridization reactions unless indicated specifically.

2.1 Agarose Gel Electrophoresis

1. Standard equipment of running agarose gels (electrophoresis tank, power supply, gel tray, and combs).

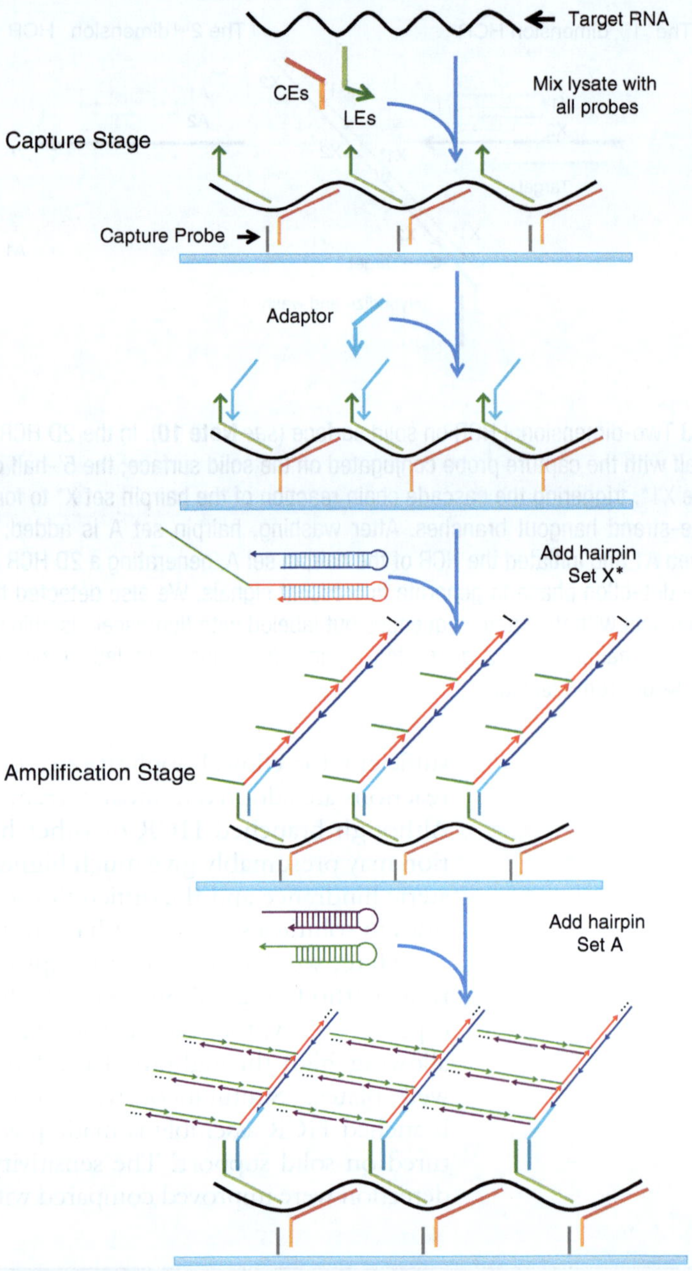

Fig. 4 Schematic illustration of the direct RNA detection assay. For each target RNA, the capture hybridization method is done with a series of oligonucleotide probes (CEs and LEs), each of these probes containing a target-specific sequence which is complementary to a different region of the target, and an additional "tail" sequence that is independent of the target sequence but can interact with either the solid support (CEs) or the adaptor (LEs). The CEs and LEs sandwich the target RNA through target-specific hybridizations. The common CE tail sequences hybridize to the capture probe conjugate on the surface of each well in 96-well plate and capture the associated target RNA onto the plate. The adaptors bind to the common LE tails and initiate the on-site 2D HCR after the addition of hairpins. Figure adapted from [5] with permission from Elsevier

2. UV transilluminator.
3. Resolving gel buffer and running buffer (1× TBE): 90 mM Tris, 89 mM boric acid, and 2.0 mM EDTA, pH 8.0.

   Add about 100 mL water to a 1 L graduated cylinder. Weigh 10.89 g Tris, 5.50 g boric acid, and 0.75 g EDTA, and then transfer to the cylinder. Add water to a volume of 900 mL. Mix and adjust pH with HCl. Make up to 1 L with water. Store at room temperature.
4. 1% Agarose with 1 ng/mL ethidium bromide (*see* **Note 1**).
5. DL2000 DNA marker: Store at 4 °C.
6. Loading buffer.

2.2 RNA Detection

1. Tabletop centrifuge with plate adapter.
2. Heat incubator.
3. Fluorescence plate reader.
4. Tinfoil sealing film.
5. Dissolving buffer: TE.
6. Stock hybridization buffer (20× SSC): 3 M NaCl, 0.3 M sodium citrate, store at 4 °C.
7. Hybridization buffer: 5× SSC (diluted from 20× SSC), store at 4 °C.
8. Wash buffer: 0.1× SSC containing 0.3 g/L lithium dodecyl sulfate, store at 4 °C.
9. Capture 96-well plate functionalized with a 22-base DNA sequence (capture probe, Fig. 4), (Diacurate), store at 4 °C (*see* **Note 2**).
10. Blood sample with *P. falciparum*, store at −20 °C (*see* **Note 3**).
11. Lysis mixture (Diacurate), store at room temperature.
12. 50 mg/mL Proteinase K, store at −20 °C (*see* **Note 3**).
13. Hairpin sets: DNA sequences designed independently using NUPACK [5] (http://www.nupack.org).
14. 100 μM oligonucleotide probes in TE: include CEs and LEs, designed by Diacurate to specifically capture malaria RNA, store at −20 °C. Dilute all probes to 10 μM with hybridization buffer before use (*see* **Note 4**).
    - CE (Capture Extender).

      CE probe contains two regions: region I on 3′-end has the complementary sequence with capture probe and region II on 5′-end can hybridize with target RNA. A "TTTTT" sequence is inserted between the two regions.
    - LE (Label Extender).

LE probe contains two regions: region I on 3′-end hybridizes with adaptor probe and region II on 5′-end has complementary sequence with target RNA. A "TTTTT" sequence is inserted between the two regions.

- A1 5′*ACA CA* GAA AGG GAA ACG ***AGT CC*** CGT TTC CCT TTC.
- A2 5′CGT TTC CCT TTC ***TGT GT*** GAA AGG GAA ACG *GGA CT*.
- X1* 5′CGT TTC CCT TTC *TGT GT* TTTTT *ATA TCC* CTC GCC GAA TCC TAG ACT ***CAA AGT*** AGT CTA GGA TTC GGC GAG.

  X1*′ consists of two regions: region I on 3′-end has the same sequence with probe X1 and region II on 5′-end is a hangout branch that can initiate the polymerization of hairpin set A. A "TTTTT" sequence is inserted into the two regions.
- X2 5′ AGT CTA GGA TTC GGC GAG ***GGA TAT*** CTC GCC GAA TCC TAG ACT *ACT TTG*.
- Italic sequence indicates the toehold region, bold typeface indicates loop sequences, and underline indicates stem sequences.

15. Target oligo: 5′ AGT CTA GGA TTC GGC GAG ***GGA TAT*** TTTTT CTC TTG GAA AGA AAG TG.

    The target oligo can hybridize on its 3′-half with the capture probe on a solid support; the 5′-half of "Target oligo" can initiate the polymerization of hairpin set X*. A "TTTTT" sequence is inserted between the two regions.
16. Adaptor: AGT CTA GGA TTC GGC GAG *GGA TAT* TTTTT ATG CTT TGA CTC AGA AAA CGG TAA CTT C.
17. 4× SYBR Green I: dilute with 4× SSC before use, store at 4 °C (*see* **Note 5**).

3 Methods

Carry out all procedures at room temperature unless otherwise specified.

3.1 Preparation

1. All hairpin probes are heated to 95 °C for 2 min and then allowed to cool to room temperature for 1 h before use.

3.2 Linear HCR by 1.5% Agarose Gel Electrophoresis

1. Mix 7 μL of water, 1 μL of probe A1, and 1 μL of probe A2, 1 μL of target with different concentrations and 1 μL of water for control, incubate at room temperature for 4 h.

2. Add loading buffer to each of the reactions, mix thoroughly and centrifuge for several seconds.
3. Carefully load DL2000 DNA marker and samples into different lanes of the gel.
4. Run agarose gels at 150 V for 40 min, until the sample line reaches approximately 75–80% of the way down the gel.
5. Visualize the gels at UV light (Fig. 2).

3.3 Two-Dimensional HCR for an Oligo Target on Solid Surface (See Fig. 3)

1. Dilute target probe (Target oligo) to 10, 1, 0.1, 0.01, and 0.001 nM respectively in hybridization buffer (*see* **Note 6**).
2. Add 100 μL of above solutions to individual capture wells.
3. Seal the wells with tinfoil sealing film and incubate at 46 °C for 1 h (*see* **Note 7**).
4. Tear the film, then quickly decant the wells and wash with 300 μL of wash buffer for three times, with quick decanting in between (*see* **Note 8**).
5. Centrifuge the plate upside-down for 1 min at 600 × *g* after the final wash and decant.
6. Mix 120 μL of hybridization buffer, 15 μL of probe X1*, and 15 μL of probe X2 (*see* **Notes 9** and **10**).
7. Add 100 μL of above mixture to the capture plate, seal and incubate at room temperature for 2 h.
8. Repeat wash steps (**steps 4** and **5**).
9. Mix 120 μL of hybridization buffer, 15 μL of probe A1, and 15 μL of probe A2.
10. Add 100 μL of above mixture to the capture plate, seal and incubate at room temperature for 2 h.
11. Repeat wash steps (**steps 4** and **5**).
12. Add 150 μL of 4× SYBR Green I in 4× SSC buffer to the reaction wells, and then incubate at room temperature for 15 min in the dark place.
13. Quantify the resulting fluorescence with a plate reader.

3.4 Direct RNA Detection in Blood Sample (See Fig. 4, Note 11)

1. Thaw the blood sample with *P. falciparum* at 4 °C before use (*see* **Note 3**).
2. Lyse 10 μL of thawed blood sample with 50 μL of lysis mixture, 85 μL of water, and 2 μL of 50 mg/mL proteinase K at 60 °C for 1 h with vigorous shaking.
3. Mix the lysate with 1.5 μL of respective CEs and LEs probes (*see* **Note 12**).
4. Add 100 μL of the above mixture to the capture plate, then seal the wells with tin foil and incubate at 58 °C overnight without shaking (*see* **Note 13**).

5. Quickly decant the wells and wash with 300 μL of wash buffer for three times, with quick decanting in between.
6. Centrifuge the plate upside-down for 1 min at 600 × *g* after the final wash and decant.
7. Add 100 μL of 1 μM adaptor probe in hybridization buffer to the wells.
8. Seal the wells with tinfoil sealing film and incubate at 46 °C for 1 h.
9. Tear the film, then quickly decant the wells and wash with 300 μL of wash buffer for three times, with quick decanting in between.
10. After the final wash, centrifuge the plate upside-down for 1 min at 600 × *g* to remove residual liquid.
11. Mix 120 μL of hybridization buffer, 15 μL of probe X1*, and 15 μL of probe X2 (the final concentration of each hairpin probe is 1 μM).
12. Add 100 μL of the above mixture to the capture plate and seal the wells, incubate at room temperature for 2 h.
13. Repeat wash steps (**steps 9** and **10**).
14. Mix 120 μL of hybridization buffer, 15 μL of probe A1, and 15 μL of probe A2.
15. Add 100 μL of above mixture to the capture plate, seal and incubate at room temperature for 2 h.
16. Repeat wash steps (**steps 9** and **10**).
17. Add 150 μL of 4× SYBR Green I in 4× SSC buffer to the reaction wells, and then incubate at room temperature for 15 min in the dark place.
18. Quantify the resulting fluorescence with a plate reader.

4 Notes

1. Ethidium bromide should be kept away from light in room temperature. As EtBr is a known mutagen, be careful when operate with this chemical and during the process of agarose gel electrophoresis.
2. Take the plate out from refrigerator and warm to room temperature before use. The capture plate is sealed with plastic film, so tear the film of specific wells where you want to add your sample to.
3. Blood sample and Proteinase K should be thawed at 4 °C and operated on ice throughout the procedure.

4. Centrifuge the lyophilized oligonucleotides at 600 × *g* for 2 min and carefully add TE to the tube after centrifuging to dissolve the probes to 100 μM with vigorous shaking.
5. SYBR Green I must be protected from light and stored at 4 °C. The concentration of SYBR Green I stock solution is 10,000× and the optimal final concentration is 4×. Dilute directly from stock solution before use.
6. Target probe (Target oligo) consists of two regions that linked with a 'TTTTT' structure. The first region located on the 3′-end can hybridize with capture probe in the 96-well plate, and the second region on the 5′-end triggers the cascade chain reaction of two hairpin sets.
7. Make sure the wells are completely sealed so the solution in the well won't evaporate during incubation.
8. Unlike previously reported HCR systems in solution, our assay is an on-site amplification technology. Leaky hairpins will be washed off without contributing to the background. Thus, we can maximize polymerization without being overly concerned about leakiness by using shorter hairpins, improving the sensitivity and making hairpin design easier.
9. Mix the hairpin probes just before use. Probe X1* and probe X2 are paired with final concentration of 1 μM. Cascade chain reaction will be triggered in the presence of target probe.
10. Hairpin probe X1* has an additional hangout sequence at the 5′-end that does not form double strand structure after polymerization, but instead, act as the initiating probe for the second step of polymerization.
11. The direct RNA detection assay is comprised of two parts: the first part is a specific RNA capture, and the second part is signal amplification through 2-dimensional hybridization chain reactions. Multiple LE probes per target could be a first step to signal amplification by providing many initial probes for the subsequent detection. Adaptor probes that simultaneously hybridize to the tails of LEs mediate the capture and amplification systems. The introduction of adaptor probe enables independent designing of DNA probes in capture stage and detection stage, making the assay more flexible. As a result, the method can be used to detect any suitable RNA or DNA targets by designing new target-specific CEs and LEs probes, while keeping the same hairpin sets used in this study. Also, a sequence of "TTTTT" is inserted into the two binding regions of adaptor probe, resulting in better signal amplification than that without "TTTTT".
12. Multiple oligonucleotide probes capture the RNA on 96-well plate in the capture stage. Based on our previous study, this

design offered cooperative hybridization between the capture probes on the solid surface and the multiple CEs binding to one target, presumably leading to a stronger capture than single CE–capture probe interaction [5], which makes sure that only the specific target can be captured to the solid surface and further trigger downstream cascade reactions.

13. As the hybridization is taken place at a temperature (58 °C) that is higher than the annealing temperature of single CE or LE probe, single probes or hairpins will be washed off and cannot be stably immobilized on the solid surface to cause nonspecific absorption. Leaky hairpin polymerization products without binding to the target will be washed off as well without causing background.

References

1. Dirks RM, Pierce NA (2004) Triggered amplification by hybridization chain reaction. Proc Natl Acad Sci U S A 101(43):15275–15278. doi:10.1073/pnas.0407024101
2. Choi HM, Beck VA, Pierce NA (2014) Next-generation in situ hybridization chain reaction: higher gain, lower cost, greater durability. ACS Nano 8(5):4284–4294. doi:10.1021/nn405717p
3. Dong J, Cui X, Deng Y, Tang Z (2012) Amplified detection of nucleic acid by G-quadruplex based hybridization chain reaction. Biosens Bioelectron 38(1):258–263. doi:10.1016/j.bios.2012.05.042
4. Spiga FM, Bonyar A, Ring B, Onofri M, Vinelli A, Santha H, Guiducci C, Zuccheri G (2014) Hybridization chain reaction performed on a metal surface as a means of signal amplification in SPR and electrochemical biosensors. Biosens Bioelectron 54:102–108. doi:10.1016/j.bios.2013.10.036
5. Xu Y, Zheng Z (2016) Direct RNA detection without nucleic acid purification and PCR: Combining sandwich hybridization with signal amplification based on branched hybridization chain reaction. Biosens Bioelectron 79:593–599. doi:10.1016/j.bios.2015.12.057

Chapter 13

In Situ Detection of MicroRNA Expression with RNAscope Probes

Viravuth P. Yin

Abstract

Elucidating the spatial resolution of gene transcripts provides important insight into potential gene function. MicroRNAs are short, singled-stranded noncoding RNAs that control gene expression through base-pair complementarity with target mRNAs in the 3′ untranslated region (UTR) and inhibiting protein expression. However, given their small size of ~22- to 24-nt and low expression levels, standard in situ hybridization detection methods are not amendable for microRNA spatial resolution. Here, I describe a technique that employs RNAscope probe design and propriety amplification technology that provides simultaneous single molecule detection of individual microRNA and its target gene. This method allows for rapid and sensitive detection of noncoding RNA transcripts in frozen tissue sections.

Key words MicroRNAs, In situ hybridization, RNA detection, Immunohistochemistry, RNAscope

1 Introduction

In situ nucleic acid hybridization (ISH) is an essential technique in biology to define the cellular distribution of nucleic acids within frozen tissues. Typically, a labeled segment of complementary DNA or RNA nucleic acid strand (probe) is used to localize a nucleic acid sequence of interest. First described in the 1950s by Gall and colleagues in studies of *Drosophila* salivary glands and *Xenopus* oocytes, ISH has become an important first step to defining gene function [1–3].

With advances in technology and popularity in transcriptome sequencing, our understanding of RNA biology has greatly expanded, and now includes both coding and noncoding classification. One subclass of noncoding RNAs are small, single-stranded regulatory RNAs termed microRNAs. Two different enzyme complexes are responsible for processing microRNA transcripts into their functional single-stranded unit of ~22-nt. In the nucleus, an RNAse III-like enzyme, Drosha, cleaves microRNA transcripts to produce a ~65-nt stem-loop structure. Once transported into the

Imre Gaspar (ed.), *RNA Detection: Methods and Protocols*, Methods in Molecular Biology, vol. 1649, DOI 10.1007/978-1-4939-7213-5_13, © Springer Science+Business Media LLC 2018

cytoplasm, these precursor microRNAs are cleaved by the essential enzyme Dicer, into single-stranded functional units [4].

MicroRNAs were originally identified from *C. elegans* genetic screens for defects in developmental transitions [5, 6]. Within the last 7–10 years, there has been an explosion of studies documenting the expanded roles of miRNAs, which now include organogenesis, stem cell maintenance and brain and cardiac morphogenesis and tissue regeneration [7–12]. MicroRNAs regulate gene expression at the posttranscriptional level, by binding to the 3′ UTR of target mRNAs and inhibiting protein translation. An individual microRNA is capable of regulating a large set of target genes, ranging from 10 to 500 target mRNAs [13, 14]. Thus, to understand microRNA biology, one must ideally detect both microRNA and target gene.

Detection of microRNAs is confounded by both its small size and low expression levels. The use of locked nucleic acid probes targeting the mature sequence has been successfully used to detect highly abundant microRNAs [15]. However, often these protocols are incompatible with simultaneous detection of a potential target gene. Advanced Cell Diagnostics (ACD) has recently pioneered probe design and amplification strategies that maximize signal intensity while suppressing background noise (Fig. 1, https://acdbio.com/). Here, I describe this approach on adult zebrafish heart cryosections detecting microRNA-101a and *fosab* mRNA.

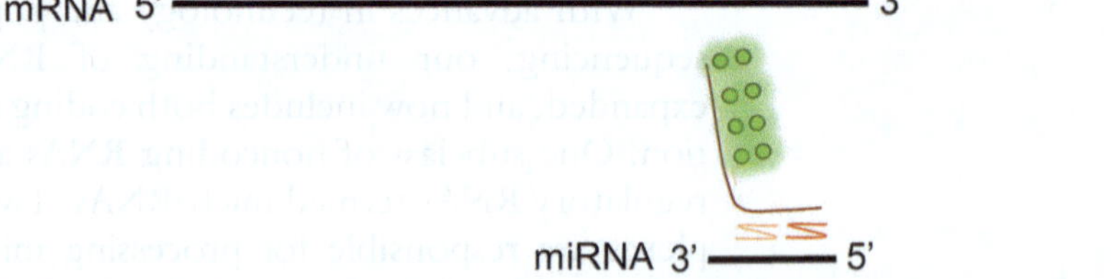

Fig. 1 Experimental paradigm of a multiplex RNAscope amplification of ISH signal. Sections are hybridized with multiple probes designed for two different target genes. Through a series of amplification chemistry with propriety technology, each probe signal is amplified. Addition of chromogenic substrates enables detection of RNAs

2 Materials and Solutions

All solutions are prepared using reverse osmosis water and filter-sterilized prior to use. Aliquots are kept frozen at −20 °C and working solutions are stored at 4 °C.

2.1 Reagents and Buffers

1. 100% ethanol.
2. 3% hydrogen peroxide.
3. 100% CitriSolv (Fisher Scientific): CitriSolv is a d-limonene-based solvent and clearing agent used for histology. It dries slowly without residue and will not harden or shrink tissue. It is a safer alternative to xylene.
4. 1× PBS: To make 1 L of PBS, dissolve 8.0 g NaCl, 0.2 g of KCl, 1.44 g of $Na_2HPO_4{\cdot}2H_2O$, and 0.2 g of KH_2PO_4 in 800 mL water. Adjust pH to 7.4 with 10 N NaOH. Add water to a final volume of 1 L.
5. RNAScope kit with AMP1-6, AP-Fast Red, and HRP-Green buffers for chromogenic detection of the targets (ACDbio).
6. Custom probes against the target RNA species (ACDbio, *see* **Note 1**).
7. Tissue fixation buffer: 4 w/v % PFA, 4 w/v % sucrose, 0.12 mM $CaCl_2$, 0.08 M Na_2HPO_4 and 0.02 M NaH_2PO_4 pH 7.4. Prepare 100 mL stock solutions of 1 M $CaCl_2$, 0.2 M Na_2HPO_4, and 0.2 M NaH_2PO_4, with the latter two solutions at pH 7.4. Combine 8 g sucrose, 24 μL 1 M $CaCl_2$, 77 mL 0.2 M Na_2HPO_4, and 23 mL 0.2 M NaH_2PO_4 into ~180 mL of water. Slowly heat and stir the solution in a hood. Gradually add 8 g PFA until completely dissolved. Adjust the volume to 200 mL with water and filter-sterilize with a 0.45 μm bottle top filter unit. Store aliquots of 15 mL at −20 °C indefinitely. Working solution is stored at 4 °C and is good for 1 week.
8. Sucrose equilibration buffer: 30 w/v % sucrose solution in water. Weigh 150 g sucrose and dissolve to a final volume of 500 mL of water. Gentle heating and stirring aids in dissolving sucrose. Filter-sterilize with a 0.45 μm bottle top filter unit. Store aliquots in 50 mL and freeze at −20 °C.
9. Antigen retrieval buffer: To unmask epitopes made unavailable by fixation, make a solution composed of 10 mM citric acid, 0.05% Tween 20. Add 1.92 g anhydrous citric acid in a 1 L beaker and dissolve it in water. Adjust the pH to 6.0 with 10 N NaOH. Add 0.5 mL 100% Tween 20 and adjust volume to 1 L with water. Store at room temperature for 3 months. Proper pH is critical for efficient antigen retrieval.
10. Tissue Freezing Medium (TFM) (Polysciences): A support matrix for embedding tissues for cryosectioning that freezes

tissue quickly without promoting crystallization even at the most extreme temperatures.

11. Wash buffer (1× PBT): To make 1 L of PBT, add 1 mL of Triton-X-100 to 1 L of PBS (*see* **Note 2**). Store at room temperature indefinitely.
12. Red detection reagent: For the detection of alkaline phosphatase (AP), solutions AP-Fast Red B and A of the RNAScope kit must be combined in a 1:60 ratio in a separate centrifuge tube to create a working solution (*see* **Note 2**).
13. Green detection reagent: To detect horseradish peroxidase (HRP, make a 1:50 ratio mixture of HRP-Green-B to HRP-Green-A of the RNAScope kit in a separate centrifuge tube (*see* **Note 3**).
14. EcoMount mounting medium (Biocare).

2.2 Equipment

1. Cryostat equipped with a microtome blade (e.g., Leica 1860UV cryostat).
2. Plastic embedding molds.
3. Iridectomy scissors.
4. Dumont #5 fine-tip forceps.
5. 1.5 mL Eppendorf tubes.
6. Superfrost/Plus microscope slides (e.g., Fisherbrand).
7. Drierite desiccant.
8. Coplin jars.
9. 500 mL beaker.
10. Aluminum foil.
11. Thermometer.
12. Heating plate.
13. Hydrophobic pen.
14. Paper towels.
15. Modified Tupperware (*see* Fig. 2 and **Note 4**).
16. Serological pipettes.
17. Hybridization incubator.

3 Methods

3.1 Tissue Preparation

The zebrafish heart is a two-chambered organ with one atrium and ventricle (Fig. 3). For orientation purposes during histology, it is important to extract the heart with the chambers and the outflow tract together. All methods are carried out at room temperature unless otherwise specified.

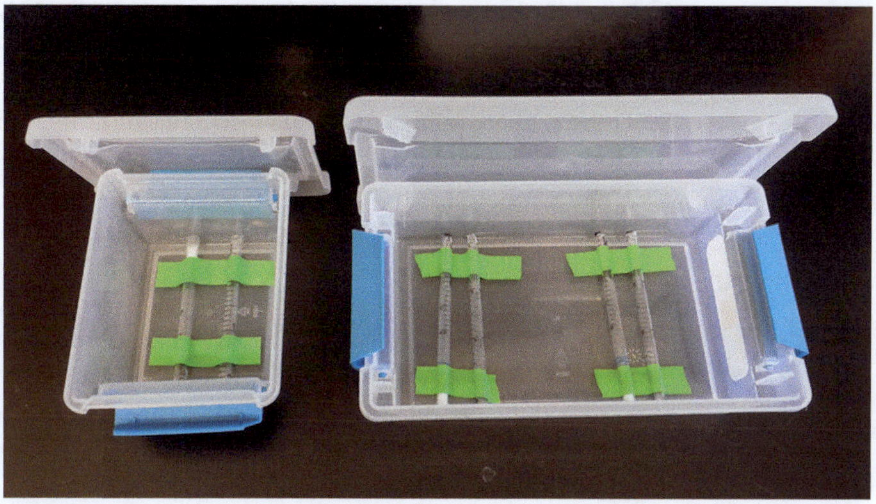

Fig. 2 A humidified chamber for ISH incubations. Serological pipettes are cut to the appropriate length and attached to the bottom surface of the Tupperware with laboratory tape. The distance between the each set of pipettes is adjusted to accommodate the width of the slide. Water soaked paper towels are placed inside the chamber to create a moist environment for incubations

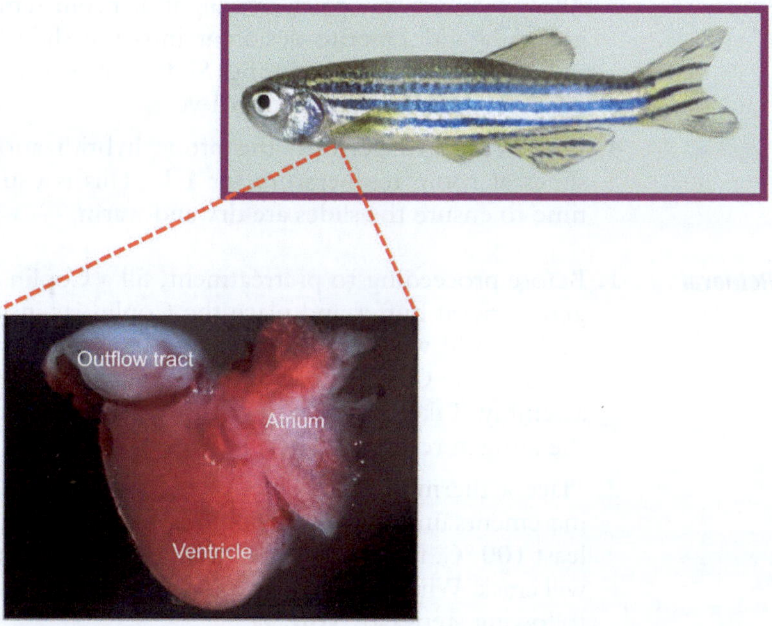

Fig. 3 The adult zebrafish heart. The zebrafish heart is a two-chambered organ. When extracting the organ for histology, it is important to keep the outflow tract, atrium, and ventricle as one unit to serve as landmarks in histology

1. Hearts are dissected from adult zebrafish using iridectomy scissors and Dumont #5 fine-tip forceps and placed directly into 1.5 mL centrifuge tube containing 1 mL of tissue fixative. Fix hearts at room temperature for 1 h. Alternatively, keep hearts in fixative overnight at 4 °C (*see* **Note 5**).

2. Remove fixative solution and replace with 1 mL 30% sucrose. Allow tissues to equilibrate for 2 h to overnight at 4 °C (*see* **Note 6**).
3. Transfer the equilibrated tissue into a plastic embedding mold. Orient the tissue with the ventricle flat on the bottom of the mold.
4. With a Kimwipes, absorb as much of the sucrose solution as possible.
5. Slowly add Tissue Freezing Medium to cover the entire tissue, filling the mold to about ¾ of the way full. If necessary, reposition the tissue with Dumont #5 forceps.
6. Freeze the embedded tissue at −80 °C for at least 2 h prior to proceeding. Tissues may be stored frozen in blocks indefinitely (*see* **Note 7**).

3.2 Tissue Sectioning

1. Before sectioning, equilibrate the tissue block to the temperature of the cryostat chamber, at −25 °C, for 30 min.
2. Section tissues at a thickness of 10 μm on cryostat and transfer sections onto Superfrost/Plus microscope slides (*see* **Note 8**).
3. Allow sections to air-dry overnight at room temperature. We generally add Drierite desiccant into the slide box with the slides to facilitate faster drying. Slides can be used immediately or stored at −20 °C for up to 1 year.
4. When ready to perform the probe hybridization, warm the slides at room temperature for 1 h. This is usually sufficient time to ensure the slides are dry and warm.

3.3 Antigen Retrieval

1. Before proceeding to pretreatment, fill a Coplin jar with antigen retrieval buffer and place the Coplin jar inside a 500 mL beaker. Add water to the beaker until the water level reaches ~3/4 of the Coplin jar. Using aluminum foil, cover the entire assembly. Take care to ensure that water is not splashing into the antigen retrieval buffer.
2. Place a thermometer inside the Coplin jar and apply heat in increments until the citrate solution reaches a temperature of at least 100 °C. If the solution is heated too quickly, the Coplin jar will crack. While the antigen buffer is heating, proceed with the following steps (*see* **Note 9**).
3. Outline the tissue sections on each slide with a hydrophobic pen in order to create a barrier to hold incubation solutions. Be cautious to not touch the tissues. Allow the ink barrier to dry completely.
4. Rinse the slides in 1× PBS for 5 min in a Coplin jar. Gently move slides up and down to remove Tissue Freezing Medium.

5. Remove slides from Coplin jar and gently tap on a stack of paper towels to remove excess PBS. It is not necessary to completely dry slides before proceeding.
6. With slides on the bench, add 500 μL of 3% hydrogen peroxide solution to each slide. Incubate for 10 min at room temperature (*see* **Note 10**).
7. Rinse twice with filtered water in Coplin jars. It is best to transfer the slides from one Coplin jar to another instead of decanting the water.
8. With a pair of blunt forceps, transfer the slides into the boiling antigen retrieval buffer. Incubate for 10 min (*see* **Note 11**).
9. Transfer the slides directly into a clean Coplin jar with water. Gently agitate by moving slides up and down four times.
10. Repeat this wash by moving slides into a clean Coplin for a total of three washes.
11. Wash slides in 100% ethanol. Gently move slides up and down five times. Transfer slides to a stack of paper towels and allow to air-dry ~2 min.
12. Redraw a barrier around the tissues with the hydrophobic pen. It is important to ensure that there are no gaps in the barrier.

3.4 Hybridization

Hybridization and amplification of the in situ RNA signal is performed in accordance with ACD suggested protocol with few modifications. Two critical factors during this process are: (1) maintaining hybridization temperature at 40 °C and (2) keeping slides moist between washes and/or incubation steps (*see* **Note 12**).

1. Place slides directly onto the serological pipettes within the humidified chamber and add 4–6 drops of ACD Protease Plus solution onto the sections. Add more drops if necessary to cover all sections (*see* **Note 13**).
2. Cover the chamber with the Tupperware lid and close tightly. Incubate at 40 °C for 30 min (*see* **Note 3**; Fig. 3).
3. Wash slides with water in a Coplin jar. Provide gently agitation by moving slides up and down four times. Repeat wash for a total of three washes.
4. Conjugated RNA probe is provided in a dropper bottle from ACD. Add a sufficient amount of the probe mixture to cover all sections on slides (*see* **Note 13**). Incubate at 40 °C for 2 h. If necessary, this incubation may be extended to up to 6 h (*see* **Note 14**).
5. Wash slides with 1× PBT buffer in Coplin jars at room temperature. Agitate by moving slides up and down during the 2 min wash. Repeat this wash for a total of three times.

6. Remove slides from Coplin jars and gently tap on a stack of paper towels to displace excess liquid.
7. Place slides in your humidified chamber and add sufficient AMP 1 solution to cover all sections. Incubate at 40 °C for 30 min.
8. Wash slides as described in **step 5**. Remove excess water by tapping slides on a stack of paper towels.
9. Place slides into the humidified chamber and add AMP 2 solution to cover all sections. Incubate at 40 °C for 15 min.
10. Wash slides in Coplin jars using the same procedure as described in **step 5**.
11. Place washed slides in humidified chamber and add sufficient AMP 3 solution to cover all sections. Cover the chamber with the Tupperware lid and Incubate at 40 °C for 30 min.
12. Wash slides with 1× PBT buffer in Coplin jars at room temperature. Agitate by moving slides up and down during the 2 min wash. Repeat this procedure for a total of three washes.
13. Remove excess liquid by tapping slides on a stack of paper towels.
14. Return washed slides to humidified chamber and add enough AMP 4 solution to cover all sections. Close the chamber and incubate for 15 min at 40 °C.
15. Wash slides with 1× PBT buffer in Coplin jars at room temperature. Agitate by moving slides up and down during the 2 min wash. Repeat this wash for a total of three times.
16. Remove excess liquid by tapping slides on a stack of paper towels.
17. Return washed slides to humidified chamber and add enough AMP 5 solution to cover all sections. Close chamber and incubate for 30 min at room temperature.
18. Wash slides with 1× PBT buffer in Coplin jars at room temperature. Agitate by moving slides up and down during the 2 min wash. Repeat this wash for a total of three times.
19. Remove excess liquid by tapping slides on a stack of paper towels.
20. Place washed slides into humidified chamber and add enough AMP 6 solution to cover all sections. Cover the chamber and incubate for 15 min at room temperature.
21. Wash slides with 1× PBT buffer in Coplin jars at room temperature. Agitate by moving slides up and down during the 2 min wash. Repeat this wash for a total of three times.
22. Remove excess liquid by tapping slides on a stack of paper towels.

3.5 Signal Detection and Mounting

Detection of amplified in situ signal is dependent on the desired chemistry and probe design. Here we describe the detection of conjugated HRP-based green probe directed against microRNA-101a and AP-based Fast Red probes designed to detect *fosab* mRNA (Fig. 4). Red and green signals are detected sequentially.

1. Add ~200 μL of the freshly prepared Red working solution directly to the sections and incubate for 30 min at room temperature. Cover the entire humidified chamber with aluminum foil to keep the slides in the dark.
2. Tap slides on a stack of paper towels to remove as much of the detection solution as possible. Transfer slides directly into a Coplin jar with 1 × PBT buffer. Agitate by moving slides up and down during the 2 min wash. Repeat this wash for a total of three times.
3. Remove excess liquid by tapping slides on a stack of paper towels.
4. For detection of HRP-Green, add ~200 μL of freshly prepared Green working solution to slides. Cover the humidified chamber and wrap with aluminum foil. Incubate for 10 min at room temperature.
5. Tap slides on a stack of paper towels to remove as much of the detection solution as possible.
6. Transfer slides directly into a Coplin jar with reverse osmosis water. Agitate by moving slides up and down during the 2 min wash. Repeat this wash for a total of three times.
7. Remove excess liquid by tapping slides on a stack of paper towels (*see* **Note 15**).
8. Dry slides in a hybridization oven at 60 °C for 30 min. It is critical that slides are completely dry.
9. Immerse slides into 100% CitriSolv in a Coplin jar.

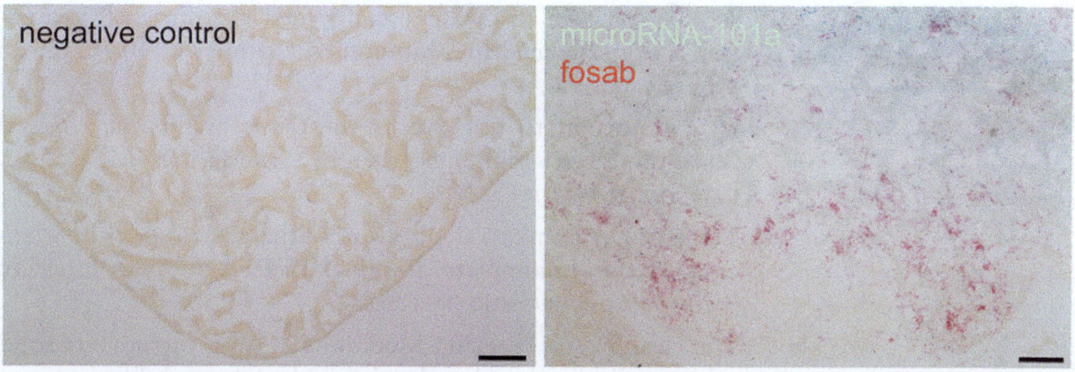

Fig. 4 RNAscope ISH detection of microRNA-101 and fosab with RNAscope probes. Uninjured and regenerating zebrafish hearts were extracted and cryosectioned at 10 μm and hybridized with either a negative control or microRNA-101a (*green*) or fosab (*red*) probes. Scale bar = 50 μm

10. Remove slides and immediately coverslip with 3–5 drops of EcoMount mounting medium. Be cautious to avoid trapping air bubbles, as they may interfere with image capture.
11. Allow slides to dry on the bench for 5–10 min.
12. Image slides on brightfield to capture signal. Figure 4 shows representative images of microRNA-101a (green) and *fosab*, a target gene (red) on adult zebrafish regenerating hearts [11].

4 Notes

1. Custom RNA probes are designed by Advanced Cell Diagnostics with propriety RNAscope amplification technology. This technology is based on synthesis of multiple RNA probes to hybridize across a target sequence. Upon hybridization, the signal is enhanced through a series of amplification steps and detected either by fluorescence or by chromogenic enzymatic reactions (Fig. 1). In this example, I ordered a probe directed against 500-bp of the microRNA primary sequence and 800-bp of a target gene.
2. Advanced Cell Diagnostics (ACD) provides a 50× concentrated wash buffer that needs to be diluted with water to 1× working concentration. We routinely use 1× PBT as a substitute and have not observed changes to signal intensity
3. In my experience ~200 μL is sufficient to cover all sections. Mix contents by pipetting up and down 3–5 times before use. Once the working solution is made, it must be used within 15 min.
4. To create a hybridization chamber, select an appropriate size Tupperware with a secure lid. Cut 5 mL serological pipettes to the length of a Tupperware and attach them onto the bottom surface in parallel at a distance that will support your slides, with laboratory tape (Fig. 2). During hybridization, it is important that you place water soaked paper towels in the bottom of the Tupperware to create a humidified chamber. All incubations are performed in a humidified chamber at 40 °C unless otherwise noted.
5. It is important to not overfix the tissue as this may induce nonspecific cross-linking, thereby interfering with RNA detection.
6. The tissue will sink to the bottom of the centrifuge tube when equilibration is complete. Residual 4% PFA does not interfere with subsequent tissue processing.
7. The clear Tissue Freezing Medium will turn into a white solid substance indicating that the tissue block has solidified.
8. It is critical to use Superfrost/Plus coated slides as sections adhere more efficiently.

9. Fixation can alter protein biochemistry such that the epitope of interest becomes masked and is no longer assessable by the probe. Citrate boiling is one technique of antigen retrieval in which the masking of an epitope is reversed and epitope–probe binding is restored.
10. This treatment will inactive endogenous phosphatase activity and reduce background staining.
11. It is important to maintain the temperature above 100 °C to ensure successful antigen retrieval.
12. ACD sells a hybridization oven that addresses the aforementioned concerns very well. However, I find that using a modified Tupperware and a standard incubator works equally as well.
13. I find that using a P1000 pipette tip to spread the solution is helpful to achieve complete coverage.
14. We have not noticed changes to signal intensity or background noise with longer incubation times
15. At this point, you can counterstain the slides with 50% hematoxylin solution using standard staining protocols. However, in our experience, counterstains have a tendency to dampen signal intensity.

Acknowledgment

This work was supported by Institutional Development Awards (IDeA) from the National Institute of General Medical Sciences of the National Institutes of Health under grant numbers P20-GM104318 and P20-GM103423, and an American Heart Association Scientist Development Grant (11SDG7210045) to V.P.Y.

References

1. Callan HG (1986) Lampbrush chromosomes. Mol Biol Biochem Biophys 36:1–252
2. Gall JG (2016) The origin of in situ hybridization – a personal history. Methods 98:4–9
3. Lampasona AA, Czaplinski K (2016) RNA voyeurism: a coming of age story. Methods 98:10–17
4. Denli AM, Tops BB, Plasterk RH, Ketting RF, Hannon GJ (2004) Processing of primary microRNAs by the microprocessor complex. Nature 432(7014):231–235
5. Lee RC, Feinbaum RL, Ambros V (1993) The C. elegans heterochronic gene lin-4 encodes small RNAs with antisense complementarity to lin-14. Cell 75(5):843–854
6. Reinhart BJ, Slack FJ, Basson M, Pasquinelli AE, Bettinger JC, Rougvie AE, Horvitz HR, Ruvkun G (2000) The 21-nucleotide let-7 RNA regulates developmental timing in Caenorhabditis elegans. Nature 403(6772):901–906
7. Zhao Y, Samal E, Srivastava D (2005) Serum response factor regulates a muscle-specific microRNA that targets Hand2 during cardiogenesis. Nature 436(7048):214–220
8. Forstemann K, Tomari Y, Du T, Vagin VV, Denli AM, Bratu DP, Klattenhoff C, Theurkauf WE, Zamore PD (2005) Normal microRNA maturation and germ-line stem cell maintenance requires Loquacious, a double-stranded RNA-binding domain protein. PLoS Biol 3(7): e236

9. Giraldez AJ, Cinalli RM, Glasner ME, Enright AJ, Thomson JM, Baskerville S, Hammond SM, Bartel DP, Schier AF (2005) MicroRNAs regulate brain morphogenesis in zebrafish. Science 308(5723):833–838
10. Hatfield SD, Shcherbata HR, Fischer KA, Nakahara K, Carthew RW, Ruohola-Baker H (2005) Stem cell division is regulated by the microRNA pathway. Nature 435(7044):974–978
11. Beauchemin M, Smith A, Yin VP (2015) Dynamic microRNA-101a and Fosab expression controls zebrafish heart regeneration. Development 142(23):4026–4037
12. Yin VP, Lepilina A, Smith A, Poss KD (2012) Regulation of zebrafish heart regeneration by miR-133. Dev Biol 365(2):319–327
13. Giraldez AJ, Mishima Y, Rihel J, Grocock RJ, Van Dongen S, Inoue K, Enright AJ, Schier AF (2006) Zebrafish MiR-430 promotes deadenylation and clearance of maternal mRNAs. Science 312(5770):75–79
14. Kloosterman WP, Plasterk RH (2006) The diverse functions of microRNAs in animal development and disease. Dev Cell 11 (4):441–450
15. Silahtaroglu AN, Nolting D, Dyrskjot L, Berezikov E, Moller M, Tommerup N, Kauppinen S (2007) Detection of microRNAs in frozen tissue sections by fluorescence in situ hybridization using locked nucleic acid probes and tyramide signal amplification. Nat Protoc 2 (10):2520–2528

Chapter 14

Padlock Probes to Detect Single Nucleotide Polymorphisms

Tomasz Krzywkowski and Mats Nilsson

Abstract

Rapid development of high-throughput DNA analyzation methods has enabled global characterization of genetic landscapes and aberrations in study subjects in a time and cost effective fashion. In most methods, however, spatial tissue context is lost since sample preparation requires isolation of nucleic acids out of their native environment. We hereby present the most recent protocol for multiplexed, in situ detection of mRNAs and single nucleotide polymorphisms using padlock probes and rolling circle amplification. We take advantage of a single nucleotide variant within conserved *ACTB* mRNA to successfully differentiate human and mice cocultured cells and apply presented protocol to genotype PCDH X and Y homologs in human brain. We provide a method for automated characterization and quantitation of target mRNA in single cells or chosen tissue area. mRNA of interest, harboring a polymorphism, is first reverse-transcribed to cDNA. Allele specific padlock probes are hybridized to the cDNA target and enzymatically circularized maintaining a physical link with the parent mRNA molecule. Lastly, circularized probes are replicated in situ, using rolling circle amplification mechanism to facilitate detection.

Key words Padlock probe, Single nucleotide polymorphism, SNP, mRNA genotyping, In situ, Single molecule, Single-cell

1 Introduction

Successful prevention of many diseases and accurate therapy guidance relies on the ability to correctly identify high-risk genetic aberrations that contribute to disease formation and progress. At present, most commonly practiced methods that allow for multiplexed mutation screening include primer extension [1, 2] and high-throughput new generation deep sequencing methods (NGS) [3]. Direct visualization of molecules, including mRNAs, overcomes the loss of information dictated by cell population averaging and provides perspective into spatiotemporal regulation of biological processes, often governed by single-cell, stochastic events. Recent developments of traditional fluorescent RNA in situ hybridization (FISH) including mRNA barcoding [4], different variants of in situ amplification like branched DNA (bDNA) FISH [5] and hybridization chain reaction (HCR) [6] have made

Imre Gaspar (ed.), *RNA Detection: Methods and Protocols*, Methods in Molecular Biology, vol. 1649, DOI 10.1007/978-1-4939-7213-5_14, © Springer Science+Business Media LLC 2018

quantitation and localization of mRNAs feasible. Although various FISH approaches facilitate assessment of mRNA levels in individual cells, single nucleotide polymorphism (SNP) genotyping remains challenging since accurate sequence recognition is negatively correlated with length of probes hybridizing to the target mRNA.

This chapter introduces an up-to-date protocol and outlines the list of notes and hints to conduct successful multiplexed mRNA SNP detection in situ, using *p*ad*l*ock *p*robes (PLP) and *r*olling *c*ircle *a*mplification (RCA). PLP is a continuous, single stranded DNA oligonucleotide with two terminal, target-specific arms (3′ arm and 5′ arm) connected by a DNA linker [7]. During target base pairing, PLP arms hybridize juxtaposed leading to probe circularization. Unlike traditional FISH approaches that solely rely on probe-mRNA hybridization, accuracy of SNP recognition with PLPs is driven by a mismatch-sensitive, thermostable DNA ligase. Differentiation between correct and incorrect target is practically binary and results in sealing the nick between probe arms rendering the probe circularized [8].

The PLP-based SNP detection protocol comprises series or enzymatic reactions, beginning with mRNA to cDNA reverse transcription (RT). To maximize the conversion rate, target specific RT primer contains several chemically modified DNA bases where 2′-O, 4′-methylene bridge locks the ribose pucker in C3′ endo conformation [9]. Such "locked" nucleic acids DNA analogues (LNAs) display much higher RNA/DNA hybridization affinity than conventional nucleic acids. After the newly synthesized cDNA strand is chemically cross-linked with formaldehyde, original mRNA molecule is degraded by ribonuclease H (RNase H is an endoribonuclease that specifically targets and hydrolyzes RNA phosphodiester bonds in RNA/DNA heteroduplexes). Part of the mRNA that was hybridized to LNA bases within a RT primer is protected from degradation thus cDNA remains physically linked with the original target mRNA (Fig. 1). Next, allele-specific PLPs are hybridized and ligated to respective cDNA targets using *Thermus thermophilus* DNA ligase (*Tth* ligase) (*see* **Note 1**). Depending on the total length of hybridizing arms (typically between 30 and 40 bp) PLP makes ~2–4 full helical turns and becomes physically threaded around or "locked" to the cDNA. This stability allows for application of more stringent washing to ensure removal of incorrectly hybridized probes. Unidirectional rolling circle amplification of the target cDNA is initiated by the ϕ29 DNA polymerase and templated by the target bound PLP [10]. DNA polymerase continuously extends the cDNA yielding 10^2–10^3 tandem copies of the original template (PLP) [11] maintaining physical contact with the original mRNA molecule (Fig. 1).

The single stranded *RCA* *p*roduct (RCP) coils up spontaneously into ~500 nm size DNA blob which is "decorated" with fluorophore-conjugated oligonucleotides (detection/decorator

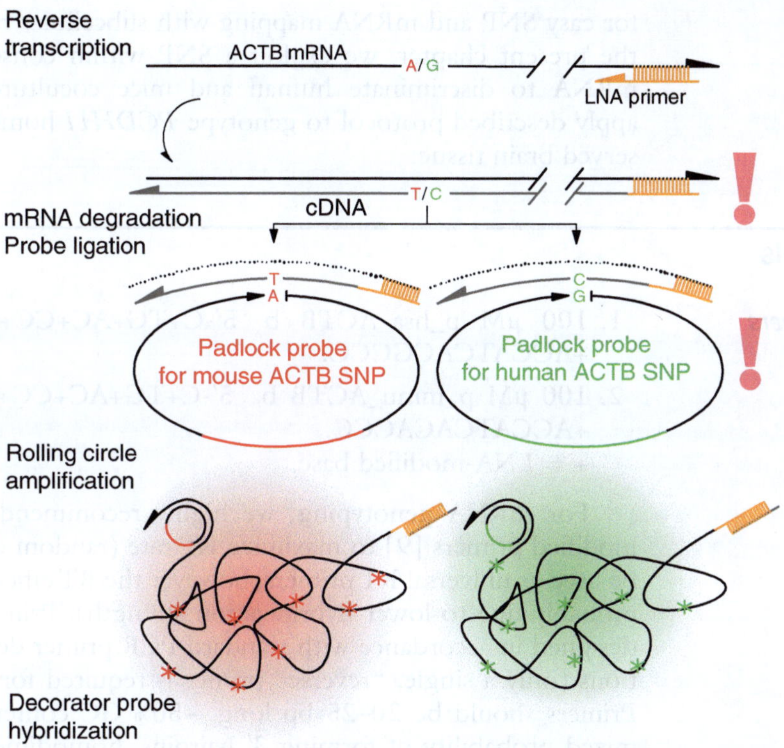

Fig. 1 Outline of in situ RNA genotyping using padlock probes and target primed rolling circle amplification. An mRNA molecule harboring a sequence variant (ACTB in the presented example) is converted into cDNA by reverse transcription, using target-specific LNA primer. Subsequently, the mRNA is degraded using RnaseH, padlock probes for the respective alleles hybridize to the target sequence and circularize upon DNA ligation. RCA generates a single-stranded DNA concatamer, which collapses into a typically half micrometer-sized DNA ball and contains hundreds of tandem repeated sequences that are complementary to the original padlock probe. Finally, fluorescently labeled decorator probes hybridize with their complementary motifs on the RCA product. Different DNA backbones of PLP for human and mouse allele allow efficient discrimination of RCP using unique, differently fluorophore-conjugated decorator probes (*green* and *red*). *Exclamation mark*: points in the protocol when the experiment can be paused (*see* **Note 23**)

oligonucleotides), complementary to the motif repeated in the amplicon. Local concentration of decorator probes creates a bright speckle readily differentiated from the background with standard EPI fluorescent microscope, even in samples with relatively high autofluorescence. By customizing PLP linker sequence, unique recognition motifs can be designed for each SNP or mRNA to be detected in the experiment. Multiplexing capacity is therefore limited to the number of decorator probes that can be differentiated with the available microscope system.

Each RCP represents a successful event of PLP hybridization, ligation, and amplification; thus, each observed signal represents a single SNP, making mRNA genotyping with padlock probes both qualitative and quantitative. Robustness and specificity of the protocol, combined with straightforward RCP localization allow user for easy SNP and mRNA mapping with subcellular resolution. In the present chapter, we exploit a SNP within conserved *ACTB* mRNA to discriminate human and mice cocultured cells and apply described protocol to genotype *PCDH11* homologs in preserved brain tissue.

2 Materials

2.1 RT Primers

1. 100 μM p_hsa_ACTB b 5′-C+TG+AC+CC+AT+GC+CC +ACCATCACGCCC.
2. 100 μM p_mmu_ACTB b 5′-C+TG+AC+CC+AT+TC+CC +ACCATCACACCC.

   + = LNA-modified base.

For mRNA genotyping, we highly recommend using LNA modified primers [9] to maximize RT rate (random decamers can be used as universal RT primers, however the RT efficiency is lower probably due to lower hybridization strength). Primer should be designed in accordance with standard PCR primer design instructions (only a single, "reverse" primer is required for in situ RT). Primers should be 20–25 bp long, ~50% GC content and minimized probability of forming 3′ hairpins, homodimer or heterodimer. Online tools like primer3 [12] are a great start as they suggest the best "right" RT primer with all thermodynamic parameters accounted for. LNA substitutions have to be suggested by the researcher and we typically substitute 3–7 DNA bases with LNA derivatives (*see* **Note 2**). Primer hybridization site should be downstream (3′) and as close to the SNP as possible (*see* **Note 3** and Supplementary Note 3 in [13]). Once LNA primer sequence is determined, it should be evaluated for potential off target RT priming to prevent formation of false-positive cDNA and signals. Blastn against RefSeq mRNA database is recommended http://blast.ncbi.nlm.nih.gov/Blast.cgi. Figure 2 shows the *ACTB* mRNA in human and mouse with SNP highlighted, illustrating typical design example for both LNA primer and allele specific PLPs.

2.2 Padlock Probes and Decorator Probes

1. 2 μM hsa_ACTB 5′ GCCGGCTTCGCGGGCGA AAAAAAAAAAA *CCTCAATGCACATGTTTGGCTCC* AAAAAAAAAAA ACGGCGCCGGCATGTGCAA**G**.
2. 2 μM mmu_ACTB 5′ GCCGGCTTCGCGGGCGA AAAAAAAAAAA *CCTCAATGCTGCTGCTGTACTAC* AAAAAAAAAAA ACGGCGCCGGCATGTGCAA**A**.

   Underline = target complementary arms, italic = detection probe complementary sequence.

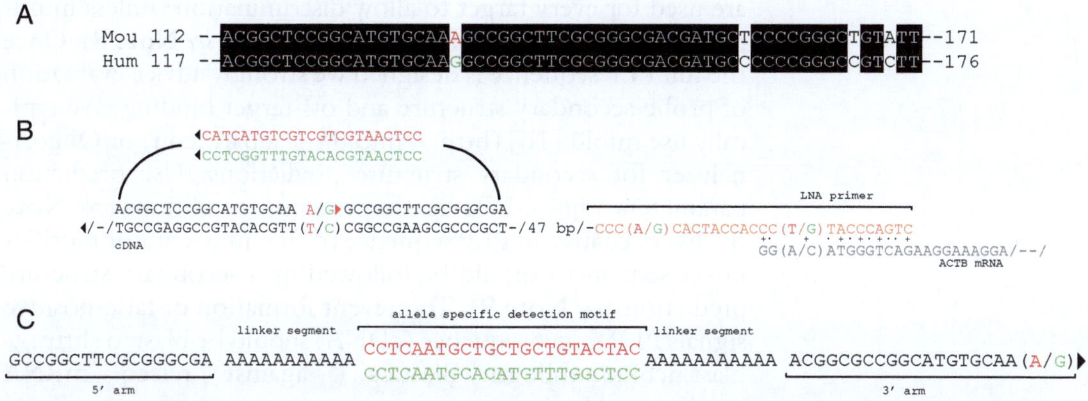

Fig. 2 Padlock probe and LNA-primer design for human and mouse *ACTB* mRNA. (**a**) Pairwise alignment of mouse and human *ACTB* mRNAs with a SNP detected highlighted. (**b**) LNA primers (*orange*; LNA bases: *bold*; SNPs: *green* for human, *red* for mouse) hybridize with *ACTB* mRNA. During the ligation step, mRNA is degraded from mRNA/cDNA heteroduplex except where it was bound to LNA bases in RT primer. Padlock arms for matching allele hybridize with the target, while discriminative 3′ base (symbolized as a *triangle* at the end of the 3′ arm) is located over the interrogated T/C SNP. Reporter sequence (*green*/*red*) is amplified and used later for detection. (**c**) 5′→ 3′ full-length sequence of the padlock probe, with different parts of the probe indicated

Manual design of PLPs is possible and it can be done relatively quickly after gaining experience. However, to avoid making errors or for designing of multiple probes we recommend using open-source, automated ProbeMaker software [14] (*see* **Note 4**). PLP arms are designed to span SNP target in the target cDNA. For optimal and efficient SNP genotyping, we advise to adjust probe arms to that their melting temperatures (T_m) exceed ligation temperature by 5–10 °C considering ligation reaction conditions (*see* **Note 5**). If multiple probes are used in the same experiment, Tm of all PLPs arms should be similar to ensure comparable performance and minimize the background. As a rule of thumb, PLP backbone should be longer than total length of both arms (target) by approximately 10 bp (according to our experience, shortened backbones result in lower detection yield, probably because cDNA needs to "bend" to accommodate both PLP arms). Discriminating nucleotide of the PLP (hybridizing to the SNP) should always be placed on the 3′ arm terminus [15] (in bold, *see* **Note 6**) and phosphate group on the 5′ terminus to allow for probe ligation. PLP can be ordered pre-phosphorylated by a vendor of choice (pre 5′-phosphorylated oligonucleotides, synthesized by IDT (http://idtdna.com) using Ultramer® chemistry work well in our hands) or phosphorylated manually according to the standard phosphorylation protocol (*see* **Note 7**). After RCA, the "decorator motif" amplified in the RCP is utilized to visualize the amplification product through hybridization of a fluorescently labeled decorator oligonucleotides thus different motifs are used for every target to allow discrimination (unless multiple SNPs are to be visualized simultaneously, *see* **Note 8**). Once the full PLP sequence is designed we strongly advice evaluation of probe secondary structure and off-target binding. We typically use mfold [16] (http://mfold.rna.albany.edu) or OligoAnalyzer for secondary structure predictions. Use prediction parameters adjusted for ligation reaction conditions (*see* **Note 5**). Every change in PLP sequence (different decorator motif or linker sequence) should be followed by a secondary structure prediction (*see* **Note 9**). To prevent formation of false-positive signals PLP target sequence (cDNA) should be blasted (http://blast.ncbi.nlm.nih.gov/Blast.cgi.) against refseq mRNA database.

3. 10 μM hsa_DO 5′ CCTCAATGCACATGTTTGGCTCC.
4. 10 μM mmu_DO 5′ CCTCAATGCTGCTGCTGTACTAC. Decorator probes (*see* **Note 8**).

2.3 Reagents

All enzymes should be stored at −20 °C. Other reagents are stored at room temperature (RT) unless stated different.

1. 40 U/μL RNase Inhibitor (e.g., DNA Gdansk RIBOPROTECT).
2. TRANSCRIPTME—RnaseH-reverse transcriptase 200 U/μL and buffer (DNA Gdansk, *see* **Note 10**).
3. 40 U/μL Tth DNA ligase (e.g., GeneCraft).
4. 5 U/μL RNaseH.
5. 10 U/μL Phi29 DNA polymerase and buffer (e.g., Thermo Scientific).
6. Biological sample: cultured cells (alive) or tissue (fresh or fresh frozen). Fixed or freshly frozen tissues should be stored at −80 °C.
7. Diethylpyrocarbonate (DEPC). Stored at 4 °C (*see* **Note 11**) *Known carcinogen. Work under a fume hood, wear gloves.*
8. RNase and DNase inactivators.
9. 70, 85, and 99.5 v/v % ethanol.
10. Xylene.
11. 0.1 M hydrochloric acid.

    Highly corrosive. Work under fume hood with rubber PVC gloves and eyes protection.
12. 3.7% formaldehyde in PBS (*see* **Note 12**).

    Known carcinogen. Use nitrile gloves, eyes protection and handle powder in the chemical fume hood.
13. Pepsin, lyophilized powder, 2.500 U/mg protein (*see* **Note 13**).
14. 0.25%trypsin–EDTA.

15. SlowFade® Gold Antifade Mountant (Thermo Scientific) or equivalent mounting medium (*see* **Note 14**).
16. 100 mM Hoechst 33342.
Store stock Hoechst 33342 solution at −20 °C. Working solutions (protected from light) can be stored at 4 °C for a couple of months.

2.4 Buffers and Solutions

All concentrated buffers are provided together with enzymes and stored according to vendor specification. Custom-made buffers should be prepared from DEPC-treated PBS or ddH_2O (*see* **Note 15**) and can be kept at RT for up to one month. We also recommend aliquoting DEPC-ddH_2O and DEPC-PBS into smaller volumes (50 mL) to minimize contamination.

1. Phosphate buffered saline (1× PBS): 137 mM NaCl, 10 mM sodium phosphate, 2.7 mM KCl, and DEPC-ddH_2O pH 7.4.
2. Washing buffer (PBS-T): 0.05% Tween 20 in 1× PBS.
3. Saline-sodium citrate buffer (20× SSC): 3 M NaCl and 300 mM trisodium citrate in DEPC-ddH_2O pH 7.
4. RT-mix (50 μL per chamber): 5 μL 10× RT-reverse transcriptase buffer, 1 μL 40 U/μL RNase Inhibitor, 0.5 μL 20 μg/μL BSA, 2.5 μL 10 mM dNTPs mix, 0.5 μL 100 μM LNA primer or 2.5 μL 100 μM random decamers, and TRANSCRIPTME-reverse transcriptase (we typically use 5 U/μL for cell lines and 20 U/μL for tissue sections). Fill up to 50 μL with DEPC-ddH_2O.
5. Ligation mix (50 μL per chamber): 5 μL 10× Tth ligase buffer, 2.5 μL 2 μM padlock probe(s), 4 μL 5 U/μL RNase H, 0.5 μL 20 μg/μL BSA, 2.5 μL 1 M KCl, 10 μL 100% formamide, and 1.25 μL 200 U/μL TTh ligase. Fill up to 50 μL with DEPC-ddH_2O
6. RCA mix (50 μL per chamber): 5 μL 10× Phi29 DNA polymerase buffer, 1.25 μL 10 mM dNTPs mix, 0.5 μL 20 μg/μL BSA, 5 μL 50% glycerol, and 5 μL 10 U/μL Phi29 DNA polymerase. Fill up to 50 μL with DEPC-ddH_2O.
7. Hybridization mix (50 μL per chamber): 5 μL 20× SSC, 10 μL 100% formamide, 0.5 μL 10 μM Decorator probe(s), 1.5 μL, and 100 mM Hoechst 33342. Fill up to 50 μL with DEPC-ddH_2O.

2.5 Equipment

1. Diamond pen.
2. Forceps.
3. 37 °C incubator.
4. 45 °C incubator.

5. Humidity chamber (e.g., empty slide or tips box with paper towel soaked with water).
6. Fluorescence microscope.
7. Secure-Seal hybridization chambers (Grace Bio-Labs) (*see* **Note 16**).
8. Coverslips (*see* **Note 17**).
9. Coplin jars.
10. Adhesive microscopy slides (e.g., Menzel Gläser SuperFrost®).
11. 150 mm × 25 mm culture dish used for cell fixation.

3 Methods

3.1 General Guidelines and Controls

RNases can maintain enzymatic activity even after prolonged autoclaving. Therefore, special measures should be taken when working with RNA.

1. We recommend designating bench area dedicated to RNA work only.
2. All bench surfaces, pipettes, or glassware should be treated with commercially available RNase- and DNAse-inactivating agents. We usually wipe benches with 100% ethanol after such treatment.
3. Sterile, disposable, free plasticware (pipette tips, slide boxes, tubes, and flasks) work best in our hands and ensure RNase-free conditions.
4. We recommend validating efficiency and specificity of new PLPs on synthetic DNA targets.
5. Ligation fidelity can be monitored in vitro as a high molecular weight band on denaturing PAGE gel (linear PLPs and templates migrate faster than circularized PLPs).
6. RCA can be monitored in vitro (templates provide the 3′ –OH group as a RCA primer, just as cDNA in regular protocol) by staining RCPs with either intercalating dyes (SYBR dyes) or decorator probes and visualized under a microscope (5–10 μL of stained RCA mix can be mounted on a microscope slide) or q-PCR system.
7. As a biological positive and negative controls, cell lines with/without target of interest provide a good model to evaluate assay specificity and sensitivity.
8. In accordance with good research practice, different day replicates are recommended, since variation in handling slides or in cell lines growth may influence the final result.

9. Finally, we recommend frequent assessment of the whole genotyping procedure on well-established system (*see* **Note 18**) to ensure maximum detection sensitivity.

3.2 Specimen Pretreatment

3.2.1 Adherent Cells

1. Culture cells in a flask until 80–90% confluence.
2. Wash cells twice with one PBS, add dislodge cells by adding 1 mL/25 cm^2 0.25% trypsin–EDTA for a 1–2 min. Aspirate Trypsin-EDTA and monitor cells trypsinization under a bright-field microscope.
3. Resuspend the cells in appropriate culturing medium (FBS in a medium inactivates the trypsin).
4. Place sterile microscope slides in a 150 mm × 25 mm culture dish and add ~22 mL of medium to cover the slides. Cells adhere to most type of slides.
5. Carefully, seed total 3 mL of suspended cells directly on the slides (*see* **Note 19**).
6. Carefully, transfer the closed dish into the incubator and incubate the cells under appropriate conditions to allow them to adhere to the microscope slides (*see* **Note 20**).
7. Wash the slides twice with cold 1× PBS and transfer the slides to a sterile Coplin jar or sterile slides transport box.
8. Fix the cells with 3.7% formaldehyde at room temperature for 20 min.
9. Discard the formaldehyde and wash the slides twice with 1× PBS (*see* **Note 21**).
10. Gradually dehydrate and pre-permeabilize the cells in ethanol series (70, 85, and 99.5% each step for 3 min).
11. Air-dry the slides and store at −80 °C (long-term storage) or −20 °C (up to 2 weeks).
12. Use pencil to note a fixation date, cell type and number of the passage on the slide's writable area (*permanent markers may dissolve upon contact with ethanol and cause severe autofluorescence*).
13. Upon use, thaw slides at room temperature.
14. Attach Secure-Seal chamber(s) and rehydrate the cells by adding 1× PBS-T into the chamber (*see* **Note 22**).
15. Remove DEPC-PBS-T and permeabilize the cells by adding 0.1 M HCl for 5 min.
16. Remove HCl and wash the cells twice with PBS-T (*see* **Note 23**).

3.2.2 Freshly Frozen Tissue

1. Tissue sections, mounted on microscope slides (*see* **Note 24**) are stored at −80 °C until use.

2. Thaw slides at RT.
3. Fix the tissue in 1× PBS for 45 min. Use a sterile Coplin Jar or perform fixation directly in the chamber (depending on specimen area).
4. Wash once with 1× PBS for 5 min.
5. (Optional) Permeabilize the tissue by incubating with pre-warmed pepsin in 0.1 M HCl at 37 °C. Digestion with 250 U/mL (0.1 mg/mL) pepsin for 5 min is a good starting point in our experience. Optimal conditions need to be identified for the respective specimen (*see* **Notes 13** and **25**).
6. Wash once with 1× PBS for 5 min.
7. Dehydrate the tissue section in the ethanol series (70, 85, and 99.5% ethanol, 3 min each).
8. Air-dry and mount Secure-Seal chambers (if fixation was conducted in the Coplin Jar).
9. Rehydrate the tissue by in 1× PBS-T.

3.2.3 Formalin-Fixed, Paraffin-Embedded (FFPE) Tissue

1. Tissue sections, mounted on microscope slides (*see* **Note 24**) are stored at −80 °C until use.
2. Thaw samples at RT.
3. Dewax samples by submerging slides in a solvent series in separate Coplin jars:

   15 min xylene, 10 min xylene, 2 × 2 min 100% ethanol, 2 × 2 min 95% ethanol, 2 × 2 min 70% ethanol, 5 min DEPC-H_2O, 5 min 1× PBS.
4. Permeabilize the tissue by incubating with pre-warmed pepsin in 0.1 M HCl at 37 °C (*see* **Notes 13** and **25**).
5. Wash with 1× PBS for 5 min.
6. Postfix the tissue in 3.7% formaldehyde for 10 min at room temperature.
7. Wash twice with 1× PBS for 5 min.
8. Dehydrate the tissue section in the ethanol series (70, 85, and 99.5% ethanol, each step for 3 min).
9. Air-dry the slides and mount Secure-Seal chambers.
10. Rehydrate the tissue with 1× PBS-T.

3.3 SNP Detection Protocol

Below, we present an updated protocol for mRNA SNP genotyping in cell lines, fresh frozen and FFPE tissue sections. We operate at 50 μL reaction volumes that can be adjusted if necessary (use larger Secure-Seal chambers, *see* **Note 16**). To minimize introduction of contaminants into chambers, work in as clean environment as possible—place a clean paper towel between lab bench and the slide; change gloves and washing buffers frequently. At temperatures above room temperature, Secure-Seal chamber inlets should be covered with standard PCR film to prevent evaporation of reaction mix. To further reduce evaporation, all incubations and reactions are performed in a humidified box.

3.3.1 Reverse Transcription

1. Apply the RT mix to the chamber.
2. Cover the chamber inlets and incubate the slides at 45 °C. The optimal incubation time needs to be determined experimentally. One hour long RT when using LNA-modified target-specific primers is usually sufficient. RT with random decamers is typically conducted overnight.
3. Wash the chamber once with 1× PBS-T.

3.3.2 Postfixation

Postfixation is an important step during which the newly synthesized cDNA strand cross-links to the adjacent chemical groups of proteins side chains.

1. We typically fix cell culture for 10 min and tissue sections for up to 45 min at room temperature in 3.7% formaldehyde (*see* **Note 12**). Postfixation time should also be optimized for every specimen.
2. Wash the chamber twice with 1× PBS-T.
3. At this point, the protocol can be halted and samples stored for a couple of days at 4 °C in 1× PBS.

3.3.3 mRNA Degradation, Padlock Probe Hybridization and Ligation

1. Apply the Ligation mix to the chamber.
2. Cover the chamber inlets and incubate the slides at 37 °C for 30 min.
3. Transfer the slide to 45 °C incubator for 45 min (*see* **Note 26**).
4. Wash the chamber twice with 1× PBS-T.
5. At this point, the protocol can be halted and samples stored for a couple of days at 4 °C in 1× PBS.

3.3.4 Rolling Circle Amplification

1. Apply the RCA mix into the chamber.
2. Cover the chamber inlets and incubate the slides at 37 °C for ≥1 h (*see* **Note 27**).
3. Wash the chamber twice with 1× PBS-T.
4. At this point, the protocol can be halted and samples stored for a couple of days at 4 °C in 1× PBS.

3.3.5 Decorator Probe Hybridization and Nuclei Counterstaining

From this step onward, light-sensitive compounds will be used. Protect decorator probes as well as DNA intercalating dyes from prolonged, direct light exposure. Samples should be kept in a dark (covered) during and after incubations.

1. Apply the Hybridization mix to the chamber.
2. Incubate the slide at RT for ~20 min (*see* **Note 28**).
3. Wash the chamber twice with 1× PBS-T.
4. Mark the position of the chamber on the backside of the slide with a diamond pen and remove the Secure-Seal chamber (*see* **Note 29**).
5. Passing the specimen through an ethanol series (70, 85, and 99.5% ethanol, each for 3 min) dehydrates sample, cleans the slide from contaminants and removes any glue residues.
6. Air-dry, apply Slow-Fade medium on the coverslip (or specimen directly) and gently, mount the specimen (*see* **Note 30**).
7. Use pencil to note date and type of experiment on the slide's writable area (***permanent markers may dissolve upon contact with ethanol***)
8. Signal and specimen are stable for a long time (blue-yellow visible light spectrum dyes for at least one year, far red dyes are less stable) when kept at 4 °C and protected from light.

3.3.6 Image Acquisition and Analysis

1. Mounted slides can be imaged immediately. Conventional wide-field epifluorescence microscopes are usually sufficient to image RCA products in tissue sections and cells.
2. Make sure to use filters appropriate for nuclear staining and fluorophores used in the experiment (*see* **Note 31**).
3. Depending on the level of detail desired, use appropriate objective (we typically use 10×, 20× and 40× high numerical aperture objectives).
4. To ensure accurate signal quantitation during image analysis, carefully adjust exposure time for a signal channels (abundant RCPs might be hard to segment if overexposed).
5. We typically acquire 5–10 μm thick z-stack of images to capture RCPs in all focal planes (if possible, preview the specimen to define" first" and" last" stack corresponding to RCPs present in the lowest and highest stack and capture all images within this range).
6. Z-Stacks with multiple focal planes are combined to a single 2D maximum intensity projection (MIP) image that can be used in image analysis. We recommend taking images from several positions for each experiment (depending on cell density) to account for cell-to-cell expression variations.
7. Open-source programs such as CellProfiler [17] or ImageJ can be used for image analysis (*see* **Note 32**).

3.4 Anticipated Results

1. Every discrete, fluorescent detected signal originates from a single, successful detection of the SNP (Fig. 3).

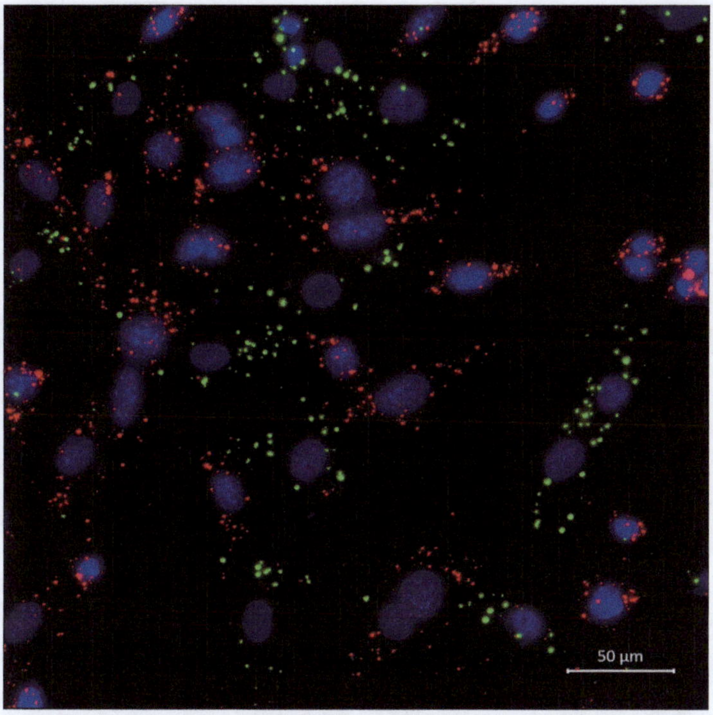

Fig. 3 In situ SNP detection of *ACTB* mRNA in human BjhTERT and mouse MEF cell line. Detection of *ACTB* SNP in human and mouse cell lines is represented as color-coded RCPs. *Green*: Detection of human *ACTB*; *Red*: Detection of mouse *ACTB* SNP. Size of the scale bar in the *inset* 50 μm

2. All RCPs should be localized to the cytoplasm, however RCPs can occasionally dissociate out of cells and generate RCPs on the glass slide (such signals should be disregarded).
3. RCPs can be easily distinguished from cellular autofluorescence by their defined brightness, shape and size.
4. Finally, RCP signal should be visible only in a single, fixed fluorescence channels (checking other channels is a good measure to evaluate signal specificity). Figure 3 shows *ACTB* mRNA genotyping in human BJhTERT and mouse MEF fibroblast cells. Mutation within the gene allowed designing two PLPs, which arms differ by only a single, 3′ terminal base (G/A in human/mouse). When PLPs for both alleles were used together in cocultured cells, SNP specific signals were exclusively generated in either human or mouse cells.
5. We applied the protocol to genotype and differentiate *PCDH X* and *Y* chromosome homologs in human embryo medulla oblongata section. This study provided detailed insight into the spatial distribution of PCDH11X/Y and NLGN4X/Y in human developing nervous tissue. Observed patterns suggested the significant development complexity in the male central nervous system even if single cell expression data was

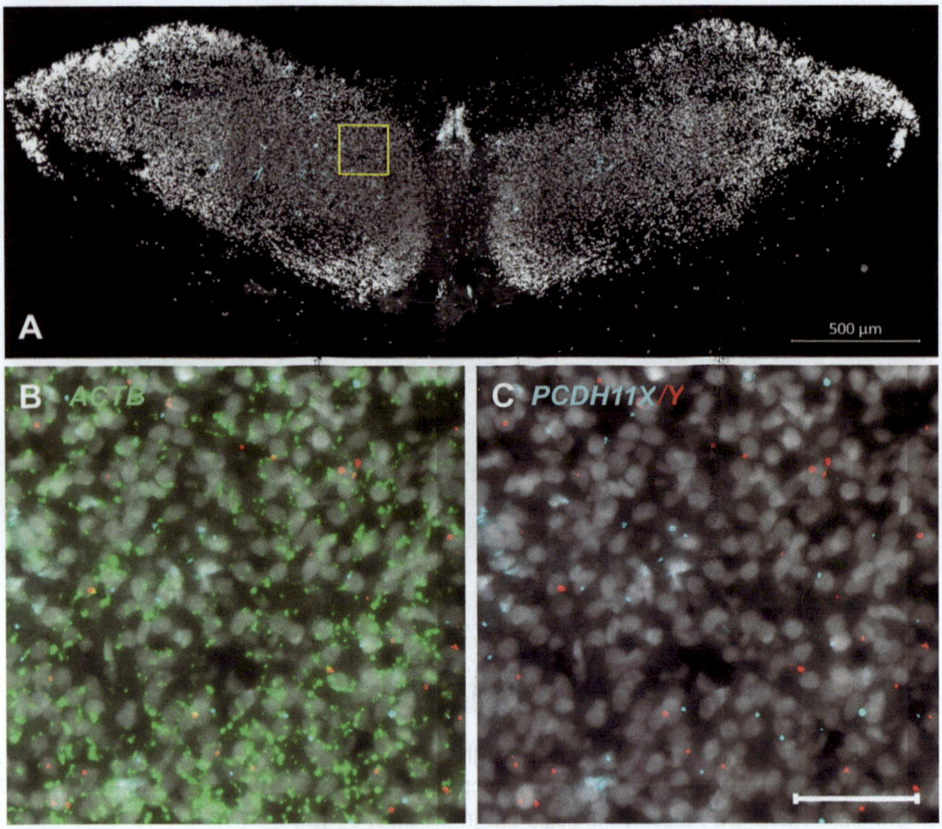

Fig. 4 In situ *PCDH11X/Y* SNP detection in male, Medulla oblongata embryo. Human male Medulla oblongata section is simultaneously hybridized with a single *ACTB* PLP, five probes for X chromosome *PCDH11* homolog and five for Y chromosome homolog of *PCDH11*. (**a**) Full Medulla oblongata section scan. *Inset* highlighted in *yellow* enlarged in **b**, **c**. (**b**) The same region of the section with staining for *ACTB* (*green*) and (**c**) *PCDH11X/Y* (*cyan* and *red* respectively). Size of the scale bar in the *inset* 50 μm. Probe sequences and experiment setup published in [18]

limited (*see* **Note 33**). Figure 4 shows a selected example of *PCDH X/Y* genotyping. Whole study including technical aspects and results is published in [18].

4 Notes

1. *Tth* DNA ligase exhibits significantly greater thermal stability and ligation specificity over traditional DNA ligases [15] and it works very well for SNP genotyping in our hands. Greater polymorphism discrimination specificity is generally observed when the discriminating nucleotide is present on the 3′ end of the PLP arm [15]. Alternative DNA ligases can be used for conventional mRNA detection and genotyping [13] or

mitochondrial DNA genotyping [19], however the SNP detection specificity should be evaluated experimentally for every target with appropriate controls (*see* Subheading 3.1).

2. LNA modified primers can be purchased from EXIQON (http://www.exiqon.com) with standard desalting as an adequate purification method. Use TE buffer (pH 8.0) to resuspend oligonucleotides to 100 μM stock. Accurate LNA bases substitution is crucial for optimal primer functionality. Too few LNA substitutions will have a minimal impact of the primer melting point, while too many may stabilize any predicted hairpins, homodimers, or off-target binding. Once primer sequence is known, we strongly advice computing primer secondary structure using any web-based secondary structure prediction tools. The OligoAnalyzer 3.1 tool from Integrated DNA Technologies (IDT), http://eu.idtdna.com/calc/analyzer works good in our hands as many parameters that usually influences DNA folding can be specified (RT conditions presented in this chapter are the following: LNA primer concentration 1 μM; monovalent and divalent cation concentration 50 mM and 4 mM respectively; and RT temperature 45 °C). LNA bases should be introduced as close to the primer 5′ end as possible, interspaced with one or more conventional DNA bases. Consecutive LNAs, evenly distributed in the primer sequence or 3′ end concentrated LNA substitutions should be avoided [20]. Based on secondary structure and homodimer formation predictions, LNAs should not be introduced in regions where dimer or hairpin formation probability is high. If no such positions are available, another primer from the primer3 list can be evaluated.
3. In our experience, longer distance between the SNP and the primer hybridization site negatively influences successful cDNA conversion and results in lowered signal amount (*see* Supplementary Note 3 and Fig. 2 in [13]). Primer can be positioned as close as 5 bp downstream SNP as long as PLP hybridization site is not overlapping with LNA bases (LNA bases likely prevent efficient PLP displacement during RCA).
4. ProbeMaker is available at http://probemaker.sourceforge.net/for both OSX and Windows platforms. Program allows importing single or batch cDNA targets in FASTA format for automated PLP design for each allele-specific variant (SNP within the sequence should be annotated as [N/N]). It is up to the user to define PLP arms binding properties like Tm and backbone sequence (including linkers and decorator probe motif). Since ProbeMaker does not correct the PLP arms T_m for the formamide presence so we typically set PLP arms T_m at 55–60 °C. Guide with examples how to use the software is available online.

5. Asymmetric arms are acceptable. We discourage designing overextended PLP arms. Even though the PLP can still generate detectable signal, high T_m can prevent probe melting if a PLP for another allele is transiently bound (in such case, the ligation will not take place and SNP will be "lost"). Monovalent and divalent ion concentrations, formamide, PLP concentration, and temperature influence the Tm of PLP arms. The hybridization and ligation reaction conditions exemplified in this chapter are the following: 75 mM monovalent ions, 10 mM divalent ions, 20% formamide, 0.1 μM padlock probe concentration, and 45 °C.
6. PLP backbone consists of "decorator motif" and linkers (Fig. 2c). Decorator motif in the PLP backbone and the corresponding decorator oligonucleotide have identical sequences (20–25 bp). While decorator motifs are unique for every mRNA or SNP target, linkers can be propagated as long as PLP secondary structure is not compromised. We typically use series of adenine nucleotides to match correct backbone length as they have minimal effect on PLP secondary structure (other bases are acceptable but can have greater impact on the structure).
7. PLP phosphorylation protocol: 10 μM final PLP concentration (can be changed if desired); 0.2 U/μL of T4 PNK kinase (e.g., Thermo Scientific); 1× PNK kinase buffer; 1 mM ATP and H_2O in a final volume of 50 μL. Mix should be incubated at 37 °C for 30 min, followed by enzyme inactivation at 65 °C for 20 min. Phosphorylated PLPs are stored at −20 °C for few months. Though we use large excess of primers and PLPs in both RT and ligation step of this protocol, slow retention of oligonucleotides on the wall of the tubes, combined with frequent freezing–thawing can influence probes performance. We recommend realiquoting oligonucleotides older than 6 months.
8. Decorator oligonucleotides can be conjugated with a fluorophore on any terminus. We routinely use 6-Carboxyfluorescein, Texas Red, Cyanine or Alexa Fluor® dyes. Decorator probe sequence should be blasted (http://blast.ncbi.nlm.nih.gov/Blast.cgi.) against refseq mRNA database to minimize nonspecific oligo binding and autofluorescence build-up.
9. Though PLPs are surprisingly tolerant to secondary structures, one should select PLP with the highest free energy. Small internal loops or hairpins are acceptable but should be avoided in the region involving recognition of the target sequence (can limit hybridization of PLP arms) or within decoration motif (can hinder hybridization of the decorator probe). Shortening

an arm where a strong hairpin is present can aid hybridization and can be compensated by extending another arm. Finally, if PLPs are used in a set, formation of heterodimers or probability of probe-on-probe ligation should be evaluated. We commonly use "two-state melting" engine to evaluate multiple probe hybridization probability (http://unafold.rna.albany.edu/?q=DINAMelt/Two-state-melting).

10. Enzymes (including reverse transcriptase, RNase inhibitor, phi29 polymerase and RnaseH) from well-known vendors like NEB or Fermentas (Thermo Fisher Scientific) performs equally well in our hands.

11. DEPC inhibits RNases present in water, buffers or labware by irreversible, covalent modification of selected amino acids [21]. Following DEPC treatment, all solutions should be autoclaved to break down DEPC residue. Less dangerous alternatives to DEPC, such as DMPC, can be considered.

12. We recommend using freshly prepared formaldehyde solutions in DEPC-PBS. Working solutions can be prepared form either 37% methanol-stabilized stock solution or from paraformaldehyde powder. Aliquots of 3.7% formaldehyde in DEPC-PBS in 1 mL (used during the experiment) and 15 mL (for cell fixation) can be stored at −20 °C. Do not freeze and thaw.

13. Lyophilized pepsin batches may vary in activity, even from the same supplier. Every new batch of pepsin should be tested on established model. We typically detect housekeeping gene (*ACTB* or *GAPDH*) in the same tissue type to evaluate detection reproducibility.

14. Far-red dyes are more susceptible to photobleaching. SlowFade® Gold Antifade Mountant works best in our experience.

15. Keep 0.1 v/v % DEPC in PBS or ddH_2O for at least 1 h at 37 °C (or overnight at RT), followed by autoclaving.

16. Secure-Seal chambers come in different sizes, shapes and depths. For experiments performed on fixed cell lines, we typically use ~50 μL chambers (round, 9 mm diameter, and 0.8 mm deep). For larger cell areas or larger tissue specimen, 100 μL or 350 μL chambers can be used.

17. To achieve optimal optical resolution, cover glass thickness needs to be adjusted for the microscope setup used.

18. For low copy number transcripts we typically genotype *KRAS* codon *12* SNPs [13] and for abundant mRNAs, *ACTB* in human/mouse fibroblasts can be detected.

19. Unlike in many in situ hybridization methods, where cells are grown or coverslips, we grow cells on the microscope slides directly. We found slides to be more resistant to breaking (especially when applying and pealing the silicone chambers

off) and on the slides, multiple experiments can be run in parallel (up to eight), making the work more convenient. Also, multiple consecutive tissue section can be placed on a single slide to facilitate fast experimenting and imaging.

20. Optimal seeding conditions should be identified experimentally for every cell line. For cells with large cytoplasm, 3 mL suspension is usually enough to create homogenous cell layer on each slide. Cells with smaller cytoplasm can be seeded at higher density. In our experience, overnight incubation allows cells to adhere to slides efficiently. Extended incubation can result in cell proliferation on-slide (too dense or clumped cells are difficult to analyze by image analysis software).
21. Formaldehyde, HCl and formamide should be disposed in accordance to local lab regulations.
22. Tween 20, as a surfactant, coats the chambers, facilitates buffer exchange and prevents formation of" dead spaces" inside the chamber. As a detergent, it can provoke bubble formation. We recommend adding a buffer into a chamber when slide is slightly tilted (gravity helps to fill up the chamber evenly).
23. Experiment can be halted at this point if necessary. In such case, replace 1× PBS-T with 1× PBS and keep the slide at 4 °C for up to 24 hours. We recommend proceeding with the protocol without interruptions until the cDNA is synthesized to minimize risk for mRNA degradation.
24. Any slides that enhance adhesion of tissue sections can be used. (SuperFrost Plus® from Menzel-Gläser work very well in our hands). Sections should be as thin as possible (preferably few cell layer). Thinner sections are more prone to break and fold during cutting. We commonly use 10 μm thick sections.
25. The fixation time needs to compromise optimal fixative diffusion and minimize loss of tissue content. We recommend pragmatic evaluation of fixation parameters (consecutive sections should be used for each condition). Fixation time may vary for different tissues and different specimen thicknesses. Housekeeping gene is a good candidate for such optimization studies. Use conditions showing maximal signal amount.
26. RNase H has the highest activity at 37 °C. It degrades RNA from mRNA/cDNA heteroduplex during the first 37 °C incubation. After 30 min, sample is transferred to 45 °C which is the optimal temperature for the *Tth* ligase. Addition of formamide into the mix lowers dsDNA stability (T_m of PLP arms/ cDNAduplex). This enables extension of PLP arms that strengthens probe "locking" on cDNA and gives a good balance between arms melting and specific binding.

27. The optimal temperature for phi29 polymerase is 37 °C. We typically conduct RCA for 1 h. If RCA is performed for several hours (or over-night) at 37 °C, RCPs may start to fragment what could interfere with accurate signal counting. If large RCPs are desired (thick tissue sections or those with high autofluorescence), we advise doing RCA at RT over-night. Such approach will generate very large but compact RCPs.
28. In a multiplexed reaction (when more than one decorator probe is used), we recommend incubation at 37 °C for 30 min (to minimize nonspecific binding of the oligonucleotides). In such case, cover the chamber inlets to prevent mix evaporation.
29. A double edge razor can be used to facilitate complete removal of the Secure-Seal chamber.
30. We typically apply 5–7 μL of mounting medium for single, 50 μL Secure-Seal chamber. Remove excess of the medium by gently pressing the slide against a coverslip (excess of medium will be absorbed by the paper towel).
31. We use a Zeiss Axioplan II Epifluorescence microscope equipped with a 100-W mercury lamp and a Hamamatsu, C4742-95 CCD camera. The following filter setup provides good wavelength separation and minimal crosstalk between different channels. 38HE (Zeiss) for imaging GFP/FITC/FAM dyes; SP102v2 (Chroma) for imaging Cy3 (minimal crass talk with 38HE filter); SP103v2 (Chroma) for imaging Cy3.5/TexasRed; SP104v2 (Chroma) for imaging Cy5; 49007 (Chroma) for imaging Cy7/Alexa 7.5 dyes.
32. CellProfiler is a great, user-friendly tool to aid biologists in image processing and analysis. With respect to presented protocol, CellProfiler offers scripts for cell segmentation (definition of the nucleus and the cytoplasm), RCP segmentation, and assignment of RCPs to individual cells or fluorescence measurements. All scripts can be implemented in automated pipeline, allowing for batch image processing. An example script for cell and RCP identification is available at CellProfiler website http://www.cellprofiler.org. Briefly, gray scale TIFF images (offering highest resolution, JPEG images are processed faster and can also be used) from individual fluorescence channels are loaded into the pipeline. Cells are segmented to nuclei and cytoplasm and RCPs are identified and related to neighboring cells. Finally, number of RCPs for each cell is exported as a .csv file, which can be used for post-analysis processing.
33. The overall sensitivity of mRNA detection in situ in cell lines using the protocol provided has been estimated to be ~30% [13]. This efficiency parameter is valid for high quality samples

(freshly prepared cell lines) and might be lower in fresh frozen material or older, more degraded FFPE sections. mRNA detection efficiency can be further decreased if more than five LNA primers are used simultaneously (most likely due to formation of strong primer–primer hybrids, enhanced by the presence of LNA analogs). Pragmatic primer evaluation to minimize probability of primer dimer formation in the whole probe pool could possibly decrease this effect. Assuming that protocol was executed as presented and all reagents where fresh, fully active (enzymes) and RNase free, we still observe signal variation for certain mRNAs with similar RPKM values. If mRNA is highly structured in the target SNP region, redesigning an LNA primer (or using random decamers) can aid RT. Relatively low sensitivity of the in situ protocol can limit single cell mRNA detection for low expression level transcripts. Though we routinely detect single cell SNP in mRNAs with RPKM ~9 like *KRAS* ([22] Fig. S1, [23] Supplementary Table 1) successful qualitative or quantitative analyses of mRNA might not require single cell resolution. Measurement of *KRAS* codon 12 and 13 point mutations ratio in selected whole-specimen area was shown to be a good approach for molecular diagnostic scoring in colorectal cancer [22]. In [18], allele signal density estimation plots were used to elucidate expression pattern differences in specimen where single cell data was limited.

References

1. Kitpipit T, Tobe SS, Kitchener AC et al (2012) The development and validation of a single SNaPshot multiplex for tiger species and subspecies identification – implications for forensic purposes. Forensic Sci Int Genet 6:250–257
2. Huang C, Chang M, Huang M, Lee F (2011) Rapid identification of Lactobacillus plantarum group using the SNaPshot minisequencing assay. Syst Appl Microbiol 34:586–589
3. Diekstra A, Bosgoed E, Rikken A et al (2015) Translating Sanger-based routine DNA diagnostics into generic massive parallel ion semiconductor sequencing. Clin Chem 61:154–162
4. Lubeck E, Cai L (2012) Single-cell systems biology by super-resolution imaging and combinatorial labeling. Nat Methods 9:743–748
5. Player AN, Shen LP, Kenny D et al (2001) Single-copy gene detection using branched DNA (bDNA) in situ hybridization. J Histochem Cytochem 49:603–612
6. Choi HMT, Beck V, Pierce N (2014) Next-generation in situ hybridization chain reaction: higher gain, lower cost, greater durability. ACS Nano 8:4284–4294
7. Nilsson M, Malmgren H, Samiotaki M et al (1994) Padlock probes: circularizing oligonucleotides for localized DNA detection. Science 265:2085–2088
8. Nilsson M, Banér J, Mendel-Hartvig M et al (2002) Making ends meet in genetic analysis using padlock probes. Hum Mutat 19:410–415
9. Petersen M, Wengel J (2003) LNA: a versatile tool for therapeutics and genomics. Trends Biotechnol 21:74–81
10. Fire A, Xu SQ (1995) Rolling replication of short DNA circles. Proc Natl Acad Sci U S A 92:4641–4645
11. Banér J, Nilsson M, Mendel-Hartvig M, Landegren U (1998) Signal amplification of padlock probes by rolling circle replication. Nucleic Acids Res 26:5073–5078
12. Untergasser A, Cutcutache I, Koressaar T et al (2012) Primer3-new capabilities and interfaces. Nucleic Acids Res 40:1–12
13. Larsson C, Grundberg I, Söderberg O (2010) In situ detection and genotyping of individual mrna molecules. Nat Methods 7:395–397

14. Stenberg J, Nilsson M, Landegren U (2005) ProbeMaker: an extensible framework for design of sets of oligonucleotide probes. BMC Bioinformatics. doi:10.1186/1471-2105-6-229
15. Luo J, Bergstrom DE, Barany F (1996) Improving the fidelity of Thermus thermophilus DNA ligase. Nucleic Acids Res 24:3071–3078
16. Zuker M (2003) Mfold web server for nucleic acid folding and hybridization prediction. Nucleic Acids Res 31:3406–3415
17. Carpenter AE, Jones TR, Lamprecht MR et al (2006) CellProfiler: image analysis software for identifying and quantifying cell phenotypes. Genome Biol 7:R100
18. Johansson MM, Lundin E, Qian X et al (2016) Spatial sexual dimorphism of X and Y homolog gene expression in the human central nervous system during early male development. Biol Sex Differ 7:5
19. Larsson C, Koch J, Nygren A et al (2004) In situ genotyping individual DNA molecules by target-primed rolling-circle amplification of padlock probes. Nat Methods 1:227–232
20. Levin JD, Fiala D, Samala MF et al (2006) Position-dependent effects of locked nucleic acid (LNA) on DNA sequencing and PCR primers. Nucleic Acids Res 34:1–11
21. Wolf B, Lesnaw JA, Reichmann ME (1970) A mechanism of the irreversible inactivation of bovine pancreatic ribonuclease by diethylpyrocarbonate. Eur J Biochem 13:519–525
22. Grundberg I, Kiflemariam S, Mignardi M et al (2013) In situ mutation detection and visualization of intratumor heterogeneity for cancer research and diagnostics. Oncotarget 4:2407–2418
23. Ke R, Mignardi M, Pacureanu A et al (2013) In situ sequencing for RNA analysis in preserved tissue and cells. Nat Methods 10:857–860

Chapter 15

Quantifying Gene Expression in Living Cells with Ratiometric Bimolecular Beacons

Yantao Yang, Mingming Chen, Christopher J. Krueger, Andrew Tsourkas, and Antony K. Chen

Abstract

Molecular beacons (MBs), a class of oligonucleotide-based probes, have enabled researchers to study various RNA molecules in their native live-cell contexts. However, it is also increasingly recognized that, when delivered into cells, MBs have the tendency to be sequestered into the nucleus where they may generate false positive signals. In an attempt to overcome this issue, MBs have been synthesized with chemically modified oligonucleotide backbones to confer greater biostability. Alternatively, strategies have been developed to minimize nuclear entry. In the latter approach, we have combined functional elements of MBs with functional elements of siRNAs that facilitate nuclear export to create a new RNA imaging platform called ratiometric bimolecular beacons (RBMBs). We showed that RBMBs exhibited long-term cytoplasmic retention, and hence a marginal level of false positive signals in living cells. Subsequent studies demonstrated that RBMBs could sensitively and accurately quantify mRNA transcripts engineered to contain multiple tandem repeats of an MB target sequence at the single-molecule level. In this chapter, we describe the synthesis of RBMBs and their applications for absolute quantification and tracking of single mRNA transcripts in cells.

Key words Molecular beacons, Ratiometric bimolecular beacons, RNA imaging, Single-molecule detection, Single-molecule fluorescence in situ hybridization, Quantifying gene expression

1 Introduction

Subcellular trafficking and localization of specific RNA molecules are crucial activities underlying many cellular processes, including embryonic development, stem cell differentiation and neuronal growth. Knowledge of RNA activities at the single-molecule level is expected to facilitate understanding of the roles that RNAs play in the regulation of these processes. One tool that has been used to study RNA in living cells is the molecular beacon [1], a class of antisense stem-loop-forming oligonucleotide (oligo) probe with the ends labeled with a fluorophore and a quencher. In the unhybridized state, self-complementation of the stem sequences at the

Imre Gaspar (ed.), *RNA Detection: Methods and Protocols*, Methods in Molecular Biology, vol. 1649, DOI 10.1007/978-1-4939-7213-5_15, © Springer Science+Business Media LLC 2018

ends of the probe places the quencher in close proximity to the reporter fluorophore to quench its fluorescence. Hybridization of the target RNA to the loop domain disrupts the stem, opening the probe and thus separating the quencher from the reporter, which can then emit a detectable fluorescence upon excitation.

Although many studies have used MBs for RNA detection, increasing evidence has indicated that MBs have limited accuracy and sensitivity for detecting intracellular RNAs, owing to their tendency to be sequestered in the nucleus and generate false-positive signals [2–4]. Additionally, because heterogeneous delivery can result in large variations in intracellular MB fluorescence, it could be difficult to determine whether high MB fluorescence reflects high RNA expression or uneven MB delivery. To overcome these drawbacks, we developed a novel probe called ratiometric bimolecular beacons (RBMBs) (Fig. 1) [5]. RBMBs are synthesized by hybridizing a stem-loop forming oligo and a linear oligo. The stem-loop forming oligo has a short and a long stem arm, and is singly labeled with a reporter dye at its short-arm terminus (5′-end). The longer arm is perfectly complementary to the linear oligo, which is labeled with a quencher and a reference dye at the two termini. Hybridization of the two oligos brings the quencher into close proximity to the reporter dye to quench its fluorescence, while the reference dye is situated at the far end of the hybrid and thus remains unquenched. Additionally, hybridization results in a 2-base (UU) single-stranded overhang at the 3′-end of the duplex, which resembles an siRNA molecule that is efficiently exported from the nucleus by exportin [5].

We have shown that, similar to MBs, RBMBs can emit a distinguishable fluorescence signal when hybridized to target RNA. Additionally, the siRNA-like domain functions to facilitate nuclear export and the unquenched reference dye allows for measurements of reporter fluorescence to adjust for cell-to-cell variability in RBMB delivery. Furthermore, we have shown that these attributes render RBMBs more effective for sensitive and accurate intracellular measurements of RNA (Fig. 1), as compared with conventional MBs [5, 6]. In this chapter, we describe synthesis of RBMBs and the assessment of their utility for quantifying single engineered mRNA transcripts in cells.

2 Materials

2.1 Plasmids

1. pLenti-d2EGFP-96mer encodes a transcript harboring 96 tandem repeats of the 50-base sequence: 5′-<u>CAGGAGTTGTGTTTGTGGACGAAG</u>AGCACCAGCCAGCTGATCGACCTCGA-3′ downstream of the d2EGFP coding sequence. The underlined sequence is the MB target sequence. When cotransfected with viral packaging vectors into a cell

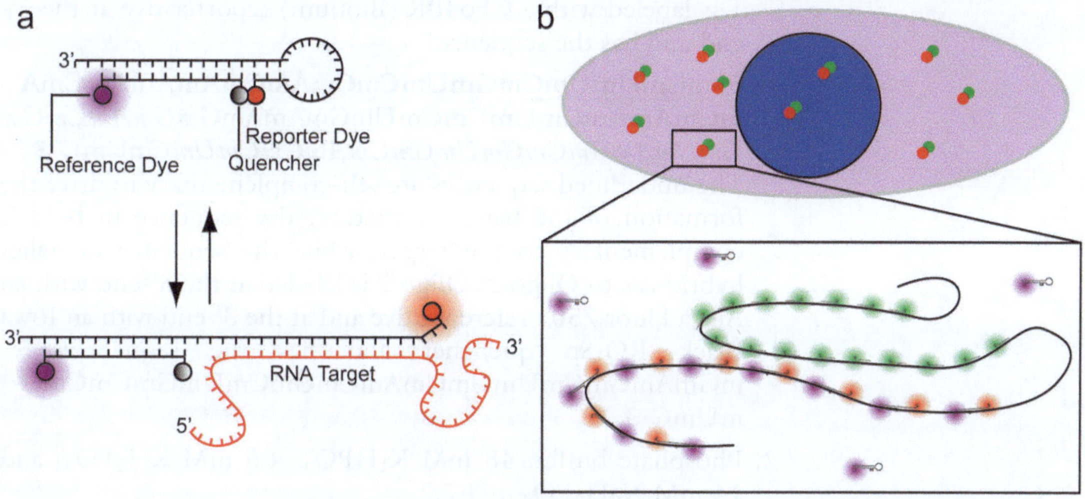

Fig. 1 Design of RBMBs and strategies for assessing RBMB performance for imaging single RNA transcripts in cells. (**a**) RBMBs are stem-loop-forming oligo probes that possess a reporter dye, quencher, and reference dye. In the absence of target, the reporter dye is held in close proximity to the quencher, leading to a low-fluorescent state. Hybridization of the loop-domain to a complementary target separates the reporter dye and quencher, resulting in the restoration of fluorescence. The reference dye remains unquenched independent of the RBMB configuration. The 3′-UU overhang drives nuclear export. (**b**) To assess the ability of RBMBs to quantify RNA copy numbers in single cells, smFISH is performed following microporation of RBMBs. In this figure, smFISH probes are *green*, while RBMB reporter dye and reference dye are *red* and *magenta*, respectively. Hybridization of the RBMBs or smFISH probes to each RNA transcript results in *bright punctate spots* within cells when imaged by conventional fluorescence microscopy. Colocalization of RBMB signal with the smFISH signal indicates successful detection of the RNA transcript by RBMBs (Reproduced from [6] with permission from Oxford University Press)

suitable for virus packaging, such as the HEK-293FT cell, the resulting lentiviral particles containing the pLenti-d2EGFP-96mer gene can be efficiently introduced into host cells (*see* **Notes 1** and **2**).

2.2 Cell Culture

1. HEK-293FT.
2. HT1080 cells.
3. HT1080-d2EGFP-96mer cells that stably express engineered d2EGFP mRNA transcripts with 96 tandem repeats in the 3′-untranslated region (UTR), generated by infecting HT1080 cells with pLenti-d2EGFP-96mer.
4. Minimum essential media (MEM) supplemented with 1% Pen/Strep and 10% FBS.
5. Phenol red-free solution of 0.25% trypsin and 1 mM EDTA.
6. 1× phosphate-buffered saline (PBS), free of Mg^{+2} and Ca^{+2}.

2.3 RBMB Design and Synthesis

1. Oligonucleotides: RBMBs are composed of two 2′-O-methyl RNA oligonucleotides, denoted as Oligo 1 and Oligo 2. Oligo

1 is labeled with a CF640R (Biotium) reporter dye at the 5′-end and has the sequence **5′-mCmUmUmCmGmUmCmCmAmCmAmAmAmCmA mCmAmAmCmUmCmCmU**mGmAmAmG*mGmAmCmGm GmCmAmGmCmGm UmGmCmAmGmCm Um*CmUmU-3′. The underlined sequences are self-complementary to drive the formation of the hairpin structure, the sequence in bold is complementary to the target, while the sequence in italics hybridizes to Oligo 2. Oligo 2 is labeled at the 5′-end with an Alexa Fluor 750® reference dye and at the 3′-end with an Iowa Black RQ-Sp quencher and has the sequence: 5′-mGmAmGmCmUmGmCmAmCmGmCmUmGmCmCmG-mUmC-3.

2. Phosphate buffer: 48 mM K_2HPO_4, 4.5 mM KH_2PO_4, and 14 mM NaH_2PO_4, pH 7.2.
3. Prep grade gel filtration column (e.g., GE Healthcare Superdex 200 or Superdex 75).
4. 10,000 MW cutoff centrifugal device (e.g., Millipore Microcon YM-10).
5. UV-Vis spectrophotometer.

2.4 Microporation

1. Microporation system (e.g., Thermo Fisher Neon transfection system).
2. MEM without antibiotics, supplemented with 10% FBS and 1× GlutaMAX (Thermo Fisher).
3. Resuspension buffer R (Thermo Fisher).
4. Electroporation buffer (Thermo Fisher).
5. Electroporation Gold Tips (10 μL size) (Thermo Fisher).
6. Electroporation tube (Thermo Fisher).
7. Microporator Pipette (Thermo Fisher).
8. 8-well chambered cover glass (e.g., Nalgene Nunc Lab-Tek).

2.5 Single-Molecule Fluorescence In Situ Hybridization

1. Nuclease-free water.
2. 4 w/v % paraformaldehyde diluted in 1× PBS.
3. 70 v/v % ethanol, prepared from anhydrous ethanol.
4. 2× SSC.
5. Wash buffer (2× SSC, 10 v/v % formamide).
6. Hybridization buffer (10 w/v % dextran sulfate, 2× SSC, 10 v/v % formamide).
7. d2EGFP FISH probes [7], a set of singly labeled probes that are complementary to different regions of the d2EGFP coding sequence.
8. Parafilm.

2.6 Microscope and Imaging Software

1. An inverted widefield fluorescence microscope. We use an Olympus IX81 motorized inverted fluorescence microscope equipped with a back-illuminated EMCCD camera (Andor), a SOLA Light Engine (Lumencor, Beaverton, OR), and a CoolLED pE-100 (740 nm, CoolLED pE-100).
2. 100× NA 1.45 objective lens.
3. Filter set for DAPI (D350/50, D460/50, 400dclp).
4. Filter set for TMR (HQ545/30, HQ620/60, Q570lp).
5. Filter set for CF640R (ET620/60x, ET700/75 m, T660lpxr).
6. Filter set for Alexa®750 (HQ710/75, HQ810/90 m, Q750lp).
7. FIJI software.
8. MATLAB (Version R2014b 64-bit, MathWorks) software.

3 Methods

3.1 RBMB Synthesis

1. Hybridize Oligo1 and Oligo2: Combine equimolar quantities of Oligo 1 and Oligo 2 in phosphate buffer in a low-retention microcentrifuge tube so that the final concentration of each oligo is 50 μM and the total volume is at least 50 μL.
2. Incubate the oligos on a thermomixer at 37 °C overnight.
3. Separate RBMBs from unhybridized oligos using a Superdex 200 or a Superdex 75 prep-grade column.
4. Concentrate purified RBMBs using a Microcon YM-10 centrifugal device (*see* **Note 3**).
5. Determine the final concentration of the RBMBs on a spectrophotometer.

3.2 Cellular Delivery of RBMBs

RBMBs are delivered into cells by microporation, a physical method that previous studies show enables efficient cellular delivery of oligo probes with high viability [5]. Specific procedures to perform microporation are described below:

1. Seed cells in tissue culture flasks in MEM growth media without phenol red or antibiotics and let them grow overnight. Be cautious not to exceed 70% confluence on day 2 (*see* **Note 4**).
2. Aspirate the media from the cells.
3. Rinse the cells with 5 mL of 1× PBS for about 2 min.
4. Aspirate the PBS and add 1 mL of phenol red-free trypsin/EDTA.
5. Following incubation of the cells in trypsin for several minutes, add media to neutralize the enzyme (*see* **Note 5**).
6. Gently resuspend the cells in media.

7. Transfer 1–1.5 mL of the cell suspension to a microcentrifuge tube.
8. Pellet the cells by centrifugation at 200 × *g* for 5 min at 4 °C.
9. Aspirate the media and gently resuspend the cell pellet in 1 mL of 1× PBS.
10. Count cells.
11. Pellet the appropriate number of cells for microporation at 400 × *g* for 5 min at 4 °C. 30,000 HT1080 cells are used for each microporation (*see* **Note 6**).
12. Aspirate the PBS and resuspend the cell pellet in resuspension buffer R at 30,000 cells per 10 μL.
13. Add 1 μL of RBMBs per 10 μL of cells such that the final concentration of RBMB is between 0 and 5 μM and pipette gently to mix (*see* **Notes 7** and **8**).
14. Microporate 10 μL of the cell suspension (approximately 30,000 cells) with 2 pulses for 25 ms at 950 V (*see* **Note 9**).
15. Following microporation, transfer microporated cells to a microcentrifuge tube prefilled with 1.5 mL of cell culture media.
16. Pellet the cells at 400 × *g* for 5 min at 4 °C.
17. Aspirate the media, being careful not to disturb the cell pellet. Add 1.5 mL of cell culture media to resuspend the pellet.
18. Repeat **steps 16–17** another two times (*see* **Notes 10** and **11**).
19. After the last wash, resuspend the cells in 250 μL of cell culture media.
20. Seed the cells in one well of an 8-well chambered cover glass until the samples are ready for imaging.

3.3 Single-Molecule Fluorescence In Situ Hybridization

1. Carefully pipette out the media from each seeded well of the 8-well chambered cover glass.
2. Gently wash the cells thrice with 250 μL 1× PBS.
3. After carefully pipetting out the PBS, in each well gently add 250 μL of 4% PFA prewarmed at 37 °C.
4. Incubate the cells in 4% PFA for 10 min at room temperature.
5. After removing PFA, add 300 μL 1× PBS to each well. Incubate the sample in 1× PBS for about 5 min. Carefully pipette out the 1× PBS and repeat this step once more.
6. After removing 1× PBS, add 400 μL of 70% ethanol.
7. Close the lid and wrap the chambered cover glass with Parafilm to minimize evaporation.
8. Store the chamber, protected from light, at 4 °C for at least 16 h.
9. Carefully pipette out the 70% ethanol.

10. Gently add 350 μL wash buffer, incubate for 5 min, and carefully pipette off. Repeat this step once more.
11. After gently removing the wash buffer, add 250 μL of the TAMRA-labeled d2EGFP mRNA smFISH probe sets (250 nM in hybridization buffer).
12. Store the chambered cover glass in a humidified chamber at 37 °C for at least 16 h covered with Parafilm to minimize evaporation.
13. Gently pipette out the d2EGFP smFISH probes.
14. Quickly and gently wash each well by adding 350 μL of wash buffer and incubating for 5 min. Repeat this step twice. Do not allow wells to dry between washes.
15. After the second wash, gently add 400 μL wash buffer to each well. Incubate the samples in a humidified chamber at 37 °C for 30 min. The samples should be covered with tinfoil.
16. Aspirate wash buffer and wash the samples twice with 2× SSC. Do not let the samples dry between washes.
17. Aspirate the 2× SSC and gently add another 300 μL of 1× PBS to each sample. It is optional to include DAPI in the 1× PBS.
18. Image in three dimensions with 0.25 μm increments in the *z*-direction using Cy5 and TRITC filter sets.
19. Save the images for each channel as a TIFF image stack.

3.4 RNA Identification and 3D-Colocalization Analysis

1. Open the image stack for each channel using Fiji [8] (*see* **Note 12**).
2. Draw a Region of Interest (ROI) around the cell of interest using the 'Freehand selections' tool.
3. Open the ROI Manager by selecting Analyze > Tools > ROI Manager.
4. Add the ROI to the list in the ROI Manager. This ROI will be used after completing **steps 5–9**. Be sure not to close the ROI Manager toolbar before obtaining the spot information for both Cy5 and TRITC channels. (This ROI is used for both channels and should not be deleted.)
5. Enhance particulate objects by selecting Process > Subtract Background and set the rolling ball radius to 2.0 pixels. Make sure to process all images (*see* **Notes 13** and **14**).
6. Identify particles in an image sequence by selecting Plugins > LoG3D. Set sigma X and sigma Y to 1, and sigma Z to 0. Select 'Process slice per slice' (*see* **Notes 15–17**).
7. Filter out particles contributed by background noise by selecting Image > Adjust > Threshold. Adjust the threshold value so the puncta resemble the bright spots visible in each slice of the original image sequence. Apply the threshold, with background pixels set to NaN, to all images in the stack.

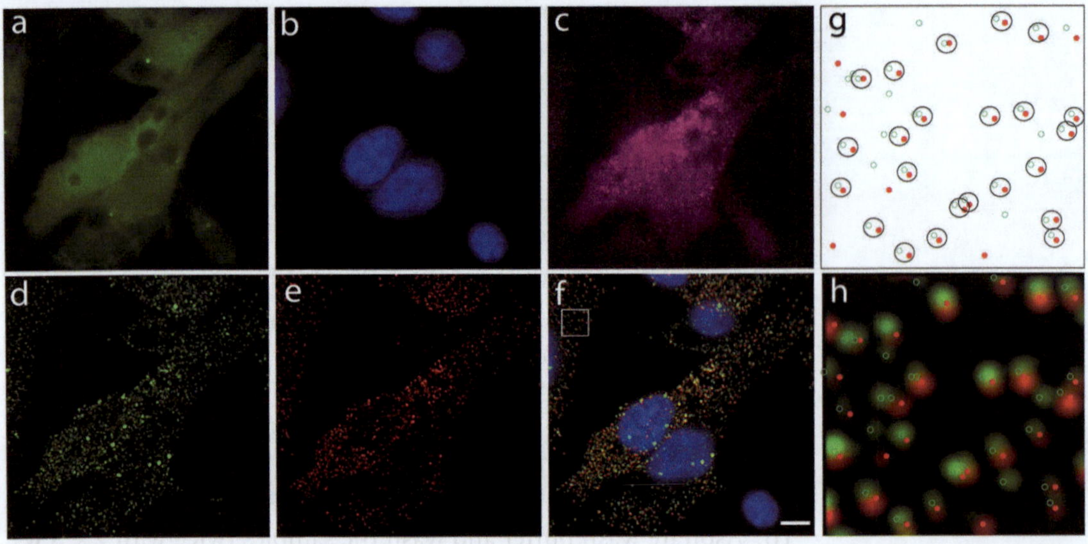

Fig. 2 Fluorescent images and analysis of engineered RNA transcripts in cells following RBMB delivery and smFISH. Five hours following microporation of HT1080-d2EGFP-96mer cells with 0.8 μM RBMBs, the cells were then fixed and smFISH was performed. (**a**) GFP, (**b**) DAPI, (**c**) RBMB-reference dye (Alexa 750), (**d**) smFISH probes (TMR) and (**e**) RBMB-reporter dye (CF640R) were acquired. Images (**a**) and (**b**) are widefield fluorescence images. Images (**c**), (**d**) and (**e**) are maximum intensity projection images. (**f**) A merged image of (**b**), (**d**) and (**e**). Individual spots in the smFISH and RBMB-reporter channels were identified and used to calculate the percent colocalization between the two signals using custom a MATLAB program. (**g**) Example MATLAB analysis that identifies smFISH (*green*) and RBMB-reporter (*red*) signals within the region outlined by a *white box* in panel (**f**). Identified colocalization events are enclosed within *black circles*. (**h**) MATLAB output overlaid on the source micrograph (Reproduced from [6] with permission from Oxford University Press)

8. Apply the preset ROI (**step 4**) to the filtered stack by selecting it from the ROI manager. The ROI should appear in each slice of the image sequence.
9. Identify single spots within the ROI by selecting Plugins > Macros > FindStackMaxima. Find the local maxima in each slice by selecting Find Maxima and setting the Noise Tolerance to 10, Output Types to Single Points and Exclude Edge Maxima (*see* **Notes 18** and **19**).
10. Invert the local maxima stack by selecting Edit > Invert. Select "Yes" to process all images (*see* **Note 20**).
11. Repeat **steps 1–10** for the second fluorescence channel using the same ROI saved in the ROI manager.
12. In each processed stack, use a custom MATLAB program to identify which 2D local maxima are 3D local maxima to remove overcounted maxima and for 3D colocalization analysis (*see* **Notes 21** and **22**). Representative fluorescence images and analysis results are shown in Figs. 2 and 3.

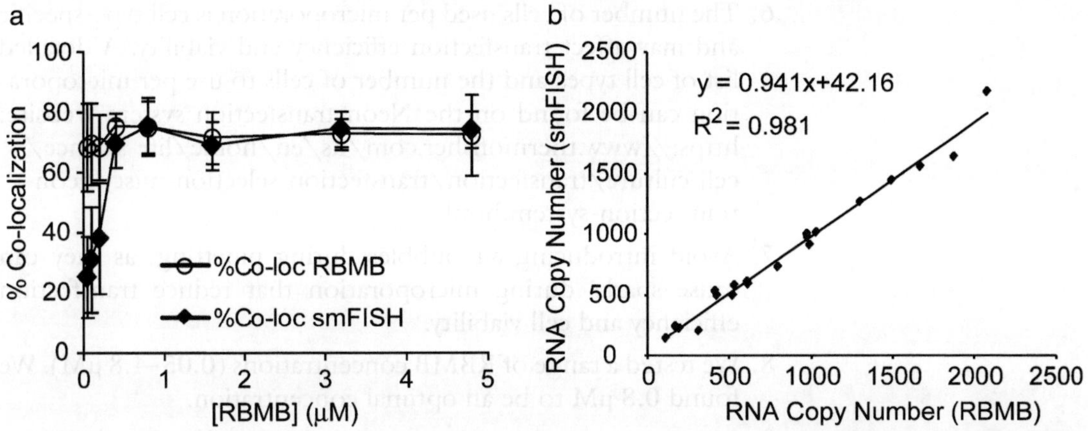

Fig. 3 Quantifying single RNA transcripts using RBMBs and smFISH. Five hours following microporation of HT1080-d2EGFP-96mer cells with 0.05–4.8 μM of RBMBs, the cells were fixed and permeabilized and smFISH was performed to assess the accuracy of MBs for detecting single RNA transcripts. (**a**) A custom Matlab program was written to analyze the percentage of MB signals that were colocalized with smFISH signals (*open circles*) and the percentage of smFISH signals that were colocalized with MB signals (*black diamonds*) in individual cells. Data represent mean ± standard deviation of at least 20 cells. (**b**) RNA copy number per cell reported by smFISH and RBMBs (0.8 μM) are in good agreement (Reproduced from [6] with permission from Oxford University Press)

4 Notes

1. Due to the multiple tandem repeats, the plasmid should be amplified in *Escherichia coli* MAX Efficiency Stbl2 (Thermo Fisher) at 30 °C to minimize recombination, which results in sequence deletion and/or rearrangement.
2. d2EGFP is a destabilized protein derived from EGFP and has a shorter intracellular lifetime than EGFP. Genes encoding EGFP or other proteins can also be used, provided that the protein expression can easily be detected at the single-cell level and that the mRNA sequence is unique in the cell type of interest to allow for the design of specific smFISH probes for single-molecule RNA detection.
3. Avoid centrifugation at speeds >5000 × *g* to avoid the RBMBs drying on the membrane.
4. To achieve high transfection efficiency and cell viability, it is important to ensure that cells are not >70% confluent prior to microporation.
5. Enough time should be given for trypsinization to ensure all of the cells are detached from the surface. Avoid vigorously pipetting the cells.

6. The number of cells used per microporation is cell type-specific and may affect transfection efficiency and viability. A detailed list of cell types and the number of cells to use per microporation can be found on the Neon transfection system website: https://www.thermofisher.com/us/en/home/life-science/cell-culture/transfection/transfection-selection-misc/neon-transfection-system.html.
7. Avoid introducing air bubbles during pipetting, as they can cause sparks during microporation that reduce transfection efficiency and cell viability.
8. We tested a range of RBMB concentrations (0.05–4.8 μM). We found 0.8 μM to be an optimal concentration.
9. Microporation parameters are cell type-specific. See the Neon Transfection system website (*see* **Note 6**) for optimal parameters for each cell line.
10. As untransfected probes can emit background fluorescence and thus hamper sensitive RNA detection, make sure that after microporation cells are sufficiently washed with enough media to remove untransfected probes.
11. After each wash, it is recommended not to aspirate out all of the wash media as this may increase the risk of aspirating out the loose cell pellet formed by the small number of cells.
12. Analysis can be performed using Fiji or ImageJ. Fiji can be downloaded at: http://fiji.sc/. ImageJ can be downloaded at: https://imagej.nih.gov/ij/.
13. In our experience, setting rolling ball radius to 2.0 pixels can enhance particulate objects without losing too much information for RNA analysis. It is recommended that the operator tries different parameters to find an optimal value.
14. The background can be further subtracted from the ROI of the cell. This could be achieved by measuring the pixel intensity of several ROIs outside the cell and subtracting the average value from the total image using Process > Math > Subtract.
15. The LoG3D plugin can be downloaded from: http://bigwww.epfl.ch/sage/soft/LoG3D/. To activate, place the file in the 'Plugin' folder of Fiji or ImageJ.
16. Images appear in gray scale after processing.
17. In our experience, these parameters enable accurate identification of particulate objects. It is recommended that the operator tries different parameters to find optimal values for their microscopy setup.
18. The 'FindStackMaxima.ijm' file should be placed in the Macros folder of Fiji or ImageJ. Install the file by selecting Plugins >

Macros > Install. The plugin can be obtained from: https://imagej.nih.gov/ij/macros/.

19. In our experience, setting noise tolerance to 10 enables better identification of spots. The operator should optimize this parameter for his or her images.
20. Be sure to invert the "black spot" images into "white spot" images and save as image sequence format before MATLAB analysis.
21. Based on our smFISH results, we determined that discrete RNA transcripts are distributed at least 1 pixel from one another. Therefore, a script should be written to compare the intensity of each local maximum in each slice with the intensity of the neighboring pixels in that slice and the slices immediately above and below (nine pixels in the slice above, eight surrounding pixels in the same slice, and nine pixels in the slice below). Each 3D maximum represents a single RNA transcript. The operator should determine how many slices and pixels are needed for RNA analysis within his or her images.
22. After RNA transcripts in smFISH and MB images are determined as described in **Note 21**, a script should be written to assess the extent of colocalization between the two signals in 3D. In our program, an MB to smFISH colocalization event occurs if within a $3 \times 3 \times 3$ voxel of a smFISH local maximum there is an MB 3D local maximum. Vice versa, we consider an smFISH to MB localization event if within a $3 \times 3 \times 3$ pixel cube of an MB local maximum there is a smFISH local maximum. A $5 \times 5 \times 5$ pixel cube can also be used for comparison between MB and smFISH signals to reduce the impact of local maxima uncertainty.

Acknowledgments

This project was supported by grants from the National Basic Research Program of China (2016YFA0100702), the National Natural Science Foundation of China (81371613) and the Beijing Natural Science Foundation (7162114).

References

1. Tyagi S, Kramer FR (1996) Molecular beacons: probes that fluoresce upon hybridization. Nat Biotechnol 14(3):303–308. doi:10.1038/nbt0396-303
2. Chen AK, Behlke MA, Tsourkas A (2007) Avoiding false-positive signals with nuclease-vulnerable molecular beacons in single living cells. Nucleic Acids Res 35(16):e105. doi:10.1093/nar/gkm593
3. Mhlanga MM, Vargas DY, Fung CW, Kramer FR, Tyagi S (2005) tRNA-linked molecular beacons for imaging mRNAs in the cytoplasm of living cells. Nucleic Acids Res 33 (6):1902–1912. doi:10.1093/nar/gki302

4. Tyagi S, Alsmadi O (2004) Imaging native beta-actin mRNA in motile fibroblasts. Biophys J 87 (6):4153–4162. doi:10.1529/biophysj.104.045153
5. Chen AK, Davydenko O, Behlke MA, Tsourkas A (2010) Ratiometric bimolecular beacons for the sensitive detection of RNA in single living cells. Nucleic Acids Res 38(14):e148. doi:10.1093/nar/gkq436
6. Zhang X, Song Y, Shah AY, Lekova V, Raj A, Huang L, Behlke MA, Tsourkas A (2013) Quantitative assessment of ratiometric bimolecular beacons as a tool for imaging single engineered RNA transcripts and measuring gene expression in living cells. Nucleic Acids Res 41(15):e152. doi:10.1093/nar/gkt561
7. Raj A, van den Bogaard P, Rifkin SA, van Oudenaarden A, Tyagi S (2008) Imaging individual mRNA molecules using multiple singly labeled probes. Nat Methods 5(10):877–879. doi:10.1038/nmeth.1253
8. Schindelin J, Arganda-Carreras I, Frise E, Kaynig V, Longair M, Pietzsch T, Preibisch S, Rueden C, Saalfeld S, Schmid B, Tinevez JY, White DJ, Hartenstein V, Eliceiri K, Tomancak P, Cardona A (2012) Fiji: an open-source platform for biological-image analysis. Nat Methods 9 (7):676–682. doi:10.1038/nmeth.2019

Chapter 16

Optimizing Molecular Beacons for Intracellular Analysis of RNA

Mingming Chen, Yantao Yang, Christopher J. Krueger, and Antony K. Chen

Abstract

Conventional molecular beacons (MBs) have been used extensively for imaging specific endogenous RNAs in living cells, but their tendency to generate false-positive signals as a result of nuclease degradation and/or nonspecific binding limits sensitive and accurate imaging of intracellular RNAs. In an attempt to overcome this limitation, MBs have been synthesized with various chemically modified oligonucleotide backbones to confer greater biostability. We have recently developed a new MB architecture composed of 2′-O-methyl RNA (2Me), a fully phosphorothioate (PS) modified loop domain and a phosphodiester stem (2Me/PS_{LOOP} MB). We showed that this new MB exhibits a marginal level of false-positive signals and enables accurate single-molecule imaging of target RNA in living cells. In this chapter, we describe detailed methods that led us to conclude that, among various PS-modified configurations, the 2Me/PS_{LOOP} MB is an optimal design for intracellular RNA analysis.

Key words Molecular beacons, 2′-O-methyl RNA, Phosphorothioate, Ratiometric imaging, Live-cell RNA detection, False-positive signals

1 Introduction

Over the past decade, it has become evident that cell fate and disease evolution are significantly influenced by the expression level, trafficking, and localization of specific RNA molecules. To better understand the molecular mechanisms underlying these processes, a variety of oligonucleotide-based probes have been developed to enable direct visualization of RNA molecules in living cells. One widely employed technology is the molecular beacon (MB), a stem-loop-forming oligonucleotide probe labeled with a fluorescent reporter and a quencher at the two termini [1]. In the nonhybridized state, the fluorophore is held in close proximity to the quencher as a result of self-annealing of the short arm sequences that form the stem domain. Hybridization of the loop domain to target RNA transcripts opens the stem, restoring MB fluorescence

Imre Gaspar (ed.), *RNA Detection: Methods and Protocols*, Methods in Molecular Biology, vol. 1649, DOI 10.1007/978-1-4939-7213-5_16, © Springer Science+Business Media LLC 2018

by separating the fluorophore from the quencher. This simple yet effective design has led to widespread use of MBs to study the intracellular behavior of various RNA molecules including mRNA [2–11], viral RNA [12–15], and noncoding RNA [16, 17].

Although MBs are considered useful tools for RNA analysis in various cellular contexts, it has become increasingly evident that in the cellular environment, MBs—particularly when synthesized with a backbone composed of DNA or 2′-O-methyl RNA (2Me)—are frequently sequestered in the nucleus where they are degraded by nucleases and/or bound nonspecifically by stem-loop binding proteins. Either of these activities can cause separation of the fluorophore and the quencher, leading to generation of background fluorescence that can hamper accurate and sensitive detection of target RNA.

One strategy to reduce MB nonspecific opening has been to chemically modify the backbone to confer greater biostability. We have recently shown the feasibility of this approach by incorporating 2Me MBs with phosphorothioate (PS) internucleotide linkages in place of conventional phosphodiester bonds. By optimizing the number and location of PS modifications within the MB backbone, we found that the design with full PS modification of the loop and no PS modification of the stem (2Me/PS_{LOOP} MB) exhibits the least false-positive signals in cells. In this chapter, we describe the ratiometric imaging methods that were utilized to evaluate the extent of nonspecific interactions of MB architectures with different PS configurations in living cells (*see* **Note 1**). We expect the methods introduced here can be easily adapted for analysis of intracellular performance of MB architectures with other chemistries, and should benefit the design of more advanced MBs or MB-based probes for in vivo applications.

2 Materials

2.1 Cell Culture

1. HeLa cells.
2. Dulbecco's Modified Eagle's Medium (DMEM) without phenol red and without antibiotics, supplemented with 10% FBS and 1× GlutaMAX (Thermo Fisher).
3. Phenol red-free solution of 0.25% trypsin and 1 mM EDTA.

2.2 Oligonucleotides

1. **2Me**

   5′-<u>mGmUmCmAmC</u>mCmUmCmAmGmCmGmUmAmAmGmUmGmAmUmGmUmC<u>mGmUmGmAmC</u>-3′.
2. **2Me/PS_{STEM}**

   5′-<u>mG*mU*mC*mA*mC</u>*mCmUmCmAmGmCmGmUmAmAmGmUmGmAmUmGmUmC*<u>mG*mU*mG*mA*mC</u>-3′.

3. **2Me/PS$_{10\text{-LOOP}}$.**

   5′-mGmUmCmAmCmC*mUmC*mAmG*mCmG*mUmA*-mAmG*mUmG*mA*mUmG*mUmC*mGmUmGmAmC-3′.

4. **2Me/PS$_{\text{LOOP}}$**

   5′-mGmUmCmAmCmC*mU*mC*mA*mG*mC*mG*mU*-mA*mA*mG*mU*mG*mA*mU*mG*mU*mC*mGmUmG-mAmC-3′.

5. **2Me/PS$_{\text{ALT}}$**

   5′-mG*mUmC*mAmC*mC*mUmC*mAmG*mCmG*mU-mA*mAmG*mUmG*mA*mUmG*mUmC*mGmU*mGmA *mC-3′.

6. **2Me/PS$_{\text{FULL}}$**

   5′-mG*mU*mC*mA*mC*mC*mU*mC*mA*mG*mC*mG* mU*mA*mA*mG*mU*mG*mA*mU*mG*mU*mC*mG*m U*mG*mA*mC-3′.

Underlined letters indicate the MB stem. m represents 2′-O-methyl RNA modification. * represents PS linkage modification. The MBs are labeled with a Cy5 fluorophore at the 5′-end and an Iowa Black® RQ-Sp quencher at the 3′-end. These MBs are complementary to luciferase RNA but not to endogenous RNAs in mammalian cells, and under ideal circumstances will remain closed, and thus quenched, in the cellular environment.

1. Luciferase target RNA oligonucleotides

   5′-GUCAGGACAUCACUUACGCUGAGUUU-3′.

2.3 Microporation

1. Microporation system (e.g., Thermo Fisher Neon transfection system).
2. 1× phosphate buffered saline (PBS), Mg^{2+}- and Ca^{2+}-free.
3. Resuspension buffer R (Thermo Fisher).
4. Electroporation buffer (Thermo Fisher).
5. Electroporation Gold Tips (10 μL size) (Thermo Fisher).
6. Electroporation tube (Thermo Fisher).
7. 8-well chambered cover glass (e.g., Nalgene Nunc Lab-Tek).
8. 10 μg/mL fibronectin.
9. Refrigerated microcentrifuge.

2.4 Microinjection

1. Microinjection system (e.g., Eppendorf Femtojet and Injectman NI 2).
2. Microinjection capillary (e.g., Eppendorf Femtotip I).
3. Microloader.

4. Microinjection buffer: 48 mM K_2HPO_4, 4.5 mM KH_2PO_4, and 14 mM NaH_2PO_4, pH 7.2 (*see* **Note 2**).
5. Glass bottom dish (e.g., MatTek).
6. Microcentrifuge.

2.5 Microscope and Imaging Software

1. An inverted widefield fluorescence microscope. We use an Olympus IX83 Motorized inverted fluorescence microscope equipped with a back-illuminated EMCCD camera (Andor) and an MT-20E excitation source (Olympus).
2. 40× 0.95NA objective lens.
3. Filter set for DAPI, EGFP, and TAMRA (e.g., Olympus MT20).
4. Filter set for Cy5 (ET620/60×, ET700/75m, and T660lpxr).
5. Filter set for IRDye® 800 (ET710/75x, ET810/90m, and T760lpxr).
6. cellSens Dimension image acquisition software (Olympus).
7. Fiji software for image analysis (*see* **Note 3**).

2.6 Synthesis of Reference-Dye Labeled Dextran

1. IRDye®800-NHS ester (e.g., Li-Cor) dissolved in anhydrous DMSO or DMF (*see* **Note 4**).
2. Reaction buffer: 50 mM sodium borate buffer, pH 8.
3. Elution buffer: 48 mM K_2HPO_4, 4.5 mM KH_2PO_4, and 14 mM NaH_2PO_4, pH 7.2.
4. Aminodextran solution: 2.5 mg/mL aminodextran (MW = 10 kDa) in reaction buffer.
5. Gel chromatography columns (e.g., GE Healthcare NAP-5).
6. Centrifugal filter units, YM-3, 3000 MW cutoff (e.g., Millipore Centricon).
7. Microcentrifuge.
8. Thermomixer.
9. Spectrophotometer (e.g., Thermo Scientific BioMate 3S).

2.7 Preparation of Microemulsion Samples

1. Span-80.
2. Tween 80.
3. Mineral oil.
4. 5 mL glass vials.
5. 50 mL tubes.
6. Magnetic stirrer and magnetic stir bars.
7. Glass pipettes.
8. Electronic scale.

3 Methods

3.1 Preparation of Reference Dye-Labeled Controls (See Note 5)

1. React 150 μL of the aminodextran solution with IRDye®800 NHS ester at a dye to dextran molar ratio of 3:1 on a thermo-mixer overnight at 25 °C.
2. Purify the mixture on a NAP-5 gel chromatography column in elution buffer to remove unbound dye.
3. Concentrate the purified IRDye® 800-labeled dextran solution on a centrifugal filter unit.
4. Determine the concentration of IRDye® 800 fluorophore using a spectrophotometer.

3.2 Establishment of In Vitro Standard Measurements

To create aqueous bubbles containing the IRDye® 800-labeled dextran and nonhybridized or fully hybridized MBs, water-in-oil emulsions are prepared as follows:

1. On an electronic scale, tare a 50 mL tube. Add 24 g of mineral oil.
2. Use a glass pipette to add 447 mg of Span 80 dropwise to the mineral oil.
3. Vortex the mixture vigorously.
4. Use a glass pipette to add 54 mg of Tween 80 dropwise to the mineral oil.
5. Vortex the mixture vigorously.
6. Add a 3 mL aliquot of the above mixture to a glass vial and stir using a magnetic stirrer. Make sure the stir bar is immersed under the emulsion surface. Stir for 1–2 min (*see* **Note 6**).
7. While stirring, add 1–2 μL of aqueous samples containing 5 μM nonhybridized or fully hybridized MBs (previously hybridized with 20 μM RNA target for at least 1 h at room temperature) and 1 μM IRDye®800-labeled dextran in a dropwise fashion into the emulsion to form water-in-oil microemulsion bubbles.
8. After stirring for 3–5 min more, take an aliquot of the emulsion sample and place it on a glass bottom dish for microscopy imaging and analysis (*see* **Note 7**).
9. Example images of microemulsion bubbles acquired using the Cy5 and IRDye®800 filter sets are shown in Fig. 1 (*see* **Notes 8** and **9**).

Microporation is used to deliver MB–dextran mixtures because it enables rapid and highly efficient delivery of MBs and dye-labeled

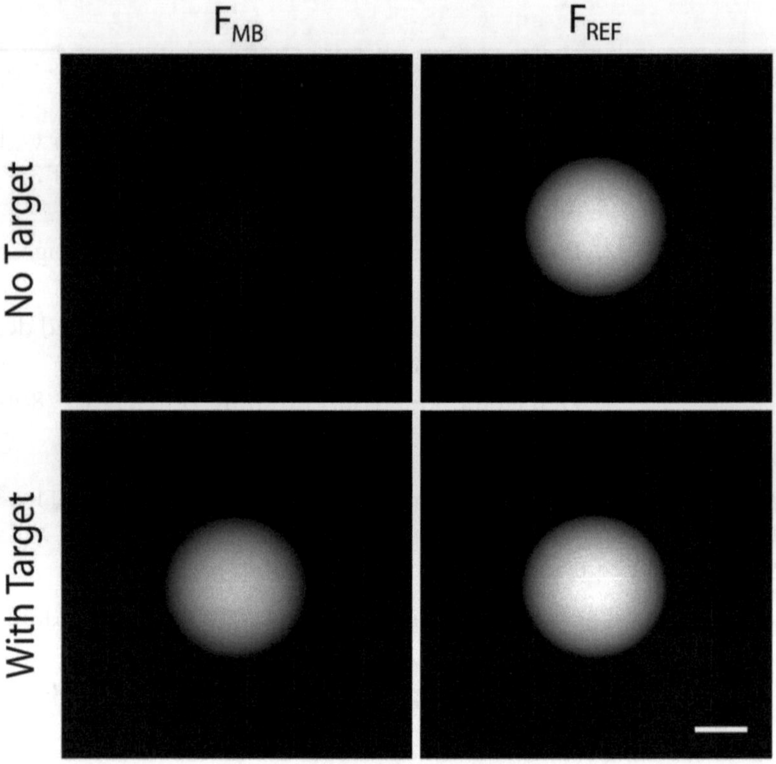

Fig. 1 Fluorescent microscopy images of microemulsion bubbles containing MBs and IRDye®800-labeled dextran in the presence and absence of nucleic acid targets. F_{MB} denotes the fluorescence signals of the MB. F_{REF} denotes the fluorescence signal of the dextran (scale bar, 10 μm)

3.3 Cellular Delivery of MB–Dextran Mixtures

dextran into a large number of cells with high viability, as shown in our previous studies [18, 19].

Day 1:

1. Coat the glass-bottom wells of the 8-well chambered cover glass by adding 250 μL fibronectin per well and incubate overnight at 37 °C (*see* **Note 10**).
2. Seed cells in T-25 flasks in DMEM growth media without phenol red or antibiotics so that they will be no more than 70% confluent on the day of the experiment (*see* **Note 11**).

Day 2:

3. Aspirate the media from the cells and rinse the flask using 5 mL of prewarmed 1× PBS for about 2 min.
4. Aspirate the PBS and add 1 mL of phenol red-free trypsin–EDTA; incubate for 1 min at room temperature.
5. Aspirate half of the trypsin (about 500 μL) and incubate at 37 °C to detach all of the cells from the flask surface (*see* **Note 12**).

6. Neutralize the remaining trypsin by adding 4.5 mL of cell culture media.
7. Resuspend the cells by pipetting gently. Make sure there are no cell clumps (*see* **Note 13**).
8. Transfer 1 mL of the cell suspension to a 1.5 mL microcentrifuge tube and pellet the cells by centrifugation at 200 × *g* for 5 min at 4 °C.
9. Aspirate the media and gently resuspend the cell pellet in 1 mL of 1× PBS. Make sure there are no cell clumps.
10. Count the cells.
11. Pellet the required number of cells necessary for microporation (generally 50,000 cells per microporation) by centrifugation at 400 × *g* for 5 min at 4 °C (*see* **Note 14**).
12. Aspirate the PBS and resuspend the cell pellet in resuspension buffer R at 5000 cells per μL.
13. Add 2 μL of sample containing MBs and IRdye® 800-labeled dextran for every 10 μL of cells such that the final concentrations of MBs and IRDye® 800-labeled dextran are 5 and 1 μM, respectively.
14. Gently mix the cells with MBs by pipetting.
15. Microporate 10 μL of the cell suspension (roughly 50,000 cells) at 1005 V with a 35 ms pulse width and 2 pulses total (*see* **Note 14**).
16. Transfer microporated cells to a microcentrifuge tube prefilled with 1.5 mL of cell culture media.
17. Pellet the cells at 400 × *g* for 5 min at 4 °C.
18. Aspirate the media. Be careful not to disturb the cell pellet. Add 1.5 mL of cell culture media to resuspend the pellet.
19. Repeat **steps 15** and **16** another two times (*see* **Notes 15** and **16**).
20. After the last wash, resuspend the cells in ~250 μL of cell culture media.
21. Seed the cells into a well of the 8-well chambered cover glass previously coated with fibronectin (*see* **Note 17**).
22. Image the cells using both Cy5 and IRDye® 800 filter sets at various time points.

3.4 Total Fluorescence Quantification

1. Using Fiji, open a pair of images corresponding to the Cy5 and IRDye®800 fluorescence of one object (microemulsion bubble or cell).
2. Stack the Cy5 and IRDye® 800 images using the function Image > Stacks > Convert Images to Stack.

3. Use the "Freehand Selections" tool to draw a Region of Interest (ROI) around the object. This object ROI is automatically applied to the two channel images within the same stack.
4. For each image, measure the total integrated fluorescence density and area (in pixels) within the object ROI by selecting Analyze > Measure.
5. For each image, measure the average background fluorescence by first drawing at least four ROIs in areas just outside the cell of interest (*see* **Note 18**).
6. Measure the mean fluorescence density (per pixel) within each "background" ROI by selecting Analyze > Measure.
7. Compute the average of the mean fluorescence densities of the four background ROIs. Multiply this value by the area of the object ROI to obtain the total background fluorescence intensity within the object ROI.
8. Subtract the total background fluorescence intensity from the total integrated fluorescence density within the object ROI to obtain the total object fluorescence.

3.5 Quantifying the Extent of Nonspecific Opening

1. After quantifying fluorescence signals of MBs and IRDye®800 in water-in-oil emulsions and cells using the method described above, nonspecific opening of MBs can be determined as previously described [18], according to the following equation:

$$\%\text{MBs opened} = \frac{R_{\text{cell}} - R_{\text{BUBBLES,CLOSED}}}{R_{\text{BUBBLES,OPENED}} - R_{\text{BUBBLES,CLOSED}}} \times 100\%.$$

2. R_{CELL} is the fluorescence ratio of living cells, and $R_{\text{BUBBLES,CLOSED}}$ and $R_{\text{BUBBLES,OPENED}}$ are the fluorescence ratios of aqueous bubbles prepared from water-in-oil emulsions of unhybridized MBs (0% opened) and MBs hybridized with excess complementary targets (100% opened), respectively (*see* **Note 19**).
3. Example images and resulting analysis that led us to conclude that the 2Me/PS$_{\text{LOOP}}$ MB is the most stable configuration among the different PS modified and non-PS modified MBs are shown in Fig. 2 (*see* **Note 20**).

3.6 Assessing the Intracellular Functionality of MBs

While 2Me/PS$_{\text{LOOP}}$ MBs exhibit the least nonspecific signals, it is also important to confirm that they still retain the ability to hybridize to target RNA in living cells. This can be achieved by microinjecting excess fully complementary synthetic target RNA molecules into cells previously microporated with MBs. If the MBs are functionally active, injection of complementary target RNA should lead to an increase in MB fluorescence.

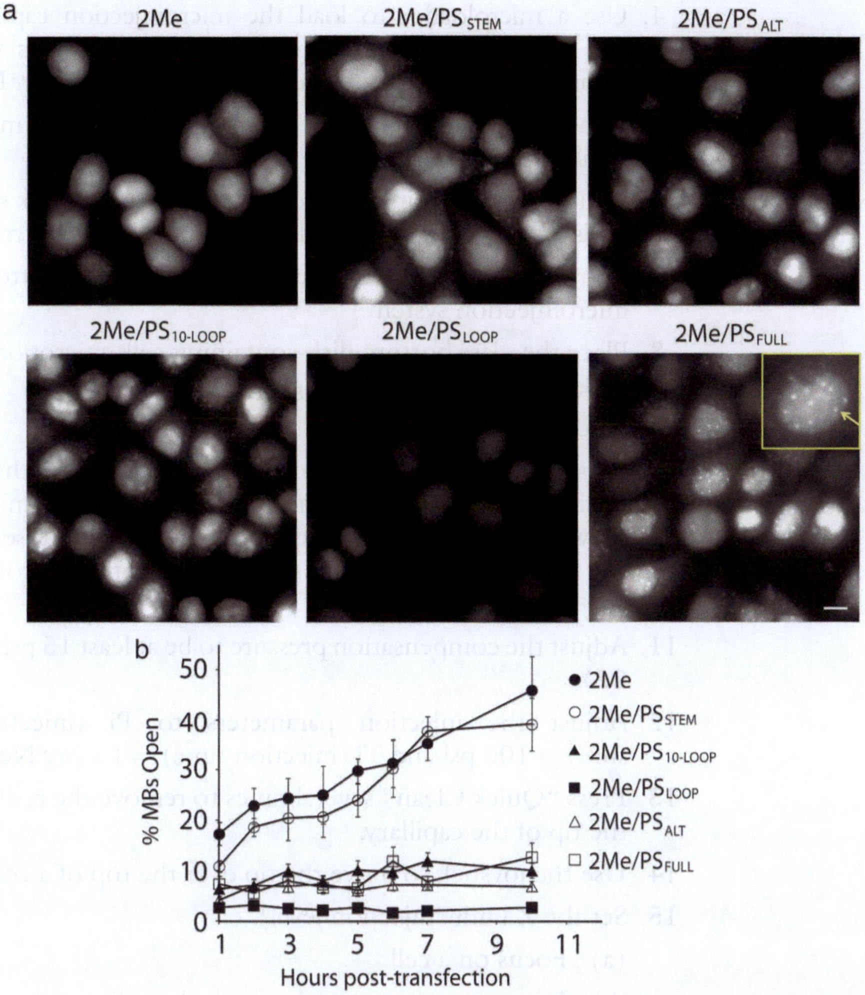

Fig. 2 Nonspecific opening of non-PS-modified and PS-modified anti-luciferase MBs in living cells. (**a**) Representative images of MBs in HeLa cells, acquired at 10 h following microporation. The MBs are not complementary to any known endogenous RNAs. The *inset* shows an expanded segment of the image. The *arrow* points to a bright punctum in the nucleus. (**b**) Quantification of the extent of MB nonspecific opening over time. Each data point represents the mean ± standard error from at least 40 cells (scale bar, 10 μm) (Reproduced from [19] with permission from Elsevier)

1. Seed HeLa cells microporated with 5 μM MBs so that they will be 10–30% confluent at the time of experiment.
2. Dilute the target RNA stock to a concentration of 20 μM in Microinjection buffer.
3. Centrifuge the target RNA sample for 20 min at 21,000 × *g*. Use the supernatant for injection as the pelleted debris can clog the microinjection capillary.

4. Use a microloader to load the microinjection capillary with 3–5 μL of the diluted target RNA solution. This volume is sufficient for injection of several hundreds of cells. (*see* **Note 21**).
5. Attach the filled microinjection capillary to the microscope capillary holder.
6. Position the microinjection capillary with the joystick so that the tip is at the center of the field of the view on the microscope.
7. Retract the capillary by pressing the "Home" button on the microinjection system.
8. Place the glass bottom dish containing cells microporated with MBs on the microscope stage.
9. Find the focal plane.
10. Press the "Home" button to return the tip to the original position at the center of the field of view on the microscope. Slowly lower the microinjection capillary to immerse the tip in the media (if not already immersed) but not yet in contact with any cells.
11. Adjust the compensation pressure to be at least 15 psi (*see* **Note 22**).
12. Adjust the injection parameters to Pi (injection pressure) = 100 psi and Ti (injection time) = 1 s (*see* **Note 23**).
13. Press "Quick Clean" several times to remove the residual air in the tip of the capillary.
14. Use the joystick to move the tip over the top of a cell.
15. Set the Z limit (injection level).
    (a) Focus on a cell.
    (b) Lower the tip to gently touch the cell membrane until a gentle wave traverses through the cell from the site of injection.
    (c) Press the "Limit" key on the InjectMan NI 2 control board to set the height as the Z limit.
16. Raise the capillary approximately 20–30 μm above the cell so that it can be moved freely without touching the cells, with the tip still visible. This height corresponds to the search level.
17. Move the microinjection capillary out of the field of view using the joystick.
18. Focus on a cell to be injected.
19. Acquire fluorescent images of the cell using the Cy5 filter set.
20. Use the joystick to move the capillary over the top of the imaged cell.

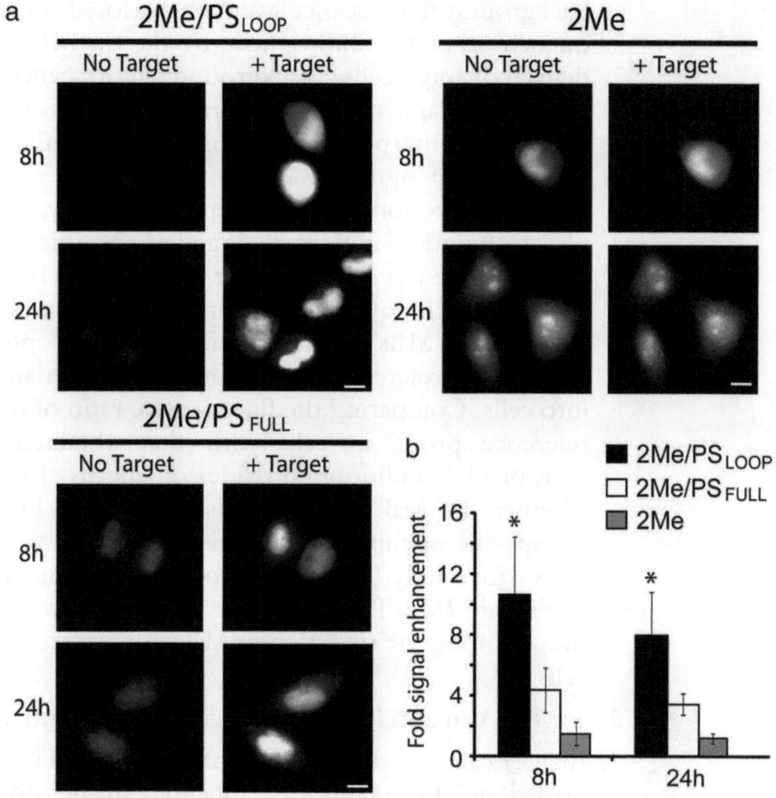

Fig. 3 Functionality analysis of 2Me/PS$_{LOOP}$ and conventional MBs in living cells. Excess complementary RNA targets were injected into HeLa cells 8 and 24 h after microporation of 2Me/PS$_{LOOP}$, 2Me/PS$_{FULL}$ or 2Me MBs. Fluorescent images were acquired before and within several minutes after injection. (**a**) Representative pre- and post-injection fluorescent images. (**b**) 2Me/PS$_{LOOP}$ MBs exhibited greatest signal enhancement upon injection of the targets into the cells. Each data point represents the mean ± SD from at least 10 cells. *Asterisk* represents significant differences ($p < 0.05$) from both 2Me/PS$_{FULL}$ and 2Me MBs (scale bar, 10 μm) (Reproduced from [19] with permission from Elsevier)

21. Press the injection button on the joystick to inject the target RNA solution into the cell. The needle will return to the search level after injection.
22. Move the microinjection capillary out of the field of view using the joystick.
23. Acquire fluorescent images of the cell at different time points using the Cy5 filter set. Example images are shown in Fig. 3 (*see* **Note 24**).

4 Notes

1. To quantify the extent of MB nonspecific opening in living cells, it is important to recognize that MBs can still emit

background fluorescence even when closed, due to imperfect quenching of the fluorophore by the quencher. Thus, when delivered into cells, background fluorescence, nonspecific interactions and cell-to-cell variations in MB delivery can all lead to misinterpretation of high uptake efficiency for MB hybridization signals. Ratiometric imaging, a technique that enables correction for unequal probe delivery, has been a popular technique for normalizing delivery efficiency in studies involving the use of fluorescent probes [20]. In the context of MBs, the technique generally involves comparing the fluorescence of the MBs of interest with that of an optically distinct, nonreactive reference probe both in solution and codelivered into cells. Comparing the fluorescence ratio of the MBs to the reference probe in cells with that obtained under well-controlled conditions provides a means for determining whether the MB signal in cells reflects MB hybridization or nonspecific opening. Using this approach, we identified that MBs with a fully PS-modified loop domain and a phosphodiester stem (2Me/PS_{LOOP} MB) exhibit the lowest false-positive signal among the six different MB architectures tested in living cells.

2. 1× PBS can also be used as microinjection buffer.
3. ImageJ can also be used for image analysis. Fiji can be downloaded at: http://fiji.sc/. ImageJ can be downloaded at: https://imagej.nih.gov/ij/.
4. Dissolve the appropriate quantity of IRDye® 800 NHS ester in 1–5 μL anhydrous DMSO or DMF before reacting with the aminodextran. Make sure the volume of DMSO or DMF is less than 10% of the total reaction mixture.
5. To identify MB backbone architectures that are stable in living cells, it is critical to use an appropriate control that enables normalization of cell-to-cell variations in delivery. As the backbone stability is unknown, it might not be appropriate to use an unquenched beacon as a normalization reference, as both quenched and unquenched beacons may be degraded. Alternatively, we have previously used fluorescently labeled dextran as a means for normalizing cellular delivery of molecular beacons [18, 19]. The IRDye®800-labeled dextran used in this study has a molecular weight (~10 kDa) similar to the MBs, therefore the two molecules were expected to exhibit similar transfection efficiencies when codelivered into cells by microporation, as shown in our previous results [18, 19].
6. Microemulsion bubbles form a range of sizes, and increasing the stirring speed can reduce bubble sizes.
7. Experimenters should try to image bubbles that have sizes similar to cells.

8. It is recommended to allow microemulsion bubbles to settle on the imaging surface (cover glass) for a few minutes before imaging.
9. Microemulsions should be prepared fresh and used within hours after preparation.
10. We have found that after microporation HeLa cells adhere faster and spread better on fibronectin-coated surfaces.
11. To achieve high transfection efficiency and cell viability, it is critical that cells are well-spread and not overconfluent before being subjected to microporation experiments.
12. The flask should be incubated at 37 °C for enough time to ensure all of the cells are fairly rounded and detached from the surface.
13. To achieve high cell viability after microporation, gentle pipetting to minimize shearing of the cells is critical.
14. The number of cells used per microporation is cell line-specific and may affect the transfection efficiency and viability. A detailed list of cell line and the number of cells to use per microporation can be found on the Neon transfection system website: https://www.thermofisher.com/us/en/home/life-science/cell-culture/transfection/transfection-selection-misc/neon-transfection-system.html.
15. It is important to wash the cells with enough media to remove unincorporated probes, which can contribute to background fluorescence and hamper accurate fluorescence quantification.
16. After every wash, it is recommended not to aspirate out the entire wash media as this risks aspirating out the loose cell pellet formed by the small number of cells.
17. Prior to seeding the cells into the 8-well chamber, it is critical to wash the chamber with 1× PBS to remove unbound fibronectin. Residual unbound fibronectin can inhibit cell attachment to the surface.
18. We have found that taking multiple ROIs around the object gives more accurate assessment of the total background fluorescence within the object ROI.
19. To determine the extent of MB nonspecific opening in cells based on fluorescence measurements acquired in solution and in cells, it is important to make sure that the emission properties of the MB reporter dye and reference dye are the same in solution and in living cells. We have previous published a method for assessing the sensitivity of commercially available fluorophores to changes in environment [20].
20. We have shown that the 2Me/PS_{LOOP} MB was the most stable configuration in other cell types including HEK-293, Jurkat,

and primary BJ cells. Additionally, we also found that for a different set of MBs with unique stem and loop lengths and sequences, the 2Me/PS_{LOOP} architecture also generated the least false-positive signals in HeLa cells. This suggests that the advantages of the 2Me/PS_{LOOP} design can be generalized to other MB sequences.

21. Slowly pipette the target RNA solution into the microinjection capillary to avoid creating air bubbles.
22. Compensation pressure is necessary to avoid entry of cell culture media into the injection capillary, which can dilute the probe.
23. Injection parameters and amounts vary with different cell types. The operator should adjust as necessary.
24. Figure 3 shows injection of excess targets at 8 and 24 h post-microporation of the 2Me/PS_{LOOP} MBs, 2Me MBs, or 2Me/PS_{FULL} MBs. Signal enhancement was significantly greater for the 2Me/PS_{LOOP} MBs as compared with 2Me and 2Me/PS_{FULL} MBs. Thus, in addition to being less susceptible to nonspecific opening, the 2Me/PS_{LOOP} MB was also more functionally active in living cells.

Acknowledgments

This project was supported by grants from the National Basic Research Program of China (2016YFA0100702), the National Natural Science Foundation of China (81371613), and the Beijing Natural Science Foundation (7162114).

References

1. Tyagi S, Kramer FR (1996) Molecular beacons: probes that fluoresce upon hybridization. Nat Biotechnol 14(3):303–308. doi:10.1038/nbt0396-303
2. Bratu DP, Cha BJ, Mhlanga MM, Kramer FR, Tyagi S (2003) Visualizing the distribution and transport of mRNAs in living cells. Proc Natl Acad Sci U S A 100(23):13308–13313. doi:10.1073/pnas.2233244100
3. Santangelo PJ, Nix B, Tsourkas A, Bao G (2004) Dual FRET molecular beacons for mRNA detection in living cells. Nucleic Acids Res 32(6):e57. doi:10.1093/nar/gnh062
4. Tyagi S, Alsmadi O (2004) Imaging native beta-actin mRNA in motile fibroblasts. Biophys J 87(6):4153–4162. doi:10.1529/biophysj.104.045153
5. Drake TJ, Medley CD, Sen A, Rogers RJ, Tan W (2005) Stochasticity of manganese superoxide dismutase mRNA expression in breast carcinoma cells by molecular beacon imaging. Chembiochem 6(11):2041–2047. doi:10.1002/cbic.200500046
6. Vargas DY, Raj A, Marras SA, Kramer FR, Tyagi S (2005) Mechanism of mRNA transport in the nucleus. Proc Natl Acad Sci U S A 102(47):17008–17013. doi:10.1073/pnas.0505580102
7. Chen AK, Behlke MA, Tsourkas A (2007) Avoiding false-positive signals with nuclease-vulnerable molecular beacons in single living cells. Nucleic Acids Res 35(16):e105. doi:10.1093/nar/gkm593
8. Chen AK, Behlke MA, Tsourkas A (2008) Efficient cytosolic delivery of molecular beacon conjugates and flow cytometric analysis of

target RNA. Nucleic Acids Res 36(12):e69. doi:10.1093/nar/gkn331

9. Wu Y, Yang CJ, Moroz LL, Tan W (2008) Nucleic acid beacons for long-term real-time intracellular monitoring. Anal Chem 80 (8):3025–3028. doi:10.1021/ac702637w
10. Rhee WJ, Bao G (2009) Simultaneous detection of mRNA and protein stem cell markers in live cells. BMC Biotechnol 9:30. doi:10.1186/1472-6750-9-30
11. Catrina IE, Marras SA, Bratu DP (2012) Tiny molecular beacons: LNA/2′-O-methyl RNA chimeric probes for imaging dynamic mRNA processes in living cells. ACS Chem Biol 7 (9):1586–1595. doi:10.1021/cb300178a
12. Santangelo P, Nitin N, LaConte L, Woolums A, Bao G (2006) Live-cell characterization and analysis of a clinical isolate of bovine respiratory syncytial virus, using molecular beacons. J Virol 80(2):682–688. doi:10.1128/JVI.80.2.682-688.2006
13. Wang A, Salazar AM, Yates MV, Mulchandani A, Chen W (2005) Visualization and detection of infectious coxsackievirus replication using a combined cell culture-molecular beacon assay. Appl Environ Microbiol 71(12):8397–8401. doi:10.1128/AEM.71.12.8397-8401.2005
14. Wang W, Cui ZQ, Han H, Zhang ZP, Wei HP, Zhou YF, Chen Z, Zhang XE (2008) Imaging and characterizing influenza A virus mRNA transport in living cells. Nucleic Acids Res 36 (15):4913–4928. doi:10.1093/nar/gkn475
15. Yeh HY, Yates MV, Mulchandani A, Chen W (2008) Visualizing the dynamics of viral replication in living cells via Tat peptide delivery of nuclease-resistant molecular beacons. Proc Natl Acad Sci U S A 105(45):17522–17525. doi:10.1073/pnas.0807066105
16. Kloc M, Wilk K, Vargas D, Shirato Y, Bilinski S, Etkin LD (2005) Potential structural role of non-coding and coding RNAs in the organization of the cytoskeleton at the vegetal cortex of Xenopus oocytes. Development 132 (15):3445–3457. doi:10.1242/dev.01919
17. Yang L, Lin C, Liu W, Zhang J, Ohgi KA, Grinstein JD, Dorrestein PC, Rosenfeld MG (2011) ncRNA- and Pc2 methylation-dependent gene relocation between nuclear structures mediates gene activation programs. Cell 147(4):773–788. doi:10.1016/j.cell.2011.08.054
18. Chen AK, Behlke MA, Tsourkas A (2009) Subcellular trafficking and functionality of 2′-O-methyl and 2′-O-methyl-phosphorothioate molecular beacons. Nucleic Acids Res 37(22):e149. doi:10.1093/nar/gkp837
19. Zhao D, Yang Y, Qu N, Chen M, Ma Z, Krueger CJ, Behlke MA, Chen AK (2016) Single-molecule detection and tracking of RNA transcripts in living cells using phosphorothioate-optimized 2′-O-methyl RNA molecular beacons. Biomaterials 100:172–183. doi:10.1016/j.biomaterials.2016.05.022
20. Chen AK, Cheng Z, Behlke MA, Tsourkas A (2008) Assessing the sensitivity of commercially available fluorophores to the intracellular environment. Anal Chem 80(19):7437–7444. doi:10.1021/ac8011347

Chapter 17

Live Imaging of Nuclear RNPs in Mammalian Complex Tissue with ECHO-liveFISH

Dan Ohtan Wang

Abstract

Multiplex RNA detection with fluorescence microscopy offers high spatial and temporal resolution required for addressing complex behaviors of RNA in living cells. Using chemically engineered linear oligonucleotide probes that emit fluorescence upon hybridization to target RNA, we have devised an imaging method suitable for studies of the dynamic regulation of nuclear RNPs, an important and yet poorly understood cellular pathway of gene expression. This new method labels specific sequences of RNA components in RNPs and thus avoids overexpression of fluorescent marker proteins that may result in entangled experimental results. Using this method, we observe in living brain tissue spatially constrained nuclear RNA foci under dynamic regulation in response to cellular transcriptional activity with individual cell heterogeneity.

Key words Exciton, Fluorescence microscopy, RNA labeling, Thiazole orange, In vivo electroporation

1 Introduction

In the nucleus, local concentrations of specific RNA and proteins (RNPs: ribonucleoprotein complex) have long been observed at subnuclear organelles such as nucleoli, nuclear speckles, paraspeckles, Cajal bodies, histone locus bodies, and promyelocytic leukemia bodies. These functionally specialized regions are prominent under light and electron microscopy and have shown close relation to nuclear structures and organization [1–4]. Concentrated nuclear RNPs have also been observed at pathological ribonuclear foci that contain abnormally lengthened repeat sequences in degenerative human muscular and nervous systems [5, 6]. Furthermore, nuclear-retained noncoding RNA species have emerged to be critical regulators of chromatin structure, transcription activity, RNA processing and modification, gene expression dynamics, and nuclear compartmentalization [7–10]. These discoveries impose new demands for a suitable imaging method that labels specific RNA species and effectively monitors nuclear RNPs with

Imre Gaspar (ed.), *RNA Detection: Methods and Protocols*, Methods in Molecular Biology, vol. 1649, DOI 10.1007/978-1-4939-7213-5_17, © Springer Science+Business Media LLC 2018

spatiotemporal resolution for understanding and manipulating nuclei in both physiological and pathological states.

Newly developed fluorescent probes with tunable photochemical properties (light absorption, quantum yields, emission fluorescence wavelengths, emission fluorescence intensities, etc.) are useful in live-cell imaging. When property changes are dependent on interaction with target nucleic acids (e.g., through hybridization), detecting such changes enables quick, simple, and reliable nucleic acid detection in both fixed and live cells [11–18]. Along the same line, we have previously applied exciton-based fluorescent probes to detect DNA and RNA molecules in situ with multiple color choices [19–26].

These linear, 13–50 nt oligonucleotide probes carry a single thymine or cytosine residue labeled with a homodimer of thiazole orange (TO, 4-[3-methyl-2,3-dihydro(benzo-1,3-thiazole)-2-methyllidene]quinolinium iodide. The TO homodimer-thymine/cytosine was named "D514" (*d*oubly labeled dyes with an excitation maximum at *514* nm) and the oligonucleotide probes were named "ECHO" (*e*xicton-*c*ontrolled *h*ybridization-sensitive fluorescent *o*ligonucleotide probes). The excitonic interaction between the two TO dyes strongly inhibits photon release, resulting efficient photoquenching [27, 28]. Upon hybridization, bis-intercalation of TO into the double stranded nucleic acids both substantially reduces the interchromophoric interaction and restricts the energy-loss rotation around the methine bond of TO, resulting in robust fluorescent emission from both TO dyes (*see* **Note 1**) [29]. Taking advantage of the excellent on–off ratio of ECHO probes, we previously developed fluorescent in situ hybridization method (ECHO-FISH) in fixed cells, a wash-free protocol that largely reduces the labor time associated with conventional FISH method [19, 30, 31].

In this chapter, we describe a step-by-step protocol of ECHO-liveFISH to label nuclear U3 snoRNA and 28rRNA in living mouse brains. The method section is divided into four parts: (1) designing effective sequence of ECHO probes and in vitro characterization; (2) introducing ECHO probes to living mouse brain; (3) tissue slice preparation for imaging; (4) typical results, control experiments, and troubleshooting. Detailed protocols for ECHO probe synthesis have been described and we encourage readers interested in synthesizing the probes in their own lab capacity to find useful information from those sources [21, 30, 31].

2 Materials

2.1 Probes

1. ECHO-liveFISH probes targeting poly(A) RNA, 28S rRNA or U3 snoRNA (Fig. 1a, d) (e.g., GeneDesign, Inc.)

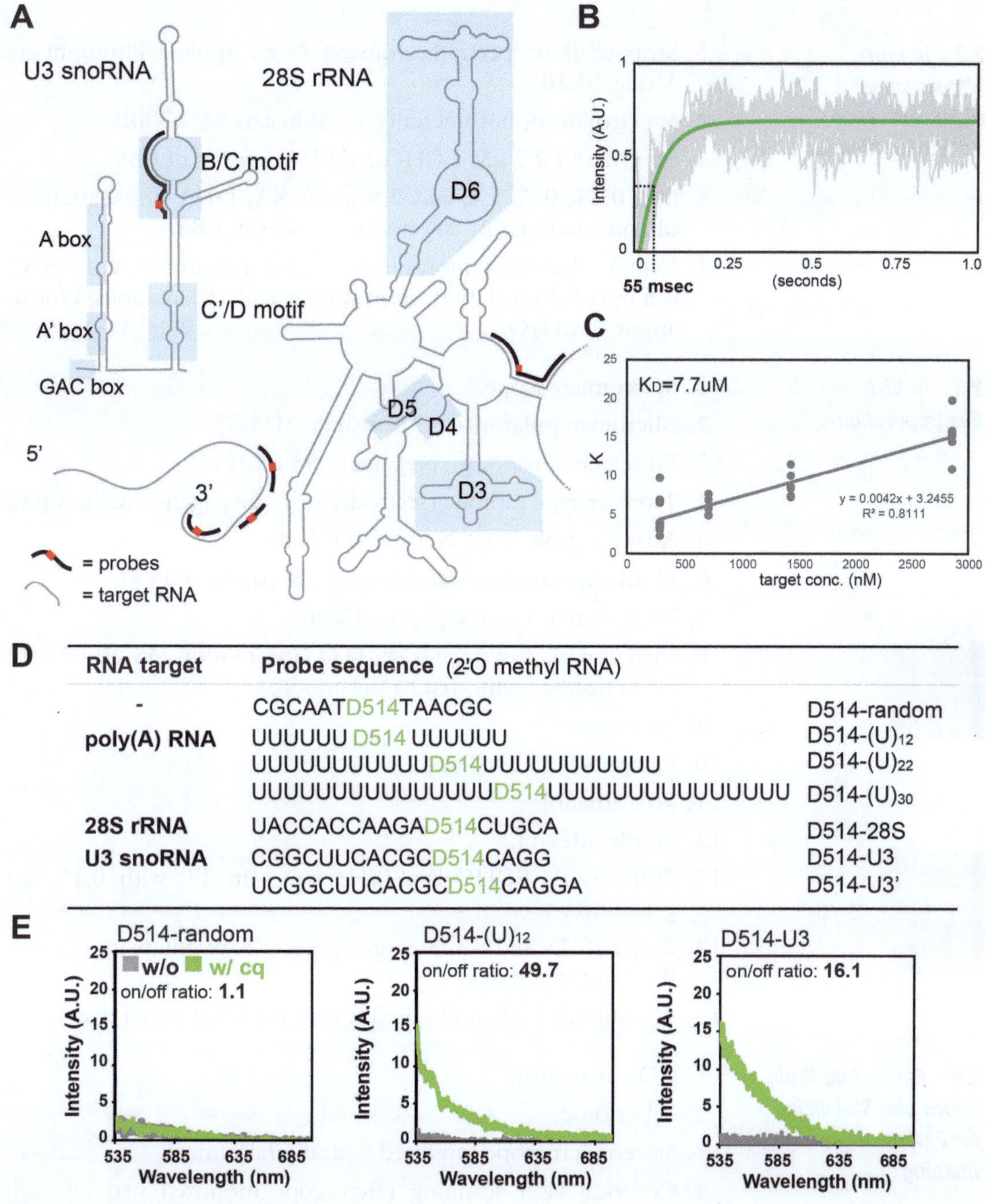

Fig. 1 Designing effective sequences of ECHO probes and in vitro characterization. (A) Transcript-specific ECHO probes (*black lines*) generated against U3 snoRNA, 28S rRNA, and poly(A) RNA (*gray lines*). Known functional motifs are labeled along the transcripts (*masked areas*). (B) Kinetics of hybridization-dependent fluorescence activation measured with stopped-flow technique. Fluorescence intensity of 50 nM probes (y) vs. the lapsed time (sec) after mixing with 1400 nM target oligonucleotides (x). *Gray lines*: 10 trials of measurements; *green line*: fitted curve ($y = ce^{-K}x + A$). Fluorescence activation occurs within tens of milliseconds in the presence of target oligonucleotides. (C) K value representing effective affinity plotted against the concentrations of target oligonucleotides. $K = K_{on}[\text{target oligo}] + K_{off}$, $K_D = K_{off}/K_{on}$. (D) A list of RNA targets and probe sequences; (E) Spectral measurement (535–700 nm) of poly(A) and U3 snoRNA probes (0.2 μM) in the absence (*gray*) or presence (*green*) of complimentary DNA oligonucleotide solutions (0.2 μM). D514-random was mixed with $d(A)_{23}$ where no fluorescence activation was observed (on–off ratio of 1.1). (Reproduced from [24] with permission from Oxford University Press)

2.2 In Vitro Measurement of Fluorescence Activation Kinetics

1. Stopped-flow spectrafluorometer (e.g., Applied Photophysics Model SX20).
2. Spectrofluorophotometer (e.g., Shimazu RF-5300PC).
3. 50 nM and 0.2 μM ECHO-liveFISH probes in PBS.
4. 0.2, 0.35, 0.7, 1.4, and 2.8 μM RNA/DNA single-stranded oligonucleotides of targeted sequences in PBS.
5. Dilution buffer in fluorophotospectral measurement: 4× SSC, 0.5 mM EDTA, 10% dextran sulfate, and 10% deionized formamide in dH_2O.

2.3 In Vivo Electroporation

1. Stereomicroscope.
2. Micromanipulator (e.g., Narishige MM-3).
3. Pulse generator (e.g., Nepagene NEPA21).
4. Tweezer-type cathode electrodes (e.g., Nepagene CUY650P3).
5. Adaptor cable (e.g., Nepagene C118).
6. Hook-type anode electrode (e.g., Nepagene C117).
7. Foot switch (e.g., Nepagene C200).
8. Injection microsyringe (e.g., ITO Corporation MS-NE05 with 33-G needle connected to the anode).
9. Tweezers.
10. Gauge 27–33 needles.
11. 70% ethanol.
12. Sterile ddH_2O.
13. 200 ng/μL ECHO-liveFISH probes in TE with 0.1% fast green dye.
14. 3 μg/μL DsRed2-B23 expressing DNA plasmids in TE with 0.1% fast green dye.
15. 200 ng/μL Cy5-d$(T)_{30}$ in TE with 0.1% fast green dye.

2.4 Generating Brain Slices and Stabilizing for Fluorescence Imaging

1. CO_2 incubator.
2. Vibratome.
3. Stereomicroscope mounted with digital camera.
4. Confocal laser scanning microscope mounted with live-cell imaging chamber (e.g., Olympus FV1000).
5. Temperature and CO_2 control modules.
6. ϕ27 mm-hole glass-bottom dish.
7. Slice anchor.
8. Artificial cerebrospinal fluid (ACSF):124 mM NaCl, 3 mM KCl, 26 mM $NaHCO_3$, 2 mM $CaCl_2/2H_2O$, 1 mM $MgSO_4/7H_2O$, 1.25 mM KH_2PO_4, and 10 mM D-glucose, bubbled with a gas mixture of 95% O_2/5% CO_2 before use.

9. 3% agarose/ACSF.
10. Collagen solution (e.g., Nitta Gelatin Cellmatrix).

2.5 Image and Statistical Analysis

1. Image J (NIH).
2. Imaris (Bitplane).
3. Prism (Graphpad).

3 Methods

3.1 Designing Effective Sequence of ECHO Probes and In Vitro Characterization

3.1.1 Design Probe Sequence

Target-dependent emission efficiency of ECHO probes can be attenuated by (1) intramolecular interaction; (2) intermolecular self-dimerization; (3) mismatching base pairs; and (4) tertiary structures of target RNA [21]. Additional considerations are taken to avoid nonspecific detection by (1) sequence-homology search against genome sequence of the target species (e.g., mm9) to guarantee single identity of detection sequence; (2) avoid functional structures of the RNA to be detected and select single-stranded region predicted by popular RNA structure prediction programs (e.g., M-fold, RNA-fold, Fig. 1a).

1. Intramolecular and intermolecular interaction within the probe sequences can be predicted using NABiT software [21]; Probe sequences with maximal scores of 100 are usually chosen for synthesis.
2. Unique target identity with 100% complementarity can be confirmed by BLAST search against the RefSeq database.

3.1.2 Kinetic Measurement Using Stopped Flow Assay

Probe characterization in vitro, including kinetics analysis and on–off ratio measurement, is used to predict how well a probe may work in vivo.

1. Kinetic measurements are carried out at room temperature using a stopped-flow spectrafluorometer equipped with dual photomultiplier tubes. Excitation 500 nm, filter set 520/35 nm (Fig. 1b).
2. 50 nM probes and 0.35, 0.7, 1.4, or 2.8 μM target RNA/DNA are dissolved separately in PBS, and mixed by firing the injection piston with simultaneous collection of fluorescence intensity as a function of time with millisecond resolution.
3. At least six experiments are performed and the average fluorescence intensity value is plotted against time.
4. Activation time is the time required for the probe fluorescence to reach half maximal intensity ($t_{1/2}$) (Fig. 1b).
5. Kinetics of hybridization-dependent fluorescence activation is fitted to $y = ce^{-K}x + A$; y: fluorescence intensity; x: lapsed time

after mixing. K is then plotted against the concentrations of target oligonucleotides (Fig. 1c).

6. To calculate effective affinity K_D, linear regression is conducted on all K values, $K = K_{on}$[target oligo] + K_{off}. K_D is calculated as $K_D = K_{off}/K_{on}$ (Fig. 1c).

3.1.3 Validating Sequence Selectivity In Vitro

1. To measure sequence-selective fluorescence activation, fluorescence spectra of the probes in the absence or presence of complementary RNA/DNA is measured using a cuvette with a 1-cm path length.
2. ECHO-liveFISH probes are mixed with equimolar (non)complementary RNA/DNA molecules in Dilution buffer by vortexing, incubated up to 5 min in Eppendorf tubes and transferred to cuvette for measurement (Fig. 1e).
3. The mixture is excited at 514 nm (1.5 nm bandwidths) and fluorescence emission between 535 and 700 nm is recorded.
4. A sum of fluorescence intensity between 535 and 700 nm is calculated as the probe on–off ratio (on: with complementary sequence; off: without complementary sequence) (Fig. 1e).

3.2 Introducing ECHO Probes into Living Mouse Brain

Intact cell membranes are not permeable to ECHO probes, which are water-soluble and hydrophilic; therefore, delivery methods assisting ECHO probes to cross plasma membrane are required. Thus, we have adapted an in vivo electroporation protocol that has been used to deliver DNA plasmids into postnatal mouse cerebella to study migration behavior of cerebellar cells [32]. This method allows oligonucleotide probes to be directly injected into the brain regions through a needle electrode (an anode, Fig. 2a, *see* **Note 2**).

1. Connect a pair of tweezer-type cathode electrodes with an adaptor cable, hook-type anode electrode and a foot switch to electrical pulse generator, and prepare sterilized surgery tools (*see* **Note 3**).
2. Stabilize an injection microsyringe with a 33-G needle connected to the anode onto the micromanipulator at a tilting angle of 60° (Fig. 2a).
3. Anesthetize a postnatal day 7–9 ICR mouse by covering its entire body with ice for 6–7 min (*see* **Note 4**).
4. Rinse the injection needle connected to the microsyringe by drawing in and pushing out 70% EtOH and sterile water alternatively three times and load 0.8 μL of sample solutions (ECHO probes or DNA plasmids) into the injection syringe by pulling back the plunger.
5. Lay the anesthetized mouse on a surgery booster and mount with a Band-Aid (Fig. 2b).

6. Clean the head–neck region of the anesthetized mice with a piece of Kimwipes wet with 70% ethanol then sterile 1× PBS immediately before surgery and carry out the subsequent surgical procedures under a stereomicroscope.
7. Cut open the skin of the head–neck region along the midline to expose the underlying muscle and skull.
8. Disconnect the muscle fibers with tweezers and drill through the skull with a gauge 27–30 needle for injection.
9. Lower the loaded injection needle through the drilled hole into the interlobular space between lobule V and VI of the mouse cerebellum to a depth of 0.5 mm (*see* **Note 5**);
10. Gently inject the sample solution into the interlobular space (*see* **Note 6**);
11. Hand-hold a pair of tweezer-type cathode electrodes and gently press against the sides of the occipital region (*see* **Note** 7).
12. Step on the foot switch to deliver electrical pulses (6 pulses of 70 V for 50 ms-duration with 150 ms-intervals for postnatal day 7 mouse, Fig. 2c) (*see* **Note 8**).
13. Stitch up the cut.
14. Revive the mice by keeping them on a 37 °C warm plate for more than 3 hours to recover and return to the litter until further experiments.

3.3 Tissue Slice Preparation for Imaging

A conventional inverted confocal fluorescence microscope can be used to observe nuclear RNP foci in individual cells given that the working distance of the lens can reach the healthy electroporated cells located next to the injection sites. To allow accessibility, we slice the cerebellum into thin sections that can be mounted onto glass bottom dishes for high-resolution single photon confocal fluorescence imaging.

1. Quickly dissect out mouse brains and transfer into ice-cold artificial cerebrospinal fluid (ACSF) bubbled with a gas mixture of 95% O_2/5% CO_2. Fluorescence can be observed in injected brain regions using a stereomicroscope (Fig. 2d)
2. Embed the brains in 3% agarose/ACSF gel and section horizontal or sagittal slices (300 μm-thickness) using a vibratome slicer.
3. Select slices under stereomicroscope and mount 2–3 slices onto a ϕ27 mm-hole glass-bottom dish, press slices down by a slice anchor if necessary and soak in ACSF (Fig. 3a).
4. Mount the dishes on to the stage of confocal microscope equipped with closed chamber with 95% O_2/5% CO_2 influx.
5. Acquire single snapshot images and time-lapse images on a confocal microscope (*see* **Note 9**).

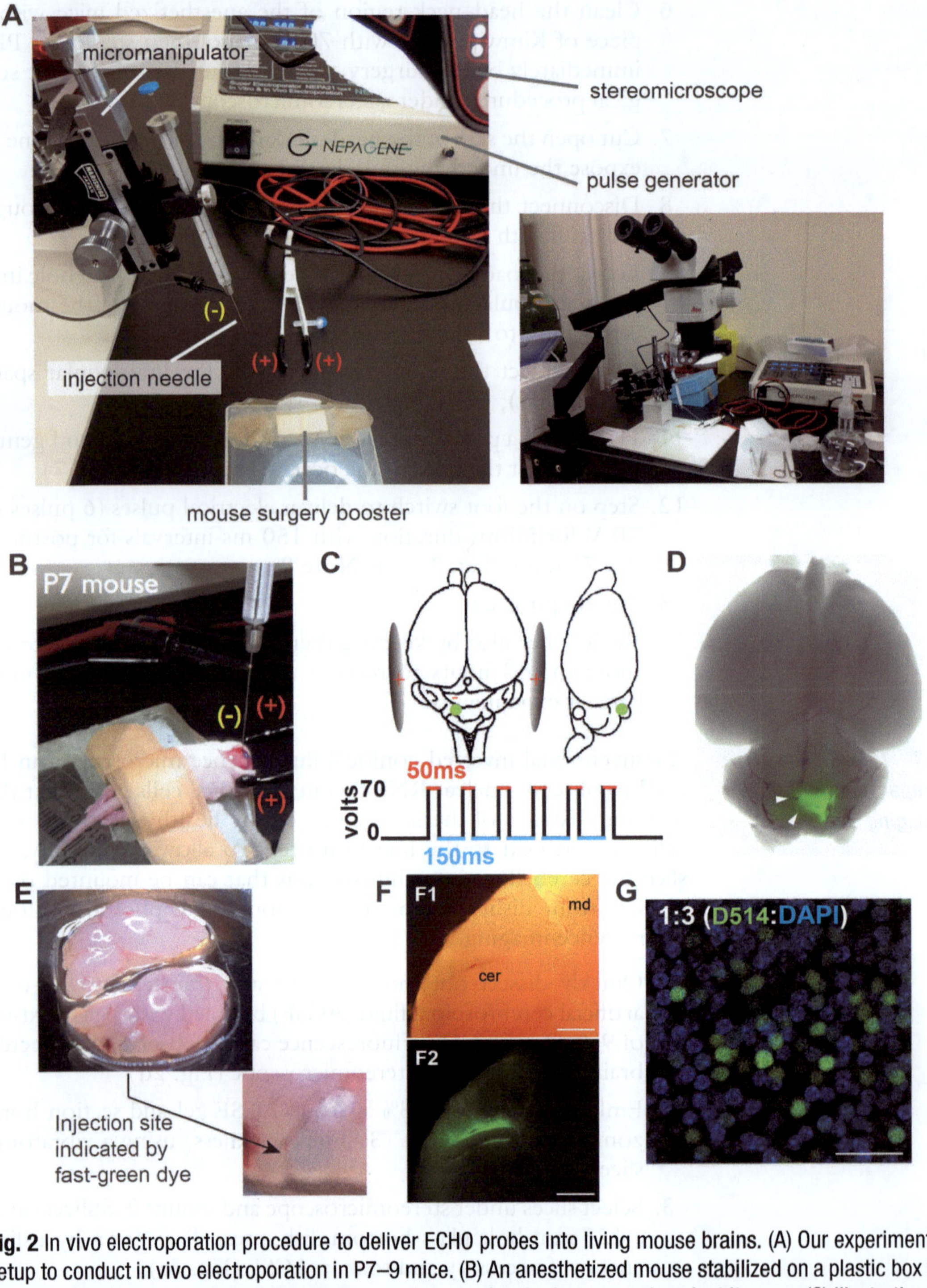

Fig. 2 In vivo electroporation procedure to deliver ECHO probes into living mouse brains. (A) Our experimental setup to conduct in vivo electroporation in P7–9 mice. (B) An anesthetized mouse stabilized on a plastic box lid with injecting/anode needle in its cerebellum and a pair of cathodes pressed against its ears. (C) Illustration of injection site and current delivery procedure. *Green dots*: probe injection sites (0.5 mm from the surface). (D) Dorsal fluorescence view of P7 cerebellum electroporated with D514-d(T)$_{12}$. (E) An open-book configuration of electroporated brain on a glass-bottom dish to reveal the inner layers of cells. Fast green dye can be used to

6. (Optional) Add drug solutions directly into the soaking ACSF when necessary.
7. Alternatively, the brains can be cut along the sagittal midline into halves and "collagen-glue" them onto glass bottom dish in an open-book configuration. Individual speckles could be observed in both sectioned slices and also in the halved hemispheres (Fig. 2e, f).
8. Probe delivery efficiency can be estimated after a post-fixed DAPI staining that labels all nuclei (Fig. 2g).

3.4 Typical Results, Validation, Control Experiments, and Troubleshooting

3.4.1 Typical Results

1. As demonstrated in Fig. 3b, a large population of granule cells in the external and inner granule layer of the cerebellum becomes fluorescent due to RNA hybridization. Magnified images reveal at single nuclear level foci-like fluorescence patterns similar for U3 snoRNA and 28S rRNA that are both localized to nucleoli, but distinct from poly (A) RNA concentrated at nuclear speckles (Fig. 3c). At this resolution, sub-nucleolar localization of U3 snoRNA and 28S rRNA cannot be distinguished (Fig. 3c). However the two RNA species have been shown to segregate after transcription inhibition, suggesting localization on distinct and independent RNPs [33, 34].
2. Time-lapse imaging can be performed up to hundreds of frames within several hours due to the molecular stability of the probes and photostability of the dyes. Movement tracking of individual foci by monitoring changes in their position and fluorescence intensity over time revealed that all three nuclear RNPs are stable with little positional change detected (Fig. 3f).

3.4.2 Estimating Sensitivity and Specificity of the liveECHO-Probes

1. To understand whether the detection of focal concentration of the target RNA is dependent on the hybridization-sensitive property of the ECHO probes, we electroporated Cy5-d$(T)_{30}$ in parallel experiments as described in Subheading 3.2.
2. Both probes hybridize to poly(A) tails; however, Cy5-d$(T)_{30}$ probes result in noisy background and diffuse fluorescence in nuclei, whereas readily distinguishable nuclear speckles with concentrated poly(A) fluorescence were observed upon electroporation of D514-$(U)_{22}$. Thus, in vivo poly(A)

Fig. 2 (continued) guide high-magnification imaging; (F) Bright-field (F1) and fluorescence (F2) views in a pair of sagittally partitioned brain halves. The fluorescence is most intense in cells adjacent to the interlobular space. Cer, cerebellum; md, midbrain. (G) Confocal images of permeabilized cerebellar slices stained with DAPI (blue) after *in vivo* electroporation of ECHO probe (*green*). One out of every three cells in the affected regions contained D514 fluorescence, indicating that delivery efficiency is roughly 30%. Images were acquired by LSM780 (Zeiss). Scale bars: 1 mm (F) and 20 μm (G). (Panels F and G are reproduced from [24] with permission from Oxford University Press)

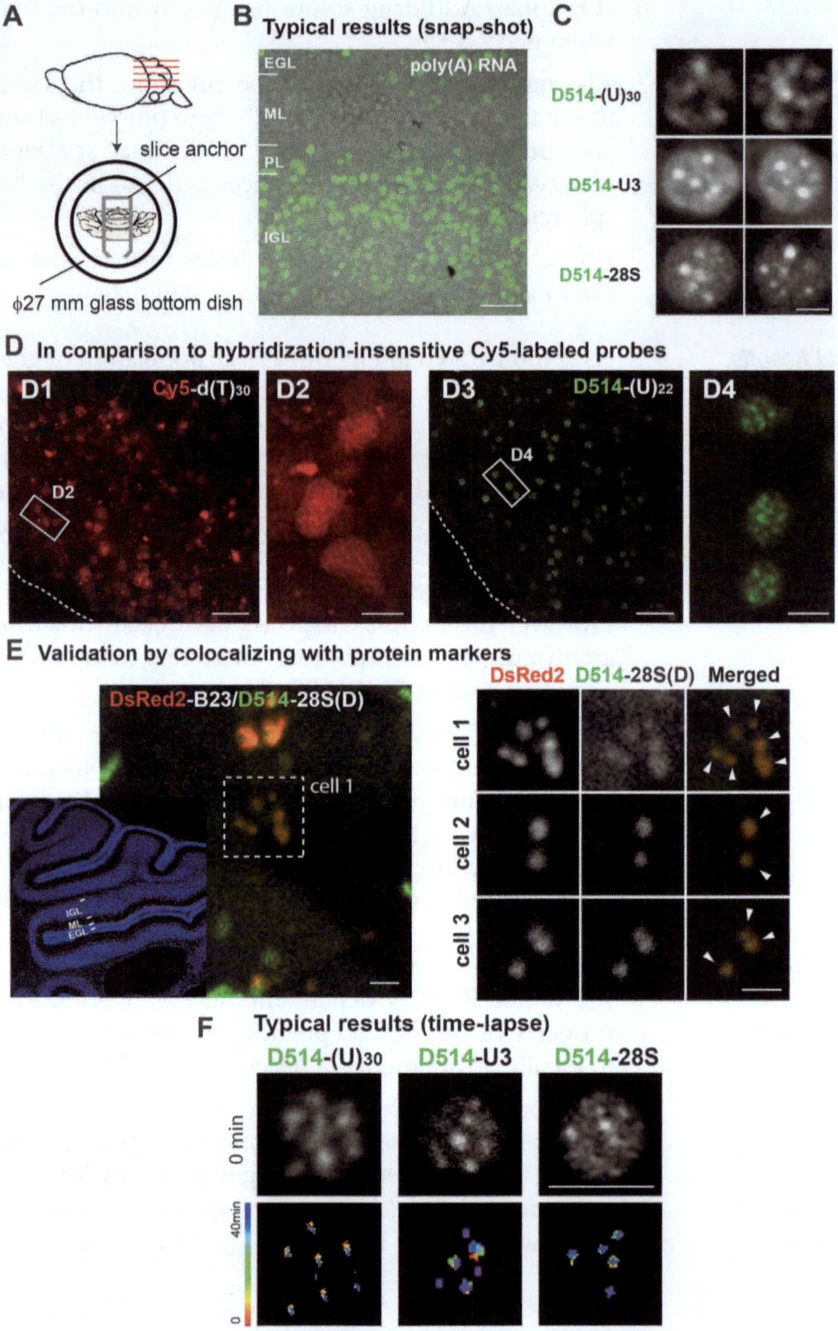

Fig. 3 Target recognition specificity validated by fluorescence patterns and colocalization with protein markers. (A) Preparation illustration of acute brain slices from electroporated mice. (B) A representative confocal image from acute cerebellar slices prepared soon after *in vivo* electroporation. EGL: external granule layer, ML: molecular layer, PL: Purkinje layer, IGL: inner granule layer. (C) Confocal images of poly(A), U3 snoRNA, and 28S rRNA in individual cerebellar granule cells after electroporation. Intranuclear foci containing target RNA concentrations are readily distinguished. The number and shape are consistent with foci of nuclear speckles and nucleoli. (D) Electroporation of oligonucleotide probes labeled with a conventional dye Cy5-d(T)$_{30}$

concentration at nuclear speckles cannot be recapitulated with a Cy5-dye labeled oligonucleotide DNA probe, indicating that both hybridization-sensitive and nuclease-resistant properties of ECHO probes are important for effective detection of target RNA (Fig. 3d).

3. To validate detection specificity, colocalization with protein markers is used. Protein B23 has been identified as a marker protein for nucleoli and when it is tagged with DsRed2, its color separation from D514 RNA probes allows simultaneous detection of U3 snoRNA (or 28S rRNA) and B23 proteins.
4. DsRed2-B23 is expressed in the cerebellum from a DNA expression plasmid using the in vivo electroporation protocol (Subheading 3.2).
5. Colocalized fluorescence signals indicate target-specific detection by ECHO-liveFISH (Fig. 3e). Likely, these foci are physically associated with chromatin and nuclear matrix that restricts their random movement.

3.4.3 Assaying Possible RNA Interference Caused by ECHO-liveFISH

siRNA and microRNA target RNA through hybridization, raising the possibility of probe-induced degradation and translation repression of the target RNA.

1. To test this possibility, we perform qRT-PCR to assess the expression level of the target RNA before and after electroporation of the probes.
2. Additionally, since 28S rRNA play critical roles in global protein production, we assess functional interference on 28S rRNA by measuring gross protein production rate in imaged cells (e.g., puromycin labeling as described in SUnSET assay [35]).

Fig. 3 (continued) showed high fluorescence background at both intracellular and extracellular locations (D1). No distinguishable intranuclear structures were resolved by Cy5-d(T)$_{30}$ labeling (D2). In contrast, D514-(U)$_{22}$ reveals robust fluorescence in nuclei with relatively low background (D3); At higher magnification, D514-(U)$_{22}$ reveals readily distinguishable poly(A) nuclear speckles in individual cerebellar cells (D4). (E) Colocalization between DsRed2-B23 and D514-28S at the nucleoli of electroporated granule neurons. P10 mice expressing DsRed2-B23 DNA plasmids were perfused and cerebellar slices were processed for DsRed2 and DAPI staining simultaneously with D514-28S hybridization. Overlapping DsRed2 and D514 fluorescence at nuclear foci indicates colocalization between B23 proteins and 28S rRNA (white arrowheads). Scale bars: 20 μm (B, D1, 3) and 2.5 μm (C, D2, 4, E). Imaging parameter: 405 nm excitation and 425–475 nm detection, 488 nm excitation and 500–600 nm detection, and 635 nm excitation and 561 nm excitation and 566–703 nm detection were used to monitor DAPI, EGFP, and DsRed2 fluorescence respectively. Images were acquired in 0.994 μm pixel size and 8 μs pixel dwell times. Image stacks of 32.48-μm depth were taken at z-step intervals of 2.32 μm. (F) Time-lapse confocal imaging of poly(A), U3 snoRNA, and 28S rRNA foci in cerebellar granule cells. *Top,* a snapshot of nuclear foci imaged in acute cerebellar slices after in vivo electroporation. *Bottom,* Track-line presentations of individual fluorescent foci over time (progressing from red to blue, 0–40 minutes). (Panels B, C, E, and F are reproduced from [24] with permission from Oxford University Press)

4 Notes

1. The interaction of TO also adds thermal stability to the probe: DNA/RNA duplex, indicated by a 7–9 °C increase in T_m for 13 nt probes.
2. The needle is reusable. For conducting multiple experiments, rinse with 70% ethanol and H_2O whenever changing sample solution and at the beginning and the end of experiments.
3. Tools can be sterilized by autoclave or alternatives.
4. Fresh ice should be used and frequently changed during the experiment. Slushy ice can result in lower reviving rate, possibly due to better heat transfer efficiency.
5. We use a marker pen to mark 1 mm from the tip of the injection needle for depth indication.
6. The fast green dye should not spread too fast and too far into the neighboring area. Oligonucleotide probe solution may spread faster than plasmid solution because of lower viscosity.
7. Bleeding should be avoided but it happens during the procedure. We use a piece of Kimwipes to get rid of blood or any excessive liquid from the animal's head before proceeding with the surgery.
8. A twitch of muscle on the head is normally observed due to the passing current. The twitching itself is harmless but it causes repositioning of the injecting needle and further damage to the brain tissue. We try to avoid such damage by retracting the needle/anode electrode to the surface of the skin immediately after electroporation.
9. Electroporated ECHO probes automatically accumulate in the nucleus. This phenomenon has been observed for in cultured cells and for other short linear oligonucleotide probes.

References

1. Fong KW, Li Y, Wang W, Ma W, Li K, Qi RZ, Liu D, Songyang Z, Chen J (2013) Whole-genome screening identifies proteins localized to distinct nuclear bodies. J Cell Biol 203(1):149–164. doi:10.1083/jcb.201303145. jcb.201303145 [pii]
2. Sutherland HG, Mumford GK, Newton K, Ford LV, Farrall R, Dellaire G, Caceres JF, Bickmore WA (2001) Large-scale identification of mammalian proteins localized to nuclear sub-compartments. Hum Mol Genet 10(18):1995–2011
3. Misteli T (2007) Beyond the sequence: cellular organization of genome function. Cell 128(4):787–800. doi:10.1016/j.cell.2007.01.028. S0092-8674(07)00126-2 [pii]
4. Sleeman JE, Trinkle-Mulcahy L (2014) Nuclear bodies: new insights into assembly/dynamics and disease relevance. Curr Opin Cell Biol 28:76–83. doi:10.1016/j.ceb.2014.03.004. S0955-0674(14)00029-5 [pii]
5. Mankodi A, Lin X, Blaxall BC, Swanson MS, Thornton CA (2005) Nuclear RNA foci in the heart in myotonic dystrophy. Circ Res 97(11):1152–1155. doi:10.1161/01.RES.0000193598.89753.e3. 01.RES.0000193598.89753.e3 [pii]

6. Lee YB, Chen HJ, Peres JN, Gomez-Deza J, Attig J, Stalekar M, Troakes C, Nishimura AL, Scotter EL, Vance C, Adachi Y, Sardone V, Miller JW, Smith BN, Gallo JM, Ule J, Hirth F, Rogelj B, Houart C, Shaw CE (2013) Hexanucleotide repeats in ALS/FTD form length-dependent RNA foci, sequester RNA binding proteins, and are neurotoxic. Cell Rep 5 (5):1178–1186. doi:10.1016/j.celrep.2013.10.049. S2211-1247(13)00648-7 [pii]
7. Mao YS, Sunwoo H, Zhang B, Spector DL (2011) Direct visualization of the co-transcriptional assembly of a nuclear body by noncoding RNAs. Nat Cell Biol 13(1):95–101. doi:10.1038/ncb2140. ncb2140 [pii]
8. Quinodoz S, Guttman M (2014) Long noncoding RNAs: an emerging link between gene regulation and nuclear organization. Trends Cell Biol 24(11):651–663. doi:10.1016/j.tcb.2014.08.009. S0962-8924(14)00146-9 [pii]
9. Chaumeil J, Le Baccon P, Wutz A, Heard E (2006) A novel role for Xist RNA in the formation of a repressive nuclear compartment into which genes are recruited when silenced. Genes Dev 20(16):2223–2237. doi:10.1101/gad.380906. 20/16/2223 [pii]
10. Hacisuleyman E, Goff LA, Trapnell C, Williams A, Henao-Mejia J, Sun L, McClanahan P, Hendrickson DG, Sauvageau M, Kelley DR, Morse M, Engreitz J, Lander ES, Guttman M, Lodish HF, Flavell R, Raj A, Rinn JL (2014) Topological organization of multichromosomal regions by the long intergenic noncoding RNA Firre. Nat Struct Mol Biol 21 (2):198–206. doi:10.1038/nsmb.2764. nsmb.2764 [pii]
11. Tyagi S, Kramer FR (1996) Molecular beacons: probes that fluoresce upon hybridization. Nat Biotechnol 14(3):303–308. doi:10.1038/nbt0396-303
12. Biggins JB, Prudent JR, Marshall DJ, Ruppen M, Thorson JS (2000) A continuous assay for DNA cleavage: the application of "break lights" to enediynes, iron-dependent agents, and nucleases. Proc Natl Acad Sci U S A 97 (25):13537–13542. doi:10.1073/pnas.240460997. 240460997 [pii]
13. Whitcombe D, Theaker J, Guy SP, Brown T, Little S (1999) Detection of PCR products using self-probing amplicons and fluorescence. Nat Biotechnol 17(8):804–807. doi:10.1038/11751
14. Holland PM, Abramson RD, Watson R, Gelfand DH (1991) Detection of specific polymerase chain reaction product by utilizing the 5′–3′ exonuclease activity of Thermus aquaticus DNA polymerase. Proc Natl Acad Sci U S A 88(16):7276–7280
15. Prigodich AE, Randeria PS, Briley WE, Kim NJ, Daniel WL, Giljohann DA, Mirkin CA (2012) Multiplexed nanoflares: mRNA detection in live cells. Anal Chem 84(4):2062–2066. doi:10.1021/ac202648w
16. Hovelmann F, Gaspar I, Ephrussi A, Seitz O (2013) Brightness enhanced DNA FIT-probes for wash-free RNA imaging in tissue. J Am Chem Soc 135(50):19025–19032. doi:10.1021/ja410674h
17. Sato S, Watanabe M, Katsuda Y, Murata A, Wang DO, Uesugi M (2015) Live-cell imaging of endogenous mRNAs with a small molecule. Angew Chem Int Ed Engl 54(6):1855–1858. doi:10.1002/anie.201410339
18. Asanuma H, Akahane M, Niwa R, Kashida H, Kamiya Y (2015) Highly sensitive and robust linear probe for detection of mRNA in cells. Angew Chem Int Ed Engl 54(14):4315–4319. doi:10.1002/anie.201411000
19. Wang DO, Matsuno H, Ikeda S, Nakamura A, Yanagisawa H, Hayashi Y, Okamoto A (2012) A quick and simple FISH protocol with hybridization-sensitive fluorescent linear oligodeoxynucleotide probes. RNA 18(1):166–175. doi:10.1261/rna.028431.111. rna.028431.111 [pii]
20. Wang DO, Okamoto A (2012) ECHO probes: Fluorescence emission control for nucleic acid imaging. J Photoch Photobio C 13 (2):112–123. doi:10.1016/j.jphotochemrev.2012.03.001
21. Ikeda S, Kubota T, Kino K, Okamoto A (2008) Sequence dependence of fluorescence emission and quenching of doubly thiazole orange labeled DNA: effective design of a hybridization-sensitive probe. Bioconjug Chem 19(8):1719–1725. doi:10.1021/bc800201m
22. Ikeda S, Kubota T, Yanagisawa H, Yuki M, Okamoto A (2009) Synthesis of exciton-controlled fluorescent probes for RNA imaging. Nucleic Acids Symp Ser (Oxf) 53:155–156. doi:10.1093/nass/nrp078. nrp078 [pii]
23. Ikeda S, Okamoto A (2008) Hybridization-sensitive on-off DNA probe: application of the exciton coupling effect to effective fluorescence quenching. Chem Asian J 3(6):958–968. doi:10.1002/asia.200800014
24. Oomoto I, Suzuki-Hirano A, Umeshima H, Han YW, Yanagisawa H, Carlton P, Harada Y, Kengaku M, Okamoto A, Shimogori T, Wang DO (2015) ECHO-liveFISH: in vivo RNA labeling reveals dynamic regulation of nuclear RNA foci in living tissues. Nucleic Acids Res 43 (19):e126. doi:10.1093/nar/gkv614. gkv614 [pii]

25. Ohmachi M, Fujiwara Y, Muramatsu S, Yamada K, Iwata O, Suzuki K, Wang DO (2016) A modified single-cell electroporation method for molecule delivery into a motile protist, Euglena gracilis. J Microbiol Methods 130:106–111. doi:10.1016/j.mimet.2016.08.018. S0167-7012(16)30228-7 [pii]
26. Hayashi G, Yanase M, Takeda K, Sakakibara D, Sakamoto R, Wang DO, Okamoto A (2015) Hybridization-sensitive fluorescent oligonucleotide probe conjugated with a bulky module for compartment-specific mRNA monitoring in a living cell. Bioconjug Chem 26 (3):412–417. doi:10.1021/acs.bioconjchem.5b00090
27. Schins JM, Agronskaia A, de Grooth BG, Greve J (1999) Orientation of the chromophore dipoles in the TOTO-DNA system. Cytometry 37(3):230–237. doi:10.1002/(SICI)1097-0320(19991101)37:3<230::AID-CYTO10>3.0.CO;2-# [pii]
28. Kasha M (1963) Energy Transfer Mechanisms and the Molecular Exciton Model for Molecular Aggregates. Radiat Res 20:55–70
29. Nygren J, Svanvik N, Kubista M (1998) The interactions between the fluorescent dye thiazole orange and DNA. Biopolymers 46(1):39–51. doi:10.1002/(SICI)1097-0282(199807)46:1<39::AID-BIP4>3.0.CO;2-Z [pii]
30. Wang DO, Okamoto A (2015) Visualization of nucleic acids with synthetic exciton-controlled fluorescent oligonucleotide probes. Methods Mol Biol 1262:69–87. doi:10.1007/978-1-4939-2253-6_5
31. Wang DO, Okamoto A (2015) ECHO-FISH for gene transcript detection in neuronal and other cells and subcellular compartments. In: Hauptmann G (ed) In Situ Hybridization Methods. NeuroMethods 99:559–584. doi:10.1007/978-1-4939-2303-8_30
32. Umeshima H, Hirano T, Kengaku M (2007) Microtubule-based nuclear movement occurs independently of centrosome positioning in migrating neurons. Proc Natl Acad Sci U S A 104(41):16182–16187. doi:10.1073/pnas.0708047104. 0708047104 [pii]
33. Jacob MD, Audas TE, Uniacke J, Trinkle-Mulcahy L, Lee S (2013) Environmental cues induce a long noncoding RNA-dependent remodeling of the nucleolus. Mol Biol Cell 24 (18):2943–2953. doi:10.1091/mbc.E13-04-0223. mbc.E13-04-0223 [pii]
34. Burger K, Muhl B, Harasim T, Rohrmoser M, Malamoussi A, Orban M, Kellner M, Gruber-Eber A, Kremmer E, Holzel M, Eick D (2010) Chemotherapeutic drugs inhibit ribosome biogenesis at various levels. J Biol Chem 285 (16):12416–12425. doi:10.1074/jbc.M109.074211. M109.074211 [pii]
35. Schmidt EK, Clavarino G, Ceppi M, Pierre P (2009) SUnSET, a nonradioactive method to monitor protein synthesis. Nat Methods 6 (4):275–277. doi:10.1038/nmeth.1314. nmeth.1314 [pii]

Chapter 18

In Vivo Visualization and Function Probing of Transport mRNPs Using Injected FIT Probes

Jasmine Chamiolo, Imre Gaspar, Anne Ephrussi, and Oliver Seitz

Abstract

Fluorogenic hybridization methods, such as the use of FIT probes, enable the in vivo detection of specific mRNAs transcribed from their endogenous, genetically nonmodified loci. Here, we describe the design, synthesis and injection of nuclease resistant FIT probes into developing *Drosophila* oocytes to detect endogenous localizing mRNAs as wells as to probe function of structural RNA elements.

Key words mRNA transport, Injection, Live cell imaging, Locked nucleic acid, Forced intercalation of thiazole orange

1 Introduction

To understand the steps of the complex and highly dynamic biogenesis of mRNPs, such as transcription, splicing, nuclear export, cytoplasmic localization, translation, storage, and decay, the mRNAs of interest have to be studied in vivo in their subcellular context and in relation to their continuously changing binding partners that compose the RNPs.

Hybridization techniques using conventional chromogenic or fluorescent labels require a differentiation step—mostly washes—to discriminate between target and background. Since such steps cannot be performed in vivo, probes with different types of fluorogenic labeling were developed over the past 2 decades: molecular beacons [1], hybridizing RNA aptamers [2], exciton-controlled hybridization-sensitive fluorescent oligonucleotide (ECHO [3, 4]) and forced intercalation of thiazole orange (FIT) probes [5]. Through different mechanisms these probe molecules substantially increase their fluorescence upon hybridization to target (responsiveness) and when delivered to target cells—e.g., by microinjection—they enable the detection of endogenous transcript molecules with great specificity. FIT probes rely on the fluorescence

Imre Gaspar (ed.), *RNA Detection: Methods and Protocols*, Methods in Molecular Biology, vol. 1649, DOI 10.1007/978-1-4939-7213-5_18, © Springer Science+Business Media LLC 2018

increase of the DNA intercalating dye thiazole orange (TO) [5]. TO is used as a base surrogate, replacing one of the internal bases of a target specific oligonucleotide. Upon mismatch-free duplex formation, the dye is positioned in a high viscosity microenvironment that limits conformational changes of TO and thus activates its fluorescence. Injecting as few as three different nuclease resistant FIT probes we successfully visualized the *oskar* transport mRNAs in developing *Drosophila* oocytes [6]. A locked nucleic acid (LNA) nucleotide adjacent to TO uniquely increased the brightness of the FIT probes and also contributed to nuclease resistance. We obtained identical results of the RNP transport properties to that measured with the reference *oskar-MS2(10x)*, MCP-EGFP (or MCP-mCherry) transgenic system. The different fluorescence spectrum of TO, EGFP (or mCherry) fluorescent proteins allowed the simultaneous visualization of two labels. Moreover, targeting the localization element (LE) of the endogenous transcript that is responsible for the intra-ooplasmic, kinesin-1 mediated transport of *oskar* mRNPs [7] with a FIT-probe, we could recapitulate the findings of our previous transgenic mutagenesis analysis while having visual confirmation on the probe–target duplex formation and the consequent disruption of the LE [6]. Here, we describe the probe design, synthesis and analysis procedures as well as the microinjection and live cell imaging of developing *Drosophila* oocytes.

2 Materials

Prepare all solutions using ultrapure water (ddH_2O) and analytical grade reagents unless indicated otherwise. Follow waste disposal and safety regulations as indicated on the reagent storing containers.

2.1 Probe Synthesis

1. 3′-Spacer-C3-CPG (1 μmol, pore size 500 Å) and 2’O-Me-RNA phosphoramidites (e.g., Link Technologies).
2. DNA phosphoramidites (e.g., Thermo Fisher Scientific).
3. LNA phosphoramidites (e.g., Exiqon).
4. Dye monomer synthesis is carried out according to F. Hövelmann et al., *Chem.Sci.*, **2016**, *7*, 128–135.
5. DNA synthesizer (e.g., MerMade-4 Synthesizer Bioautomation).
6. Activator: Hyacinth DMT-solution (e.g., emp Biotech).
7. Capping A: THF–lutidine–acetic anhydride (e.g., emp Biotech).
8. Capping B: 10 v/v % 1-methylimidazole in THF (e.g., emp Biotech).

9. Oxidizer: 0.02 M iodine in pyridine, water, and THF 0.4:9:90.6 (v:v:v) (e.g., emp Biotech).
10. Deblock: 6% dichloroacetic acid in methylchloride (e.g., emp Biotech).
11. Acetonitrile for DNA-synthesis.

2.2 Probe Cleavage and Purification

1. 32% aqueous ammonia.
2. 55 °C rocking heating block.
3. 80% aqueous AcOH.
4. *i*ProOH.
5. 3 M ammonium acetate.
6. Reversed Phase HPLC e.g., 1105 HPLC System from Gilson, using a Waters X-Bridge BEH130 C18 column (10 × 150 mm, 5 μm) at a flow rate of 8 mL/min at 55 °C. (DMTr-on: 15–40% B in 10 min, A = 0.1 M triethylammonium acetate, aq. pH 7.4; B = MeCN) (DMTr-off: 5–20% B in 10 min, A = 0.1 M triethylammonium acetate, aq. pH 7.4; B = MeCN)

2.3 Probe Analytic

1. Reversed Phase HPLC, e.g., 1105 HPLC System from Gilson by using a Waters X-Bridge BEH130 C18 column (4.6 × 250 mm, 5 μm) at a flow rate of 1.5 mL/min at 55 °C.

   (10–50% B in 10 min, A = 0.1 M triethylammonium acetate, aq. pH 7.4; B = MeCN).
2. MALDI-TOF mass spectra (e.g., Shimadzu Axima Confidence) in positive mode with HPA matrix (1:1 mixture of = 50 mg/mL 3-hydroxypicolinic acid in MeCN/H2O 1:1 and 50 mg/mL diammonium citrate in MeCN/H2O 1:1).

2.4 In Vitro Measurements

1. Fluorescence spectrometer (e.g., Varian Cary Eclipse).
2. 10 mm quartz cuvettes.
3. Phosphate buffer: 10 mM Na_2HPO_4, 100 mM NaCl, pH 7.0.
4. UV–Vis spectrometer (e.g., Varian Cary Bio 100).
5. ATTO 520 and ATTO 590 fluorescent dyes.

2.5 Fly Husbandry and Dissection

Flies are raised on standard cornmeal agar at 25 °C in humidified incubators.

1. Dissection microscope station equipped with CO_2 pistol and pad to anesthetize the flies.
2. 100% ethanol.
3. Dissection buffer (BRB80): 80 mM PIPES–KOH, pH 6.9, 1 mM EGTA, and 2 mM $MgCl_2$.

4. Oocyte cultivation medium: full S2 cell culture medium, 15% fetal bovine serum (FBS), 1% streptomycin–penicillin antibiotics mixture. Store at 4 °C until required. Before use, freshly add 200 μg/mL human insulin.
5. A 3–9 well dissection plate.
6. Dumont #4 and #5 forceps.
7. Two sharpened tungsten needles mounted in handles.
8. Voltalef 10S halocarbon oil.

2.6 Wash-Free Fluorescent in Situ Hybridization (FISH)

1. Fixative: Mix a vial (10 mL) of 16% electron-microscopy grade paraformaldehyde (PFA) with 30 mL sterilized phosphate buffered saline (PBS, pH 7.4). Filter through 0.22 micron size particle filter to remove aggregates and other impurities. Store at 4 °C until required.
2. Wash buffer (IBEX): 10 mM Tris–HCl, pH 7.5–7.7, 100 mM KCl, 1 mM EDTA, 0.3 v/v % Triton-X-100.
3. 20 mg/mL Proteinase K (optional).
4. Ethylene carbonate (EC, optional).
5. 1.5 mL Eppendorf tubes.
6. Tube stands.
7. Micropipettes.
8. Nutator.
9. 37 °C rocking heating block.
10. 22 mm × 22 mm coverslips, nontreated glass slides to mount the specimen.
11. Coverslip sealant, e.g., transparent, nonglittering nail polish.

2.7 Microinjection

1. Eppendorf Femtotips II microinjection capillaries or
2. Borosilicate capillaries (e.g., Sutter 100-50-10).
3. Micropipette puller with 3 × 3 mm box filament to make microinjection capillaries (e.g., Sutter P-97).
4. Injection buffer (IB): 10 mM Tris–HCl, pH 7.5–7.7, 100 mM KCl, 1 mM $MgCl_2$.
5. Table-top microcentrifuge.
6. Microinjector (e.g., Eppendorf Femtojet series).
7. Incised plastic slide (*see* Fig. 1b) and 22 × 22 mm coverslips.
8. Double sided tape.
9. Thin plastic sheet, e.g., used X-ray films.
10. 2–3 mm wide strips of Whatman 3 mm filter paper.

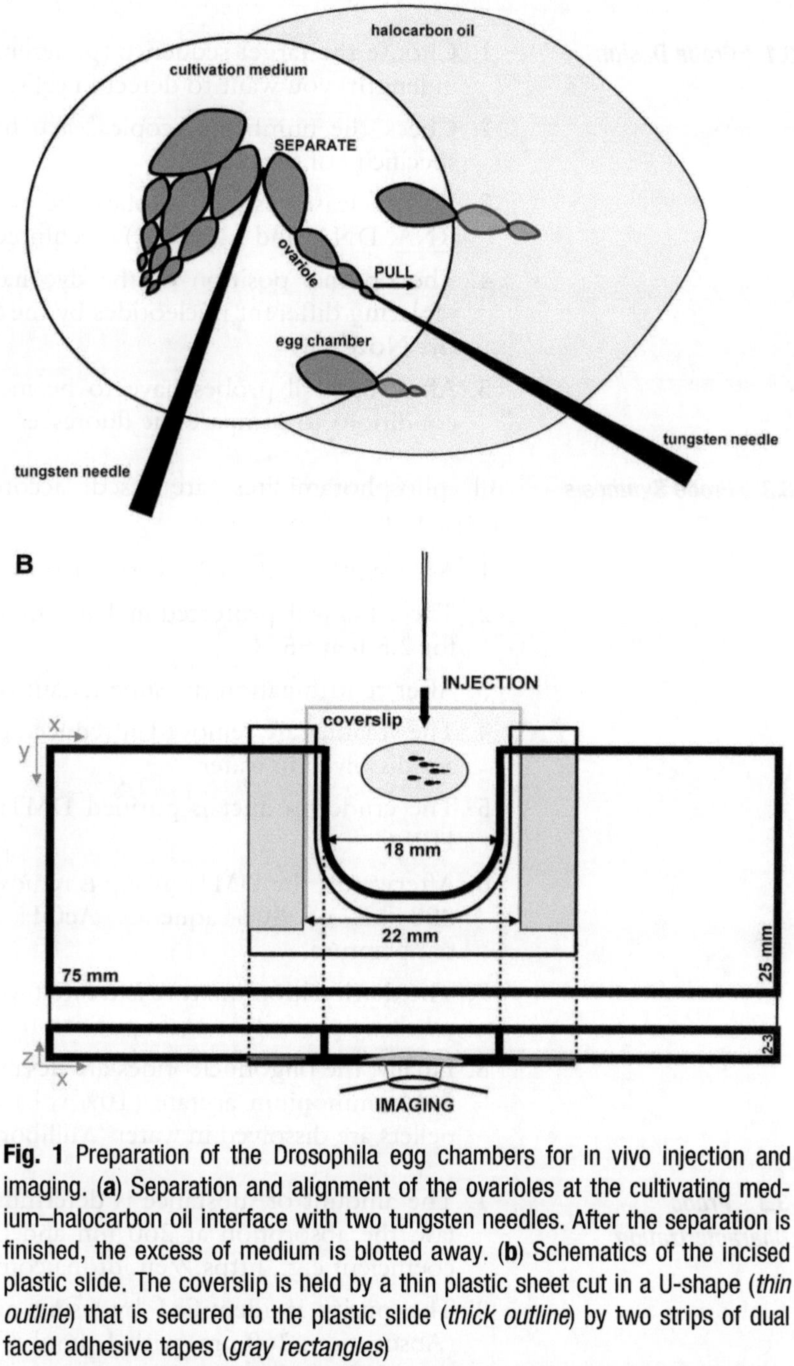

Fig. 1 Preparation of the Drosophila egg chambers for in vivo injection and imaging. (**a**) Separation and alignment of the ovarioles at the cultivating medium–halocarbon oil interface with two tungsten needles. After the separation is finished, the excess of medium is blotted away. (**b**) Schematics of the incised plastic slide. The coverslip is held by a thin plastic sheet cut in a U-shape (*thin outline*) that is secured to the plastic slide (*thick outline*) by two strips of dual faced adhesive tapes (*gray rectangles*)

2.8 Microscopy

1. Inverted confocal microscope with a low and a high magnification, high numerical aperture objective (e.g., Leica SP8 with a 20× dry and a 63× 1.4 NA objective).
2. 488 nm or 514 nm laser lines to excite thiazole orange (TO).
3. 561 nm or 594 nm laser lines to excite quinoline blue (QB).

3 Methods

3.1 Probe Design

1. Choose the target sequence (preferentially 18–20 nucleotides in length) you want to detect in cells.
2. Check the number of copies, structure, singularity, and the specificity of the target.
3. For nuclease resistant probes the combined use of 2′O-Me RNA, DNA, and LNA (X_L) is required.
4. The optimal position of the dye has to be determined by replacing different nucleotides by the Ser(TO/QB)-monomer (*see* **Note 1**).
5. Afterwards all probes have to be measured under the same conditions to compare the fluorescence properties.

3.2 Probe Synthesis

All phosphoramidites are used according to manufactures instructions.

1. After synthesis the CPGs are dried under reduced pressure.
2. The CPG is deprotected in 1 mL of aqueous ammonia (32%) for 2.5 h at 55 °C.
3. After centrifugation the supernatant is collected.
4. The volatiles are removed at reduced pressure and the residues are dissolved in water.
5. The crude product is purified DMTr-on by preparative RP-HPLC.
6. Afterwards, the DMTr group is removed upon treatment with 300 μL of 80% aqueous AcOH for 15 min at room temperature.
7. The detritylation mixture is treated with 1 mL *i*PrOH and the resulting precipitate again purified by RP-HPLC DMTr-off.
8. Finally, the oligonucleotides are desalted by precipitation with 3 M ammonium acetate (10% vol.) and 1 mL *i*PrOH. The pellets are dissolved in water (Millipore) and analyzed.

3.3 Probe Characterization

1. The amount of substance is determined using Lambert–Beer law, the absorption at 260 nm and the calculated extinction coefficient e.g., https://eu.idtdna.com/calc/analyzer.
2. The purity is identified by using analytical RP-HPLC-UV (Absorption: 260 nm) with 1 nmol of probe solved in water.
3. Characterization by MALDI-TOF mass spectrometry requires approx. 1 nmol of probe in HPA matrix for measurements in the positive detection mode.

3.4 In Vitro Analysis of the Probes

1. The fluorescence and absorption measurements are performed with a 0.5 μM concentration of probe in 10 mm quartz cuvettes with phosphate buffer.
2. For melting analysis the absorbance at 260 nm is monitored during a thermal cycle (three times 25–90 °C in 0.5 °C/min) with 1 eq of RNA target.
3. First the fluorescence of the probe is measured in the single stranded state, without the target nucleic acid (*see* **Note 2**):

   TO: $\lambda_{ex} = 485$ nm, $\lambda_{em} = 500$–700 nm
   QB: $\lambda_{ex} = 560$ nm, $\lambda_{em} = 575$–750 nm
4. Next, the absorption spectrum (700–220 nm, 1 nm steps) of the probe is measured in the same cuvette.
5. 5 eq RNA is added and the fluorescence measurement (**step 3**) is repeated for the double stranded probe.
6. **Steps 3–5** are performed three times per analyzed probe and three times for the blank.
7. The average of three fluorescence measurements (corrected with the fluorescence of the blank and normalized by the absorption at 260 nm) is calculated for both the single and double stranded states of the probes.
8. The responsiveness of the probes is calculated with following formula:

$$I/_{I_0} = \frac{I_{ds}}{I_{ss}}$$

   The readout of the fluorescence intensity: TO = 535 nm, QB = 605 nm.
9. The extinction coefficient for every wavelength is calculated as follows:

$$\varepsilon(x) = \frac{Abs_x \cdot \varepsilon_{(260)}}{Abs_{(260)}}$$

10. The quantum yields of the single and double stranded states are evaluated measuring the ATTO dyes (ATTO 520 for TO and ATTO 590 for QB) under the same conditions as the probes using following formula:

$$QY = \frac{\int I_{Probe} \cdot Abs_{ref} \cdot QY_{ref}}{\int I_{ref} \cdot Abs_{probe}}$$

11. The brightness is calculated as follows:

$$\mathrm{Br} = \left(\frac{\mathrm{QY} \cdot \varepsilon_{(485/560)}}{1000}\right)$$

12. The probes with the best fluorescence properties (brightness and responsiveness) are tested by wash-free FISH (Subheading 3.6) and subsequently may be used for in vivo visualization.

3.5 Fly Dissection

1. Take a clean 3–9 well dissection plate, fill one well with 100% ethanol (~2 mL) and two wells with BRB80.
2. Anesthetize the females prepared for dissection (*see* **Note 3**) and transfer them into the well containing the ethanol, make sure they sink to the bottom (*see* **Note 4**).
3. After 30 s incubation, transfer them to the next well with BRB80 and wash away the ethanol. Transfer them to the third well.
4. Using the #4 forceps crush the thorax of the fly and hold it steady. With the #5 forceps gently grab the ventral side of the abdomen close to the posterior tip and with a firm movement open it.
5. Isolate the bulging ovaries and separate them from the rest of the internal organs (e.g., intestines, Malpighian tubules, and the oviduct).
6. Carry on with the wash-free FISH or the microinjection protocols.

3.6 Wash-Free FISH

To test whether the probes have access to the target RNA in its native context and conformation and that the RNA–probe duplex yields sufficient signal–noise ratio a rapid wash-free FISH is carried out.

1. Transfer the dissected ovaries into 500 μL fixative in a 1.5 mL Eppendorf tube and nutate them for 20 min at room temperature (*see* **Note 5**).
2. Remove the fixative and wash with 1 mL IBEX three times 10 min.
3. Add 500 μL fresh IBEX prewarmed to 37 °C and place the ovaries into the heating block (*see* **Note 6**).
4. Add the probe(s) to be tested at a 10–50 nM per probe concentration.
5. Rock for 20 min with 1200–1400 RPM at 37 °C.
6. Take the tube out of the heating block and allow the ovaries to settle at room temperature.

7. Transfer the ovaries in a 20–25 μL volume to a nontreated clean glass slide with a P200 micropipette using a cut 200 μL pipette tip
8. With the two tungsten needles separate the ovarioles and flatten the ovaries as much as possible.
9. Gently lay on top of the ovaries a 22 × 22 mm coverslip, make sure that there are no air bubbles trapped between the two glass surfaces.
10. Seal the edges of the coverslip with nail polish or with some other coverslip sealant.
11. Image with a high magnification, high NA objective that would be used for the in vivo imaging and analyze the intensity of the specific signal relative to the cytoplasmic background. The observed contrast is a good proxy to estimate the in vivo performance of the microinjected probes. The following modifications of the wash-free FISH protocol will address the different mechanisms that may underlie the insufficient contrast.

3.6.1 Troubleshooting: Inefficient Target Recognition

To make the target segment of the mRNA more available for hybridization, insert these steps between **steps 2** and **3** of Subheading 3.6 (*see* **Notes 6** and 7).

1. Add 1 μg/mL Proteinase K to the last IBEX wash and digest the specimen for 5 min.
2. Remove the digestion solution and add 500 μL IBEX + 0.05% SDS preheated to 92 °C. Incubate for 5 min at 92 °C.
3. Add 1 mL ice-cold IBEX and chill on ice for 1–2 min. Proceed with Subheading 3.6, **step 3** but extend the incubation time to 60 min.

3.6.2 Troubleshooting: High Degree of Unspecific Activation of the Probe Fluorescence

To test the degree of unspecific activation of the probe make the following modifications of Subheading 3.6 (*see* **Notes** 7 and **8**):

1. Supplement the IBEX used for hybridization at Subheading 3.6, **step 3** with 15% EC.
2. After the hybridization (Subheading 3.6, **step 5**) wash the specimen with 1 mL IBEX once at 37 °C and then twice at 25 °C for 3 × 10 min.

3.7 Preparation of Microinjection Capillaries (Optional)

The aim is to obtain Type B microinjection capillaries with 5–7 mm long taper and 0.5–1.0 μm tips. On a Sutter P-97 micropipette puller with 3 × 3 mm box filament we use the following program:

1. Determine the RAMP temperature of one glass capillary from the batch.
2. Enter the following one line program into the puller:

HEAT: RAMP + 5 C, PULL = 35, VEL = 75, TIME = 130 d, PRESSURE = 500

3. Insert a new capillary and execute the program.
4. Repeat **step 3** to obtain at least 8–10 microinjection capillaries.

3.8 Microinjection and Imaging

1. Prepare the probe(s) for injection by diluting to 2–10 μM/probe final concentration in injection buffer.
2. Spin down at min 12,000 × *g* for 2 min to remove any aggregates or impurities that may clog the microinjection needle. Transfer the supernatant to a clean tube.
3. Load 1.5–2.5 μL of the prepared mixture into a microinjection needle and mount it into the capillary holder of the injector system (*see* **Note 9**).
4. Place a drop of Voltalef 10S oil onto a clear coverslip and focus it with a low magnification objective on the stage of the injection microscope.
5. Using the micromanipulator of the injection system, plunge the tip of the microinjection needle into the oil (*see* **Note 10**).
6. Apply some positive pressure to the needle and observe the flow of the injection solution. Adjust the pressure and duration of the injection until a small (diameter ~20–50 μm) liquid bubble forms under the oil. Typically, this happens at 1000–1200 hPa applied for 0.2–0.3 s using a gas operated injector (*see* **Note 11**).
7. Leave the tip under the oil until preparing the specimen (*see* **Note 12**).
8. Dissect ovaries as described in Subheading 3.5 and transfer them onto one side of a clean coverslip in a small drop of BRB80 with the #5 forceps.
9. Place them into a drop of oocyte cultivating medium (5 μL) in the middle of the same coverslip for 10s.
10. Transfer them to another drop of cultivating medium close to the other side of the same coverslip. Place a drop of Voltalef 10S oil adjacent to the specimen that the two drops form an interface.
11. Using cleaned tungsten needles, separate the ovarioles from each other. Isolate those that contain oocytes in the appropriate developmental stage(s) and using the germarium as handle, pull the ovarioles to the medium–oil interface with one needle. Move the young egg chambers under the oil and leave the rest in the medium. Remove the late oogenetic stages from the ovarioles if you do not need them, as they may hinder injection of the earlier oocytes (Fig. 1a).

12. After the ovarioles are positioned, blot away the cultivating medium using the Whatman filter strips. This will also remove all the egg chambers that are not fixed under the oil. The Voltalef 10S oil will subsequently invade the space of the cultivating medium. With some filter strips, smear the edges of the oil to ensure isolation of the selected ovarioles from the atmosphere. Also, blot away the two other drops on the coverslip from **steps 8** and **9**.
13. Under the oil, arrange the ovarioles as such that the appropriate egg chambers are physically accessible for the microinjection needle (Fig. 1a)
14. Place and secure the coverslip onto the incised plastic slide (Fig. 1b).
15. To transfer the specimen to the injection microscope, first retract the needle from the storing oil droplet. Carefully place the specimen holder onto the stage and return the needle until its tip is in the oil droplet that covers the ovarioles (*see* **Note 10**).
16. Check if the microinjection settings are correct and adjust if necessary (**step 6**).
17. Focus the specimen and adjust the tip position to bring it into focus.
18. Slowly approach an egg chamber and try to move the tip into it. A successful piercing is indicated by initial distortion and sudden relaxation of the surface of the egg chamber (*see* **Note 13**).
19. Once the tip is inside the oocyte, inject with the determined pressure. Look for the injection effect. Ideally, the organelles (e.g., yolk granules) should be slightly displaced by the injection volume and fluorescence of the probes should be detected (*see* **Note 14**).
20. Either focus on another egg chamber and repeat **steps 18** and **19** or proceed with microscopy.

3.9 Imaging the Probe Labeled Transcripts

1. Transfer the specimen to the injection microscope and/or switch to the high magnification imaging objective.
2. Set up the detection settings as follows (*see* **Note 15**):
   (a) 488 nm or 514 nm laser excitation, 525 nm–575 nm detection window in case of TO labeled probes (Fig. 2).
   (b) 561 nm or 594 nm excitation, 605 nm–655 nm detection window in case of QB labeled probes.
3. We recommend acquiring images with a minimum of twofold oversampling in the lateral dimensions (x and y) and with a temporal resolution where the imaged mRNPs do not change their position by more than a particle diameter between

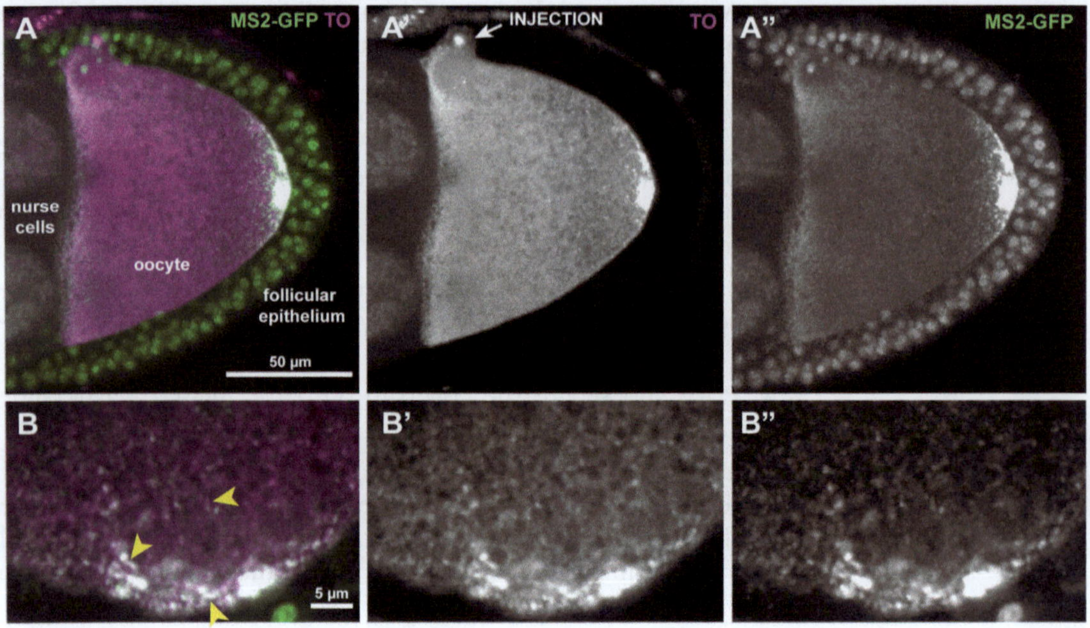

Fig. 2 Injected *Drosophila* oocyte. (**a–a"**) Z-stack of a Stage 9 oocyte expressing the *oskarMS2*-MCP-EGFP reporter system [8] (*green*, **a"**) injected with a mixture of three TO-labeled LNA mixmer FIT probes that target *oskar* mRNA (*magenta*, **a"**). The site of the injection is marked by a microdamage in the follicular epithelium (*arrow*, **a′**). Note that there is a mild aspecific accumulation of the injected probes in the nuclei (e.g., nurse cells o the left). (**b–b"**) Time projection of 4 s (12 frames) of an in vivo time lapse acquisition. *Oskar* mRNPs are highlighted with MCP-EGFP (*green*, **a"**) and TO (*magenta*, **b′**). *Yellow* arrowheads (**b**) indicate streaks drawn by mRNPs moving along a linear trajectory. These trajectories are marked by both MCP-EGFP and TO

adjacent frames. The resolution of the images we typically acquire are 120 nm × 120 nm × 800 nm at 3.3–5 Hz (*see* **Note 16**).

4 Notes

1. The dye should be at least 2–3 nucleotides away from the terminus.
2. The temperature of measurement should be the same as the temperature of the in vivo experiments.
3. To obtain ovaries that contain all developmental stages and that are not in a relative nutrient deficient state, place 5–6 females and the same number of males into a fresh vial containing the standard food and granular baker's yeast 16–24 h before dissection.
4. The hydrophobic wax covering the cuticle is dissolved by the ethanol wash preventing the fly from floating on the surface of the dissection buffer. This brief wash does not affect the viability of the ovaries.

5. Since PFA and formaldehyde covalently cross-link protein molecules to RNA and also break the sugar phosphate backbone of nucleic acids, prolonged fixation will result in greatly reduced FISH quality.
6. Since the purpose of the wash-free FISH analysis is to keep the RNA in its native state, those steps of a classical FISH protocol that make the RNA more available for hybridization (e.g., limited Proteinase K digestion and denaturation of RNA secondary structures by high temperature) are omitted.
7. These modifications will reduce the likelihood of unspecific binding. Optionally the hybridization temperature may be increased up to the melting temperature of the probes.
8. If neither of the modifications described in Subheadings 3.6.1 and 3.6.2 improve the signal to background ratio, the likely cause of the poor contrast is the insufficient activation of the probe fluorescence by duplex formation (low in situ responsiveness). While the inefficient target recognition of the probe inevitably requires that a different segment of the RNA is targeted, insufficient or unspecific activation of the probe fluorescence may be solved by moving the TO dye to another position within the same sequence.
9. In case of a gas operated injector, load the microinjection needle from the back with an Eppendorf capillary loading tip. Make sure that the solution reaches the very tip of the capillary before inserting it into the holder. Use gravity and/or firm shaking of the capillary to increase the speed of this process. When using an oil based injector, plunge the tip of the injector-mounted needle into a drop of the probe solution placed onto a coverslip under the injection microscope. Apply negative pressure to load the tip from the front.
10. Plunging the tip into any liquid drops on a glass surface requires special care to prevent damaging the needle. Turn on the transmission light of the microscope and navigate the tip into the light cone while keeping a 1–2 cm distance from the glass. Once the light gleams on the needle, gently lower it until it touches the surface of the liquid. Then switch to the eyepiece of the microscope and slowly move the needle in the lateral directions (x and y) until you find the tip. Usually, it appears rather unfocused, mostly just as a shadow. Focus to it by using the focus control of the microscope. Then by coadjusting the axial position of the micromanipulator and the microscope focus, the tip can be lowered slowly until it reaches the glass surface. When the tip is seen bending elevate its position slightly to prevent any damage to occur.
11. Sometimes the tip of the needle is sealed—especially if it was self-made—or gets clogged. If no liquid bubbles appear despite

the highest available pressure applied, position the stage such that the tip only slightly reaches trough the edge of the glass slide or coverslip. Slowly lower the needle until the tip bends and snaps onto the other side of the glass. Now, slowly raise the tip till it bends and snaps back to focus. Usually, this maneuver opens the tip without affecting the final diameter and thus the piercing ability. However, if it fails repeatedly or the tip gets broken, replace the needle and start over from **step 3** of Subheading 3.7.

12. Exposing the tip to open air for prolonged periods will cause clogging, so keep it under the oil also when preparing or imaging the specimen.

13. Assessing the axial position of the tip relative to the specimen is difficult. As a rule of thumb look for indications of physical interaction, e.g., movement, rotation and distortion of the specimen and bending of the tip. Movement of the specimen indicates too large tip diameter and a consequent inappropriate piercing pressure, whereas rotation of the egg chambers signifies too lateral hit which usually stems from a too shallow angle of the needle. Ideally, this angle is between 30 and 45° relative to the glass plane holding the specimen.

14. It may happen that the tip gets stuck in the perivitelline space between the oocyte and the follicular epithelium. In such cases, formation of a liquid bubble can be observed after injection, accompanied by an immensely bright fluorescent signal of the probe around and within the oocyte due that it gets endocytosed by the oocyte into yolk granules. Due to the intensity of this aspecific fluorescence, the affected oocyte cannot be used for mRNP imaging even after reinjection.

15. TO and QB labeled probes can be simply combined with simultaneous imaging of red (e.g., mCherry) or green fluorescent proteins (e.g., EGFP), respectively due to the differences in the emission and absorption spectra of these fluorescent molecules. Using microscopes equipped with a supercontinuum based light source (e.g., a white-light laser) TO can be combined with EGFP with minimal cross-talk using the following sequential scan setup: 470 nm laser line and 480–520 nm detection window to image GFP and 525 nm laser line and 535–575 nm detection window to image TO (Fig. 2).

16. We found no decline in *oskar* mRNP motility within the first hour after dissection. Very little or no RNP motility and a subsequent formation of larger RNP aggregates over the course of an hour are indicative of either inefficient insulin signaling or probable interference of the probe with normal mRNP biogenesis. To distinguish between the two scenarios, repeat the microinjection and imaging using cultivation medium made with a fresh aliquot of human insulin.

References

1. Tyagi S, Kramer FR (1996) Molecular beacons: probes that fluoresce upon hybridization. Nat Biotechnol 14(3):303–308. doi:10.1038/nbt0396-303
2. Sato S, Watanabe M, Katsuda Y, Murata A, Wang DO, Uesugi M (2015) Live-cell imaging of endogenous mRNAs with a small molecule. Angew Chem 54(6):1855–1858. doi:10.1002/anie.201410339
3. Oomoto I, Suzuki-Hirano A, Umeshima H, Han YW, Yanagisawa H, Carlton P, Harada Y, Kengaku M, Okamoto A, Shimogori T, Wang DO (2015) ECHO-liveFISH: in vivo RNA labeling reveals dynamic regulation of nuclear RNA foci in living tissues. Nucleic Acids Res 43 (19):e126. doi:10.1093/nar/gkv614
4. Okamoto A (2011) ECHO probes: a concept of fluorescence control for practical nucleic acid sensing. Chem Soc Rev 40(12):5815–5828. doi:10.1039/c1cs15025a
5. Kohler O, Jarikote DV, Seitz O (2005) Forced intercalation probes (FIT probes): thiazole orange as a fluorescent base in peptide nucleic acids for homogeneous single-nucleotide-polymorphism detection. Chembiochem 6 (1):69–77. doi:10.1002/cbic.200400260
6. Hovelmann F, Gaspar I, Loibl S, Ermilov EA, Roder B, Wengel J, Ephrussi A, Seitz O (2014) Brightness through local constraint–LNA-enhanced FIT hybridization probes for in vivo ribonucleotide particle tracking. Angew Chem 53(42):11370–11375. doi:10.1002/anie.201406022
7. Ghosh S, Marchand V, Gaspar I, Ephrussi A (2012) Control of RNP motility and localization by a splicing-dependent structure in oskar mRNA. Nat Struct Mol Biol 19(4):441–449. doi:10.1038/nsmb.2257
8. Zimyanin VL, Belaya K, Pecreaux J, Gilchrist MJ, Clark A, Davis I, St Johnston D (2008) In vivo imaging of oskar mRNA transport reveals the mechanism of posterior localization. Cell 134(5):843–853. doi:10.1016/j.cell.2008.06.053

Chapter 19

Visualizing RNA in Live Bacterial Cells Using Fluorophore- and Quencher-Binding Aptamers

Murat Sunbul, Ankita Arora, and Andres Jäschke

Abstract

To elucidate the roles, dynamics, and regulation of RNAs, it is vital to be able to visualize the RNA of interest (ROI) in living cells noninvasively. Here, we describe a novel live-cell RNA imaging method using fluorophore- and quencher-binding aptamers, which can be genetically fused to the ROI. In this method, new membrane permeable and nonfluorescent fluorophore–quencher conjugates were utilized, and we showed that their fluorescence increases dramatically upon binding to fluorophore- or quencher-binding aptamers. This phenomenon allowed for labeling the ROI with many different colored fluorophores and also dual-color imaging of two distinct RNAs in live bacteria. Our approach uses small RNA tags and small molecule fluorophores for labeling, thereby minimal perturbation on the function and dynamics of the RNA of interest is expected.

Key words Aptamer, Fluorophore, RNA imaging, RNA localization, Contact quenching, Live cell imaging, Fluorescence microscopy, RNA trafficking, SRB-2 aptamer, DNB aptamer

1 Introduction

RNAs play a diverse role in living cells ranging from translation of proteins, regulation and silencing of genes to catalysis of chemical reactions and epigenetics. A mainstay of regulatory processes has been spatial and temporal localization of various RNAs to specific sites, which can be dictated by local or external stimuli [1]. Additionally, it has been previously shown that even bacterial cells spatially organize their RNAs to promote functional compartmentalization in the absence of organelles [2, 3]. Furthermore, localization of bacterial RNAs can regulate their interactions with nucleases and control their stability [4]. However, methods to visualize RNAs and study their dynamics in live bacterial cells are still in their infancy [5]. Therefore, in order to completely understand the function of RNAs in living cells, new approaches for imaging of intracellular RNA are of vital importance.

Imre Gaspar (ed.), *RNA Detection: Methods and Protocols*, Methods in Molecular Biology, vol. 1649, DOI 10.1007/978-1-4939-7213-5_19, © Springer Science+Business Media LLC 2018

1.1 RNA Imaging Using a Fluorophore-Binding Aptamer

Previously, several short RNA sequences (aptamers) which bind to specific fluorogenic dyes [6–8] or fluorophores such as fluorescein [9], sulforhodamine B [9], and tetramethylrhodamine [10] have been developed by using Systematic Evolution of Ligands by EXponential enrichment (SELEX) [11]. However, they were not utilized for in vivo RNA imaging applications, presumably due to low affinity, toxicity of the dyes or high background fluorescence in cells. Lately, an aptamer, named Spinach, that binds to 3,5-difluoro-4-hydroxybenzylidine imidazolinone (DFHBI, a mimic of the GFP chromophore) has been developed and utilized for in vivo RNA imaging [12]. Spinach was genetically encoded to the target RNAs, which become fluorescent upon binding of the fluorogenic dye to the aptamer. The first generation Spinach aptamer and DFHBI dye were further coevolved to create more stable, smaller and brighter aptamers [13–15]. In addition to Spinach and its derivatives, recently an aptamer, named Mango, with an extremely high affinity to thiazole orange was discovered and appeared to be a promising tool for in vivo RNA imaging [16].

We recently developed a new method combining contact-quenched (*see* **Note 1**) fluorogenic dyes and a fluorophore-binding aptamer to image RNA in live bacterial cells (*see* Fig. 1a) [17]. This method was based on noncovalent interactions between a fluorophore-binding aptamer (SRB-2, sulforhodamine B binding aptamer) and a fluorophore (SR, sulforhodamine B) (*see* Fig. 2a). In this approach, the fluorescence of SR was diminished drastically by

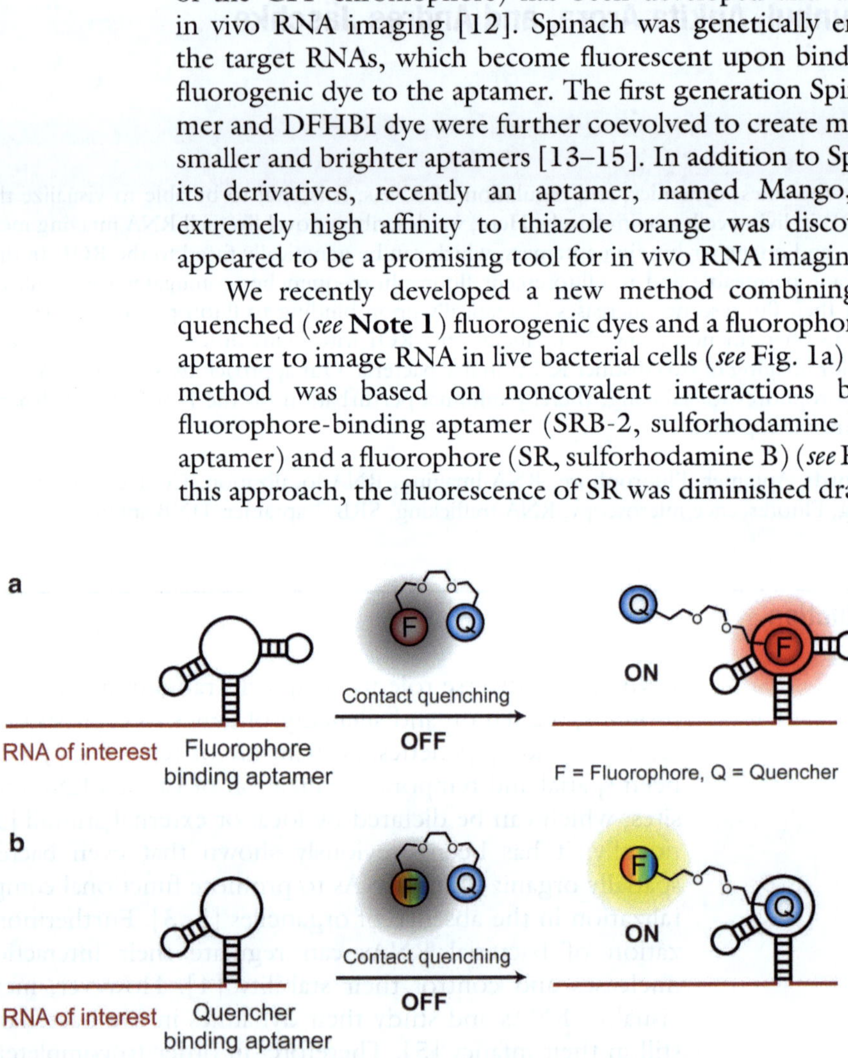

Fig. 1 Principles of the RNA imaging method based on fluorophore- and quencher-binding aptamers. The contact-quenched fluorophore–quencher conjugates (OFF) light up upon binding to (**a**) a fluorophore binding aptamer, or (**b**) a quencher-binding aptamer (ON). RNA of interest can be fused to one of the fluorescence enhancing aptamers and imaged in the presence of the fluorophore–quencher conjugate. F denotes any fluorophore and Q denotes a contact quencher. Images were reproduced from [22] with permission from Oxford University Press

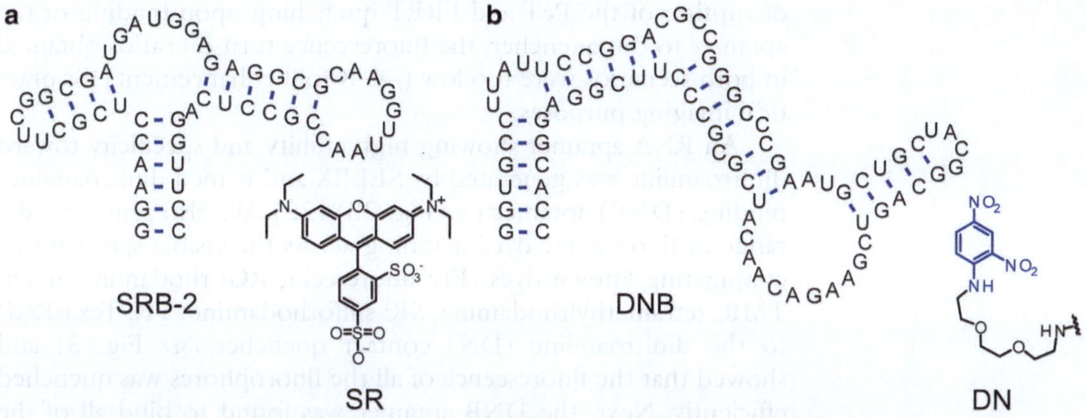

Fig. 2 Mfold-predicted secondary structure of (**a**) SRB-2 aptamer and its binding partner sulforhodamine B (SR); (**b**) DNB aptamer and its binding partner dinitroaniline (DN)

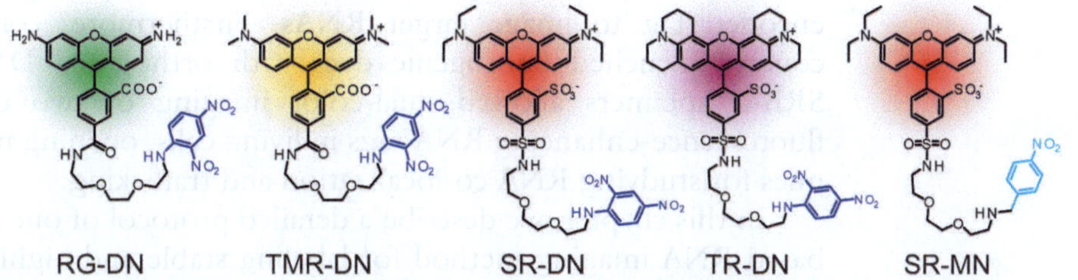

Fig. 3 Structures of the fluorogenic dyes used in this protocol. Dinitroaniline (DN) and p-nitrobenzylamine (MN) are the contact quenchers used in this protocol. *RG-DN* rhodamine green dinitroaniline, *TMR-DN* tetramethylrhodamine dinitroaniline, *SR-DN* sulforhodamine dinitroaniline, *TR-DN* TexasRed dinitroaniline, *SR-MN* sulforhodamine p-nitrobenzylamine

attaching a very efficient contact quencher (DN, dinitroaniline) via a triethylene glycol linker to yield a fluorogenic dye (SR-DN) (*see* Fig. 3). The fluorescence of SR-DN increases ~100-fold upon binding to SRB-2 and this dye was utilized to image an RNA of interest genetically fused to SRB-2 in live bacteria.

1.2 RNA Imaging Using a Quencher-Binding Aptamer

Even though fluorophore-binding aptamers are invaluable tools for in vivo RNA imaging, a new aptamer has to be generated for each fluorophore with a different structure and a different color, which hampers the multiplexing potential of fluorophore-binding aptamers. Therefore, we decided to generate an RNA aptamer that binds to dinitroaniline, which was previously discovered as a general and very efficient contact quencher [17, 18]. With this quencher-binding aptamer, RNA of interests can be easily labeled with different colored fluorophores (*see* Fig. 1b). It is worth mentioning that quencher-binding aptamers were previously developed; however, the quenching mechanism of the fluorescent dyes were based on PeT [19, 20] or FRET [21] quenching. Due to incomplete disruption of the PeT and FRET quenching upon binding of the aptamer to the quencher, the fluorescence turn-on ratios obtained in both scenarios were too low (~4–8-fold enhancement) for practical imaging purposes.

An RNA aptamer showing high affinity and specificity toward dinitroaniline was generated by SELEX and named dinitroaniline-binding (DNB) aptamer (*see* Fig. 2b) [22]. We also synthesized a range of fluorogenic dyes spanning across the visible spectrum by conjugating known dyes (FL: fluorescein, RG: rhodamine green, TMR: tetramethylrhodamine, SR: sulforhodamine, TR: TexasRed) to the dinitroaniline (DN) contact quencher (*see* Fig. 3) and showed that the fluorescence of all the fluorophores was quenched efficiently. Next, the DNB aptamer was found to bind all of the fluorogenic dyes, leading to fluorescence increase ranging from 5- to 73-fold in vitro and in living cells (*see* **Note 2**). When expressed in *E. coli*, the DNB aptamer could be labeled and visualized with different colored fluorophores and can be used as a genetically encoded tag to image target RNAs. Furthermore, combining contact-quenched fluorogenic dyes with orthogonal DNB and SRB-2 aptamers allowed dual-color imaging of two different fluorescence-enhancing RNA tags in living cells, opening new avenues for studying RNA co-localization and trafficking.

In this chapter, we describe a detailed protocol of our aptamer based RNA imaging method for labeling stable and highly abundant RNAs in live bacterial cells. Moreover, using tandem repeats of the DNB aptamer, we explain how to label GFP mRNA in live *E. coli*.

2 Materials

2.1 Plasmid Construction

1. pET28 plasmid (Agilent Technologies).
2. pET-tRNA (*see* **Note 3**).
3. pET-SRB2 and pET-DNB containing SRB-2 and DNB aptamers, respectively (*see* Fig. 4b, **Note 3**).
4. pET-SRB2-DNB (*see* Fig. 4c, **Note 4**).
5. pET-GFP-4xDNB and pET-GFP-8xDNB (*see* Fig. 4d, **Note 5**).
6. Oligonucleotide primers one of which should contain SRB-2 or DNB sequence (*see* **Note 6**) (SRB-2: 5′ *GGAACCTCGCTTCGGCGATGATGGAGAGGCGCAAGG TTAACC GCCTCAGGTTCC*; DNB: 5′ *GGTGCCTTATT CCGGACGCCGGGCCCGAATGCTGCT ACGGCAGTCG AAGACAACATCGCGCCCTTCGGAGGCACC*).
7. dNTP (10 mM of each nucleotide) solution.
8. Thermostable DNA polymerase.

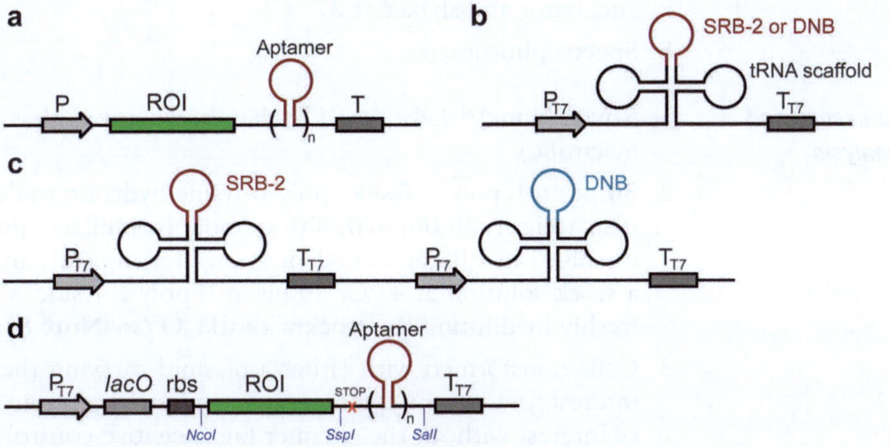

Fig. 4 (**a**) A general scheme for a cloning vector for RNA imaging with aptamers. P denotes a promoter sequence, ROI denotes an RNA of interest and T denotes a terminator sequence. A single copy or tandem repeats of the aptamer should be fused to the ROI. (**b**) Cloning vector used for expressing tRNA (negative control, pET-tRNA) or aptamers (SBR-2 or DNB) embedded in a tRNA scaffold (pET-SRB2 and pET-DNB). (**c**) Cloning vector used for expressing both SRB-2 and DNB aptamers in *E. coli* (pET-SRB2-DNB). (**d**) A general scheme of a cloning vector for mRNA imaging in bacteria (pET-GFP-4xDNB and pET-GFP-8xDNB)

9. PCR purification kit (e.g., Qiagen).
10. Restriction enzymes.
11. DNA ligase.
12. 1% agarose and gel electrophoresis equipment.
13. 10 mg/mL ethidium bromide.
14. 1× TAE (Tris–acetate–EDTA) Buffer: 40 mM Tris–acetate and 1 mM EDTA at pH 8.3.
15. Mini-prep kit (Qiagen).

2.2 Fluorophore–Quencher Conjugates

Fluorogenic dyes used in this study are not commercially available. Therefore, they have to be synthesized and purified as described in the literature [17, 23] (*see* **Note 7**).

1. 100 μM of RG-DN (in DMSO), 100 μM of TMR-DN (in DMSO), 100 μM of SR-DN (in DMSO), 100 of μM SR-MN (in DMSO), 100 μM of TR-DN (in DMSO).

2.3 Bacterial Growth

1. BL21 Star™ (DE3) (Invitrogen), and DH5α™ (Invitrogen) competent *E. coli* strains.
2. 1 M isopropyl-β-D-thio-galactopyranoside (IPTG).
3. 30 mg/mL kanamycin (Kan) in water.
4. Autoclaved Luria–Bertani (LB) media: 10 g tryptone/L, 5 g yeast extract/L, and 5 g NaCl/L.
5. LB-agar plates containing 30 μg/mL of kanamycin.
6. 50 mL baffled Erlenmeyer flask.

7. Incubator and shaker at 37 °C.
8. Spectrophotometer.

2.4 Microscopy and Image Analysis

1. 8-well Nunc™ Lab-Tek™ II chambered cover glass (Thermo Scientific).
2. 50 μg/mL poly-D-lysine: poly-D-lysine hydrobromide (molecular weight 30,000–70,000 g/mol, lyophilized powder) is dissolved in PBS at a concentration of 1 mg/mL and kept as a stock solution at 4 °C. 50 μg/mL poly-D-lysine is prepared freshly by diluting the stock with dH_2O (*see* **Note 8**).
3. Cells transformed with either a plasmid carrying the RNA of interest fused to the aptamer tag or a plasmid carrying the RNA of interest without the aptamer tag (negative control).
4. 1 M isopropyl-β-D-thio-galactopyranoside (IPTG).
5. Live-cell imaging solution (Invitrogen): 20 mM HEPES, 140 mM NaCl, 2.5 mM KCl, 1.8 mM $CaCl_2$, and 1 mM $MgCl_2$, pH 7.4. This solution is always supplemented with additional 5 mM $MgCl_2$ and 20 mM glucose and will be referred as "imaging solution" in the following sections.
6. Fluorogenic dyes: 100 μM of RG-DN, TMR-DN, SR-DN, TR-DN, and SR-MN (all of them dissolved in DMSO).
7. An automated wide field epifluorescence microscope, e.g., a Nikon TiE equipped with a Nikon 100× Plan Apo lambda oil immersion objective (NA 1.45), a cooled CCD Hamamatsu Orca-AG camera and a TokaiHit INU ZILCS incubator box.
8. Incubator and a shaker at 37 °C.
9. Fiji image analysis software (https://fiji.sc/).

3 Methods

3.1 Expression Plasmids

1. Both SRB-2 (54-nucleotide) and DNB (75-nucleotide) aptamers are quite small and do not require special scaffolds for RNA imaging in bacterial cells. Therefore, they can be directly fused to either 3′ or 5′ of the RNA of interest during PCR by using forward and reverse primers one of which carries the aptamer sequence (*see* **Note 6**).
2. Restriction sites should also be included in the primer sequence and the PCR product can be cloned into the multiple cloning site (MCS) of pET28 plasmid (*see* **Note 6**).
3. (Optional) To image less stable and low abundant RNAs tandem repeats of SRB-2 or DNB aptamers can be utilized which increases the signal-to-noise ratio (*see* **Note 5**).
4. Using appropriate restriction enzymes substitute the GFP coding sequence (cds) with the cds of the RNA of interest in pET-GFP-4xDNB or pET-GFP-8xDNB plasmids (*see* Fig. 4d, **Note 5**).

3.2 Induction of Reporter RNA Expression in Bacteria

Plasmids encoding the aptamer or aptamer tagged ROI should be transformed into an appropriate bacterial strain expressing T7 RNA polymerase. Induction with IPTG initiates the transcription of the RNA aptamers, which are imaged subsequently.

1. Transform BL21 Star™ (DE3) competent *E. coli* cells with expression plasmids (pET-tRNA, pET-SRB2, or pET-DNB) encoding for the respective aptamers or tRNA scaffold (control) (*see* **Notes 9** and **10**).
2. Plate transformed cells on LB-agar plates supplemented with 30 μg/mL of kanamycin for selection of the bacteria carrying the plasmids.
3. Pick a single colony from each plate and start overnight cultures in 5 mL of LB medium supplemented with 30 μg/mL of kanamycin by incubating the colony at 37 °C with vigorous shaking at 150 rpm.
4. Measure the optical density (OD_{600}) for a 1:10 dilution of each overnight culture.
5. Start a fresh culture using the overnight culture as a starter culture with an OD_{600} of 0.05 in a 50 mL baffled Erlenmeyer flask containing 10 mL of LB medium with 30 μg/μL kanamycin.
6. When the OD_{600} reaches 0.4, the cultures are induced with 1 mM IPTG and shaken additionally for 3 h (*see* **Note 11**) at 37 °C.

3.3 Imaging Aptamers in Bacteria

This section gives details about the steps for imaging various aptamers transcribed after the induction of pET expression vectors in live *E. coli* cells.

3.3.1 Preparation of the Poly-D-Lysine Coated Glass Chamber Slides

Bacterial cells do not adhere to glass slides. To ensure adhesion of the bacterial cells to a glass surface during imaging, the glass slides must be coated with poly-D-lysine (*see* **Note 12**).

1. 8-well Nunc™ Lab-Tek™ II chambered cover glass slides are removed from packaging and coated with 250 μL of 50 μg/mL poly-D-lysine and incubated at room temperature for 30–45 min (*see* **Note 8**).
2. The poly-D-lysine solution is aspirated and the chamber slides are washed twice with 400 μL sterile Millipore water to remove excess poly-D-lysine.
3. The excess water is aspirated after the final wash and the slides are allowed to air-dry at room temperature.

3.3.2 Preparation of E. coli for Imaging

1. Prepare *E. coli* cells expressing aptamers and tRNA scaffold (negative control) as described in Subheading 3.2.

2. Aliquot 0.2 mL of culture from each Erlenmeyer flask and transfer to a fresh microcentrifuge tube. Pellet the cells at 4400 × *g* for 1 min at room temperature.
3. Remove the LB media and suspend the cells in 1 mL of imaging solution (*see* **Note 13**).
4. Pellet the cells again at 4400 × *g* for 1 min at room temperature, remove the imaging solution, and resuspend the cells in 1 mL of imaging solution.
5. Add 0.2 mL of cell suspension to poly-D-lysine coated 8-well glass chamber slides (Subheading 3.3.1) and incubate the cells expressing SRB-2, DNB and tRNA scaffold (negative control) at 37 °C for 30–45 min to promote the adhesion of bacterial cells to the surface.
6. Gently wash the wells twice with 0.4 mL of imaging solution to remove unattached *E. coli* cells.
7. Finally add 300 μL of imaging solution containing 1 μM of the appropriate fluorogenic dye (RG-DN, TMR-DN, SR-DN, or TR-DN) and incubate the cells at 37 °C for 10 min before imaging (*see* **Note 14**).

   Cells expressing SRB-2 aptamer should be incubated with SR-DN (*see* Fig. 5a). Cells expressing DNB aptamer can be incubated with one of RG-DN, TMR-DN, SR-DN, or TR-DN. (*See* Fig. 5b). Cells expressing tRNA scaffold were incubated with RG-DN, TMR-DN, SR-DN, and TR-DN.

3.3.3 Imaging the Aptamers

1. Place the 8-well slide with immobilized *E. coli* in the prewarmed incubation chamber at 37 °C attached to the microscope.
2. Use the 100× objective under bright-field illumination to focus on adherent *E. coli* cells.
3. For the fluorescence illumination, a metal halide lamp and the following filter settings are used: for RG-DN: 470/30 nm excitation filter, 495 nm dichroic beam splitter, 525/30 nm emission filter; for TMR-DN and SR-DN: 560/40 nm excitation filter, 595 nm dichroic beam splitter, 630/60 nm emission filter; for TR-DN: 580/20 nm excitation filter, 595 nm dichroic beam splitter, 630/60 nm emission filter.
4. Focus on the well with cells expressing either SRB-2 or DNB incubated with the appropriate fluorogenic dye and determine the suitable exposure time for each dye such that one obtains highest fluorescent signal without saturating the pixels.
5. Acquire both bright-field and fluorescent illumination images.
6. Acquire the images of the bacteria carrying pET-tRNA plasmid by using exactly the same exposure time. This image will be used for comparisons.

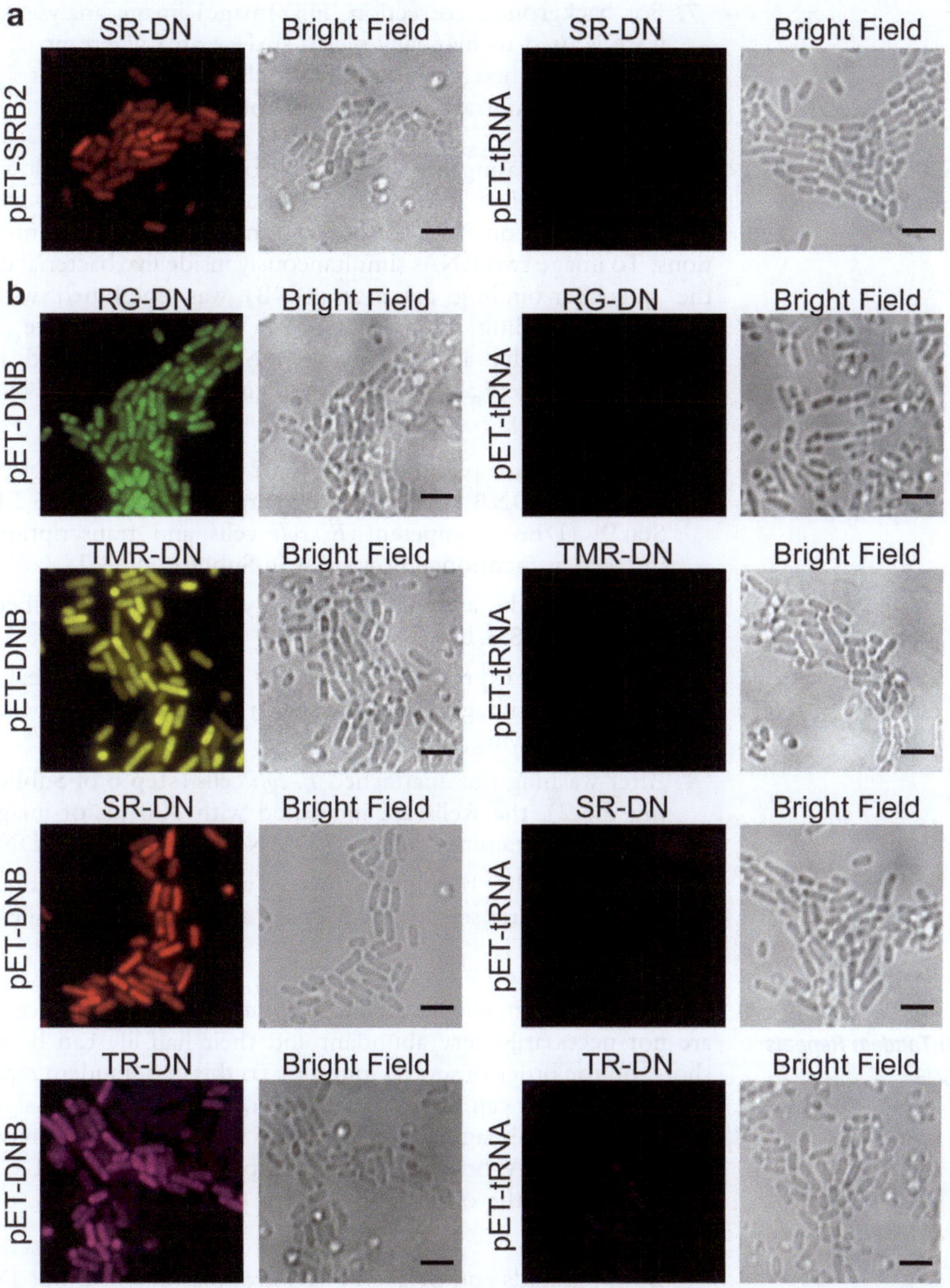

Fig. 5 (**a**) Imaging SRB-2 aptamer in live *E. coli* with SR-DN dye. Bacteria were transformed with either pET-SRB2 or pET-tRNA plasmid. Transcription was induced with IPTG and bacteria were incubated with 1 μM of SR-DN (*red*) for 10 min and imaged at 37 °C. (**b**) Imaging DNB aptamer in live *E. coli* with various fluorophore–dinitroaniline conjugates. Bacteria were transformed with either pET-DNB or pET-tRNA plasmid. Transcription was induced with IPTG. Bacteria were incubated with 1 μM of RG-DN (*green*), TMR-DN (*yellow*), SR-DN (*red*) or TR-DN (*magenta*) for 10 min and imaged at 37 °C. Scale bar, 3 μm

7. For background correction, Fiji/ImageJ image analysis software is used to manually pick a surface area where no *E. coli* cells are attached (*see* **Note 15**) and the mean fluorescent signal obtained is subtracted from the whole image.

3.4 Dual-Color RNA Imaging

There is an increasing demand to develop methods that allow for imaging of multiple RNAs simultaneously in living cells, which facilitates studies on RNA co-localizations and RNA-RNA interactions. To image two RNAs simultaneously inside live bacterial cells, the quencher-binding aptamer (DNB) was combined with a fluorophore-binding aptamer (SRB-2). Using two cognate pairs of fluorogenic dye–aptamer, namely, SR-MN–SRB-2 and RG-DN–DNB, we were able to successfully image two ROIs in live *E. coli* (*see* **Note 16**).

1. The expression plasmid (pET-SRB2-DNB) and control plasmids (pET-DNB and pET-SRB2) are transformed into BL21 Star™ (DE3) competent *E. coli* cells and transcription is induced as mentioned previously in Subheading 3.2.
2. The glass slides are coated with poly-D-lysine as described in Subheading 3.3.1.
3. *E. coli* cells expressing the SRB-2 and DNB aptamers are prepared for imaging as described in Subheading 3.3.2 except for the following steps.
4. After washing the unattached *E. coli* cells (**step 6** of Subheading 3.3.2), the wells are incubated with 300 μL of imaging solution containing 1 μM of SR-MN and 1 μM of RG-DN.
5. Aptamers in live bacteria are imaged as mentioned in Subheading 3.3.3. Filter sets for SR-MN dye are as same as the ones for SR-DN (*see* Fig. 6).

3.5 mRNA Imaging with Tandem Repeats of DNB

Imaging mRNA in live bacteria can be quite challenging since they are not necessarily very abundant and their half-life can be very short, on the order of several minutes. To this end, tandem repeats of DNB aptamer can be very advantageous to increase the signal-to-noise ratio in imaging. Here, we use two different plasmids to label GFP mRNA; one has only four repeats of DNB (pET-GFP-4xDNB), while the other one contains 8 repeats of DNB (pET-GFP-8xDNB).

1. Both plasmids and pET-GFP (control plasmid) without DNB aptamer are transformed into BL21 Star™ (DE3) competent *E. coli* cells and transcription is induced as mentioned previously in Subheading 3.2.
2. The glass slides are coated with poly-D-lysine as described in Subheading 3.3.1.

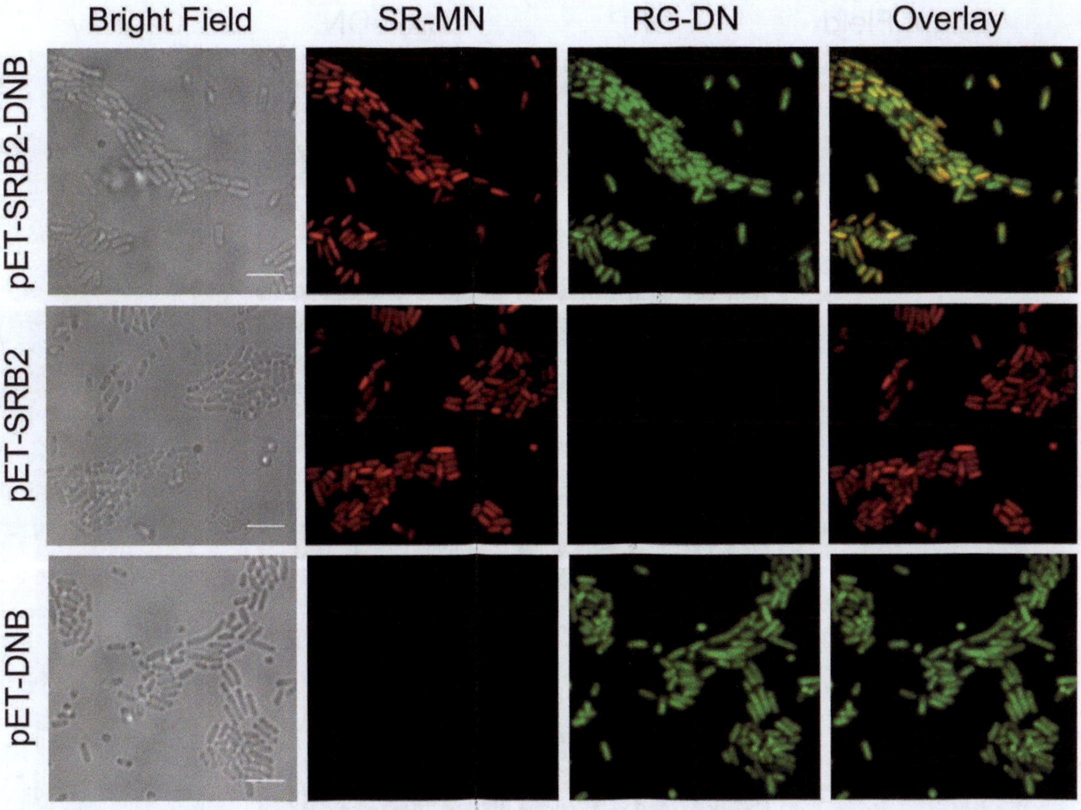

Fig. 6 Dual-color imaging of SRB-2 and DNB aptamers in live *E. coli* with RG-DN (*green*, 1 μM) and SR-MN (*red*, 1 μM). Fluorescence signal in both *red* and *green* channels was detected in cells expressing both DNB and SRB-2, while cells expressing either DNB or SRB-2 showed fluorescence only in the *green* or *red* channel, respectively. Scale bar, 5 μm. Images were reproduced from [22] with permission from Oxford University Press

3. *E. coli* cells expressing the GFP protein and GFP mRNA with or without DNB tandem repeats are prepared for imaging as described in Subheading 3.3.2 except for the following steps.
4. After washing the unattached *E. coli* cells (**step 6** of Subheading 3.3.2), the wells are incubated with 300 μL of imaging solution containing 1 μM of TMR-DN.
5. Both GFP protein and mRNA in live bacteria are imaged as mentioned in Subheading 3.3.3. Filter sets for GFP are as same as the ones for RG-DN (*see* Fig. 7 and **Note 17**).

4 Notes

1. Contact quenching is a type of static quenching where the fluorophore and quencher interact with each other to form a nonfluorescent intramolecular ground state dimer with its own distinct absorption spectrum.

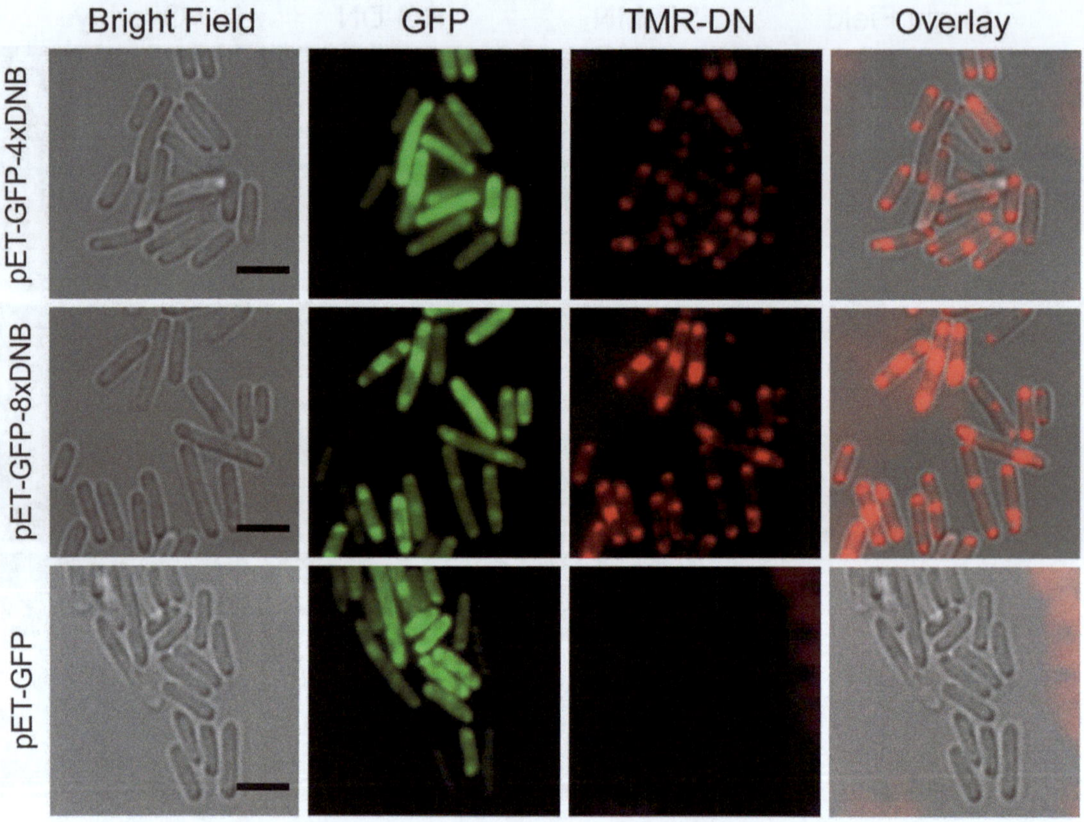

Fig. 7 Imaging of GFP mRNA in live bacteria. Bacteria were transformed with pET-GFP, pET-GFP-4xDNB, or pET-GFP-8xDNB and transcription was induced with IPTG. Bacteria were incubated with 1 μM of TMR-DN (*red*) for 10 min and both GFP protein (*green*) and GFP mRNA (*red*) were imaged at 37 °C. GFP mRNA was found to localize at the poles of the bacteria. Scale bar, 3 μm

2. The fluorogenic dyes RG-DN, TMR-DN, SR-DN, and TR-DN show 6-, 73-, 56-, and 15-fold fluorescence enhancement upon binding to DNB, respectively, and the dissociation constants are calculated to be 4.48 ± 0.60 μM, 0.35 ± 0.05 μM, 0.80 ± 0.10 μM, and 18.0 ± 1.8 μM, respectively. The best dye for DNB aptamer is TMR-DN, because it shows the highest fluorescence turn-on ratio and has the lowest K_D.
3. To image RNA in bacterial cells, we used several different plasmids in this protocol. The first plasmid (pET-tRNA) expresses only the tRNA scaffold and was used as a negative control to evaluate the background fluorescence due to non-specific binding of the fluorogenic dyes to cellular biomolecules. pET-SRB2 and pET-DNB plasmids contain SRB-2 and DNB aptamers, respectively, embedded in a tRNA scaffold placed between T7 promoter and T7 terminator sequences (*see* Fig. 4b).

4. For dual-color RNA imaging experiments, we created a single plasmid (pET-SRB2-DNB) which expresses both SRB-2 and DNB aptamers embedded in the tRNA scaffold from their own independent T7 promoter/terminator systems (*see* Fig. 4c). It is also possible to use two different plasmids for the same purpose: one transcribes an ROI-1-SRB2 fusion and the other one transcribes an ROI-2-DNB fusion. They should be compatible to each other and carry different antibiotic selection markers. Bacteria can be transformed with both of the plasmids and expression of both transcripts can be induced by the addition of IPTG.
5. We prepared two plasmids (pET-GFP-4xDNB or pET-GFP-8xDNB) for the imaging of GFP mRNA in live bacteria. In these plasmids, GFP is expressed from the T7/lacO promoter and tandem repeats (4 or 8 times) of DNB were fused to the 3′ UTR region of the GFP gene right after the stop codon (*see* Fig. 4d). Any RNA of interest can be substituted for GFP by taking advantage of the restriction enzyme sites before and after the GFP sequence. Alternatively, tandem repeats of DNB can be cut out from the plasmids (pET-GFP-4xDNB or pET-GFP-8xDNB) by using appropriate restriction enzymes, purified on an agarose gel, and used as a cassette, which can be cloned into any plasmid carrying the gene of interest.
6. Since SRB-2 and DNB aptamers are quite small in size, they can easily be fused to the ROI during PCR by using a relatively long primer. Primer design for the PCR depends if the aptamer will be fused to the 3′ or 5′ end of the ROI. For example, if the aptamer is to be fused to the 5′ end of the ROI, forward primer should contain an appropriate restriction site sequence, either SRB-2 or DNB aptamer sequence, and a gene specific sequence, respectively (5′–3′ direction). Whereas, the reverse primer should contain an appropriate restriction site sequence and a reverse complementary of the gene specific sequence, respectively (5′–3′ direction). If the aptamer is to be fused to the 3′ end of the ROI, forward primer should contain an appropriate restriction site sequence and a gene specific sequence, respectively (5′–3′ direction), whereas the reverse primer should contain an appropriate restriction site, the reverse complementary of either SRB-2 or DNB aptamer, and the reverse complementary of a gene specific sequence, respectively (5′–3′ direction). Finally, PCR product can be cloned into any vector of interest (e.g., pET vectors) by using restriction site digestion and ligation reactions.
7. The synthesized dyes for in vivo imaging applications have to be extremely pure. Since the contact-quenched dyes are nonfluorescent, any fluorescent impurity would dramatically decrease the turn-on ratios.

8. Prepare the working solution of poly-D-lysine freshly each time just before it is needed. Stock solution can be stored at 4 °C. Avoid multiple cycles of freeze–thaw of poly-D-lysine stock solution. Numerous freeze-thaw cycles decrease the efficiency of bacterial cells adhesion to the slide.
9. Other *E. coli* strains can also be used as long as they have the gene for T7 RNA polymerase expression, because all pET plasmids that we use here carry a T7 promoter/terminator system.
10. Perform a fresh transformation of the expression plasmids each time to ensure efficient expression of the desired RNA aptamers.
11. IPTG induction can be carried out for 2–4 h. In our experience, no significant increase in fluorescence intensity is observed between 2-h and 4-h induction times.
12. To ensure proper adhesion of bacterial cells it is important to use borosilicate glass chambers instead of the polymer based slides.
13. Imaging solution must be used in live-cell imaging experiments as it offers the following two advantages over LB medium: (1) components of LB medium prevent adherence of *E. coli* to poly-D-lysine-treated dishes, (2) live-cell imaging medium has a lower background fluorescence than LB medium.
14. We observed different levels of background fluorescence for different dyes. For example, we obtained the worst signal-to-noise ratio with TR-DN and the best ratio with TMR-DN.
15. Choose the area with no cells attached carefully by using bright-field illumination and avoid selecting attached bacterial cells by mistake.
16. It is important to ensure that the dye–aptamer pairs used for dual imaging are orthogonal to each other, i.e., they do not bind to each other and they do not interfere with each other. Previously, the SRB-2 aptamer was used to image RNA in live bacterial cells using the SR-DN dye. However, the combination of the DNB–SRB-2 aptamer pair with the RG-DN (green)–SR-DN (red) dye pair would not allow dual-color imaging of two different RNA molecules since the DNB aptamer would bind to both dyes due to the presence of the same DN contact quencher. Therefore, we conjugated another contact quencher, namely, p-nitrobenzylamine (MN), to yield SR-MN for exclusive labeling of the SRB-2 aptamer. This new combination, SRB-2/SR-MN and DNB/TMR-DN, allowed for simultaneous labeling of two distinct RNAs in a single live bacterium.

17. We clearly see GFP mRNA localization at the poles of bacteria. However, this data should be interpreted very carefully. pET plasmids localize at the poles, and this is the reason why we spot mRNAs (and partially GFP protein) at the poles of the bacteria.

Acknowledgments

This work is supported by Helmholtz Initiative on Synthetic Biology. Murat Sunbul thanks the Alexander von Humboldt Foundation for a postdoctoral fellowship. Ankita Arora thanks the DAAD for a doctoral fellowship.

References

1. Martin KC, Ephrussi A (2009) mRNA localization: gene expression in the spatial dimension. Cell 136(4):719–730. doi:10.1016/j.cell.2009.01.044
2. Broude NE (2011) Analysis of RNA localization and metabolism in single live bacterial cells: achievements and challenges. Mol Microbiol 80(5):1137–1147. doi:10.1111/j.1365-2958.2011.07652.x
3. Montero Llopis P, Jackson AF, Sliusarenko O, Surovtsev I, Heinritz J, Emonet T, Jacobs-Wagner C (2010) Spatial organization of the flow of genetic information in bacteria. Nature 466(7302):77–81. doi:10.1038/nature09152
4. Keiler KC (2011) RNA localization in bacteria. Curr Opin Microbiol 14(2):155–159. doi:10.1016/j.mib.2011.01.009
5. Kannaiah S, Amster-Choder O (2016) Methods for studying RNA localization in bacteria. Methods 98:99–103. doi:10.1016/j.ymeth.2015.12.010
6. Babendure JR, Adams SR, Tsien RY (2003) Aptamers switch on fluorescence of triphenylmethane dyes. J Am Chem Soc 125(48):14716–14717
7. Constantin TP, Silva GL, Robertson KL, Hamilton TP, Fague K, Waggoner AS, Armitage BA (2008) Synthesis of new fluorogenic cyanine dyes and incorporation into RNA fluoromodules. Org Lett 10(8):1561–1564. doi:10.1021/ol702920e
8. Lee J, Lee KH, Jeon J, Dragulescu-Andrasi A, Xiao F, Rao J (2010) Combining SELEX screening and rational design to develop light-up fluorophore-RNA aptamer pairs for RNA tagging. ACS Chem Biol 5(11):1065–1074
9. Holeman LA, Robinson SL, Szostak JW, Wilson C (1998) Isolation and characterization of fluorophore-binding RNA aptamers. Fold Des 3(6):423–431
10. Carothers JM, Goler JA, Kapoor Y, Lara L, Keasling JD (2010) Selecting RNA aptamers for synthetic biology: investigating magnesium dependence and predicting binding affinity. Nucleic Acids Res 38(8):2736–2747. doi:10.1093/nar/gkq082. gkq082 [pii]
11. Ellington AD, Szostak JW (1990) In vitro selection of RNA molecules that bind specific ligands. Nature 346(6287):818–822. doi:10.1038/346818a0
12. Paige JS, KY W, Jaffrey SR (2011) RNA mimics of green fluorescent protein. Science 333(6042):642–646. doi:10.1126/science.1207339
13. Strack RL, Disney MD, Jaffrey SR (2013) A superfolding Spinach2 reveals the dynamic nature of trinucleotide repeat-containing RNA. Nat Methods 10(12):1219–1224. doi:10.1038/nmeth.2701
14. Song W, Strack RL, Svensen N, Jaffrey SR (2014) Plug-and-play fluorophores extend the spectral properties of Spinach. J Am Chem Soc 136(4):1198–1201. doi:10.1021/ja410819x
15. Filonov GS, Moon JD, Svensen N, Jaffrey SR (2014) Broccoli: rapid selection of an RNA mimic of green fluorescent protein by fluorescence-based selection and directed evolution. J Am Chem Soc 136(46):16299–16308. doi:10.1021/ja508478x
16. Dolgosheina EV, Jeng SCY, Panchapakesan SSS, Cojocaru R, Chen PSK, Wilson PD, Hawkins N, Wiggins PA, Unrau PJ (2014) RNA mango aptamer-fluorophore: a bright, high-affinity complex for RNA labeling and tracking. ACS Chem Biol 9(10):2412–2420. doi:10.1021/cb500499x

17. Sunbul M, Jäschke A (2013) Contact-mediated quenching for RNA imaging in bacteria with a fluorophore-binding aptamer. Angew Chem Int Ed 52(50):13401–13404. doi:10.1002/anie.201306622
18. Sadhu KK, Mizukami S, Watanabe S, Kikuchi K (2010) Turn-on fluorescence switch involving aggregation and elimination processes for beta-lactamase-tag. Chem Commun (Camb) 46 (39):7403–7405. doi:10.1039/c0cc02432e
19. Sparano BA, Koide K (2005) A strategy for the development of small-molecule-based sensors that strongly fluoresce when bound to a specific RNA. J Am Chem Soc 127(43):14954–14955
20. Sparano BA, Koide K (2007) Fluorescent sensors for specific RNA: a general paradigm using chemistry and combinatorial biology. J Am Chem Soc 129(15):4785–4794
21. Murata A, Sato S, Kawazoe Y, Uesugi M (2011) Small-molecule fluorescent probes for specific RNA targets. Chem Commun (Camb) 47(16):4712–4714. doi:10.1039/c1cc10393h
22. Arora A, Sunbul M, Jäschke A (2015) Dual-colour imaging of RNAs using quencher- and fluorophore-binding aptamers. Nucleic Acids Res 43(21):e144. doi:10.1093/nar/gkv718
23. Sunbul M, Nacheva L, Jäschke A (2015) Proximity-induced covalent labeling of proteins with a reactive fluorophore-binding peptide tag. Bioconjug Chem 26(8):1466–1469. doi:10.1021/acs.bioconjchem.5b00304

Chapter 20

Method for Imaging Live-Cell RNA Using an RNA Aptamer and a Fluorescent Probe

Shin-ichi Sato, Kenji Yatsuzuka, Yousuke Katsuda, and Motonari Uesugi

Abstract

Live-cell imaging of mRNA dynamics is increasingly important to understanding spatially restricted gene expression. We recently developed a convenient and versatile method that uses a gene-specific RNA aptamer and a fluorescent probe to enable spatiotemporal imaging of endogenous mRNAs in living cells. The method was validated by live-cell imaging of the endogenous mRNA of β-actin. The new RNA-imaging technology might be useful for live-cell imaging of any RNA molecules.

Key words Live-cell imaging, mRNA, RNA aptamer, Small molecule, Chemical biology

1 Introduction

Imaging of RNAs in living cells is a powerful approach to understanding the intracellular dynamics of RNA and for measuring spatiotemporal gene expression. A number of different methods have been developed for detecting RNA targets [1–15]. The use of molecular beacons, the fluorescence of which changes upon binding to native mRNA targets, offers one of the best approaches [16–19]. Although such molecular beacons could provide valuable information about RNA dynamics in live cells, this method suffers from several potential drawbacks: injection of beacons into living cells might damage the cells, and poor stability of beacons in cells would produce background fluorescence. New technology for RNA imaging is needed to overcome these drawbacks. We recently developed a convenient and versatile method that permits spatiotemporal imaging of specific native RNAs in living cells [20, 21]. The method employs transfection of a plasmid encoding a gene-specific RNA aptamer, combined with a cell-permeable synthetic small molecule, BHQ1-Cy3, the fluorescence of which is restored only when the RNA aptamer hybridizes with its cognitive mRNA. Here, we describe the synthesis of BHQ1-Cy3, the selection of the

Imre Gaspar (ed.), *RNA Detection: Methods and Protocols*, Methods in Molecular Biology, vol. 1649, DOI 10.1007/978-1-4939-7213-5_20, © Springer Science+Business Media LLC 2018

BHQ1 aptamer, and the construction of RT aptamers that hybridize to the target RNA. This imaging technology could provide new insights into RNA function and gene expression.

2 Materials

All solutions are prepared using ultrapure water (purchased from Merck Millipore) and analytical grade reagents. Reagents are prepared and stored at room temperature unless otherwise indicated. Distilled water and special grade solvents and reagents are used for chemical syntheses.

2.1 Reagents

1. EAH Sepharose™ 4B resin (General Electric).
2. BHQ1 carboxylic acid (e.g., Biosearch Technologies) (*see* **Note 1**).
3. BHQ1 amine (e.g., Biosearch Technologies) (*see* **Note 1**).
4. 4-(4,6-dimethoxy-1,3,5-triazin-2-yl)-4-methylmorpholinium chloride n-hydrate (DMT-MM; e.g., Wako Pure Chemical Industries).
5. Triethylamine.
6. Acetic acid.
7. Dioxane.
8. Methanol.
9. Fmoc-β-Ala-Wang Resin (0.18 meq/g; e.g., Peptides International).
10. *N,N*-dimethylformamide (DMF).
11. Dichloromethane.
12. 20 v/v % piperidine in DMF (e.g., Watanabe Chemical Industries).
13. Fmoc-amido-dPEG4™-acid (e.g., Quanta BioDesign).
14. *N,N'*-diisopropylcarbodiimide (e.g., Wako Pure Chemical Industries).
15. 1-Hydroxybenzotriazole monohydrate (HOBt; e.g., Wako Pure Chemical Industries).
16. Cy3-NHS-ester (e.g., Lumiprobe) (*see* **Note 1**).
17. Acetic anhydride (e.g., Tokyo Chemical Industry) (*see* **Note 2**).
18. Trifluoroacetic acid (TFA; e.g., Wako Pure Chemical Industries).
19. Triisopropylsilane (e.g., Tokyo Chemical Industry).
20. *O*-(6-chlorobenzotriazol-1-yl)-*N,N,N',N'*-tetramethyluronium hexafluorophosphate (HCTU; e.g., Wako Pure Chemical Industries).

21. *N,N*-diisopropylethylamine (e.g., Wako Pure Chemical Industries).
22. PCR Master mix (e.g., Toyobo KOD FX Neo).
23. 3 M sodium acetate buffer Solution pH 5.2.
24. 70% ethanol.
25. TE Buffer Solution pH 8.0.
26. T7 in vitro transcription kit including DNase (Thermo Fisher Scientific MEGAscript® T7 Kit).
27. 10 M ammonium acetate solution pH 7.0.
28. DNA purification columns (e.g., General Electric NAP-5).
29. 0.22 μm centrifugal filter units (e.g., Merck Millipore Ultra-free®-MC, GV 0.22 μm).
30. Reverse transcription (RT) kit (e.g., Toyobo ReverTra Ace® qPCR RT Kit).
31. *Eco*RI, *Bam*HI, *Bgl*II, and *Hind*III restriction enzymes.
32. DNA Ligation Kit (e.g., Takara Bio < Mighty Mix>).
33. Competent high DH5α *E. Coli* (e.g., Toyobo Co., Ltd.).
34. Miniprep Kit (e.g., QIAGEN QIAprep Spin Miniprep).
35. Midiprep Kit (e.g., QIAGEN HiSpeed Plasmid Midi).
36. 4% PFA in 0.1 M PBS (e.g., Muto Pure Chemicals).
37. pUC19 DNA (e.g., Takara Bio).
38. pSuper.neo vector (Oligoengine).
39. LB plate containing 100 μg/mL of carbenicillin.
40. 1× PBS.
41. Hoechst 33342 (e.g., Dojindo Molecular Technologies).

2.2 Oligonucleotides

1. N60 template DNA: 5′-GAA TTC CGC GTG TGC ACA CC–N60–GTC CGT TGG GAT CCT CAT GG-3′.
2. Forward primer: 5′-GCT AAT ACG ACT CAC TAT AGG GAA TTC CGC GTG TGC ACA CC-3′ (T7 promoter sequence is underlined).
3. Reverse primer: 5′-CCA TGA GGA TCC GAA CGG AG-3′.
4. RT aptamer template 1: 5′-GAT CCC CGG AGC AAT GAT GGC **CTA GAT AAA TTC GGA** GCT TGA TCT TCA TTT TTA-3′ (RNA recognition arms are underlined, and the BHQ1-binding loop is bold).
5. RT aptamer template 2: 5′-AGC TTA AAA ATG AAG ATC AAG CTC CGA ATT TAT CTA GGC CAT CAT TGC TCC GGG -3′.

2.3 Buffers and Reaction Mixes

1. PCR mixture: 10 μL N60 template DNA, 250 μL 2× KOD FX Neo buffer, 100 μL dNTP mixture (2 mM each), 10 μL 10 μM Forward primer, 10 μL 10 μM Reverse primer, 10 μL 1.0 U/μ L KOD FX Neo, and 110 μL RNase-free water.
2. In vitro transcription mixture: 8 μL template DNA, 2 μL 10× T7 reaction buffer, 2 μL 2 mM ATP, 2 μL 2 mM CTP, 2 μL 2 mM GTP, and 2 μL 2 mM UTP, 2 μL T7 enzyme mix.
3. RT reaction mixture: 7 μL template RNA, 0.5 μL 10 μM reverse primer, 2 μL 5× RT reaction buffer, and 0.5 μL RT enzyme mix.
4. Ligation mixture: 5 ng insert DNA, 5 ng linearized plasmid DNA, RNAse free water up to 1 μL, and 1 μL Ligation mix.
5. Annealing buffer: 10 mM Tris–HCl, pH 7.6, 100 mM KCl, and 5 mM $MgCl_2$.
6. 2× SSC buffer: 300 mM NaCl, 30 mM sodium citrate, pH 7.0, 50 v/v % formamide.
7. Prehybridization buffer: 300 mM NaCl, 30 mM sodium citrate, pH 7.0, 10 w/v % dextran sulfate, 2 mM vanadyl-ribonucleoside complex, 0.02 w/v % RNase-free BSA, 40 μg *E. coli* tRNA, and 30 v/v % formamide.

2.4 Cell Culture

Cells are maintained in Dulbecco's modified Eagle medium, supplemented with 100 units/mL penicillin, 100 μg/mL streptomycin sulfate, and 10 v/v % fetal bovine serum, at 37 °C in a humidified 5% CO_2 incubator.

1. Trypsin–EDTA (0.25%), phenol red (e.g., Thermo Fisher Scientific).
2. Dulbecco's modified Eagle medium (e.g., Thermo Fisher Scientific).
3. Penicillin–streptomycin, Liquid (10,000 units penicillin; 10,000 μg streptomycin) (e.g., Thermo Fisher Scientific).
4. Fetal bovine serum (e.g., Biowest).
5. Transfection Reagent (e.g., Promega FuGENE® HD).
6. PDL-Coated Glass Bottom ViewPlate-96 (PerkinElmer).
7. CO_2 incubator.

2.5 Equipment

1. HPLC system (e.g., Shimadzu Prominence).
2. 5 μm, 20.0 × 100 mm preparative columns (e.g., GL Sciences Inertsil ODS-3).
3. Fluorescence spectrometer (e.g., PerkinElmer LS 55).
4. Spectrophotometer (e.g., Hitachi High-Technologies U-3010).
5. Tabletop high-speed microcentrifuge.

6. Thermal cycler.
7. Confocal microscope (e.g., Yokogawa Electric CV1000).
8. LC-MS system (e.g., Shimadzu LCMS-2010EV).
9. 2.2 μm, 3 × 50 mm analytical columns (e.g., Shimadzu GLC Shim-pack XR-ODS).
10. Freeze dryer (e.g., Tokyo Rikakikai FDU-1200).

3 Methods

3.1 Preparation of the BHQ1-Immobilized Resin

1. Place 0.5 mL of EAH Sepharose™ 4B resin in a Poly-Prep Chromatography Column and allow to drain by gravity.
2. Wash the resin thoroughly with at least three bed-volumes of dioxane–H_2O (3:1, v/v).
3. Add 1.0 μmol BHQ1 carboxylic acid, 0.10 mmol triethylamine, and 0.10 mmol DMT-MM in 0.5 mL of dioxane–H_2O (3:1, v/v) to the drained resin, and shake the suspension at room temperature overnight (*see* **Note 3**).
4. Drain the coupling solution from the resin, and add 0.50 mmol acetic acid and 0.50 mmol DMT-MM freshly prepared in 0.5 mL of dioxane–H_2O (3:1, v/v) to cap unreacted amino groups on the resin.
5. After overnight incubation, drain the resin and wash thoroughly with dioxane–H_2O (1:1, v/v) followed by methanol–H_2O (1:1, v/v).
6. Resuspend the resin with methanol–H_2O (1:3, v/v) to make a 50% slurry. Store at 4 °C until use.

3.2 Preparation of BHQ1-Cy3 Probe

Synthesis of the BHQ1-Cy3 probe is summarized in Scheme 1.

1. Deprotect the Fmoc group of 1.5 μmol Fmoc-β-Ala-Wang Resin (*see* **Note 4**) with 20% piperidine in DMF three times 10 min.
2. The resin is then washed with DMF three times 10 min.
3. To conjugate the linker, 7.4 μmol Fmoc-amido-dPEG4™-acid, 7.4 μmol *N,N'*-diisopropylcarbodiimide, and 7.4 μmol HOBt in 20 μL of DMF are added to the resin and incubated at room temperature for 1 h (*see* **Note 5**) to obtain Compound **1**.
4. Wash the resin with DMF three times 10 min.
5. To deprotect the Fmoc group, the resin is treated with 20% piperidine in DMF three times 10 min (*see* **Note 5**).
6. Wash the resin with DMF three times 10 min.

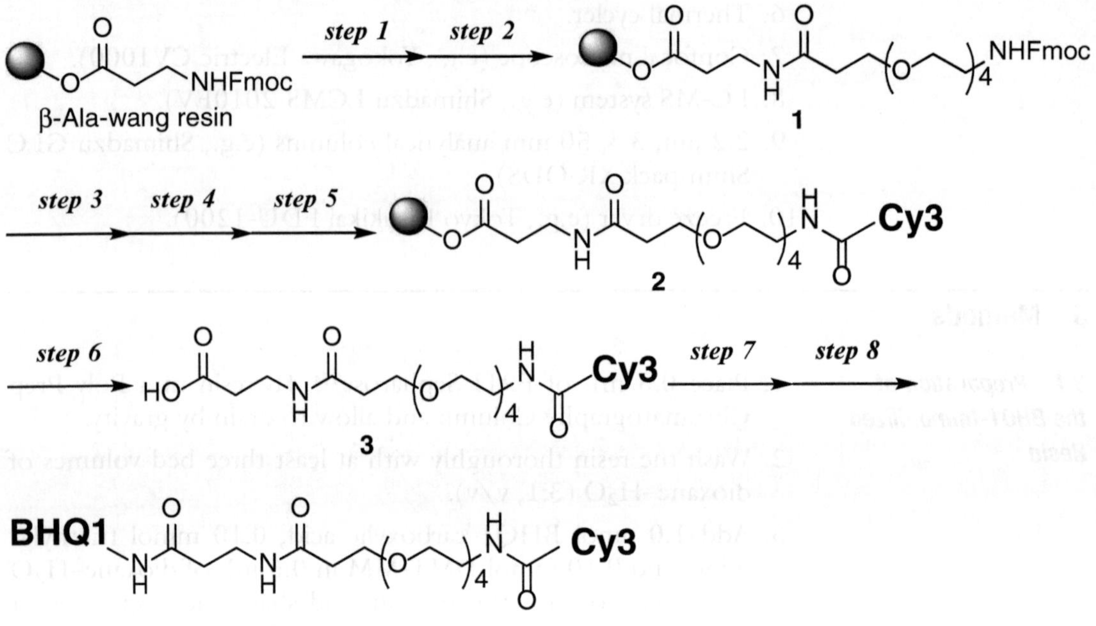

4 (BHQ1-Cy3)

Scheme 1 Synthesis of BHQ1-Cy3 probe

7. To conjugate the fluorophore, apply 3.0 μmol Cy3-NHS-ester in 36 μL DMSO to the resin, and incubate it overnight at room temperature in the dark to obtain Compound **2**.
8. The resin is then washed with DMF and dichloromethane (*see* **Note 6**).
9. To cap the unreacted amino groups on the resin, the resin is treated with 0.5 mL 25% acetic anhydride in dichloromethane at room temperature for 5 min.
10. Wash the resin with dichloromethane three times for 1 min.
11. To cleave Compound **3** from the resin, TFA–triisopropylsilane–H_2O (95:2.5:2.5) is added to the resin and incubated at room temperature for 2 h.
12. The mixture is then filtered and concentrated in vacuo, and the residue is lyophilized with H_2O.
13. To couple with BHQ1, the residue is dissolved in a solution of 3.0 μmol HCTU and 3.0 μmol HOBt in 20 μL DMSO, and is incubated at room temperature for 10 min.
14. Subsequently, 7.9 μmol *N,N*-diisopropylethylamine and 3.0 μmol BHQ1 amine are added to the dissolved residue, and the mixture is incubated overnight at room temperature in the dark to obtain Compound **4** (BHQ1-Cy3 probe) (*see* **Note 7**).

15. The mixture is diluted and acidified with 0.6 mL of 10 v/v % TFA in acetonitrile–H_2O (1:1).
16. The mixture is then purified by HPLC (0–3 min: CH_3CN, 60%; 3–23 min: CH_3CN, 60–80% at RT, $t_R = 19.2$ min). Compound **4** (BHQ1-Cy3 probe) is obtained as a dark purple amorphous solid (1.5 mg, 82% based on the resin) by lyophilization (*see* **Note 8**).
17. The BHQ1-Cy3 probe is dissolved in DMSO and stored at −30 °C.

3.3 In Vitro Selection of RNA Aptamers

In vitro selection is performed according to standard procedure for isolating RNA aptamers for BHQ1 [22–24] (Fig. 1). The selection cycle is repeated several times to enrich RNA species specifically bound to BHQ1.

3.3.1 Preparation of DNA Pools (See Note 9)

1. Prepare the PCR mixture in PCR tubes.
2. Place the tubes in a thermal cycler and perform the PCR amplification using the following program: 2-min initial denaturation at 94 °C, followed by 30 sequential cycles of

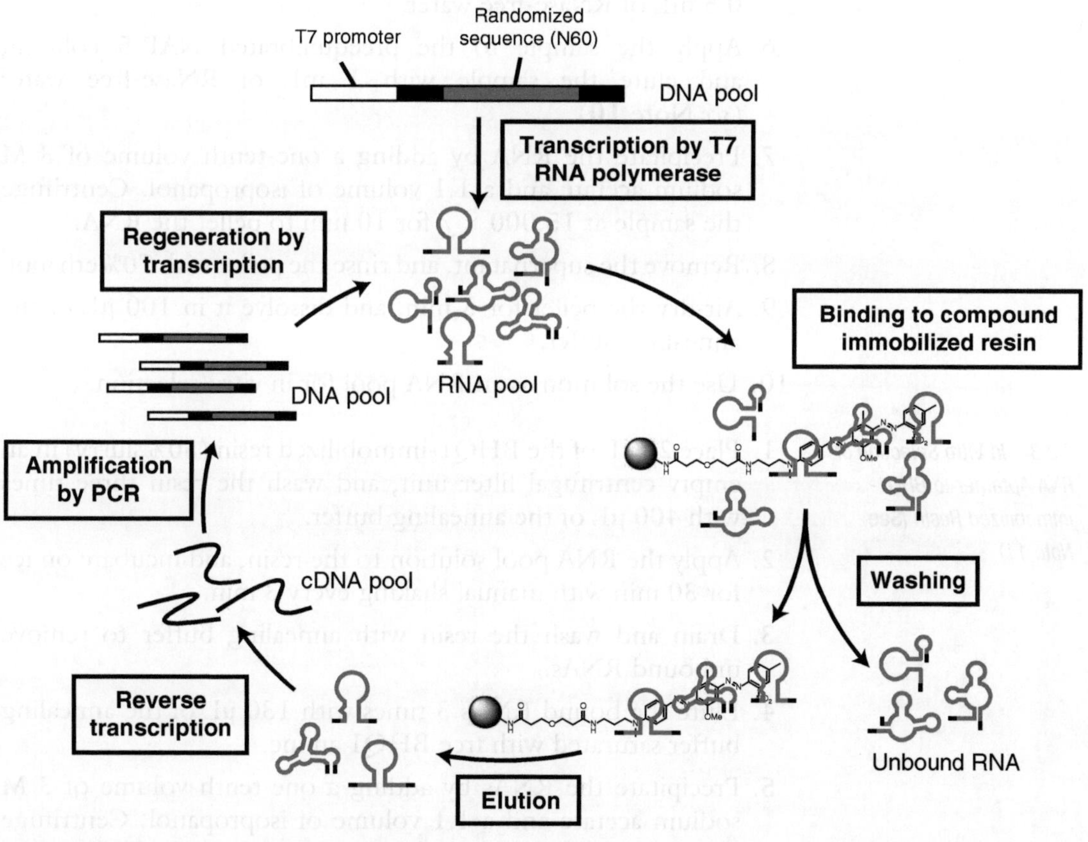

Fig. 1 In vitro selection of BHQ1 aptamers

denaturation at 98 °C for 10 s, annealing at 55 °C for 30 s, and extension at 68 °C for 30 s.

3. Precipitate the DNA by adding a one-tenth volume of 3 M sodium acetate and a 1.1 volume of isopropanol. Centrifuge the sample at 15,000 × *g* for 10 min to pellet the DNA.
4. Remove the supernatant carefully, and rinse the pellet with 70% ethanol.
5. Air-dry the pellet for 5 min, and dissolve it in a 20-μL TE buffer solution.

3.3.2 Preparation of RNA Pools

1. Prepare the in vitro transcription reaction at room temperature in the order as described in the Materials.
2. Incubate the reaction mixture overnight at 37 °C.
3. Add 1 μL of DNase and incubate for 15 min at 37 °C.
4. Precipitate the RNA by adding a one-third volume of 10 M ammonium acetate and a 1.5 volume of isopropanol. Centrifuge the sample at 15,000 × *g* for 10 min at 4 °C to pellet the RNA.
5. Thoroughly remove the supernatant, and dissolve the pellet in 0.5 mL of RNase-free water.
6. Apply the sample to the preequilibrated NAP-5 column, and elute the sample with 1 mL of RNase-free water (*see* **Note 10**).
7. Precipitate the RNA by adding a one-tenth volume of 3 M sodium acetate and a 1.1 volume of isopropanol. Centrifuge the sample at 15,000 × *g* for 10 min to pellet the RNA.
8. Remove the supernatant, and rinse the pellet with 70% ethanol.
9. Air-dry the pellet for 5 min, and dissolve it in 100 μL of the annealing buffer.
10. Use the solution as an RNA pool for in vitro selection.

*3.3.3 In Vitro Selection of RNA Aptamer to BHQ1-Immobilized Resin (**See** Note 11)*

1. Place 20 μL of the BHQ1-immobilized resin (50% slurry) in an empty centrifugal filter unit, and wash the resin three times with 400 μL of the annealing buffer.
2. Apply the RNA pool solution to the resin, and incubate on ice for 30 min with manual shaking every 3 min.
3. Drain and wash the resin with annealing buffer to remove unbound RNAs.
4. Elute the bound RNAs 3 times with 130 μL of the annealing buffer saturated with free BHQ1 amine.
5. Precipitate the RNAs by adding a one-tenth volume of 3 M sodium acetate and a 1.1 volume of isopropanol. Centrifuge the sample at 15,000 × *g* for 10 min to pellet the RNA.

6. Remove the supernatant and rinse the pellet with 70% ethanol.
7. Air-dry the pellet for 5 min, and dissolve it in a 15-μL TE buffer solution.

3.3.4 Reverse Transcription

1. Heat the eluted resin-bound RNA for 5 min at 65 °C, then immediately chill it on ice.
2. Prepare the RT reaction mixture.
3. The reaction is performed for 15 min at 37 °C, followed by heat-inactivation of RTase at 98 °C for 5 min.
4. 4.6 μL of the reverse-transcribed products are used as templates for PCR amplification to generate the DNA pool for the next round of selection (*see* **Note 12**).

3.3.5 Cloning and Sequencing of RNA Aptamers for BHQ1

1. After the final round of selection, digest the resulting DNA pool with restriction enzymes *Eco*RI and *Bam*HI.
2. Prepare the ligation mixture with *Eco*RI and *Bam*HI pUC19 plasmid as a vector.
3. Incubate the mixture for 30 min at 16 °C to allow the ligation reaction.
4. Transform DH5α with the ligation product, and heat shock at 42 °C for 1 min and spread the cells on an LB plate containing 100 μg/mL of carbenicillin.
5. Incubate the plate overnight at 37 °C.
6. Select a single colony and culture overnight at 37 °C in a liquid LB medium, and isolate the plasmid using the QIAprep Spin Miniprep Kit.
7. Sequence the insert DNA using primers in the vector backbone (*see* **Note 13**).

3.4 Preparation of RNA-Targeting Aptamer (RT-Aptamer)

The specific RNA-recognition aptamer is prepared as shown in Fig. 2.

1. The stem region is removed from the BHQ1-binding stem-loop structure [20], leading to the formation of a flexible BHQ1-binding loop that lacks in BHQ1-binding activity.
2. Two short RNA sequences (RNA targeting arms), complementary to a 24-base oligonucleotide in a target RNA sequence, are attached to the flexible structure (*see* **Note 14**).
3. The RNA-targeting arms hybridize to a target RNA, consequently forming the BHQ1-binding loop on the target mRNA (*see* **Note 15**).

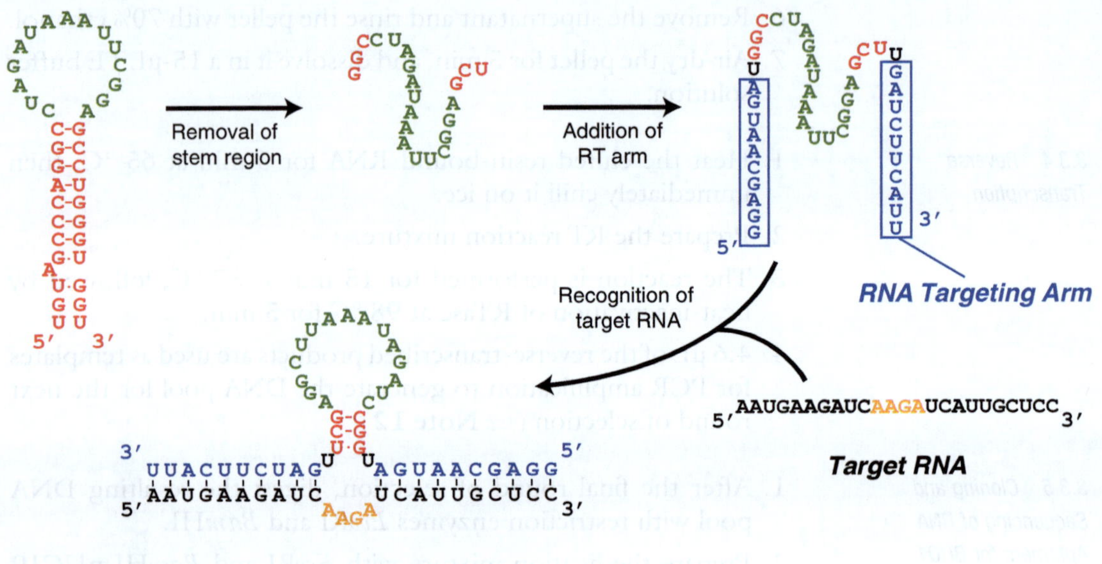

Fig. 2 Preparation of the BHQ1-binding RNA aptamer

3.5 Fluorescence In Situ Hybridization (FISH) in Fixed Cells with the RT-Aptamer [25]

1. Fix HeLa cells on 96-well plates (3.0×10^3 cells per well) with a 4% paraformaldehyde at room temperature for 15 min.
2. Wash the cells twice with PBS.
3. Permeabilize the cells with 70% ethanol overnight at 4 °C.
4. Soak the cells in 2× SSC buffer at 50 °C for 15 min.
5. Incubate the cells at 37 °C for 24 h with the prehybridization buffer.
6. Incubate the cells with 100 ng of the RT-aptamer in 2× SSC buffer (50 μL) for 2 h at 37 °C.
7. Treat the cells with 30 μL of 1 μg/mL Hoechst 33342 in 2× SSC buffer at 37 °C for 30 min.
8. Add a 30 μL of 5 μM BHQ1-Cy3 probe in 2× SSC buffer, incubate at 37 °C for 30 min, and then wash out excess probe with 2× SSC buffer.
9. Observe the target RNAs under a confocal microscope (*see* **Note 16**).

3.6 Preparation of an Expression Vector Encoding RT-Aptamer [26–28]

1. Prepare the dsDNA template encoding RT-aptamer by annealing two 100 μM synthetic ssDNA oligos, RT aptamer templates 1 and 2 [Fig. 3] in annealing buffer.
2. Prepare the ligation mixture using *Bgl*II–*Hind*III linearized pSuper.neo plasmid as a vector, and incubate the mixture for 30 min at 16 °C to allow the ligation reaction.
3. Transform DH5α with the ligation product, and heat shock at 42 °C for 1 min and spread the cells on an LB plate containing 100 μg/mL of carbenicillin.

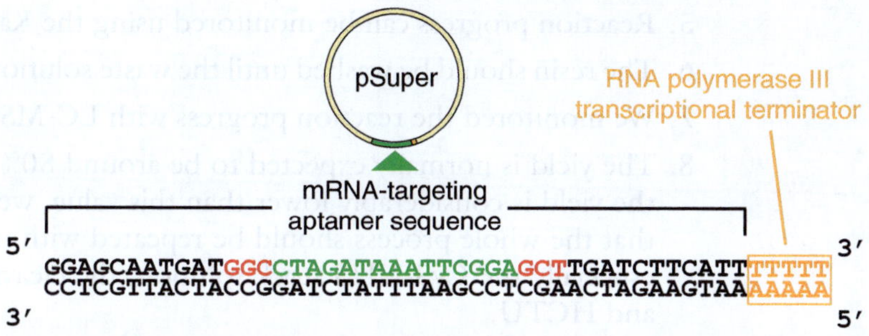

Fig. 3 Design and construction of an expression vector encoding the RT-aptamer

4. Incubate the plate overnight at 37 °C.
5. Select a single colony, and culture in a liquid LB medium overnight at 37 °C, and isolate the plasmid using a Plasmid Midi Kit.
6. Sequence the insert DNAs using a universal primer (*see* **Note 13**).

3.7 Live-Cell Imaging of mRNA Targets with RT-Aptamers

1. Seed HeLa cells on 12-well plates (1.0×10^5 cells per well) and culture overnight at 37 °C.
2. Transfect the cells with the pSuper.neo vector encoding an RT-aptamer by using FuGENE HD Transfection Reagent, according to the manufacturer's protocol [29].
3. Detach the cells from the plates with 0.25 w/v % trypsin–EDTA and reseed the cells on 96-well plates (5.0×10^3 cells per well) in the complete growth medium.
4. Culture the cells for 4 h at 37 °C in a humidified 5% CO_2 incubator.
5. Stain the cells with the BHQ1-Cy3 probe (final concentration: 5 μM) and Hoechst 33,342 (final concentration: 1 μg/mL) in the complete growth medium (50 μL) for 5 min at 37 °C in a humidified 5% CO_2 incubator (*see* **Notes 17** and **18**).
6. Wash the cells with 1× PBS, and observe the RNA targets under a confocal microscope (*see* **Note 16**).

4 Notes

1. A 5 w/v % solution in DMSO is prepared before use.
2. A 25 v/v % solution in CH_2Cl_2 is prepared immediately before use.
3. We used DMT-MM as a coupling reagent, but other water-soluble coupling reagents can be used.
4. Fmoc-β-Ala-Wang Resin should be swelled in DMF prior to the coupling reactions.

5. Reaction progress can be monitored using the Kaiser test.
6. The resin should be washed until the waste solution is colorless.
7. We monitored the reaction progress with LC-MS.
8. The yield is normally expected to be around 80% or higher. If the yield is considerably lower than this value, we recommend that the whole process should be repeated with new reagents; especially the Cy3-NHS stock solution, the cleavage solution, and HCTU.
9. In the preparation of the first DNA pool, we used a sufficient amount of DNA template to maintain the diversity of the library.
10. A NAP-5 column is equilibrated with three column-volumes of RNase-free water.
11. The percentage of RNA bound to the resin is calculated as 100 (AbsT − AbsF)/AbsT, where AbsT is the UV absorbance of the RNA solution before binding to the resin, and AbsF is that of the flow-through fraction.
12. PCR was stopped during the positive acceleration phase to avoid nonspecific amplification. The purity of PCR products was assessed by 8 w/v % native polyacrylamide gel electrophoresis.
13. M13-RV was used as a universal primer for sequencing.
14. The RNA targeting arms are connected with a three-base-pair stem segment through U mononucleotide bridges.
15. We normally single out an effective siRNA site as an RT-aptamer recognition sequence. Such a site might exhibit high levels of exposure for hybridization.
16. Cell images were taken with a vertical range of 4 μm and displayed as maximum intensity projection (MIP). Time-lapse images were taken at 20–30 s intervals for up to 10 min. The cell images were taken with a vertical range of 2 μm, and each image stack was then projected onto a single plane.
17. We added the BHQ1-Cy3 probe from a stock solution in DMSO (1 mM) and Hoechst 33,342 from a stock solution in H_2O (1 mg/mL) to wells filled with media.
18. Overstaining (>5 min) results in an increase of background fluorescence.

Acknowledgment

This work was supported in part by JSPS (26220206 to M.U. and 26440005 to S.S.), ZE Research Program, IAE (ZE27B-18 to S. S.), and an iCeMS research acceleration grant. iCeMS is supported by World Premier International Research Center Initiative (WPI), MEXT, Japan. This work was inspired by the international and interdisciplinary environments of the JSPS Core-to-Core Program, "Asian Chemical Biology Initiative." We also thank M. Nakashima for manuscript preparation.

References

1. Tyagi S (2009) Imaging intracellular RNA distribution and dynamics in living cells. Nat Methods 6:331–338
2. Lampasonaa AA, Czaplinski K (2016) RNA voyeurism: a coming of age story. Methods 98:10–17
3. Park HY, Lim H, Yoon YJ, Follenzi A, Nwokafor C, Lopez-Jones M, Meng X, Singer RH (2014) Visualization of dynamics of single endogenous mRNA labeled in live mouse. Science 343:422–424
4. Santangelo PJ, Lifland AW, Curt P, Sasaki Y, Bassell GJ, Lindquist ME, Crowe JE Jr (2009) Single molecule-sensitive probes for imaging RNA in live cells. Nat Methods 6:347–349
5. Lionnet T, Czaplinski K, Darzacq X, Shav-Tal Y, Wells AL, Chao JA, Park HY, de Turris V, Lopez-Jones M, Singer RH (2011) A transgenic mouse for in vivo detection of endogenous labeled mRNA. Nat Methods 8:165–170
6. Mili S, Moissoglu K, Macara IG (2008) Genome-wide screen reveals APC-associated RNAs enriched in cell protrusions. Nature 453:115–119
7. Shav-Tal Y, Darzacq X, Shenoy SM, Fusco D, Janicki SM, Spector DL, Singer RH (2004) Dynamics of single mRNPs in nuclei of living cells. Science 304:1797–1800
8. Paige JS, Nguyen-Duc T, Song W, Jaffrey SR (2012) Fluorescence imaging of cellular metabolites with RNA. Science 335:1194
9. Song W, Strack RL, Jaffrey SR (2013) Imaging bacterial protein expression using genetically encoded RNA sensors. Nat Methods 10:873–875
10. Strack RL, Disney MD, Jaffrey SR (2013) A superfolding Spinach2 reveals the dynamic nature of trinucleotide repeat-containing RNA. Nat Methods 10:1219–1224
11. Hövelmann F, Gaspar I, Loibl S, Ermilov EA, Roder B, Wengel J, Ephrussi A, Seitz O (2014) Brightness through local constraint—LNA-enhanced FIT hybridization probes for in vivo ribonucleotide particle tracking. Angew Chem Int Ed 53:11370–11375
12. Ozawa T, Natori Y, Sato M, Umezawa Y (2007) Imaging dynamics of endogenous mitochondrial RNA in single living cells. Nat Methods 4:413–419
13. Wang DO, Matsuno H, Ikeda S, Nakamura A, Yanagisawa H, Hayashi Y, Okamoto A (2012) A quick and simple FISH protocol with hybridization-sensitive fluorescent linear oligodeoxynucleotide probes. RNA 18:166–175
14. Wang DO, Okamoto A (2015) Visualization of nucleic acids with synthetic excitation-controlled fluorescent oligonucleotide probes. Methods Mol Biol 1262:166–175
15. Oomoto I, Suzuki-Hirano I, Umeshima H, Han YW, Yanagisawa H, Carlton P, Harada Y, Kengaku M, Okamoto A, Shimogori T, Wang DO (2012) ECHO-liveFISH: in vivo RNA labeling reveals dynamic regulation of nuclear RNA foci in living tissues. Nucl Acids Res 43: e126
16. Tyagi S, Kramer FR (1996) Molecular beacons: probes that fluoresce upon hybridization. Nat Biotechnol 14:303–308
17. Bratu DP, Cha BJ, Mhlanga MM, Kramer FR, Tyagi S (2003) Visualizing the distribution and transport of mRNAs in living cells. Proc Natl Acad Sci U S A 100:13308–13313
18. Sokol DL, Zhang X, Lu P, Gewirtz AM (1998) Real time detection of DNA-RNA hybridization in living cells. Proc Natl Acad Sci U S A 95:11538–11543
19. Tyagi S, Alsmadi O (2004) Imaging native beta-actin mRNA in motile fibroblasts. Biophys J 87:4153–4162
20. Murata A, Sato S, Kawazoe Y, Uesugi M (2011) Small-molecule fluorescent probes for specific RNA targets. Chem Commun 47:4712–4714
21. Sato S, Watanabe M, Katsuda Y, Murata A, Wang DO, Uesugi M (2015) Live-cell imaging of endogenous mRNAs with a small molecule. Angew Chem Int Ed 54:1855–1858
22. Ellington AD, Szostak JW (1990) In vitro selection of RNA molecules that bind specific ligands. Nature 346:818–822
23. Tuerk C, Gold L (1990) Systematic evolution of ligands by exponential enrichment: RNA ligands to bacteriophage T4 DNA polymerase. Science 249:505–510

24. Robertson DL, Joyce GF (1990) Selection in vitro of an RNA enzyme that specifically cleaves single-stranded DNA. Nature 344:467–468
25. Yoshimura H, Inaguma A, Yamada T, Ozawa T (2012) Fluorescent probes for imaging endogenous β-actin mRNA in living cells using fluorescent protein-tagged pumilio. ACS Chem Biol 7:999–1005
26. Tuschl T (2002) Expanding small RNA interference. Nat Biotechnol 20:446–448
27. Brummelkamp TR, Bernards R, Agami R (2002) A system for stable expression of short interfering RNAs in mammalian cells. Science 296:550–552
28. Brummelkamp TR, Bernards R, Agami R (2002) Stable suppression of tumorigenicity by virus-mediated RNA interference. Cancer Cell 2:243–247
29. http://www.promega.jp/resources/protocols/technical-manuals/101/fugene-hd-transfection-reagent-protocol/

Chapter 21

RNA Live Imaging in the Model Microorganism *Ustilago maydis*

Sabrina Zander, Kira Müntjes, and Michael Feldbrügge

Abstract

An essential feature of protein expression is the tight regulation of when and where a protein is translated from its cognate mRNA. This spatiotemporal expression is particularly important in guaranteeing the correct and efficient targeting of proteins to defined subcellular sites. In order to achieve local translation, mRNAs must be deposited at specific locations. A common mechanism is the active transport of mRNAs along the actin or microtubule cytoskeleton. To study such dynamic transport processes in vivo RNA live imaging is the method of choice. This method is based on the principle that defined binding sites for a heterologous RNA-binding protein (RBP) are inserted in the 3′ UTR of target mRNAs. Coexpression of the RBP fused to a fluorescent protein enables mRNA detection in vivo using fluorescence microscopy techniques. In this chapter we describe the well-established method of studying microtubule-dependent mRNA transport in the eukaryotic model microorganism *Ustilago maydis*. The presented experimental design and the microscopic techniques are applicable to a broad range of other organisms.

Key words mRNPs, Dual color microscopy, Endosomal mRNA transport, Early endosomes, Microtubule-dependent transport

1 Introduction

Every cell needs to determine precisely when and where mRNAs are to be translated into proteins. This spatiotemporal expression is, for example, crucial to support compartmentalization in eukaryotic cells. Therefore, it is rather common that mRNAs localize to a specific subcellular site in the cytoplasm for the precise delivery of the encoded protein [1–4]. A prominent mechanism for mRNA localization is their active transport along the cytoskeleton [5–7]. For long-distance movement mRNAs are transported along microtubules [5]. A well-studied example in eukaryotic microorganisms is endosomal mRNP transport in *U. maydis*, a trafficking process that is needed for the efficient unipolar growth of infectious hyphae

Sabrina Zander and Kira Müntjes contributed equally to this work.

Imre Gaspar (ed.), *RNA Detection: Methods and Protocols*, Methods in Molecular Biology, vol. 1649, DOI 10.1007/978-1-4939-7213-5_21, © Springer Science+Business Media LLC 2018

[8, 9]. Endosomal movement along microtubules is mediated by the concerted action of plus-end directed kinesin-3 type motor Kin3 and minus-end directed cytoplasmic dynein Dyn1/2 [10]. The associated transport machinery distributes mRNAs and ribosomes throughout the hyphae to avoid aberrant gradients of mRNAs and translated proteins around the nucleus [11, 12]. Importantly, endosomal transport of septin mRNAs and encoded proteins is needed for assembly of heteromeric septin complexes and their delivery and incorporation into higher-order filaments at the growth pole of hyphae (Fig. 1a; [13, 14]).

Key players for mRNA transport are RNA-binding proteins (RBPs) that recognize specific *cis*-acting elements within the cargo mRNAs [15, 16]. The so-called RNA zip codes can by recognized by their primary sequence or their more complex tertiary structures. Recent evidence revealed that RBPs act in concert to increase affinity and specificity for the interaction with distinct RNA elements forming a higher-order structure called mRNP (mRNA ribonucleoprotein complex; [6, 16]). The key RNA-binding protein of septin mRNA transport in *U. maydis* is Rrm4, which contains three N-terminal RNA recognition motifs for RNA-binding and a C-terminal MademoiseLLE domain for interaction with the adaptor-like protein Upa1 [8, 17–19]. The latter connects

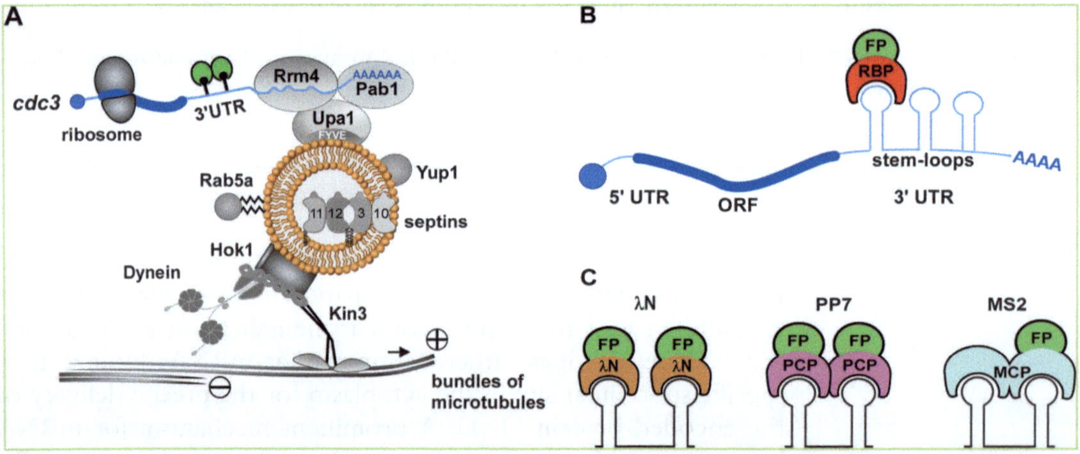

Fig. 1 Concept of RNA live imaging in *U. maydis*. (**a**) Schematic representation illustrating endosomal mRNA transport including the RNA live imaging concept in *U. maydis*. *cdc3* mRNA (*blue*) is bound by the RNA binding protein Rrm4. In the 3′ UTR stem-loops are integrated that are bound by RNA-binding proteins tagged with fluorescence proteins (*green*). The endosomal transport machinery is shown in *gray* (Rab5a, small G protein; Yup1, SNARE; Hok1, motor adaptor; Kin3, plus-end directed kinesin; septins Cdc3, Cdc10, Cdc11, and Cdc12 are indicated as bean-like shapes). (**b**) Scheme of RNA live imaging depicting mRNA with stem-loops in 3′ UTR. These stem-loops are recognized by an RNA-binding protein (RBP, *red*) coupled to a fluorescence protein (FP, *green*). (**c**) Comparison of three different systems for the visualization of mRNA: in the λN system the RBP (λN*; *brown*) binds as a monomer, in the PP7 system the RBP (PCP; *pink*) binds as a dimer and in the MS2 system the RBP (MCP; *light blue*) is already expressed as a synthetically fused tandem dimer

Rrm4-containing mRNPs with the endosomal surface by a lipid-binding FYVE domain (Fig. 1a; [9, 18]). In essence, microtubule-dependent transport of early endosomes is not only needed for endocytic sorting to the vacuole but also for long-distance transport of mRNPs and even of whole organelles such as peroxisomes [7, 20].

To study mRNA localization fluorescent in situ hybridization is a very powerful technique that has been perfected down to the single molecule level. However, in order to study dynamic processes RNA live imaging is the method of choice [21]. RNA live imaging was first applied studying mating type switching in *Saccharomyces cerevisiae* [22]. Inserting binding sites for the MS2 phage coat protein in the 3′ UTR of mating type switch repressor *ASH1* mRNA enabled live recording of mRNA transport [22]. The strategy of inserting binding sites in the 3′ UTR was also used in other systems to study, for example, the influence of RBP targeting on decay or to visualize translation in vivo [23–25]. The system has to be applied carefully because of an accumulation of intermediates of mRNA degradation containing MS2 binding sites [23, 26, 27]. However, processively moving mRNAs are less likely to be affected by such problems.

A related example is the coat protein of bacteriophage PP7 (PCP), an RNA phage of *Pseudomonas aeruginosa*, and its cognate RNA stem-loops (PBS; [28]). Importantly, the MS2 and PP7 coat proteins share only 15% sequence identity and their cognate RNA stem-loops differ from each other [28]. Thus, they can be simultaneously used to detect two mRNAs without interference. Another in vivo RNA labeling system is derived from the anti-terminator protein N from phage λ. The first 22 residues of the λN peptide are sufficient for recognition of its boxB RNA hairpins [29]. Thus, a small truncated version is used for RNA live imaging (Fig. 1c; [11, 30]). All phage proteins are fused at their C-terminus to fluorescence proteins and no alterations in RNA-binding have been described.

In this chapter we report on the use of RNA live imaging during microtubule-dependent transport in *U. maydis*. The growing hyphae of this organism provide a highly organized stereotypic microtubule cytoskeleton ideal to study transport of motile mRNAs and their interactions with mRNP composing protein molecules in vivo. We describe the experimental design and the strategy of constructing such RNA tracking systems. This includes both multimerization of the binding sites and the fluorescence proteins (FPs) to increase sensitivity. Furthermore, it is advisable to fine-tune the expression of the RBP-FP fusion to obtain optimal signal-to-noise ratios. Alternatively, we present the use of a nuclear localization signal (NLS) for the RBP-FP fusion. Thereby, unbound RBP-FPs are targeted from the cytoplasm to the nucleus, also improving the signal-to-noise ratio. Finally, we compare different phage proteins as well as the quantification of the results like velocity, amount and distances traveled.

2 Materials

2.1 Plasmids

1. pmax pona 12xTRICK 24xMS2SL (Addgene #64542, [25]).
2. Potef-mCherry-cdc3-16boxB-3′ UTR (Addgene #86465, [13]).
3. phage ubc nls ha pcp gfp (Addgene #64539, [25]).
4. Potef-λN*-3xGfp-NLS (Addgene #86780, [18]).
5. pST4 TET CMV intron renilla 24xPP7 24xMS2 (Addgene #84444, [25]).
6. phage UBC NLS-HA-2xMCP-tagRFP (Addgene #64541, [25]).

2.2 PCR Oligos and Cloning

Underlined sequence represents target specific portion of the PCR oligos, sequence highlighted in bold shows the newly introduced restriction site.

1. PP7loops-SacII-Fwd:
   5′CCG**CCGCGG**ACACGGCCGTGTATTA.
2. PP7loops-SacII-Rev:
   5′**CCGCGG**GCTGATCCACTCGAGAGATC.
3. PP7CP-BamHI-Fwd:
   5′GGC**GGATCC**ATGTCCAAAACCATCGTTCTTTCGG.
4. PP7CP-BamHI-Rev:
   5′CAT**GGATCC**TGAACGGCCCAGCGGCACAAGG TTGACG.
5. *Bam*HI and *Sac*II restriction enzymes (e.g., New England Biolabs).

2.3 Working Solutions for U. maydis [31, 32]

1. Trace elements: 0.06 w/v % H_3BO_3, 0.14 w/v % $MnCl_2 \cdot 4\ H_2O$, 0.4 w/v % $ZnCl_2$, 0.4 w/v % $Na_2MoO_4 \cdot 2\ H_2O$, 0.1 w/v % $FeCl_3 \cdot 6\ H_2O$, 0.04 w/v % $CuSO_4 \cdot 5\ H_2O$, add ddH_2O, sterile filter.
2. Salt solution: 16 w/v % KH_2PO_4, 4 w/v % Na_2SO_4, 8 w/v % KCl, 1.32 w/v % $CaCl_2$, 8 v/v % trace elements, 1 w/v % $MgSO_4$ (water free), add ddH_2O, sterile filter.
3. Vitamin solution: 0.1 w/v % thiamine hydrochloride, 0.05 w/v % riboflavin, 0.05 w/v % pyridoxine, 0.2 w/v % calcium pantothenate, 0.05 w/v % p-aminobenzoic acid, 0.2 w/v % nicotinic acid, 0.2 w/v % choline chloride, 1 w/v % *myo*-inositol, add ddH_2O, sterile filter.
4. Complete medium (CM): 0.25 w/v % casamino acids, 0.1 w/v % yeast-extract, 1.0 v/v % vitamin solution, 6.25 v/v % salt solution, 0.05 w/v % herring sperm DNA, 0.15 w/v % NH_4NO_3, add deionized water, adjust pH with 5 M NaOH to 7.0, add glucose (CM-glc) or arabinose (CM-ara) solution after autoclaving (1% f.c.).

5. Nitrate minimal medium (NM): 0.3 w/v % KNO_3, 6.25 v/v % salt solution, add deionized water, adjust pH with 5 M KOH to 7.0, add glucose (glc) or arabinose (ara) solution after autoclaving (1% f.c.).
6. Sodium citrate sorbitol solution (SCS): Solution 1: 20 mM tri-Na-citrate, 1 M sorbitol; Solution 2: 20 mM citric acid, 1 M sorbitol; add solution 2 to solution 1 until pH 5.8 is reached, autoclave.
7. Sorbitol, Tris–HCl, $CaCl_2$ (STC): 1 M sorbitol, 10 mM Tris–HCl pH 7.5, 100 mM $CaCl_2$, sterile filter.
8. STC/PEG: 40 v/v % PEG 3350 in STC-buffer.
9. RegLight: 1.0 w/v % yeast extract, 0.4 w/v % Bacto peptone, 0.4 w/v % sucrose, 18.22 w/v % sorbitol, 1.5 w/v % agar, add deionized water and autoclave.
10. Selection antibiotics, e.g., hygromycin.
11. Protoplasting solution: 12.5 mg/mL *Trichoderma* lysing enzymes in SCS, filter-sterilize through a 22 μm filter, solution has to be fresh for optimal enzymatic activity.
12. Glass reaction tubes, baffled flasks 250 and 100 mL, petri dishes, rotation wheel, shaker.

2.4 General Microscopy Equipment for Live Cell Imaging of *U. maydis* [31]

1. Wide-field fluorescence microscope. We use a Zeiss Axio Observer.Z1 in combination with a Photometrics CoolSNAP HQ2 CCD camera.
2. Laser illumination to excite GFP (488 nm/100 mW) or mCherry (561 nm/150 mW).
3. 63× NA 1.4 or 100× NA 1.3 objectives.
4. Filter sets for GFP: ET 470/40×, ET 252/50 m, T495_PXR.
5. Filter sets for mCherry: ET 560/40×, ET 630/75 m, T585lp.
6. For dual-color microscopy we use a VS-LMS4 Laser Merge-System (Visitron Systems).
7. MetaMorph (Molecular Devices, version 7) is used to control the microscopes, to process and analyze the acquired images.
8. Microscope slides with 76 × 26 mm (e.g., Marienfeld).
9. High precision coverslips with 18 × 18 mm, $D = 170 \pm 5$ μm (e.g., Zeiss).
10. 3 w/v % melted agarose (e.g., Bio-Rad).

3 Methods

An important requirement for the detection of mRNAs in fungal model systems is the stable integration of heterologous RNA stem-loops in the 3′ UTR of the mRNA of interest (Fig. 1b). This can be achieved by integration at the endogenous locus using standard techniques adapted from model fungi [32–34]. In addition, the corresponding RBP fused to a fluorescence protein needs to be coexpressed. For this, we recommend ectopic expression at defined loci in *U. maydis* [35]. To ensure high quality RNA live imaging, the expression level of the RBP has to be adjusted by using suitable promoters. To achieve a high signal-to-noise ratio during RNA live imaging, the RBP can be fused to an NLS to remove cytoplasmic background fluorescence (*see* Subheading 3.5).

3.1 Cloning of the Reporter Constructs

Mainly three systems are used to visualize mRNAs in living cells: The MS2, PP7 and λN system (Fig. 1c). All three systems were applied for the visualization of *cdc3* mRNA in *U. maydis.*

1. To visualize *cdc3* mRNA in vivo two plasmids have to be generated. The first one encodes the RBP coupled to a fluorescence protein. The other plasmid contains cognate stem-loops, integrated into the 3′ UTR of the gene of interest (GOI).
2. To construct the MS2 and PP7 containing plasmids for RNA live imaging the already available *cdc3-16boxB*/λN system was used [13, 18]. As an example, the construction of the PP7 system is described in the following steps.
3. The plasmid "pmax pona 12xTRICK 24xMS2SL" was used for the amplification of 12 PP7 stem-loops. To amplify the PP7 stem-loops the PCR oligonucleotides PP7loops-SacII-Fwd and PP7loops-SacII-Rev can be used. For cloning *Sac*II restriction sites were integrated upstream and downstream of the stem-loops. The resulting PCR product was inserted into the 3′ UTR of *cdc3* using the plasmid "Potef-mCherry-cdc3-16boxB-3′ UTR". Here, 16 copies of the boxB stem-loops were replaced with 12 copies of the PP7 stem-loops using *Sac*II restriction sites (*see* **Note 1**).
4. The PP7 RNA-binding coat protein was taken from the plasmid "phage ubc nls ha pcp gfp". For amplification of the PP7 coat protein PCR oligonucleotides PP7CP-BamHI-Fwd and PP7CP-BamHI-Rev can be used. The resulting PCR products were integrated into plasmid "Potef-λN*-3xGfp-NLS" by replacing the λN* protein using *Bam*HI restriction sites.
5. The MS2 loops and the MS2 RNA-binding protein were integrated in the same way as described for the PP7 system, using the "pST4 TET CMV intron renilla 24xPP7 24xMS2" and "phage UBC NLS-HA-2xMCP-tagRFP" plasmids as templates of the MS2 loop array and the MS2 coat protein, respectively.

3.2 Generation of *U. maydis* Strains

To generate *U. maydis* strains the desired constructs are integrated in the genome by homologous recombination. The homologous flanking regions should be about 1 kb in size [33, 34]. To this end, linearized constructs can be transformed into *U. maydis* protoplasts. A detailed visual protocol showing the procedure is available [32]. In brief:

1. Inoculate a preculture in 3 mL CM-glc and incubate it on a rotating wheel for 24 h at 28 °C.
2. Inoculate a main culture in 50 mL CM-glc and allow growth until exponential phase (OD_{600} of around 0.8).
3. Pellet cells for 5 min, 1500 × *g* and wash the pellet in 25 mL SCS.
4. Resuspend the pellet in 2 mL protoplasting solution and incubate cells for 5–20 min at room temperature (RT). Check protoplasting process under the microscope and stop process when 30–40% of the cells are round or resemble pinheads.
5. Wash 3× in 10 mL cold SCS, centrifuge at 1000 × *g* for 5 min. Keep protoplasts on ice.
6. Wash once in 10 mL cold STC and resuspend the pellet in 1 mL cold STC. Make 100 μL aliquots in prechilled tubes and freeze them at −80 °C until further use.
7. For transformation prepare two-layered selection plates (12 mL for each layer). Bottom layer contains RegLight agar with antibiotic, e.g., hygromycin (double amount of the usual concentration; 400 μg/mL), the upper layer contains only RegLight agar.
8. Thaw protoplast on ice and add 1 μL heparin (15 mg/mL) and 1 μg linearized plasmid. Incubate for 10 min on ice.
9. Add 500 μL STC/PEG to the transformation tube. Incubate for 15 min on ice.
10. Distribute cells on two transformation plates.
11. Incubate plates at 28 °C for 5–10 days.
12. To confirm correct transformants genomic DNA of potentially positive candidates are extracted and homologous integration events need to be verified by Southern blot experiments [32, 33, 35].

3.3 Live Cell Imaging of a Highly Motile RNA-Binding Protein

RBPs are key components orchestrating mRNA transport. Therefore, analyzing the localization of motile RBPs like Rrm4 from *U. maydis* can give a first insight into the transport machinery of mRNAs.

1. To visualize the subcellular localization of Rrm4, fuse the coding sequence of GFP C-terminally to Rrm4 using standard cloning techniques [33, 34] (*see* **Note 2**).
2. Do not alter the endogenous promoter to avoid overexpression artifacts. This is achieved by designing the integration event downstream of the coding sequence.

3. Transform the generated Rrm4-GFP expression plasmid into *U. maydis* as described in Subheading 3.2 [32–34].
4. Inoculate yeast-like growing cells in 3 mL of CM-glc in a glass tube and grow culture on rotation wheel for 24 h at 28 °C.
5. Inoculate a main-culture in a 250 mL baffled flask by mixing 15 μL of the 24 h culture with 30 mL CM-glc medium and incubate it for 15–20 h shaking with 200 rpm at 28 °C.
6. For induction of hyphal growth measure the optical density (OD_{600}) of the cultures and centrifuge the corresponding volume at 3500 × *g* for 5 min to obtain an OD_{600} of 0.5 in 20 mL. Wash once with NM-glc media and resuspend the cells in 20 mL of the same media to induce hyphal growth (*see* **Note 3**). Continue shaking with 200 rpm at 28 °C for 7–8 h.
7. Prepare agarose cushions by pipetting 250 μL melted agarose solution on a microscope slide. Add a second slide directly on top of the agarose drop, parallel to the lower slide. After 20–30 min one slide can be removed by carefully sliding it sideways.
8. Pipet 1 μL of the cell culture onto the cushion, distribute the cell culture by swaying the slide and let it dry for 1–2 min (*see* **Note 4**).
9. To visualize motile Rrm4-GFP signals choose appropriate illumination for the excitation of chosen fluorophores (here GFP, 488 nm excitation); use the 63× or the 100× objective and apply a laser power (100 mW) of 30–40%.
10. Record movies with 150 frames and 150 ms exposure time at each frame (*see* **Note 5**).
11. Use the camera in stream mode to visualize movement throughout the hyphae and ensure the fastest acquisition. Rrm4 shuttles bidirectionally along microtubules throughout the whole cell (Fig. 2a).
12. After recording moving Rrm4-Gfp signals, convert the movie into a kymograph, which plots traveled distance over time using the software package MetaMorph (Fig. 2a and b; *see* **Note 6**). Processively moving Rrm4-Gfp signals are depicted as diagonal lines (Fig. 2a and b).
13. The velocity of moving RBPs can be measured by marking such processive Rrm4-Gfp signals. For this, the start and end point of the moving signal is connected by a line (Fig. 2b). The travelled distance per time represents the velocity of the moving signal (Software package Metamorph, Version 7).

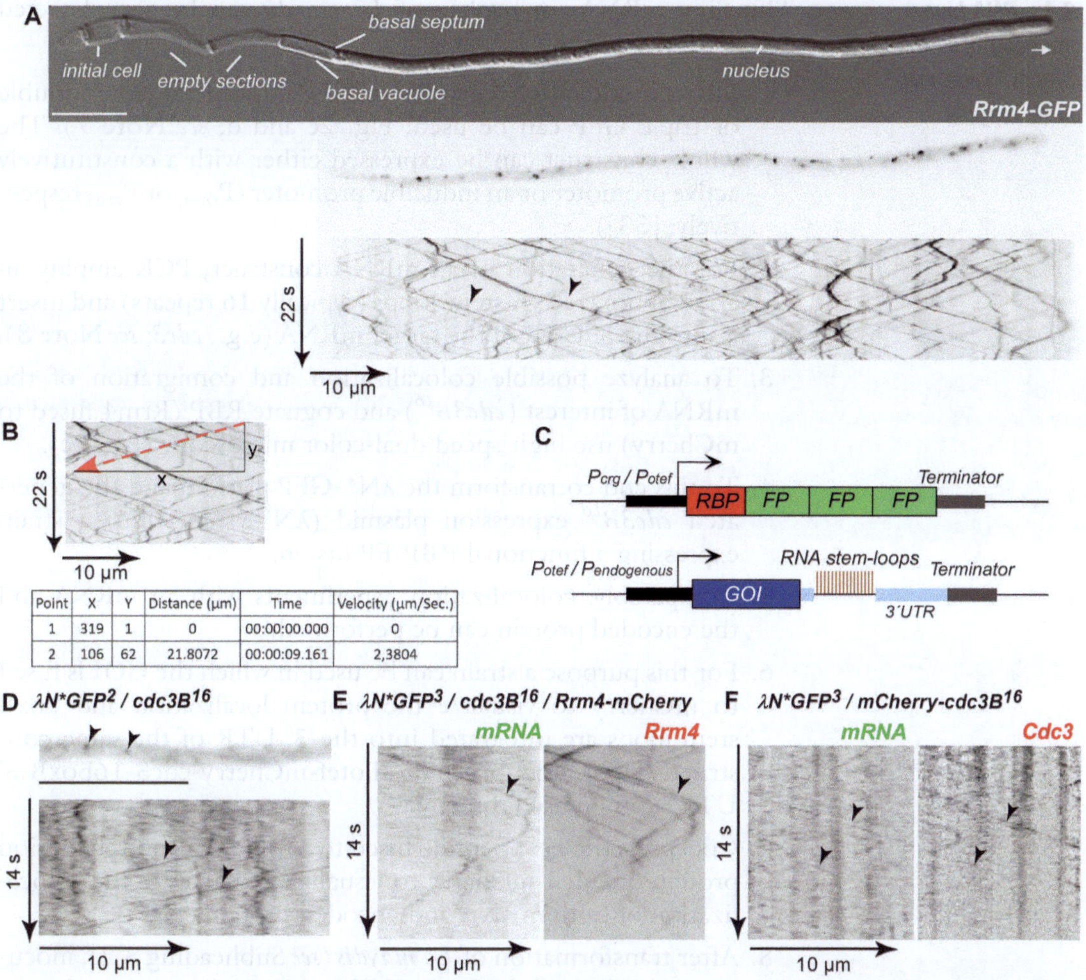

Point	X	Y	Distance (µm)	Time	Velocity (µm/Sec.)
1	319	1	0	00:00:00.000	0
2	106	62	21.8072	00:00:09.161	2,3804

Fig. 2 Components of the endosomal mRNA transport. (**a**) DIC image of an Rrm4-GFP expressing hypha 6 h after induction of hyphal growth with corresponding fluorescence image and kymograph (scale bar, 10 μm, inverted fluorescence image). (**b**) Example for velocity measurement within a kymograph generated by the software package MetaMorph (inverted image). The *red line* indicates a processively moving signal from the hyphal tip toward the basal pole (*right* to *left*). Based on the position, distance and elapsed time the velocity was calculated. (**c**) Schematic representation of the components of the RNA live imaging technique in *U. maydis*. The *upper part* shows the RNA-binding protein (RBP) fused to three fluorescence proteins (FP) expressed at the ectopic ip^{s} locus in *U. maydis*. Expression is either driven by an arabinose-inducible promoter P_{crg} or a constitutively active promoter P_{otef}. The *lower part* shows the locus of the gene of interest (GOI). Its expression is either driven by P_{otef} or by the endogenous promoter. 16 boxB stem-loops (*red*) are inserted in the 3′ UTR. (**d**) Micrograph showing *cdc3* mRNA signals in hypha indicated by *black arrowheads* (*top*, scale bar 10 μm, inverted image). Kymograph visualizes movement of *cdc3* mRNA (*black arrowheads*). (**e**) Kymographs of the described strain to visualize the colocalization of mRNA (*green*) and Rrm4 (*red*). *Arrowhead* marks colocalization event (inverted images). (**f**) Kymographs of the described strain to visualize the colocalization of mRNA (*green*) and encoded protein (*red*). *Arrowheads* mark colocalization event (inverted images)

3.4 RNA Live Imaging of Endosomal mRNA Transport in *U. maydis*

Shuttling mRNAs in hyphae of *U. maydis* can be also detected using the λN system.

1. Fuse a modified λN* peptide to GFP multimers (either double or triple GFP can be used; Fig. 2c and d; *see* **Note 7**). The whole construct can be expressed either with a constitutively active promoter or an inducible promoter (P_{otef}, or P_{crg}, respectively; [33]).
2. For the generation of an mRNA construct, PCR amplify an array of boxB RNA stem-loops (typically 16 repeats) and insert it into the 3′ UTR of the target mRNA (e.g., *cdc3*; *see* **Note 8**).
3. To analyze possible colocalization and comigration of the mRNA of interest (*cdc3B*^{16}) and cognate RBP (Rrm4 fused to mCherry) use high-speed dual-color microscopy (Fig. 2e).
4. To this end cotransform the λN*-GFP plasmid and the generated *cdc3B*^{16} expression plasmid (λN system) into a strain expressing a functional RBP-FP fusion.
5. Comparably, colocalization experiments with an mRNA and the encoded protein can be performed.
6. For this purpose a strain can be used in which the GOI is fused to mCherry to visualize the protein localization and boxB stem-loops are integrated into the 3′ UTR of the same construct. Here, the plasmid "Potef-mCherry-cdc3-16boxB-3′ UTR" can be used directly.
7. Coexpress the λN* peptide fused to a triple GFP and the fusion protein of Cdc3-mCherry to visualize and analyze the colocalization of both mRNA and encoded protein (Fig. 2f).
8. After transformation of *U. maydis* (*see* Subheading 3.2), inoculate yeast-like cells in 3 mL of CM-glc in a glass tube and grow culture on rotation wheel for 24 h at 28 °C.
9. Inoculate a main-culture in a 250 mL baffled flask with 30 mL CM-ara (1% w/v f.c.) medium (dilute the 24 h culture to ~1:1000) to induce the P_{crg} promoter or CM-glc (1% w/v f.c.) medium (dilute the 24 h culture to ~1:2000) if the expression is controlled by P_{otef} promoter. Incubate cells 15–20 h shaking with 200 rpm at 28 °C (*see* **Note 9**).
10. For induction of hyphal growth adjust OD_{600} to 0.5 in 20 mL and shift the cells to NM-ara or NM-glc media by centrifugation at 3500 × *g* for 5 min (*see* **Note 10**). Wash once in the same media and resuspend cells in 20 mL. Continue shaking at 200 rpm at 28 °C for 7–8 h (*see* dilute the 24 h culture to ~).
11. To prepare microscopy follow **steps** 7 and **8** in Subheading 3.3.
12. To visualize moving mRNA choose appropriate illumination for the excitation of chosen fluorophores (GFP or mCherry).

Use the 63× or the 100× objective and use a laser power (100 mW) of 20–30%.

13. Record movies with 150 frames and 150 ms exposure time (*see* **Note 5**).
14. Use the camera in stream mode to visualize movement and ensure the fastest acquisition.
15. After recording moving mRNAs, convert the movie into a kymograph using the software package MetaMorph (*see* **Note 6**). Comparable to the analysis of the movement of Rrm4 signals (*see* Subheading 3.3), the velocity and the distance can be measured with the software package MetaMorph (*see* **Note 11**).
16. For the simultaneous detection of mRNA and protein a two-channel imager is used. The imager splits green and red emission lights and detects them on separate regions of the CCD camera chip. Use a 100× Plan-Neofluar (NA 1.3) objective with the respective laser lines for imaging (*see* **Note 12**).
17. To ensure minimal photobleaching first determine the minimal laser power to detect the signal for each fluorophore (*see* **Note 13**). To decrease photobleaching alternating laser excitation (msAlex) together with DualView technology can be applied [31].

3.5 RNA Live Imaging Comparing Heterologous RBP with and without NLS

Signal-to-noise ratio during RNA live imaging can be improved by a fusion of the RBP to a NLS (Fig. 3a).

1. PCR amplify the λN* as described for PP7 coat protein in Subheading 3.1. Fuse it in frame to GFP multimers and an NLS sequence. Alternatively, the plasmid "Potef-λN*-3xGfp-NLS" can be used directly.
2. The desired plasmid is transformed using a strain (*see* Subheading 3.2), which harbors 16 boxB stem-loops integrated into the 3′ UTR of the *cdc3* mRNA (Fig. 3b and c).
3. The NLS ensures the targeting of the unbound RBPs into the nucleus and therefore decrease the cytoplasmic background.
4. Perform microscopic analysis as described above (Subheading 3.4, **steps 8–14**).
5. Adjust the contrast of the recorded movies with available software packages (here: MetaMorph, Version 7) to visualize moving mRNAs (Fig. 3b and c).
6. Analyze the movement of mRNAs using kymographs. Because of the lower cytoplasmic background using an RBP with an NLS, moving mRNA signals can be visualized over extended periods of time (Fig. 3b and c, right).

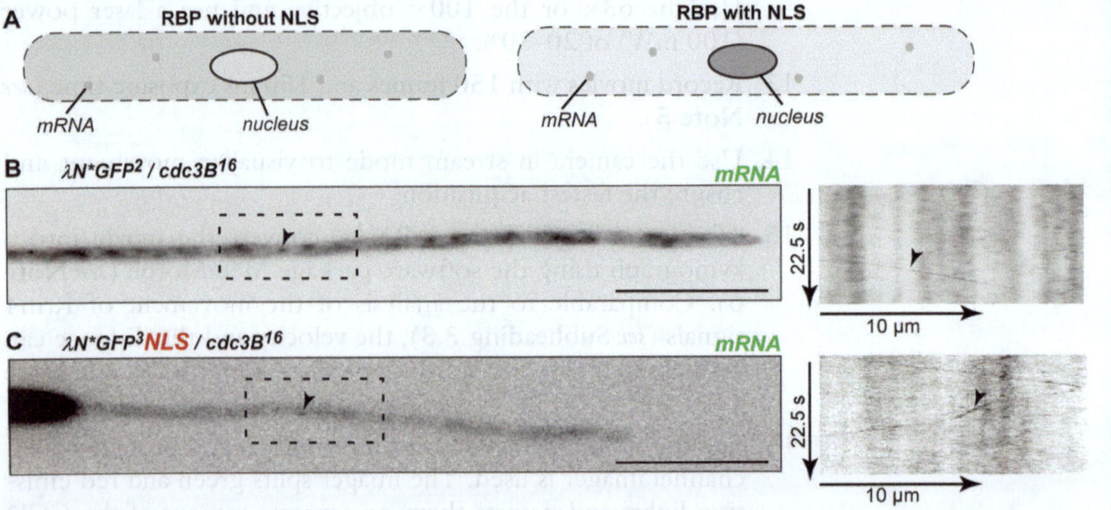

Fig. 3 mRNA visualization with and without a NLS sequence. (**a**) Schematic representation showing the difference between RNA live imaging using a RBP-FP with or without a nuclear localization sequence (NLS). (**b**) Background adjusted fluorescence image of a strain expressing λN* coupled to GFP and *cdc3* mRNA with boxB stem-loops (*left*, scale bar 10 μm). mRNA signal is marked by *arrowheads*, note the strong cytoplasmic background. Kymograph corresponding to the indicated area (*right*) visualizes movement of the corresponding signal (*inverted images*). (**c**) Background-adjusted fluorescence image of a strain expressing λN* coupled to GFP with an NLS and *cdc3* mRNA with 16 boxB stem-loops (*left*, scale bar 10 μm). *Arrowhead* marks motile mRNA. Kymograph on the right visualizes movement of the corresponding signal (inverted images)

3.6 RNA Live Imaging Comparing Different Heterologous RBPs

1. In contrast to the already described boxB loops, the PP7 stem-loop consists of 25 nucleotides (nt) and the MS2 stem-loop has a size of 19 (nt). For binding of the MS2 coat protein both of the single stranded-adenosine (A) residues at the top of the stem and the bulged-A are essential (Fig. 4a). For the PP7 stem-loop (PBS) a similar bulged-A is needed for tightest binding of the RPB (Fig. 4a).
2. The RBP deriving from the bacteriophage λ binds to the RNA stem-loops as a monomer (Fig. 1c; [36]). Twenty-two amino acids of the RBP are sufficient for binding. Furthermore, three point mutations in the λN-peptide increase the binding to the cognate boxB stem-loops (λN*; [36]).
3. The coat proteins of MS2 and PP7 bind as dimers to their RNA-binding sites (Fig. 1c; [28, 37]). In contrast to the PCP, the MCP is already expressed as a dimer which is then fused to different numbers of fluorescent proteins.
4. PCR amplify the PP7 or MS2 stem-loops (typically 12 or 24 repeats, respectively) and insert it into the 3′ UTR of the target (e.g., *cdc3)* as described in Subheading 3.1 (*see* **Notes 1** and **14**).
5. Generate a fusion of the cognate RBP fused to GFP multimers as described in Subheading 3.1.

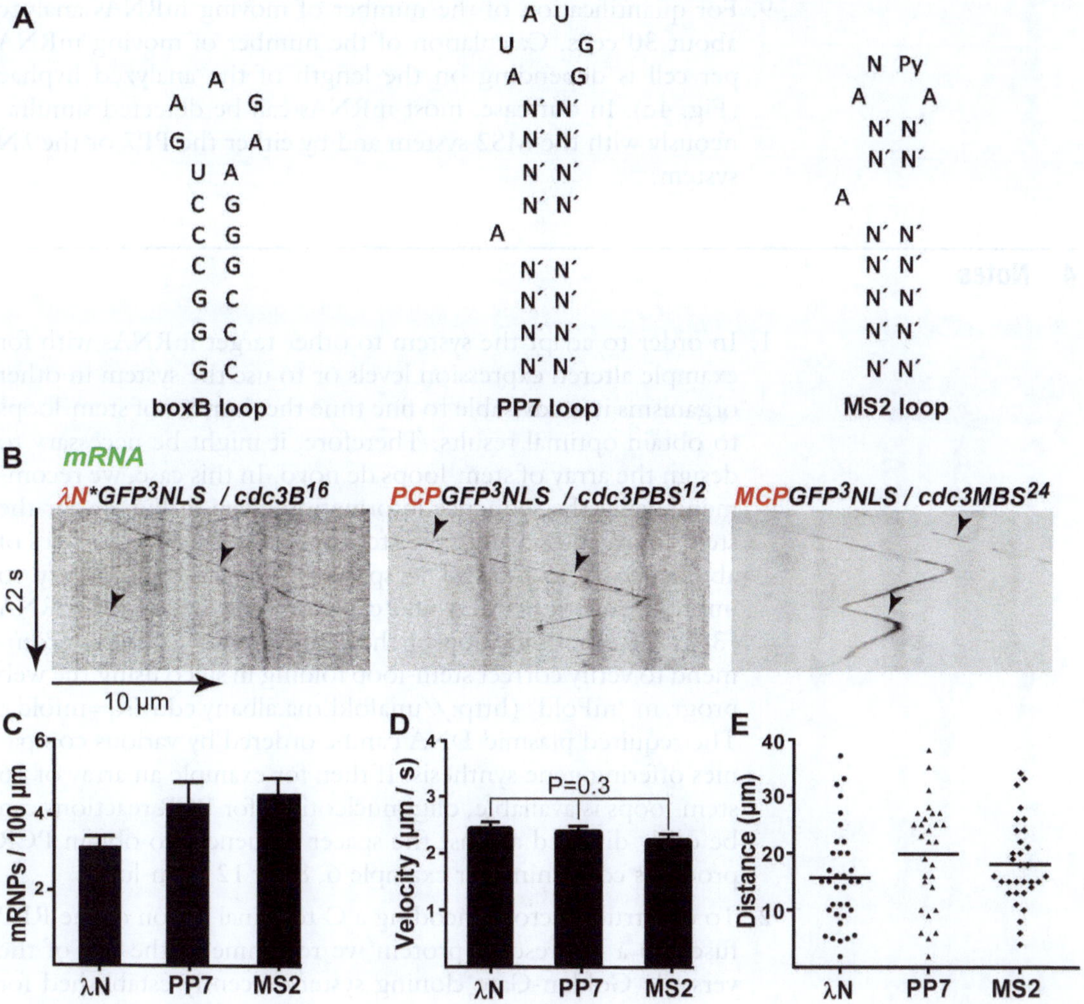

Fig. 4 Analysis of RNA live imaging of *cdc3* comparing different systems. (**a**) Schematic representation of the secondary structures of boxB, PP7, and MS2 RNA stem-loops. (**b**) Kymographs of labeled *cdc3* mRNA with three different RBPs (λN*, PCP, and MCP) fused to triple GFP and containing an NLS (inverted images). *Arrowheads* indicate moving mRNPs. Hyphae were analyzed 8 h post induction. (**c**) Number of processively moving RBP-FP signals per 100 μm of hyphae containing different systems for RNA live imaging (error bars represent s.e.m.; 30 cells were analyzed in three independent experiments). (**d**) Velocity of observed signals in hyphae containing different RNA live imaging systems (total directed signals; error bars represent s.e.m., 30 cells were analyzed in three independent experiments; one way ANOVA, $\alpha = 0.05$). (**e**) Range of movement of signals in hyphae containing different RNA live imaging systems ($n = 3$, data points and median are indicated)

6. Cotransform the desired constructs in *U. maydis* as described in Subheading 3.2 (*see* **Note 14**).
7. Perform the microscopic analysis as described above and generate kymographs (Subheading 3.4, **steps 8–15**; Fig. 4b).
8. To find the best suited setup, compare the different systems in respect to the number of detectable moving mRNAs (mRNP complexes), the velocity of moving signals and the covered distance (Fig. 4c–e).

9. For quantification of the number of moving mRNAs analyze about 30 cells. Calculation of the number of moving mRNA per cell is depending on the length of the analyzed hyphae (Fig. 4c). In our case, most mRNAs can be detected simultaneously with the MS2 system and by either the PP7 or the λN system.

4 Notes

1. In order to adapt the system to other target mRNAs with for example altered expression levels or to use the system in other organisms it is advisable to fine tune the number of stem-loops to obtain optimal results. Therefore, it might be necessary to design the array of stem-loops de novo. In this case, we recommend using the sequence information given in Fig. 4a for the stem-loops and separate the stem-loops by spacer sequences of about 40 nucleotides. The spacer sequences should vary to improve genetic stability and expression of the tagged mRNA [38]. After optimization of the sequence we strongly recommend to verify correct stem-loop folding in silico using the web program mFold (http://unafold.rna.albany.edu/?q=mfold). The required plasmid DNA can be ordered by various companies offering gene synthesis. If then for example an array of 16 stem-loops is available, oligonucleotides for PCR reactions can be easily directed against the spacer sequences to obtain PCR products containing for example 6, 8, or 12 stem-loops.
2. To construct vectors encoding a C-terminal fusion of the RBP fused to a fluorescent protein we recommend the use of the versatile Golden-Gate cloning system recently established for *U. maydis* [34]. In brief, the stop codon of the RBP of interest is replaced by a sequence encoding a short linker amino acid sequence AANAAT, which is fused to the start codon of the FP such as GFP. This is achieved by using the DNA sequence G<u>CGGCCAACGCGGCCA</u>CC [33]. Thereby an *Sfi*I restriction enzyme site (underlined) is introduced that can also be used for further standard cloning procedures, if for example the FP should be replaced by a different color (e.g., replace GFP with mCherry; [33]).
3. Switching the nitrogen source induces the hyphal growth of *U. maydis* due to the expression of the responsible genes with the *nar1* promoter [39].
4. Swaying the microscope slide ensures an appropriate distribution of cells. It is highly recommended to let the cell suspension dry on the slide to increase plane orientation of hyphae.
5. For better visualization the exposure time can be adjusted ranging from 100 to 250 ms.

6. For the generation of kymographs other software packages than MetaMorph like the Zeiss software (Zen; Zeiss) or Fiji (ImageJ, https://imagej.nih.gov/ij/ [40, 41]) are available.
7. The gene encoding the λN*-peptide carries three point mutations [36]. These mutations increase the binding to the cognate boxB stem-loops. The peptide is expressed ectopically in the *ip^s* locus of *U. maydis*. The expression of the RBP is driven by the arabinose inducible promoter P_{crg1} or the constitutively active promoter P_{otef}. Three GFP molecules are fused to the C-terminus of the λN*-peptide without a visible effect on RNA binding.
8. Sixteen boxB stem-loops are integrated in the 3′ UTR of the *cdc3* mRNA. The stem-loops are integrated 100 bp downstream of the stop codon of Cdc3, followed by 600 bp 3′UTR sequence. In between each stem-loop a linker sequence of 8 bp is used. The whole construct is integrated at the endogenous locus by homologous recombination.
9. The switch to arabinose as single carbon source for the induction of the P_{crg} results in a slower growth rate. To obtain comparable culture densities to the strains expressing the RBP with the P_{otef} promoter the ratio of inoculation has to be adjusted.
10. In comparison to strains grown in glucose containing media, a delay in filament induction can be observed in strains growing in media containing arabinose. This is due to arabinose as single carbon source.
11. Measured velocities for moving mRNAs should be comparable to the velocity of moving RBPs if they localize in identical complexes.
12. Using camera chip binning 2 can improve signal-to-noise ratio thus facilitating detection.
13. The 488 nm laser also bleaches mCherry. Therefore, laser power has to be adjusted carefully.
14. Make sure that an increased number of integrated stem-loops and therefore extended 3′ UTR do not decrease mRNA stability.

Acknowledgments

We thank lab members for critical comments on the manuscript as well as Dr. Jeffrey A. Chao for plasmids and helpful advice on RNA live imaging. Our research was in part financed by grants from the German Science Foundation through DFG-FOR2333, DFG-Fe448/8-1; DFG-Fe448/9-1, DFG-EXC1024 CEPLAS and DFG-CRC1208.

References

1. Buxbaum AR, Haimovich G, Singer RH (2015) In the right place at the right time: visualizing and understanding mRNA localization. Nat Rev Mol Cell Biol 16(2):95–109. doi:10.1038/nrm3918
2. Jansen RP (2001) mRNA localization: message on the move. Nat Rev Mol Cell Biol 2 (4):247–256
3. St Johnston D (2005) Moving messages: the intracellular localization of mRNAs. Nat Rev Mol Cell Biol 6(5):363–375
4. Martin KC, Ephrussi A (2009) mRNA localization: gene expression in the spatial dimension. Cell 136(4):719–730
5. Tekotte H, Davis I (2002) Intracellular mRNA localization: motors move messages. Trends Genet 18(12):636–642
6. Jansen RP, Niessing D, Baumann S, Feldbrügge M (2014) mRNA transport meets membrane traffic. Trends Genet 30:408–417. doi:10.1016/j.tig.2014.07.002
7. Salogiannis J, Reck-Peterson SL (2016) Hitchhiking: a non-canonical mode of microtubule-based transport. Trends Cell Biol. doi:10.1016/j.tcb.2016.09.005
8. Vollmeister E, Feldbrügge M (2010) Posttranscriptional control of growth and development in *Ustilago maydis*. Curr Opin Microbiol 13:693–699
9. Haag C, Steuten B, Feldbrügge M (2015) Membrane-coupled mRNA trafficking in fungi. Annu Rev Microbiol 69:265–281. doi:10.1146/annurev-micro-091014-104242
10. Schuster M, Sreedhar K, Fink G, Collemare J, Roger Y, Steinberg G (2011) Kinesin-3 and dynein cooperate in long-range retrograde endosome motility along a non-uniform microtubule array. Mol Biol Cell 22:3645–3657
11. König J, Baumann S, Koepke J, Pohlmann T, Zarnack K, Feldbrügge M (2009) The fungal RNA-binding protein Rrm4 mediates long-distance transport of *ubi1* and *rho3* mRNAs. EMBO J 28:1855–1866
12. Higuchi Y, Ashwin P, Roger Y, Steinberg G (2014) Early endosome motility spatially organizes polysome distribution. J Cell Biol 204 (3):343–357. doi:10.1083/jcb.201307164
13. Baumann S, König J, Koepke J, Feldbrügge M (2014) Endosomal transport of septin mRNA and protein indicates local translation on endosomes and is required for correct septin filamentation. EMBO Rep 15:94–102
14. Zander S, Baumann S, Weidtkamp-Peters S, Feldbrügge M (2016) Endosomal assembly and transport of heteromeric septin complexes promote septin cytoskeleton formation. J Cell Sci 129:2778–2792
15. Böhl F, Kruse C, Frank A, Ferring D, Jansen RP (2000) She2p, a novel RNA-binding protein tethers *ASH1* mRNA to the Myo4p myosin motor via She3p. EMBO J 19 (20):5514–5524
16. Müller M, Heym RG, Mayer A, Kramer K, Schmid M, Cramer P, Urlaub H, Jansen RP, Niessing D (2011) A cytoplasmic complex mediates specific mRNA recognition and localization in yeast. PLoS Biol 9(4):e1000611. doi:10.1371/journal.pbio.1000611
17. Becht P, König J, Feldbrügge M (2006) The RNA-binding protein Rrm4 is essential for polarity in *Ustilago maydis* and shuttles along microtubules. J Cell Sci 119:4964–4973
18. Pohlmann T, Baumann S, Haag C, Albrecht M, Feldbrügge M (2015) A FYVE zinc finger domain protein specifically links mRNA transport to endosome trafficking. Elife 4. doi:10.7554/eLife.06041
19. Feldbrügge M, Zarnack K, Vollmeister E, Baumann S, Koepke J, König J, Münsterkötter M, Mannhaupt G (2008) The posttranscriptional machinery of *Ustilago maydis*. Fungal Genet Biol 45:S40–S46
20. Guimaraes SC, Schuster M, Bielska E, Dagdas G, Kilaru S, Meadows BR, Schrader M, Steinberg G (2015) Peroxisomes, lipid droplets, and endoplasmic reticulum "hitchhike" on motile early endosomes. J Cell Biol 211(5):945–954. doi:10.1083/jcb.201505086
21. Lampasona AA, Czaplinski K (2016) RNA voyeurism: a coming of age story. Methods 98:10–17. doi:10.1016/j.ymeth.2015.11.024
22. Bertrand E, Chartrand P, Schaefer M, Shenoy SM, Singer RH, Long RM (1998) Localization of *ASH1* mRNA particles in living yeast. Mol Cell 2(4):437–445
23. Garcia JF, Parker R (2016) Ubiquitous accumulation of 3′ mRNA decay fragments in *Saccharomyces cerevisiae* mRNAs with chromosomally integrated MS2 arrays. RNA 22(5):657–659. doi:10.1261/rna.056325.116
24. Wu B, Eliscovich C, Yoon YJ, Singer RH (2016) Translation dynamics of single mRNAs in live cells and neurons. Science. doi:10.1126/science.aaf1084
25. Halstead JM, Lionnet T, Wilbertz JH, Wippich F, Ephrussi A, Singer RH, Chao JA (2015) Translation. An RNA biosensor for imaging the first round of translation from single cells

to living animals. Science 347 (6228):1367–1671. doi:10.1126/science.aaa3380

26. Haimovich G, Zabezhinsky D, Haas B, Slobodin B, Purushothaman P, Fan L, Levin JZ, Nusbaum C, Gerst JE (2016) Use of the MS2 aptamer and coat protein for RNA localization in yeast: a response to “MS2 coat proteins bound to yeast mRNAs block 5′ to 3′ degradation and trap mRNA decay products: implications for the localization of mRNAs by MS2-MCP system”. RNA 22(5):660–666. doi:10.1261/rna.055095.115
27. Garcia JF, Parker R (2015) MS2 coat proteins bound to yeast mRNAs block 5′ to 3′ degradation and trap mRNA decay products: implications for the localization of mRNAs by MS2-MCP system. RNA 21(8):1393–1395. doi:10.1261/rna.051797.115
28. Chao JA, Patskovsky Y, Almo SC, Singer RH (2008) Structural basis for the coevolution of a viral RNA-protein complex. Nat Struct Mol Biol 15(1):103–105. doi:10.1038/nsmb1327
29. Tan R, Frankel AD (1995) Structural variety of arginine-rich RNA-binding peptides. Proc Natl Acad Sci U S A 92(12):5282–5286
30. Daigle N, Ellenberg J (2007) LambdaN-GFP: an RNA reporter system for live-cell imaging. Nat Methods 4(8):633–636
31. Baumann S, Takeshita N, Grün N, Fischer R, Feldbrügge M (2015) Live cell imaging of endosomal trafficking in fungi. In: Tang BL (ed) Methods in molecular biology: membrane trafficking, vol 1270, 2nd edn. Springer, New York, pp 347–363
32. Bösch K, Frantzeskakis L, Vranes M, Kamper J, Schipper K, Göhre V (2016) Genetic manipulation of the plant pathogen *Ustilago maydis* to study fungal biology and plant microbe interactions. J Vis Exp 115. doi:10.3791/54522
33. Brachmann A, König J, Julius C, Feldbrügge M (2004) A reverse genetic approach for generating gene replacement mutants in *Ustilago maydis*. Mol Genet Genomics 272:216–226
34. Terfrüchte M, Jöhnk B, Fajardo-Somera R, Braus G, Riquelme M, Schipper K, Feldbrügge M (2014) Establishing a versatile Golden Gate cloning system for genetic engineering in fungi. Fungal Genet Biol 62:1–10
35. Loubradou G, Brachmann A, Feldbrügge M, Kahmann R (2001) A homologue of the transcriptional repressor Ssn6p antagonizes cAMP signalling in *Ustilago maydis*. Mol Microbiol 40:719–730
36. Austin RJ, Xia T, Ren J, Takahashi TT, Roberts RW (2002) Designed arginine-rich RNA-binding peptides with picomolar affinity. J Am Chem Soc 124(37):10966–10967
37. Peabody DS, Lim F (1996) Complementation of RNA binding site mutations in MS2 coat protein heterodimers. Nucleic Acids Res 24 (12):2352–2359
38. Wu B, Miskolci V, Sato H, Tutucci E, Kenworthy CA, Donnelly SK, Yoon YJ, Cox D, Singer RH, Hodgson L (2015) Synonymous modification results in high-fidelity gene expression of repetitive protein and nucleotide sequences. Genes Dev 29(8):876–886. doi:10.1101/gad.259358.115
39. Brachmann A, Weinzierl G, Kämper J, Kahmann R (2001) Identification of genes in the bW/bE regulatory cascade in *Ustilago maydis*. Mol Microbiol 42(4):1047–1063
40. Schneider CA, Rasband WS, Eliceiri KW (2012) NIH Image to ImageJ: 25 years of image analysis. Nat Methods 9(7):671–675
41. Schindelin J, Arganda-Carreras I, Frise E, Kaynig V, Longair M, Pietzsch T, Preibisch S, Rueden C, Saalfeld S, Schmid B, Tinevez JY, White DJ, Hartenstein V, Eliceiri K, Tomancak P, Cardona A (2012) Fiji: an open-source platform for biological-image analysis. Nat Methods 9(7):676–682. doi:10.1038/nmeth.2019

Chapter 22

Real-Time Fluorescence Imaging of Single-Molecule Endogenous Noncoding RNA in Living Cells

Hideaki Yoshimura and Takeaki Ozawa

Abstract

Visualizing RNA in living cells is increasingly important to facilitate accumulation of knowledge about the relation between specific RNA dynamics and physiological events. Single-molecule fluorescence imaging of target RNAs is an excellent approach to analyzing intracellular RNA motion, but it requires special techniques for probe design and microscope setup. Herein, we present a principle and protocol of an RNA visualization probe based on an RNA binding protein of the Pumilio homology domain (PUM-HD). We also describe the setup and operation of a microscope, and introduce an application to visualize telomeric repeats-containing RNA with telomeres and a telomere-related protein: hnRNPA1. This imaging technique is applicable to visualization of different RNAs, especially including repetitive sequences, in living cells.

Key words Fluorescence imaging, Single molecule, Repetitive RNA, PUM-HD

1 Introduction

Generally, RNA has been considered merely as a messenger of genetic information. However, the importance of intracellular transportation and localization of mRNA for physiological phenomena has been recognized recently [1, 2]. Furthermore, various noncoding RNAs have been implicated in many biological events, especially controlling chromatin states [3, 4]. Noncoding RNAs show such unique dynamics and localization in their functions as localization on particular loci on chromosomes and formation of particle-like structures called speckles [5, 6]. Consequently, the implication of dynamics and localization of RNAs in their functions promotes them as a target of interest in the bioimaging field. Nevertheless, live cell imaging studies of RNAs have not performed intensively, in contrast to live cell fluorescence imaging studies of various proteins, which have provided a great deal of information for elucidating the mechanisms of biological phenomena. A major reason for the delay of development on RNA imaging technique is

Imre Gaspar (ed.), *RNA Detection: Methods and Protocols*, Methods in Molecular Biology, vol. 1649, DOI 10.1007/978-1-4939-7213-5_22, © Springer Science+Business Media LLC 2018

the difficulty of labeling target RNAs in living cells. RNA cannot be fused with fluorescent or tag proteins through genetic engineering.

A possible approach to label target RNAs in living cells is to use an RNA binding protein that specifically addresses the target RNAs. In the simplest design, fusing the RNA binding protein to a fluorescent protein such as enhanced green fluorescent protein (EGFP) produces a fluorescent probe to label the target RNA. To generalize this approach, the RNA binding protein must have the following features: selective recognition of substantial length of RNA sequence to assure the specificity to the target RNA, capability of tailor-made design for variety of RNA sequences, and sufficient affinity to the target RNA sequence for labeling in living cells. Pumilio homology domain (PUM-HD) of human PUMILIO 1 is a promising candidate for RNA binding proteins to be used in RNA probes. The PUM-HD consists of eight repeated motifs, each of which binds specifically to an RNA base through formation of hydrogen bonds and van der Waals interaction [7]. Then an 8-base sequence of RNA is captured completely by a PUM-HD. A noteworthy feature of PUM-HD is its design flexibility [8]. Reported crystal structures of PUM-HD allow the production of tailor-made design of PUM-HD mutants to recognize particular RNA sequences. We previously succeeded in labeling endogenous β-actin mRNA in living cells using a PUM-HD based probe [9–11] consisting of two subunits including different PUM-HD mutants targeting particular sites in the 3′UTR of β actin mRNA and split fluorescent protein fragments. Upon these two subunits attached on a β actin mRNA at a time, fluorescent protein reconstitution is induced: the probe fluoresces. Consequently, β actin mRNA can be visualized selectively in living cells using fluorescence microscopy.

Using a PUM-HD based probe, we recently visualized single-molecule motion of telomeric repeat-containing RNA (TERRA), which is a noncoding RNA transcribed from telomeres, and which therefore contains a telomeric repeat sequence region [12]. In addition to the biological importance of TERRA and telomeres, RNAs including repetitive sequences have arisen as targets of interest in many biological fields. For instance, genes containing CAG repeats are known to introduce diseases [13]. The origin of the diseases is suspected not only as the translated proteins but also RNAs. In addition, several repetitive sequences provide markers for RNA splicing or translation [14, 15]. About 20% of the total genome is now regarded as regions of repetitive sequences. Therefore, RNAs including repetitive sequences are anticipated as important targets to elucidate functions of the RNAs and mechanisms of physiological events and diseases.

This chapter introduces the principles and methods of simultaneous visualization of TERRA, telomeres, and a telomere related protein hnRNPA1 in living cells [12]. We designed a fluorescent probe that targets the telomeric repeat region to label using a

PUM-HD mutant; PUM-HD was mutated to address the sequence UUAGGGUU (designated as mPUMt). Telomeres were visualized using a fusion protein of iRFP and TRF1. Then hnRNPA1 was fused to a SNAP tag and was conjugated with a red fluorescent dye, tetramethylrhodamine (TMR). We introduced the genes of these three probes into living cells and observed them using TIRF microscopy. Herein, we describe details of methods for this sample preparation and observation.

2 Materials

2.1 Plasmids (See Note 1)

1. NLS-GN-mPUMt-GC/pcDNA3.1(+).
2. hnRNPA1-SNAPf/pcDNA3.1(+).
3. iRFP-TRF1/pCLNCX.
4. pCL-10A1 (a packaging vector in RetroMax system).

2.2 Cell Culture and Transfection

1. Gag293 cells.
2. U2OS cells.
3. Dulbecco's modified Eagle's medium (DMEM) supplemented with 10% fetal bovine serum (FBS).
4. Trypsin solution (0.005 w/v %).
5. Opti-MEM.
6. Lipofectamine 2000 (Invitrogen Corp.).
7. Lipofectamine LTX (Invitrogen Corp.).
8. Polybrene.
9. Phosphate-buffered saline (PBS). Autoclaved before storage at room temperature.
10. Glass-bottomed dish (e.g., Asahi Glass Co., Japan).
11. CO_2 incubator.

2.3 Fluorescence Microscopy Observation

1. Hanks' balanced salt solution (HBSS).
2. MEM nonessential amino acids solution (100×, Gibco).
3. Imaging medium: HBSS containing 1× MEM nonessential amino acids solution, 2.5 g/L glucose, 2 mM glutamine, 1 mM sodium pyruvate, and 10 mM HEPES pH 7.4.
4. 1 μM SNAP-TMR.
5. Inverted fluorescence microscope, (e.g., Olympus IX81), equipped with a 100× 1.49 NA oil immersion objective, and a home-built excitation optical system.
6. EM-CCD camera (e.g., ImagEM; Hamamatsu Photonics K.K.).

7. sCMOS camera (e.g., ORCA Flash 4.0 V2; Hamamatsu Photonics K.K.).
8. MetaMorph software (Molecular Devices).
9. 488 nm laser (e.g., Coherent Sapphire 488 LP).
10. 561 nm laser (e.g., Coherent Sapphire 561 LP).
11. 647 nm laser (e.g., CUBE; Coherent Inc.).
12. Optical materials (Sigma Koki Co. Ltd., *see* Fig. 3).

    Convex lenses of $f = 10$ and 200 for beam expanders, $f = 50$ and 250 for relay lenses.
13. Optical filters (Semrock Inc.).

    FF662-FDi01 for split iRFP fluorescence and shorter ones.
    FF580-FDi01 for split TMR and GFP fluorescence.
    FF01-697/58 for iRFP fluorescence bandpass filter.
    FF01-609/54 for TMR fluorescence bandpass filter.
    FF03-525/50 for GFP fluorescence bandpass filter.

3 Methods

3.1 Design of PUM-HD-Based RNA Probes

In visualization of TERRA, a PUM-HD-based RNA recognition method and a split fluorescent protein reconstitution technique were adopted to create the TERRA probe. Before introducing details of the methods and protocols of sample preparation and observation of TERRA in living cells, we herein describe the principles of RNA recognition by PUM-HD, split fluorescent protein fragments reconstitution technique, and the probe design.

1. PUM-HD is an RNA binding domain of human PUMILIO1, an RNA binding protein that associates with 3′UTR of mRNAs to control gene translation in collaboration with other proteins such as eIF4 [16]. Hall et al. reported the crystal structure of PUM-HD (Fig. 1a) [7], and demonstrated that PUM-HD consists of eight repeated motifs (repeat 1–8). In each motif, three amino acids form hydrogen bonds and van der Waals interaction specifically to a single RNA base. Actually, PUM-HD completely recognizes an 8-base RNA sequence of UGUAYAUA (Y is C or U) (Fig. 1b and c).
2. Although no motifs that would recognize a cytosine base are present in the wild-type PUM-HD, a study using screening on a random mutagenesis method generates a motif that recognizes a cytosine base selectively [17].
3. Based on information from the crystal structure, PUM-HD can be modified e.g., by site directed PCR mutagenesis to recognize 8-base RNA sequences that are different from the original one by inducing site-specific substitution of particular amino

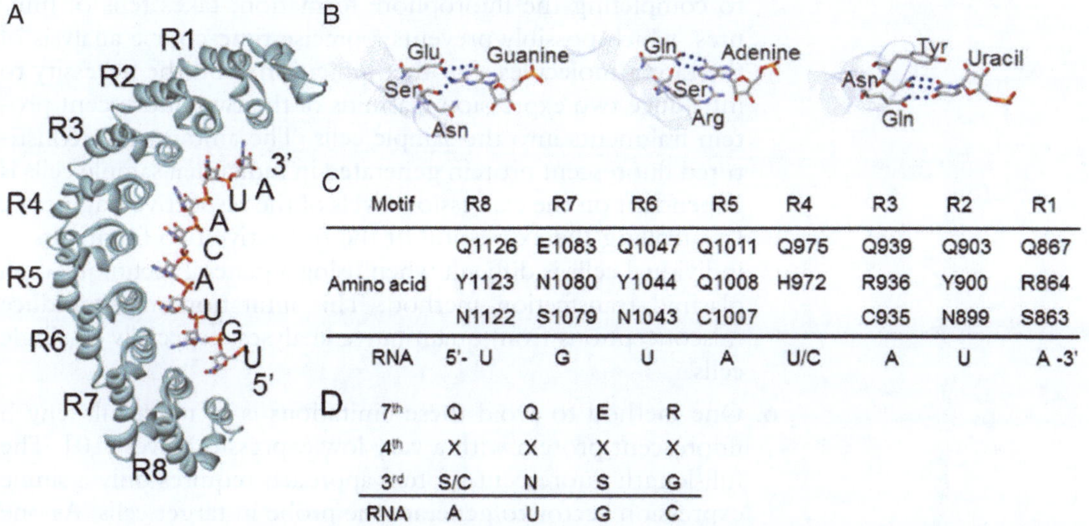

C

Motif	R8	R7	R6	R5	R4	R3	R2	R1
	Q1126	E1083	Q1047	Q1011	Q975	Q939	Q903	Q867
Amino acid	Y1123	N1080	Y1044	Q1008	H972	R936	Y900	R864
	N1122	S1079	N1043	C1007		C935	N899	S863
RNA	5'- U	G	U	A	U/C	A	U	A -3'

D

7th	Q	Q	E	R
4th	X	X	X	X
3rd	S/C	N	S	G
RNA	A	U	G	C

Fig. 1 Crystal structure and RNA recognition of PUM-HD. (**A**) Crystal structure of PUM-HD. (**B**) Enlarged images of the amino acids–RNA interacting portions. (**C**) Combination of RNA recognizing amino acids in the eight repeated motifs in the wild-type PUM-HD. (**D**) The universal code of RNA recognizing amino acids at the third, fourth, and seventh positions in the second helix, and recognized the RNA bases. X indicates positions at which any amino acid is acceptable

acids participating in RNA recognition [8] (*see* Fig. 1d and **Note 2**).

4. In addition to the usage of PUM-HD, employment of a fluorescent protein reconstitution technique is important for the design of PUM-HD based probes for sensitive and precise monitoring of the target RNA. Fluorescence from excess probes that is not binding to the target RNA causes a high background signal and disturb detection of the target RNA. The fluorescent protein reconstitution technique is a strategy to reduce the background fluorescence signal from excess probe molecules [18, 19].
5. For the fluorescent protein reconstitution technique, N-terminal and C-terminal fragments of EGFP (GN and GC, respectively) are used. When the two fragments become mutually close, they cause a reconstitution reaction to form the structure of the full-length protein and regain their original fluorescence [19–21]. Using this technique, RNA probes can be designed to emit fluorescence just upon binding to the target RNA. Although the approach on fluorescent protein reconstitution is effective to eliminate background fluorescence and to improve the signal-to-background ratio in the obtained images, this approach has some limitations. An important limitation is the slow rate of the reconstitution reaction. The reaction of the pair of EGFP fragments, from mutual approach

to completing the fluorophore formation, takes tens of minutes, which possibly prevents a precise time-course analysis of the target molecules. Another limitation is on the necessity to introduce two expression plasmids of the two fluorescent protein fragments into the sample cells. The amount of reconstituted fluorescent protein generated in individual sample cells is dependent on the expression levels of the respective fragments. Controlling the expression of the respective two fragments in individual cells is difficult when using a general technique on a plasmid transfection method. This limitation might induce misconceptions from quantitative analyses, especially of single cells.

6. One method to avoid these limitations is to use a full-length fluorescent protein with a very low expression level [10]. The full-length fluorescent protein approach requires only a single expression vector to generate the probe in target cells. As one might expect, excess probe containing a full-length fluorescent protein causes a severe background signal that critically hampers the detection of target molecules. Therefore, the use of full-length fluorescent protein-based probe requires stricter control of the expression level of the probe. Consequently, although an option to use full-length fluorescent protein based probes is applicable, a split fluorescent protein reconstitution approach is the prime choice for RNA probe design.
7. Based on these methods and principles, we designed TERRA probe, in which GN and GC were fused respectively to the N-terminal and C-terminus of mPUMt [12]. This probe alone does not emit fluorescence. In the presence of TERRA, which has a telomere repeat region, multiple probe molecules bind to the repetitive region. Then, a GN portion and GC portions in two adjacent probes cause a reconstitution reaction to yield full-length EGFP. Consequently, the present probe fluorescently labels TERRA with low background. This strategy is expected to be applicable to visualize various RNAs including repetitive sequences. The following sections present our protocols of simultaneous observation of TERRA, telomeres, and hnRNPA1 using the present TERRA probe (NLS-GN-mPUMt-GC, Fig. 2), a telomere probe (iRFP-TRF1), and hnRNPA1-SNAPf.

3.2 Cell Sample Preparation

3.2.1 Preparation of the iRFP-TRF1 Expressing U2OS Cell Line Through Viral Infection

1. Transfect the iRFP-TRF1/pCLNCX and the packaging vector into Gag293 cells using Lipofectamine 2000™.
2. Change the cultivation medium to D-MEM containing 10% FBS 8 h after transfection.
3. 24 h after transfection, take the supernatant containing viral particles and mix it with 50% fresh medium and polybrene.

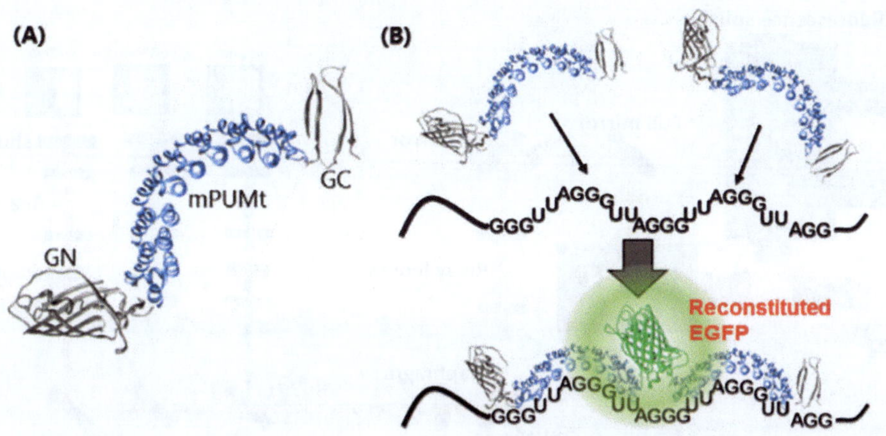

Fig. 2 Schematic of the principle of the present TERRA probe: (**A**) illustration of the probe, which consists of an N-terminal EGFP fragment (GN); a PUM-HD mutant mPUMt that is modified to recognize an RNA of a UUAGGGUU sequence, and a C-terminal EGFP fragment; (**B**) mechanism of the probe function to label TERRA selectively (Reproduced from [12] with permission from Nature Publishing Group)

4. Place the virus-containing mixture to U2OS cells for infection.
5. After 4 h of infection, remove the medium and replace it with fresh D-MEM containing 10% FBS.
6. Culture the infected U2OS cells in DMEM containing 10% FBS.

3.2.2 Preparation of the Observed Cell Sample Expressing iRFP-TRF1, hnRNPA1-SNAPf, and the TERRA Probe

1. Culture the infected U2OS cells in DMEM containing 10% FBS on glass-bottomed dishes.
2. Introduce plasmids of hnRNPA1-SNAPf/pcDNA3.1(+) and NLS-GN-mPUMt-GC/pcDNA3.1(+) into the cells using Lipofectamine LTX™ according to the manufacturer's protocol (*see* **Note 3**).
3. Incubate the cells for 4–6 h.
4. Incubate the cells with 0.1 μM SNAP-TMR for 30 min in a CO_2 incubator to stain hnRNPA1-SNAPf.
5. Wash the cells with DMEM containing 10% FBS to eliminate unbound SNAP-TMR.
6. Replace the cell cultivation medium with imaging medium.
7. Place the sample onto the microscope stage after setting the microscope for observations.

3.3 Microscopy

3.3.1 Construction and Tuning of the Microscope Setup

1. Place optical components as Fig. 3 shows *.
2. Modulate the diaphragms' height to make their centers identical to that of optical port of the microscope ($H = 193.5$ mm for an Olympus IX 81; Olympus Optical Inc.).

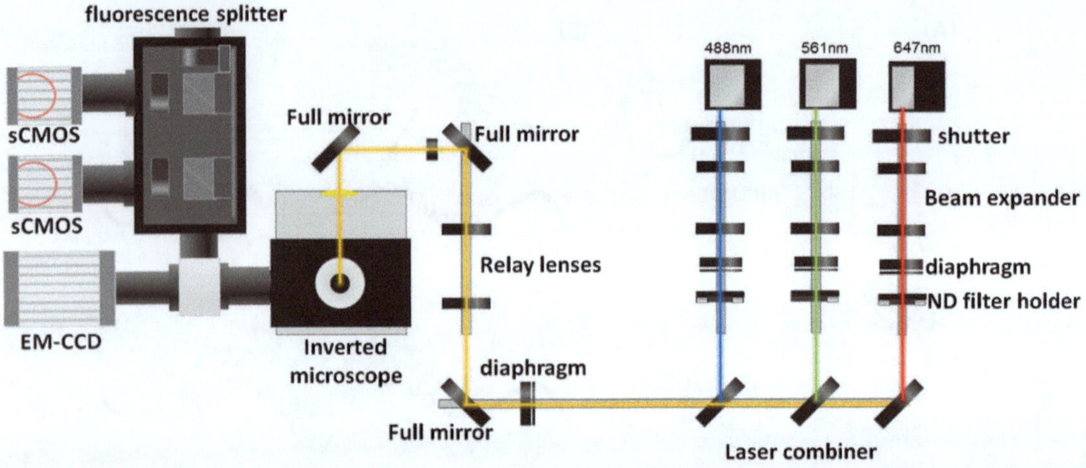

Fig. 3 Layout of optical setup in the microscope system to monitor the dynamics of TERRA, telomeres, and hnRNPA1 in living cells

3. Put all lenses off. Modulate the positions and angles of the lasers and mirrors to make the laser light pass through the centers of the respective diaphragms. Then the laser light goes horizontally at the height of the optical port of the microscope.
4. Put the expanders and relay lenses into the laser lines and optimize the line in the same manner as done before putting the lenses.
5. Tune on the cameras and modulate the laser position to be the center of the field of view.

3.3.2 Observation

1. Turn on the excitation lasers.
2. Place appropriate neutral density filters in respective laser lines.
3. Modulate the full mirror angle to have the laser light pass vertically from the objective.
4. Modulate the distance between the beam expander lenses to make the laser light be a parallel beam.
5. Place a sample glass-bottom dish on the microscope stage.
6. Make the incident angle of the laser light be large to satisfy the TIRF condition.
7. Focus at the upper surface of the glass bottom. Small particles of fluorescent substances adsorbing on the surface of the glass will be detected.
8. Modulate the incident angle of the excitation laser light to be slightly smaller to make oblique illumination (*see* **Note 4**).
9. Optimize the focus position roughly to observe intracellular RNAs. Find a cell expressing the probes at an appropriate level.

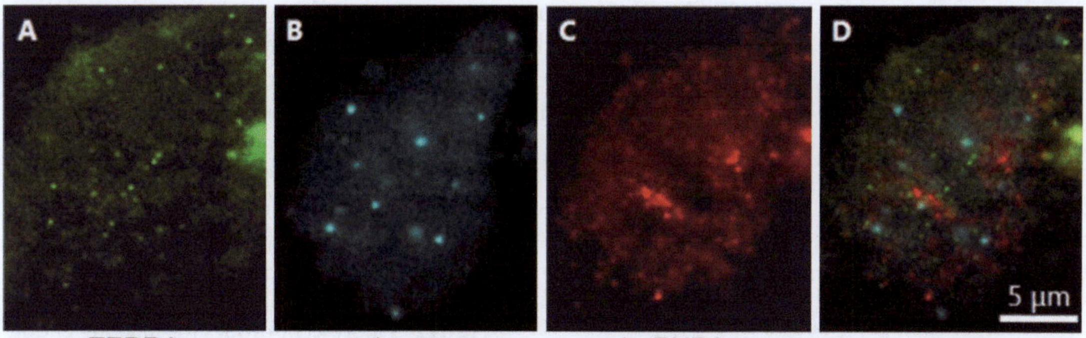

Fig. 4 Images of TERRA using present methods: (**A**) TERRA labeled with the present probe, (**B**) iRFP-TRF1 that represent telomeres, (**C**) hnRNPA1-SNAP-TMR, and (**D**) merge image

10. After finding an appropriate cell, fine-tune the focus to an appropriate position for observation of the target RNA.
11. Capture images (Fig. 4) and movies.

4 Notes

1. The expression plasmids and their sequence information are available from the authors' lab (http://www.chem.s.u-tokyo.ac.jp/~analyt/en/home_en.html).
2. The dissociation constant of PUM-HD mutants to the recognized 8-base RNA is largely dependent on the three RNA bases at the 5′ terminus [8]. When the three bases are 5′-UGU-3′, the dissociation constant is reported to be nanomolar, whereas the three-base sequence is not UGU, the dissociation constant is approximately 100 nM. In case of mPUMt, the initial three-base sequence is UUA. The dissociation constant was 100 nM.
3. A short incubation time after transfection of the TERRA probe is recommended to avoid excess expression of the probe. Given the irreversible reaction of EGFP reconstitution, overexpression of the probe molecule results in accumulation of reconstituted EGFP, which outnumbers TERRA and which generates a severe background fluorescence signal in the observation. Therefore, the expression level of the probe is expected to be sufficiently small by making the incubation time after the transfection much shorter than described in the transfection reagent manufacturer's protocols.
4. In oblique illumination, making the light thin results in a high signal-to-noise ratio because background fluorescence from out-of-focus regions is eliminated, whereas the visible field area becomes small [22]. In case of our TERRA observation, the target RNA localizes in the nucleus. We made the

diaphragm narrow to generate a thin excitation laser beam. When cytoplasmic RNA is the target of the observation, excitation light with larger diameter will be better to observe the entire cytoplasmic region simultaneously.

References

1. Martin KC, Ephrussi A (2009) mRNA localization: gene expression in the spatial dimension. Cell 136(4):719–730. doi:10.1016/j.cell.2009.01.044. S0092-8674(09)00126-3 [pii]
2. Buxbaum AR, Haimovich G, Singer RH (2015) In the right place at the right time: visualizing and understanding mRNA localization. Nat Rev Mol Cell Biol 16(2):95–109. doi:10.1038/nrm3918. http://www.nature.com/nrm/journal/v16/n2/abs/nrm3918.html#supplementary-information
3. Geisler S, Coller J (2013) RNA in unexpected places: long non-coding RNA functions in diverse cellular contexts. Nat Rev Mol Cell Biol 14(11):699–712. doi:10.1038/nrm3679
4. Heward JA, Lindsay MA (2014) Long non-coding RNAs in the regulation of the immune response. Trends Immunol 35(9):408–419. doi:10.1016/j.it.2014.07.005
5. Engreitz Jesse M, Sirokman K, McDonel P, Shishkin AA, Surka C, Russell P, Grossman Sharon R, Chow Amy Y, Guttman M, Lander Eric S (2014) RNA-RNA interactions enable specific targeting of noncoding RNAs to nascent pre-mRNAs and chromatin sites. Cell 159(1):188–199. doi:10.1016/j.cell.2014.08.018
6. Oomoto I, Suzuki-Hirano A, Umeshima H, Han Y-W, Yanagisawa H, Carlton P, Harada Y, Kengaku M, Okamoto A, Shimogori T, Wang DO (2015) ECHO-liveFISH: in vivo RNA labeling reveals dynamic regulation of nuclear RNA foci in living tissues. Nucleic Acids Res. doi:10.1093/nar/gkv614
7. Wang X, McLachlan J, Zamore PD, Hall TMT (2002) Modular recognition of RNA by a human pumilio-homology domain. Cell 110(4):501–512. doi:10.1016/S0092-8674(02)00873-5
8. Cheong CG, Hall TMT (2006) Engineering RNA sequence specificity of pumilio repeats. Proc Natl Acad Sci U S A 103(37):13635–13639. doi:10.1073/pnas.0606294103
9. Yamada T, Yoshimura H, Inaguma A, Ozawa T (2011) Visualization of nonengineered single mRNAs in living cells using genetically encoded fluorescent probes. Anal Chem 83(14):5708–5714. doi:10.1021/ac2009405
10. Yoshimura H, Inaguma A, Yamada T, Ozawa T (2012) Fluorescent probes for imaging endogenous beta-actin mRNA in living cells using fluorescent protein-tagged pumilio. ACS Chem Biol 7(6):999–1005. doi:10.1021/cb200474a
11. Yoshimura H, Ozawa T (2016) Chapter 3: Monitoring of RNA dynamics in living cells using PUM-HD and fluorescent protein reconstitution technique. In: Grigory SF, Samie RJ (eds) Methods in enzymology, vol 572. Academic Press, New York, pp 65–85. doi:10.1016/bs.mie.2016.03.018
12. Yamada T, Yoshimura H, Shimada R, Hattori M, Eguchi M, Fujiwara TK, Kusumi A, Ozawa T (2016) Spatiotemporal analysis with a genetically encoded fluorescent RNA probe reveals TERRA function around telomeres. Sci Rep 6:38910. doi:10.1038/srep38910
13. Green KM, Linsalata AE, Todd PK (2016) RAN translation—what makes it run? Brain Res 1647:30–42. doi:10.1016/j.brainres.2016.04.003
14. Matsui M, Corey DR (2016) Non-coding RNAs as drug targets. Nat Rev Drug Discov. Advance online publication. doi:10.1038/nrd.2016.117
15. Faulkner GJ, Kimura Y, Daub CO, Wani S, Plessy C, Irvine KM, Schroder K, Cloonan N, Steptoe AL, Lassmann T, Waki K, Hornig N, Arakawa T, Takahashi H, Kawai J, Forrest ARR, Suzuki H, Hayashizaki Y, Hume DA, Orlando V, Grimmond SM, Carninci P (2009) The regulated retrotransposon transcriptome of mammalian cells. Nat Genet 41(5):563–571. http://www.nature.com/ng/journal/v41/n5/suppinfo/ng.368_s1.html
16. Vardy L, Orr-Weaver TL (2007) Regulating translation of maternal messages: multiple repression mechanisms. Trends Cell Biol 17(11):547–554. doi:10.1016/j.tcb.2007.09.002
17. Filipovska A, Razif MF, Nygard KK, Rackham O (2011) A universal code for RNA recognition by PUF proteins. Nat Chem Biol 7(7):425–427. doi:10.1038/nchembio.577

18. Ozawa T, Yoshimura H, Kim SB (2013) Advances in fluorescence and bioluminescence imaging. Anal Chem 85(2):590–609. doi:10.1021/ac3031724

19. Yoshimura H, Ozawa T (2014) Methods of split reporter reconstitution for the analysis of biomolecules. Chem Rec 14(3):492–501. doi:10.1002/tcr.201402001

20. Magliery TJ, Wilson CGM, Pan W, Mishler D, Ghosh I, Hamilton AD, Regan L (2005) Detecting protein–protein interactions with a green fluorescent protein fragment reassembly trap: scope and mechanism. J Am Chem Soc 127(1):146–157. doi:10.1021/ja046699g

21. Ozawa T, Nogami S, Sato M, Ohya Y, Umezawa Y (2000) A fluorescent indicator for detecting protein–protein interactions in vivo based on protein splicing. Anal Chem 72(21):5151–5157. doi:10.1021/ac000617z

22. Tokunaga M, Imamoto N, Sakata-Sogawa K (2008) Highly inclined thin illumination enables clear single-molecule imaging in cells. Nat Methods 5(2):159–161. doi:10.1038/nmeth1171

Chapter 23

Live Imaging of mRNA Synthesis in *Drosophila*

Hernan G. Garcia and Thomas Gregor

Abstract

mRNA synthesis is one of the earliest readouts of the activity of a transcribed gene, which is of particular interest in the context of metazoan cell fate specification. These processes are intrinsically dynamic and stochastic, which makes in vivo single-cell measurements inevitable. Here, we present the application of a technology that has been widely used in single celled organisms to measure transcriptional activity in developing embryos of the fruit fly *Drosophila melanogaster*. The method allows for quantification of instantaneous polymerase occupancy of active gene loci and thereby enables the development and testing of models of gene regulation in development.

Key words Nascent mRNA, Transcription activity, Quantitative live imaging, Fluorescence, *Drosophila*

1 Introduction

The observation of mRNA transcript production in real time has become a familiar routine in the context of single cells [1–3]. This method entails the use of an mRNA tagging system in which nascent transcripts are tagged with multiple repeats of a stem loop sequence that is recognized by a cognate binding protein (Fig. 1a). The latter is constitutively expressed and fused to a fluorescent protein that can be visualized using standard live microscopy techniques [4–6]. The stem loop cluster binds multiple fluorescent proteins resulting in spatially localized fluorescence at the gene locus, which is further enhanced by each additional polymerase that is engaged in transcriptional elongation (Fig. 1b). In the fruit fly *Drosophila melanogaster*, this MS2 system has been implemented to study maternal mRNA transport in oocytes [7] and transcription [8–13].

Here we present a detailed protocol to visualize and quantify nascent transcripts in living fly embryos using the MS2 stem loop system (Fig. 1c). Details are provided regarding the transgenic

Imre Gaspar (ed.), *RNA Detection: Methods and Protocols*, Methods in Molecular Biology, vol. 1649, DOI 10.1007/978-1-4939-7213-5_23, © Springer Science+Business Media LLC 2018

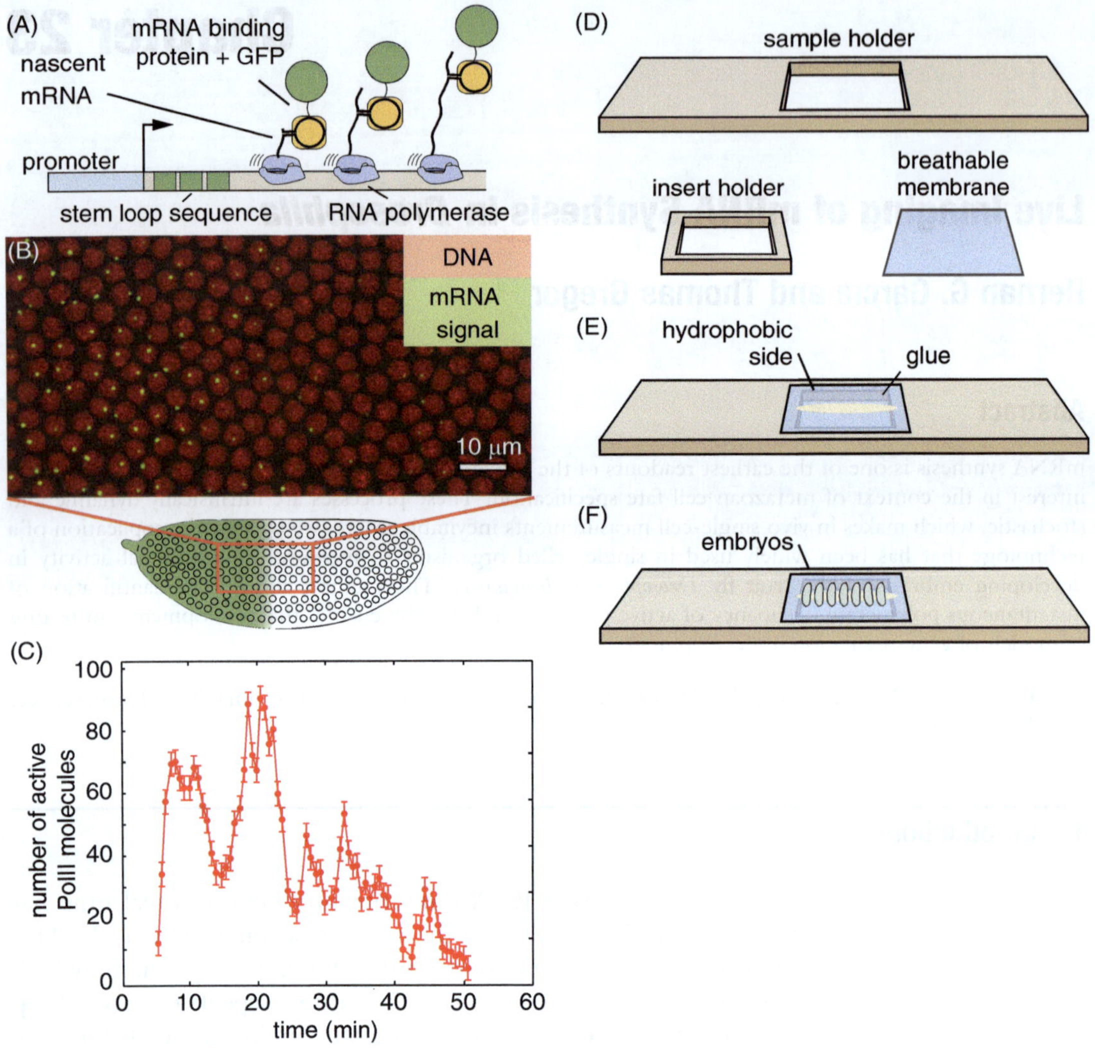

Fig. 1 Quantifying transcriptional dynamics in live fly embryos. (**a**) 24 repeats of the MS2 loop sequence are added to a gene that, when transcribed, fold into a stem loop that is recognized by an mRNA binding protein fused to GFP; fluorescence is proportional to transcriptional activity. (**b**) Typical field of view showing sites of transcription in single nuclei. (**c**) Number of actively transcribing Pol II molecules as a function of time for a labeled site of nascent transcript formation. (**d–f**) Sample holder and breathable membrane for embryo mounting

construct design and implementation, embryo handling for live imaging, and various suitable microscopy techniques. We discuss confocal and two-photon imaging conditions for optimal high quality images for quantitative analysis. In particular, we discuss calibration of the fluorescence signal to obtain absolute units in terms of numbers of actively transcribing RNA Polymerase II (PolII) molecules.

2 Materials

2.1 Creating MS2 Reporter Constructs and Transgenic Fly Lines

1. Plasmid pCR4-24XMS2SL-stable [4] (Addgene #31865) to make a new MS2 reporter construct.
2. Plasmids pIB-hbP2-P2P-lacZ-αTUb3′UTR [14] or pBφ-eve2-MS2-yellow [10] for plasmids ready for transgenesis using recombination mediated cassette exchange (RMCE) [15] or single attP site integration [16], respectively.
3. Transformation competent *E. coli* (e.g., XL1-Blue).
4. Stbl2 competent *E. coli* cells (Life Technologies 10268-019).
5. Bloomington *Drosophila* stock center fly lines #27388 (for RMCE) or #9750 (single attP).
6. In-house microinjection setup to generate transgenics or third-party company such as Bestgene (www.thebestgene.com) or Rainbow Transgenics (www.rainbowgene.com).

2.2 Fly Lines

1. yw;His2Av-mRFP;Pnos-MCP-EGFP [8] from Bloomington *Drosophila* Stock Center (#60340) expresses MCP-GFP and Histone-RFP maternally.
2. yw;P2P-MS2-lacZ and yw;P2P-lacZ reporter fly lines [8] for absolute calibration of PolII molecules actively transcribing (available upon request).

2.3 Embryo Collection

1. Apple juice agar plates.
2. Active dry yeast (e.g., Fleischmann's ADY 2192).
3. Halocarbon oil 27 (e.g., Sigma-Aldrich).
4. Absorbent paper towels.
5. 20% bleach.
6. Distilled or reverse osmosis water.

2.4 Embryo Mounting

1. Breathable Lumox Film (e.g., Sarstedt 94.6077.317).
2. Double-sided sticky tape.
3. 50 mL conical tubes.
4. 1.5 mL Eppendorf tubes.
5. Heptane.
6. Platform rocker (Nutator).
7. Tabletop centrifuge.
8. Scintillation vial.
9. Embryo and membrane slide holder (http://www.sculpteo.com/en/gallery/public/hggarcia/).
10. Dumont #5 forceps.
11. Round brush (size 0 or smaller).

2.5 Imaging

1. Custom-built 2-photon microscope [17] with a Zeiss 25× 0.8 NA objective, GaAsP photomultipliers (Hamamatsu H10770PA-40 SEL), Pockel Cell or neutral density filters to modulate laser output, Coherent power meter #1159770 and ScanImage control software (www.scanimage.org).
2. Confocal microscope with high sensitivity photodetectors. We use a Leica SP8 with 63×, 1.4 NA objective, Argon and Helium-Neon lasers, or white light laser and HyD detectors and a Zeiss 780 with 40×, 1.4 NA objective, Argon and Helium-Neon lasers and HyD detectors.
3. Diagnostic slides for measuring flat field (Chroma #92001).

2.6 Data Analysis

1. Mathworks Matlab.
2. FlyRNAQuant(github.com/PrincetonUniversity/FlyRNAQuant).

3 Methods

The protocol begins with the creation of transgenic *Drosophila* lines bearing stem loop-tagged reporter constructs. The mounting of embryos onto a custom-made slide sample holder (Fig. 1d–f) is described. This sample holder makes it possible to flatten the embryos and have as many nuclei as possible in the same focal plane. Lastly, the imaging protocol for the measurement of transcriptional activity in live embryos is described.

3.1 Creating DNA Reporter Molecules

1. Use the unique restriction sites in pIB-hbP2-P2P-lacZ-α TUb3′UTR (RMCE) or pBφ-eve2-MS2-yellow (single attP insertion) to insert new regulatory regions using regular cutting-and-pasting with restriction enzymes or using Gibson assembly (vector maps available at benchling.com/garcialab).
2. Transform the newly generated plasmids into supercompetent cells such as XL1-Blue.
3. Some of the MS2 stem loops can be lost during the previous step. Before sequencing colonies screen them by digesting candidate plasmids using restriction enzymes such as EcoRV and running an analytical gel. Send for sequencing with primers for the new insert and for the stem loops.
4. Once sequencing is confirmed transform the plasmid into Stbl2 cells for archival purposes.
5. Generate transgenic flies by injecting the newly generated vector in-house or through a company.

3.2 Embryo Glue

1. Densely pack a 50 mL conical tube with strips of double-sided sticky tape.
2. Fill the tube with heptane.

3. Mix the tube overnight on a nutator.
4. Using forceps extract the tape and pipette the heptane into Eppendorf tubes.
5. Spin down the tubes at maximum speed on a tabletop centrifuge.
6. Pipette the supernatant into a scintillation vial and store at 4 °C.

3.3 Preparing the Sample Holder for Live Imaging of Embryos

1. Identify the hydrophobic side of the permeable membrane using a sharpie: it will be harder to write on the hydrophobic side.
2. Cut a 2 cm × 2 cm square.
3. Mount the membrane into the membrane holder as shown in Figs. 1d and e.
4. Apply a thin line of glue to the membrane (Fig. 1e). This glue will avoid embryo rolling during sample preparation.

3.4 Preparing Transgenic Embryos for Live Imaging

1. Two days before imaging prepare a cage of flies containing 50–100 virgin females of His-RFP;MCP-GFP flies and 20–40 males containing the reporter construct.
2. 90 min before preparing the embryos replace the plate in the cage for a new one with fresh yeast.
3. After the waiting time replace the plate again.
4. Cover plate with halocarbon oil and image embryos using a dissecting scope with trans-illumination (*see* **Note 1**).
5. Using the forceps pick early embryos [18] and transfer them to a small (1 cm × 1 cm) cutout of a paper towel.
6. Cover with bleach for 1 min.
7. Absorb bleach using a paper towel and wash with water for 1 min.
8. Absorb water and transfer paper towel to a petri dish with water. Undamaged embryos will float.
9. Using the brush transfer embryos to a small dry cutout of paper towel.
10. Pick embryos one by one in the right orientation using the brush and place them on the glue on the membrane (Fig. 1f) (*see* **Note 2**).
11. Put two drops of halocarbon oil on embryos and place an 18 × 18 mm 1.5 cover glass on them. Let the sample sit for 1 min.
12. Inspect the samples for embryo flattening. If embryos need further flattening use the side of a Kimwipes to absorb oil from the side of the cover glass. Monitor cover glass so that it does not slide and the embryos so that they do not explode.

13. This sample can be used both on an upright or inverted microscope configuration.

3.5 Setting Up the Microscopes for Imaging

Turn on the microscope system and lasers at least 90 min before imaging the embryos.

3.5.1 2-Photon Microscopy for Live Imaging

1. Set the TiSa laser wavelength to 970 nm.
2. Set up power meter at the objective exit.
3. Park scanner or set it to a high zoom of 80 or larger and scan.
4. Measure power and use the Pockel Cell or the neutral density filter to adjust it to 10 mW.
5. Set scanning to a pixel size of 220 nm, 512 pixels per line, 256 lines per frame, and a frequency of 250 Hz per line (4 ms/line).
6. Set Z-stack continuous acquisition to ten slices separated by 1 μm, each slice is averaged three times.

3.5.2 Confocal Microscopy for Live Imaging

1. Use a 488 nm and a 561 nm laser excitation lines.
2. Set up power meter at the 10× objective exit, zoom to 40×, and find the right laser power settings for a 35 μW output power in 488 nm and 20 μW output power in 561 nm.
3. Set scanning to a pixel size of 220 nm, 512 pixels per line, 256 lines per frame and a frequency of 400 Hz per line and bidirectional scanning.
4. Set Z-stack continuous acquisition to 21 slices separated by 0.5 μm, each slice is averaged three times.

3.6 Live Imaging

1. Find the embryos using brightfield illumination and save their positions using an automated XY stage.
2. Using fluorescence illumination look for an embryo whose nuclei are just migrating to the surface.
3. Find the nuclei and transcription spots, and start a continuous acquisition by centering the Z-stack on the middle of the nuclei (*see* **Notes 3** and **4**).
4. The sample should be monitored regularly as nuclei and their transcription spots tend to drift during acquisition. Adjust the center of the Z-stack as needed throughout the experiment.
5. Once done with the acquisition take an image of the full embryo image in order to determine the correct AP position of the zoomed-in image and to check for photobleaching (*see* **Note 5**).
6. Using the same acquisition parameters of the Z-stack measure the flat field using diagnostic slides for measuring flat field (*see* **Note 6**).

3.7 Live Data Analysis

1. Download the FlyRNAQuant software and manual.
2. Install the software as instructed in the manual and download a sample data set.
3. Follow the analysis of the sample data set all the way to the end to ensure the full functionality of the code.
4. Transfer the obtained data to the analysis computer with analysis software.
5. Go through the different steps detailed in the manual for the automated analysis.
6. If single cell information is required, particular attention needs to be paid to the curation step of the traces.

3.8 Calibration of Absolute Number of Active PolII Molecules

1. Measure the transcriptional activity of reporter fly line P2P-MS2-lacZ as described in Subheadings 3.3, 3.4, 3.6, and 3.7.
2. Integrate the fluorescence over nuclear cycle 13 to infer the number of total number of mRNA molecules produced as described in [8].
3. Measure the number of mRNA molecules produced by the reporter line P2P-lacZ in the cytoplasm during mitosis 12 and mitosis 13 using the single-molecule FISH (smFISH) protocol described in Chapter 8 of this book [19].
4. Take the difference of mRNA molecules obtained by smFISH during mitosis 12 and 13 and compare it to the quantity obtained by live imaging during the same developmental stage in order to calibrate the arbitrary fluorescence units of the MS2-MCP-EGFP reporter [8].

4 Notes

1. The halocarbon oil allows to easily image through the embryos and score their developmental time.
2. Mounting a typical amount of 16 embryos usually ensures finding several in the right orientation and developmental stage for imaging.
3. If no spots or low fluorescent signal is detected, check and/or increase laser power. Image the flat field slide as a means to make sure that the detectors are working properly.
4. If the imaging conditions are checked and the problem of no/low signal persists, it is possible that the MS2 stem loops were lost during the cloning procedure. Check the number of loops as in **step 3** of Subheading 3.1 both in the transgenic fly line, and the plasmid originally used to generate this line. If necessary, rescreen bacterial colonies.

5. If pronounced photobleaching on the MCP-GFP channel is detected reduce the laser power and/or reduce the frequency of acquisition.
6. The laser power will have to be reduced dramatically so as to not saturate and damage the detectors.

Acknowledgments

The authors are grateful through support provided by the Princeton Dickie Fellowship (H.G.), the Burroughs Wellcome Fund Career Award at the Scientific Interface (H.G.), the Sloan Research Foundation (H.G.), the Searle Scholars Program (T.G. and H.G.), the Shurl & Kay Curci Foundation (H.G.), the Human Frontier Science Program (H.G.), and the National Institutes of Health Grants P50 GM071508 and R01 GM097275 (T.G.).

References

1. Fusco D, Accornero N, Lavoie B, Shenoy SM, Blanchard JM, Singer RH, Bertrand E (2003) Single mRNA molecules demonstrate probabilistic movement in living mammalian cells. Curr Biol 13(2):161–167
2. Golding I, Paulsson J, Zawilski SM, Cox EC (2005) Real-time kinetics of gene activity in individual bacteria. Cell 123(6):1025–1036
3. Larson DR, Zenklusen D, Wu B, Chao JA, Singer RH (2011) Real-time observation of transcription initiation and elongation on an endogenous yeast gene. Science 332 (6028):475–478. doi:10.1126/science.1202142. 332/6028/475 [pii]
4. Bertrand E, Chartrand P, Schaefer M, Shenoy SM, Singer RH, Long RM (1998) Localization of ASH1 mRNA particles in living yeast. Mol Cell 2(4):437–445. S1097-2765(00)80143-4 [pii]
5. Yunger S, Rosenfeld L, Garini Y, Shav-Tal Y (2010) Single-allele analysis of transcription kinetics in living mammalian cells. Nat Methods 7(8):631–633. doi:10.1038/nmeth.1482. nmeth.1482 [pii]
6. Lionnet T, Czaplinski K, Darzacq X, Shav-Tal Y, Wells AL, Chao JA, Park HY, de Turris V, Lopez-Jones M, Singer RH (2011) A transgenic mouse for in vivo detection of endogenous labeled mRNA. Nat Methods. doi:10.1038/nmeth.1551. nmeth.1551 [pii]
7. Forrest KM, Gavis ER (2003) Live imaging of endogenous RNA reveals a diffusion and entrapment mechanism for nanos mRNA localization in Drosophila. Curr Biol 13 (14):1159–1168
8. Garcia HG, Tikhonov M, Lin A, Gregor T (2013) Quantitative imaging of transcription in living Drosophila embryos links polymerase activity to patterning. Curr Biol 23 (21):2140–2145. doi:10.1016/j.cub.2013.08.054
9. Lucas T, Ferraro T, Roelens B, De Las Heras Chanes J, Walczak AM, Coppey M, Dostatni N (2013) Live imaging of bicoid-dependent transcription in Drosophila embryos. Curr Biol 23 (21):2135–2139. doi:10.1016/j.cub.2013.08.053
10. Bothma JP, Garcia HG, Esposito E, Schlissel G, Gregor T, Levine M (2014) Dynamic regulation of eve stripe 2 expression reveals transcriptional bursts in living Drosophila embryos. Proc Natl Acad Sci U S A 111(29):10598–10603. doi:10.1073/pnas.1410022111
11. Bothma JP, Garcia HG, Ng S, Perry MW, Gregor T, Levine M (2015) Enhancer additivity and non-additivity are determined by enhancer strength in the Drosophila embryo. Elife 4. doi:10.7554/eLife.07956
12. Ferraro T, Esposito E, Mancini L, Ng S, Lucas T, Coppey M, Dostatni N, Walczak AM, Levine M, Lagha M (2016) Transcriptional memory in the Drosophila embryo. Curr Biol 26(2): 212–218. doi:10.1016/j.cub.2015.11.058
13. Fukaya T, Lim B, Levine M (2016) Enhancer control of transcriptional bursting. Cell. doi:10.1016/j.cell.2016.05.025

14. Chen H, Xu Z, Mei C, Yu D, Small S (2012) A system of repressor gradients spatially organizes the boundaries of bicoid-dependent target genes. Cell 149(3):618–629. doi:10.1016/j.cell.2012.03.018
15. Bateman JR, Lee AM, Wu CT (2006) Site-specific transformation of Drosophila via phiC31 integrase-mediated cassette exchange. Genetics 173(2):769–777. doi:10.1534/genetics.106.056945. genetics.106.056945 [pii]
16. Venken KJ, He Y, Hoskins RA, Bellen HJ (2006) P[acman]: a BAC transgenic platform for targeted insertion of large DNA fragments in D. melanogaster. Science 314(5806):1747–1751. doi:10.1126/science.1134426
17. Liu F, Morrison AH, Gregor T (2013) Dynamic interpretation of maternal inputs by the Drosophila segmentation gene network. Proc Natl Acad Sci U S A 110(17):6724–6729. doi:10.1073/pnas.1220912110
18. Figard L, Sokac AM (2011) Imaging cell shape change in living Drosophila embryos. J Vis Exp (49). doi:10.3791/2503
19. Little SC, Tikhonov M, Gregor T (2013) Precise developmental gene expression arises from globally stochastic transcriptional activity. Cell 154(4):789–800. doi:10.1016/j.cell.2013.07.025

Chapter 24

Imaging Newly Transcribed RNA in Cells by Using a Clickable Azide-Modified UTP Analog

Anupam A. Sawant, Sanjeev Galande, and Seergazhi G. Srivatsan

Abstract

Robust RNA labeling and imaging methods that enable the understanding of cellular RNA biogenesis and function are highly desired. In this context, we describe a practical chemical labeling method based on a bioorthogonal reaction, namely, azide–alkyne cycloaddition reaction, which facilitates the fluorescence imaging of newly transcribed RNA in both fixed and live cells. This strategy involves the transfection of an azide-modified UTP analog (AMUTP) into mammalian cells, which gets specifically incorporated into RNA transcripts by RNA polymerases present inside the cells. Subsequent posttranscriptional click reaction between azide-labeled RNA transcripts and a fluorescent alkyne substrate enables the imaging of newly synthesized RNA in cells by confocal microscopy. Typically, 50 μM to 1 mM of AMUTP and a transfection time of 15–60 min produce significant fluorescence signal from labeled RNA transcripts in cells.

Key words Nucleotide analog, Click chemistry, Bioorthogonal reaction, Posttranscriptional chemical modification, Azide–alkyne cycloaddition reaction, RNA labeling, RNA imaging

1 Introduction

Bioconjugation strategy based on chemoselective azide–alkyne cycloaddition reaction has become a powerful tool to functionalize RNA both in vitro and in cells [1–6]. Usually alkyne group is incorporated into RNA by chemical or enzymatic methods and further functionalization is accomplished by carrying out a copper-catalyzed azide–alkyne cycloaddition (CuAAC) reaction with the desired probe containing azide group [7–15]. This is because alkyne-containing phosphoramidite and nucleotide triphosphate (NTP) substrates used in the chemical and enzymatic RNA labeling methods are highly stable and easily accessible. However, the use of cytotoxic Cu catalysts in the cycloaddition reaction between alkyne-labeled nucleic acid and azide counterpart is a major concern, especially in the study of RNA in live cells [16].

In comparison to alkyne functionality, azide group can undergo a wider variety of bioorthogonal chemical reactions namely

Imre Gaspar (ed.), *RNA Detection: Methods and Protocols*, Methods in Molecular Biology, vol. 1649, DOI 10.1007/978-1-4939-7213-5_24, © Springer Science+Business Media LLC 2018

CuAAC, strain-promoted azide–alkyne cycloaddition (SPAAC) and azide–phosphine Staudinger ligation reactions. Although the versatility of azide group has been well exploited in labeling and imaging glycans, proteins, and lipids, incorporation of clickable azide group into nucleic acids has remained a challenge [17–21]. This is largely due to the (1) instability of azide substrates under solid-phase oligonucleotide synthesis conditions and (2) lack of effective methods to incorporate metabolically stable azide group into cellular RNA. We have recently developed a tool box of azide-modified UTP analogs, which can be conveniently introduced into RNA oligonucleotides by transcription reaction [22–24]. The azide-labeled RNAs have been subsequently functionalized post-transcriptionally with several biophysical probes by using click reactions. Notably, endogenous RNA polymerases specifically incorporate one of the nucleotide analogs, 5-azidomethyl UTP (AMUTP, Fig. 1), into cellular RNA transcripts [24]. This labeling process further facilitated the visualization of newly synthesized RNA in fixed and live cells by using CuAAC and SPAAC reactions with a suitable fluorescent alkyne substrate.

Here, we provide a detailed stepwise protocol for in situ labeling of cellular RNA with azide group and subsequently image newly transcribed RNA by using click reactions. In the first part, experimental procedure to image cellular RNA in fixed cells by using CuAAC reaction is described. In the second part, we enumerate the steps to visualize azide-labeled cellular RNA in live cells by using SPAAC reaction with a fluorescent cyclooctyne probe.

2 Materials

Prepare all solutions using autoclaved water and analytical grade reagents, and store at room temperature (unless indicated otherwise). In this work, we have used fluorescent dyes Alexa 594 alkyne and Cy3 DBCO (Fig. 1). Store these probes in a deep freezer (−20 or −40 °C). In this work, we have used the transfecting agent DOTAP (*N*-[1-(2,3-dioleoyloxy)propyl]-*N*,*N*,*N*-trimethylammonium methylsulfate), which is procured from Roche. Follow our reported stepwise procedure to prepare AMUTP from 5-azidomethyluridine (AMU), which is now available with Jena Bioscience GmbH [23, 24]. Store media and reagents required for cell culture experiments at 4 °C (unless indicated otherwise).

2.1 Buffer Solutions for Cell Culture and Click Reactions

1. HEPES buffered saline (HBS): 20 mM HEPES, 150 mM NaCl, pH 7.4. Weigh 2.6 g of HEPES and 4.38 g of NaCl in a 500 mL graduated cylinder. Allow HEPES and NaCl to dissolve completely by adding 400 mL of autoclaved water. Adjust the pH of the buffer to 7.4 by dropwise addition of 1 M NaOH and finally adjust the volume to 500 mL by adding

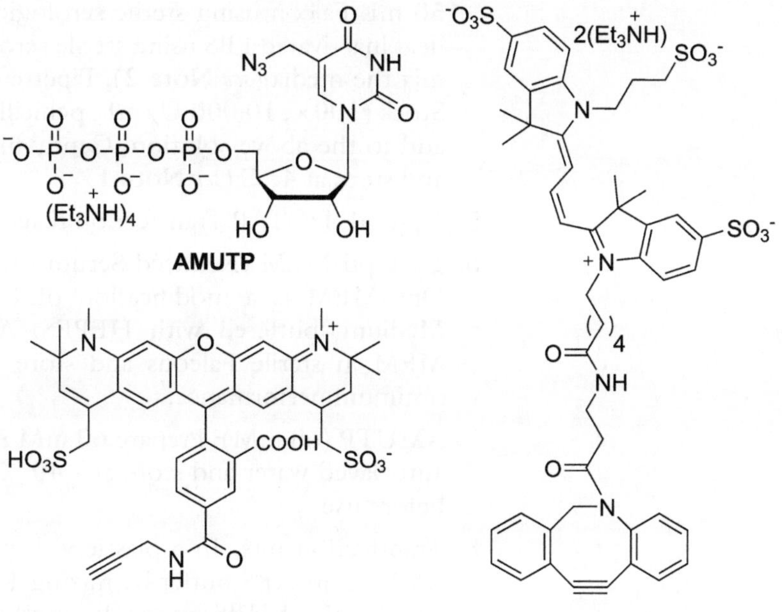

Fig. 1 Chemical structure of 5-azidomethyl UTP (AMUTP) and fluorescent dyes, Alexa 594 alkyne and Cy3 DBCO

autoclaved water. Sterilize by filtering HBS solution using sterile 0.2 μm syringe filter inside the cell culture hood and store at 4 °C (*see* **Note 1**).

2. 1× Phosphate buffer saline (1× PBS): 10 mM phosphate buffer containing 138 mM NaCl, 2.7 mM KCl, pH 7.4. Dissolve one pack of PBS powder in 1000 mL of Milli-Q water. Autoclave 1× PBS and then filter through sterile 0.2 μm syringe filter inside the cell culture hood. Aliquot 1× PBS in 50 mL sterile Falcon tubes inside biosafety cabinet and store at 4 °C (*see* **Note 1**).
3. Tris buffered saline (TBS): 50 mM Tris base, 150 mM NaCl, pH 7.4. Dissolve 6.05 g of Tris base and 8.76 g of NaCl in 1000 mL graduated cylinder containing 800 mL of autoclaved water. Adjust the pH to 7.4 by slowly adding 1 M HCl. Adjust the volume to 1000 mL by adding autoclaved water. Aliquot the TBS solution in 50 mL Falcons and store at 4 °C. Thaw TBS buffer at RT prior to further use.

2.2 Cell Culture

1. CO_2 cell culture incubator.
2. Cell culture dishes.
3. Sterile coverslips.
4. Dulbecco's modified Eagle's medium (DMEM) in 10% fetal bovine serum (FBS): Aliquot 45 mL of plain DMEM in sterile

50 mL Falcon using sterile serological pipette. Add 5 mL of heat inactivated FBS using sterile serological pipette and gently mix the media (*see* **Note 2**). Pipette out 500 μL of Pen-Strep Stock (100×, 10,000 U/mL, penicillin and streptomycin) and add to the above solution. Gently mix this cell culture media and store at 4 °C (*see* **Note 1**).

5. 1 μg/μL DOTAP Transfection reagent (Roche, *see* **Note 3**).
6. 1× Opti-MEM (Reduced Serum Minimum Essential Media): Opti-MEM is a modification of Eagle's Minimal Essential Medium, buffered with HEPES. Aliquot 40 mL of Opti-MEM in sterile Falcons and store at 4 °C in the dark for optimum performance.
7. AMUTP (60 mM): Prepare 60 mM AMUTP stock solution in autoclaved water and store at −40 °C. Thaw it on an ice bath before use.
8. Transfection mix: In a plastic vial, prepare 50 μL of 12 mM AMUTP in HBS buffer by mixing 10 μL of 60 mM AMUTP and 40 μL of HBS buffer. In another vial, prepare a 100 μL DOTAP solution containing 30 μL of 1 μg/μL DOTAP and 70 μL of HBS. Transfer the 12 mM AMUTP into the above DOTAP solution and mix well. Incubate this mix at RT for 15 min. After 15 min add 450 μL of reduced serum media (Opti-MEM) and mix gently to generate the transfection mix (final volume 600 μL and 1 mM AMUTP). Prepare this mix in biosafety cabinet and just before transfection experiments.
9. 3.7% formaldehyde: Prepare working solution of 3.7 v/v % formaldehyde solution by diluting 5 mL of 37% formaldehyde in 45 mL of 1× PBS in 50 mL sterile Falcon. Store 3.7% formaldehyde solution at room temperature.
10. Triton X 0.5% solution: Prepare a primary stock of 10 w/v % Triton X-100 solution by dissolving 1 g of Triton X-100 in 8 mL of 1× PBS in 15 mL sterile Falcon. Mix stock solution carefully by stirring the Falcon on a rotor to avoid frothing (*see* **Note 4**). Add 1× PBS to make 10 mL stock solution of 10% Triton X-100 and mix well. Prepare 50 mL working solution of 0.5% Triton X-100 by diluting 2.5 mL of 10% Triton X-100 using 45 mL 1× PBS. Store 0.5% Triton X-100 solution at room temperature.
11. Live cell imaging media: Prepare live cell imaging medium using DMEM (Life Technologies, Catalog No. 31053028) in 10% FBS. Take 45 mL of plain DMEM in a sterile 50 mL Falcon and add 5 mL of heat inactivated FBS using sterile serological pipette and gently mix the media. Store this live cell imaging medium at 4 °C (*see* **Note 1**).
12. (Optional) 2 μM actinomycin D in Opti-MEM (*see* **Note 5**).

2.3 Reagents for Azide–Alkyne Cycloaddition Reaction

1. 7.5 mM Alexa594 alkyne: Prepare a stock solution of Alexa Fluor® 594 alkyne (7.5 mM) by adding 69 μL of DMSO to 0.5 mg Alexa Fluor® 594 (Thermo Scientific). Vortex the solution, spin and store at −40 °C (*see* **Note 6**).
2. 10 mM Cy3 DBCO: Prepare a primary stock solution of 20 mM Cy3 DBCO by adding 101 μL of DMSO to 2 mg Cy3 DBCO vial (Click Chemistry Tools). Mix the content well by pipetting and vortexing and prepare 50 μL working stock of 10 mM Cy3 DBCO in DMSO. Store Cy3 DBCO stock solutions at −40 °C (*see* **Note 6**).
3. 100 mM $CuSO_4$: Freshly prepare the $CuSO_4$ solution (1 mL) using autoclaved water. Store this solution at 4 °C (*see* **Note 7**).
4. 100 mM sodium ascorbate: Freshly prepare the sodium ascorbate solution by dissolving 19.81 mg of sodium ascorbate in 1 mL of autoclaved water. Store sodium ascorbate solution at 4 °C (*see* **Note 7**).
5. CuAAC reaction mix (per coverslip): 429 μL TBS buffer, 20 μL 100 mM $CuSO_4$, 1 μL 7.5 mM Alexa594 alkyne, 50 μL 100 mM sodium ascorbate. Prepare freshly by adding ingredients in the indicated order at **step 5** of Subheading 3.2.
6. Cyclooctyne dye solution (per coverslip): 0.5 μL 10 mM Cy3 DBCO in 500 μL 1× PBS containing 0.01% Triton X. Prepare freshly at **step 2** of Subheading 3.3.
7. Triton X 0.01% solution: 0.01% Triton-X-100 in 1× PBS.
8. Washing buffer: 1× PBS containing 2 mM sodium azide.

2.4 DNA Staining Dyes

1. 1 mg/mL DAPI: Prepare DAPI stock solution using 1× PBS in an amber-colored 1.5 mL Eppendorf tube. Mix well by pipetting and vortexing until DAPI dissolves completely. Store this stock solution at 4 °C in the dark. Prepare 500 μL 2 μg/mL working solution for staining DNA freshly by diluting DAPI stock 1:500 in 1× PBS (*see* **Note 8**).
2. 20 mM Hoechst 33342: Store Hoechst 33342 stock solution at −20 °C. Thaw the vial at RT and freshly prepare 1 μM stock solution of Hoechst dye (1 mL) in the complete DMEM by serial dilution (*see* **Note 9**).

3 Methods

Reagents, buffers, media, and labware such as pipettes, plastic vials, and Falcons necessary for cell culture experiments should be sterilized by wiping with 70% aqueous ethanol and then only transferred to the cell culture hood. A flowchart of steps involved in the imaging of cellular RNA by using azide–alkyne cycloaddition reaction is provided in Fig. 2.

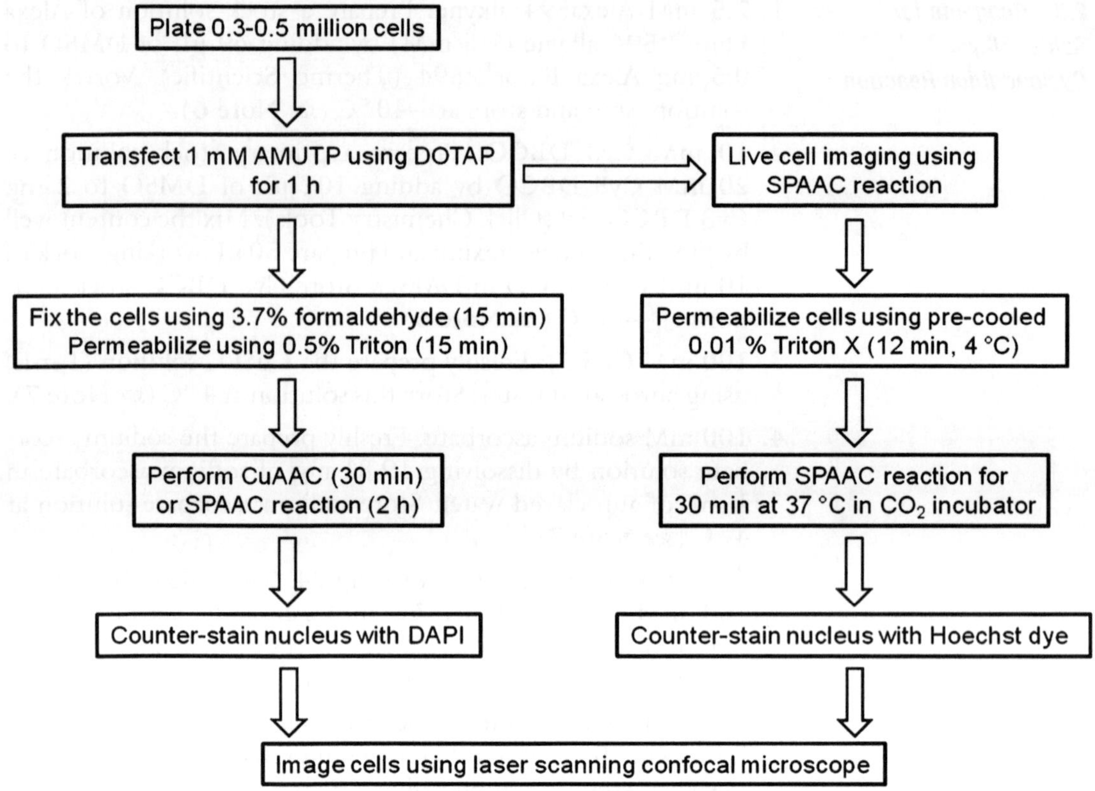

Fig. 2 Flowchart depicting the stepwise procedure for imaging newly transcribed RNA in cells by using AMUTP. Azide-labeled RNA transcripts were detected by performing CuAAC/SPAAC reactions in fixed/live cells

3.1 Transfection of AMUTP into Cells

1. Culture mammalian cells (e.g., HeLa cells) in complete DMEM supplemented with 10% FBS (*see* **Notes 1** and **3**).
2. Seed 0.3–0.5 million HeLa cells in a 6-well plate containing glass cover slips. Place the plate in the CO_2 incubator at 37 °C for 24 h (*see* **Note 10**).
3. Take a new 6-well dish and add 1 mL of HBS to each well and transfer the coverslips with cells to the individual wells with the help of a forceps.
4. Remove HBS using aspirator and carefully add 600 μL of the AMUTP transfection mix over the coverslip.
5. Gently swirl the plate in clockwise and anticlockwise direction so that the transfection mix covers the surface of the coverslip. Place the plate in the CO_2 incubator at 37 °C for 1 h (*see* **Note 3**).

3.2 Imaging RNA in Fixed Cells by Using CuAAC Reaction

1. After transfecting AMUTP for 1 h, remove the transfection mix. Wash the cells in each well three times with 1 mL of 1× PBS (*see* **Note 11**).
2. Add freshly prepared 1 mL of 3.7% formaldehyde in 1× PBS to each well and incubate for 15 min at RT.
3. Remove the fixative and wash the cells in each well twice with 1 mL of 1× PBS.
4. Add 1 mL of Triton X 0.5% solution to each well and further incubate the cells at RT for 15 min.
5. Meantime prepare fresh CuAAC reaction mix.
6. Remove the Triton X 0.5% solution (**step 4**), then wash the cells in each well with 1 mL of 1× PBS (3 × 5 min). Remove the wash solution.
7. Add 500 μL of freshly prepared CuAAC reaction mix to each well. Swirl the plate gently to insure even distribution of reaction mix over the coverslip.
8. Incubate the plate for 30 min at RT in dark.
9. Remove the reaction cocktail and wash the cells with 1 mL of wash buffer (1× PBS containing 2 mM sodium azide) for 20 min and rinse again with 1 mL of 1× PBS.
10. Counter-stain the DNA by using DAPI (500 μL, 2 μg/mL) in dark for 2 min at RT. Wash the coverslips with 1 mL of 1× PBS (*see* **Note 8**).
11. Remove the coverslip and place it upside-down on a microscope slide containing 10 μL of SlowFade Gold Antifade Mountant (Invitrogen). Seal the edges of coverslip using nail polish.
12. Image the cells using a confocal laser scanning microscope with an oil immersion objective at 40× or 63× (Fig. 3, *see* **Notes 5** and **12**).

3.3 Imaging RNA in Fixed Cells by Using SPAAC Reaction

1. Perform the steps as enumerated in Subheading 3.1 (**steps 1–6**) and 3.2 (**steps 1–4**).
2. Prepare a fresh cyclooctyne dye solution and transfer it to the coverslip.
3. Swirl the plate gently to insure even distribution of reaction mix over the coverslip.
4. Incubate the plate for 2 h at RT in dark.
5. Wash cells once with 1 mL of Triton X 0.01% solution. Further, wash the cells with 1 mL of 1× PBS for 5 min. Repeat this washing step with 1× PBS two more times.
6. Counter-stain DNA with DAPI and image the cells by following the **steps 9–12** described in Subheading 3.2 (Fig. 4).

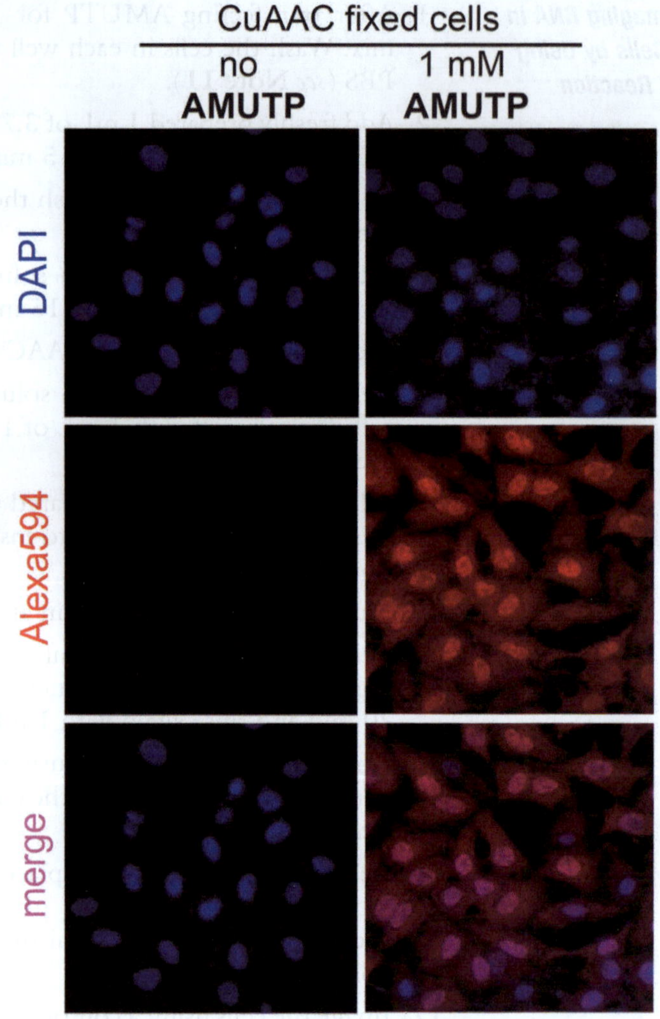

Fig. 3 Imaging cellular RNA transcription using AMUTP. Cultured HeLa cells were transfected with AMUTP (1 mM) for 1 h using DOTAP. The cells were then fixed, permeabilized and the AMU-labeling was detected by performing CuAAC reaction with Alexa 594 alkyne. This figure has been reproduced by permission of Nucleic Acids Research: Oxford Journals [24]

3.4 Imaging RNA in Live Cells by Using SPAAC Reaction

1. Culture HeLa cells in complete DMEM supplemented with 10% FBS (*see* **Note 1**).
2. Seed 0.3–0.4 million HeLa cells in a 35 mm dish having a glass surface. Place the dish in the CO_2 incubator at 37 °C for nearly 12 h (*see* **Note 13**).
3. Remove the media from the culture dish and carefully transfer 600 μL of the transfection mix containing AMUTP into the dish.

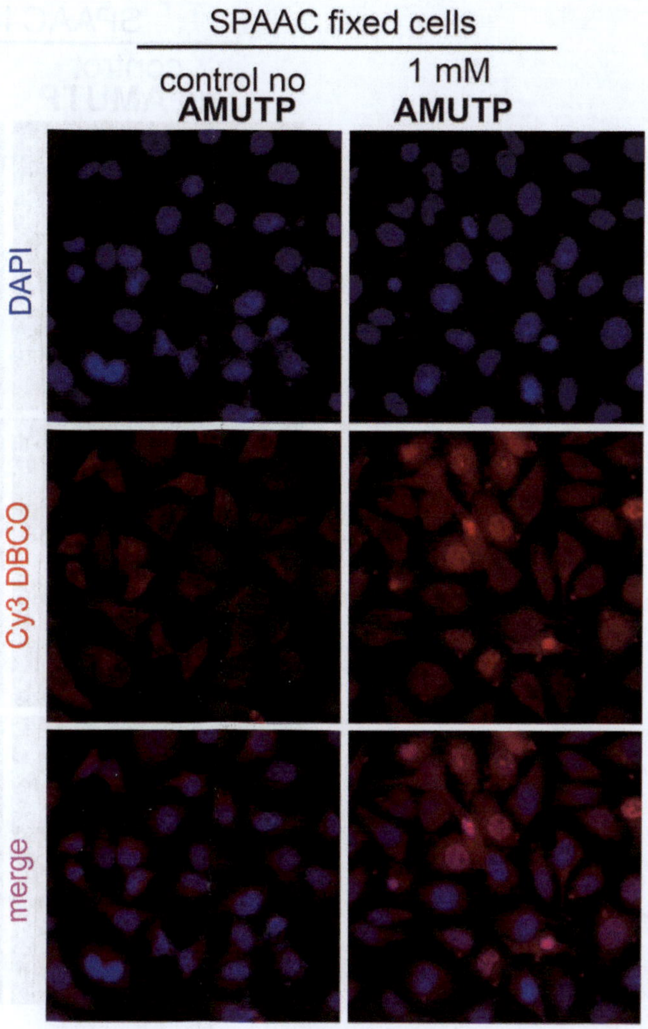

Fig. 4 Imaging cellular RNA transcription using AMUTP under copper-free SPAAC reaction conditions. HeLa cells were transfected with 1 mM of AMUTP for 1 h. The cells were fixed and stained by SPAAC reaction using Cy3 DBCO. This figure has been reproduced by permission of Nucleic Acids Research: Oxford Journals [24]

4. Gently swirl the dish in a clockwise and anticlockwise direction so that the transfection mix covers the surface of the cells. Place the dish in the CO_2 incubator at 37 °C for 1 h.
5. Wash the cells with 1 mL of complete DMEM and add 1 mL of precooled Triton X 0.01% solution containing 40 μM of cyclooctyne dye Cy3 DBCO. Incubate at 4 °C for 12 min and wash three times with 1 mL of 1× PBS.
6. Supplement the cells with 1 mL of freshly prepared DMEM containing 40 μM of Cy3 DBCO. Place the dish immediately in the incubator at 37 °C for 30 min.

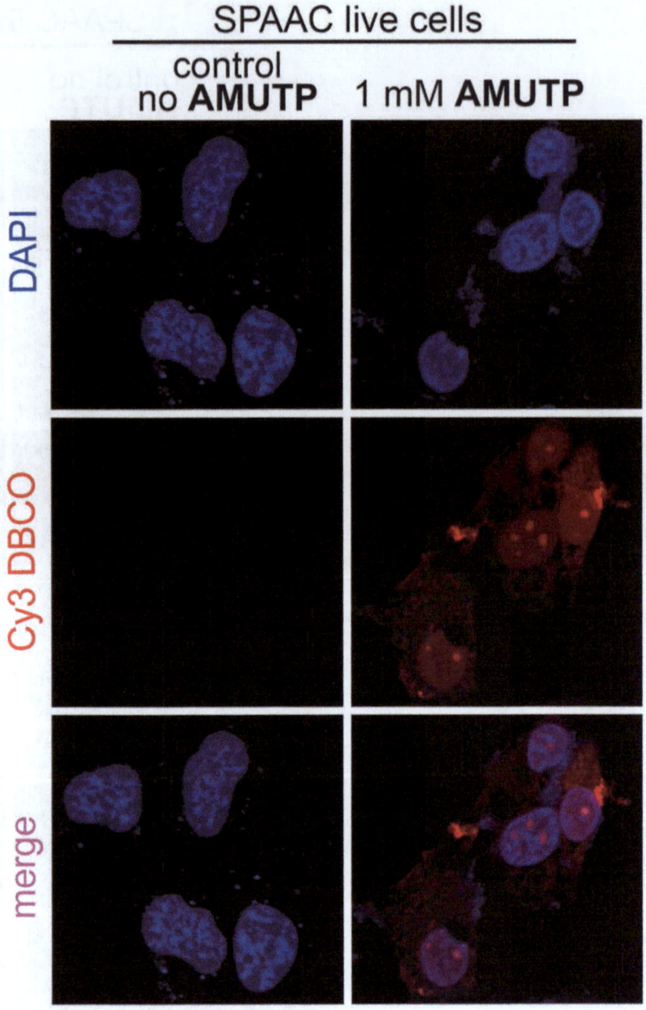

Fig. 5 Imaging cellular RNA transcription using AMUTP under copper-free SPAAC reaction conditions in live HeLa cells. The AMU-labeled RNA transcripts in live cells were detected by treating the cells with Cy3 DBCO. This figure has been reproduced by permission of Nucleic Acids Research: Oxford Journals [24]

7. After SPAAC staining (30 min at 37 °C), remove the media and add 1 mL of Hoechst 33342 (1 μM) prepared in complete DMEM to the cells. Place the dish immediately in the incubator at 37 °C for 30 min.
8. Wash the cells once with 1 mL of live cell imaging medium prepared as mentioned in Subheading 2.2.
9. Add 1 mL of live cell imaging medium and image the cells using a confocal laser scanning microscope with an oil immersion at 63× (*see* **Notes 12** and **14**, Fig. 5).

4 Notes

1. Thaw buffer solutions and media required for cell culture in the water bath (37 °C) containing autoclave water prior to use in cell culture experiments. Sterilize the bottles and Falcons by wiping with 70% ethanol, and then transfer them into the culture hood. Always open the reagents and media inside the cell culture hood only.
2. See manufacturer's instruction for thawing FBS. We usually thaw the frozen stock of the heat inactivated FBS in +4 °C fridge and complete the thawing process at room temperature. FBS should be mixed few times during the thawing process by gently swirling the bottle. Aliquot 30–40 mL and store in deep freezer. Working aliquot of FBS can be also stored at 4 °C. Avoid vigorous shaking while handling FBS as it is susceptible to frothing.
3. The following RNA labeling protocol has been developed using HeLa cells with an optimized AMUTP concentration of 1 mM and transfection time of 1 h. However, we have performed experiments to test a range of AMUTP concentrations (50 μM–4.0 mM) and transfection time (15 min–3 h) to determine the best labeling condition in HeLa cells. Apparently, longer incubation times (≥3 h) and concentrations of AMUTP ≥2 mM resulted in detectable reduction in cellular RNA staining. In our experiments, DOTAP served as a significantly better transfecting agent for AMUTP as compared to Lipofectamine. Although this protocol will also work for other cell lines, it is recommended that the concentration of AMUTP and transfection time should be optimized for each of the cell lines.
4. Detergents are main constituents of Triton X-100 and excessive shaking will lead to frothing. Therefore, handle Triton X-100 carefully to avoid cross contamination during processing cells.
5. In order to confirm the specific incorporation of AMUTP into cellular RNA, we treat the cells with actinomycin D (RNA polymerase inhibitor). Essentially, the described procedure can be adopted wherein actinomycin D (final concentration 2 μM, 600 μL) prepared in Opti-MEM is added before transfecting AMUTP. The cells are incubated in CO_2 incubator at 37 °C for 5 h. Remove 150 μL of actinomycin D solution and add freshly prepared 150 μL of transfection mix and incubate for 1 h. Perform the fixation, permeabilization, and click staining reaction as mentioned in Subheading 3.2.
6. Thaw the vials at RT and spin it well prior to use. Store the stocks of fluorescent dyes in deep freezer (−20 or −40 °C) and

avoid extended exposure to light. Dye stocks can be either wrapped using aluminum foil or stored in amber-colored tubes.

7. Degassing of DMSO and autoclaved water used for preparing the stock solutions for CuAAC reaction is recommended. However, in our cell-based experiments we have observed efficient staining by click reaction even when reagents stocks were not degassed.
8. DAPI is light sensitive therefore all stock solution preparation and staining should be performed in the dark. Store the stock solutions at 4 °C in amber-colored plastic vials.
9. Freshly prepare stock solutions in dark and inside cell culture hood. Working solution to stain DNA in live cells must be thawed in a 37 °C water bath. Make sure that the staining solution is warm (~37 °C) to avoid heat shock to cells. Store the stock solutions at −20 °C in amber-colored plastic vials.
10. Typically we use 100 mm dish for culturing the cells. Depending on the cell count prepare cell suspension in the complete media required for seeding into 6-well dish. Prior to seeding, place sterile glass coverslip in each well and pipette the media containing the cells at the center of the well. Swirl the plate very carefully to make sure uniform distribution of cells on glass coverslip. Place the plate in the incubator after properly labeling the wells. The incubation time should be optimized to get ~70% cell confluence.
11. After washing the cells with 1× PBS, the fixation, permeabilization, and click staining can be performed outside the cell culture hood.
12. Always perform a control staining reaction with cells treated with transfecting mix without AMUTP.
13. The incubation time should be optimized to get ~70% cell confluence.
14. It is best to image live cells after click and Hoechst staining immediately to minimize cell death.

Acknowledgments

This work was supported by the Department of Science and Technology, India, SERB grant (EMR/2014/000419) to S.G.S. and the Centre of Excellence in Epigenetics grant from the Department of Biotechnology, Government of India to S.G. The authors wish to thank Arun Tanpure, Progya Mukherjee, Soumitra Athavale, Rahul Jangid, and Ashwin Kelkar for discussion and help with work that has led to the optimization of these protocols.

References

1. El-Sagheer AH, Brown T (2010) New strategy for the synthesis of chemically modified RNA constructs exemplified by hairpin and hammerhead ribozymes. Proc Natl Acad Sci U S A 107 (35):15329–15334
2. Paredes E, Evans M, Das SR (2011) RNA labeling, conjugation and ligation. Methods 54(2):251–259
3. Schulz D, Rentmeister A (2014) Current approaches for RNA labeling in vitro and in cells based on click reactions. Chembiochem 15(16):2342–2347
4. Jao CY, Salic A (2008) Exploring RNA transcription and turnover in vivo by using click chemistry. Proc Natl Acad Sci U S A 105 (41):15779–15784
5. Grammel M, Hang H, Conrad NK (2012) Chemical reporters for monitoring RNA synthesis and poly (A) tail dynamics. Chembiochem 13(8):1112–1115
6. Qu D, Zhou L, Wang W, Wang Z, Wang G, Chi W, Zhang B (2013) 5-Ethynylcytidine as a new agent for detecting RNA synthesis in live cells by "click" chemistry. Anal Biochem 434 (1):128–135
7. Gramlich PM, Wirges CT, Manetto A, Carell T (2008) Postsynthetic DNA modification through the copper-catalyzed azide–alkyne cycloaddition reaction. Angew Chem Int Ed 47(44):8350–8358
8. El-Sagheer AH, Brown T (2010) Click chemistry with DNA. Chem Soc Rev 39 (4):1388–1405
9. Gramlich PM, Warncke S, Gierlich J, Carell T (2008) Click–click–click: single to triple modification of DNA. Angew Chem Int Ed 47 (18):3442–3444
10. Xu Y, Suzuki Y, Komiyama M (2009) Click chemistry for the identification of G-quadruplex structures: discovery of a DNA–RNA G-quadruplex. Angew Chem Int Ed 48(18):3281–3284
11. Paredes E, Das SR (2011) Click chemistry for rapid labeling and ligation of RNA. Chembiochem 12(1):125–131
12. Ishizuka T, Kimoto M, Sato A, Hirao I (2012) Site-specific functionalization of RNA molecules by an unnatural base pair transcription system via click chemistry. Chem Commun 48 (88):10835–10837
13. Phelps KJ, Ibarra-Soza JM, Tran K, Fisher AJ, Beal PA (2014) Click modification of RNA at adenosine: structure and reactivity of 7-ethynyl-and 7-triazolyl-8-aza-7-deazaadenosine in RNA. ACS Chem Biol 9(8):1780–1787
14. Samanta A, Krause A, Jäschke A (2014) A modified dinucleotide for site-specific RNA-labelling by transcription priming and click chemistry. Chem Commun 50 (11):1313–1316
15. Sawant AA, Mukherjee PP, Jangid RK, Galande S, Srivatsan SG (2016) A clickable UTP analog for the posttranscriptional chemical labeling and imaging of RNA. Org Biomol Chem 14:5832–5842
16. Kennedy DC, McKay CS, Legault MC, Danielson DC, Blake JA, Pegoraro AF, Stolow A, Mester Z, Pezacki JP (2011) Cellular consequences of copper complexes used to catalyze bioorthogonal click reactions. J Am Chem Soc 133(44):17993–18001
17. Weisbrod SH, Marx A (2007) A nucleoside triphosphate for site-specific labelling of DNA by the Staudinger ligation. Chem Commun 18:1828–1830
18. Neef AB, Luedtke NW (2014) An azide-modified nucleoside for metabolic labeling of DNA. Chembiochem 15(6):789–793
19. Kiviniemi A, Virta P, Lönnberg H (2008) Utilization of intrachain 4′-C-azidomethylthymidine for preparation of oligodeoxyribonucleotide conjugates by click chemistry in solution and on a solid support. Bioconjug Chem 19(8):1726–1734
20. Pourceau G, Meyer A, Vasseur J-J, Morvan F (2009) Azide solid support for 3′-conjugation of oligonucleotides and their circularization by click chemistry. J Org Chem 74 (17):6837–6842
21. Santner T, Hartl M, Bister K, Micura R (2013) Efficient access to 3′-terminal azide-modified RNA for inverse click-labeling patterns. Bioconjug Chem 25(1):188–195
22. Rao H, Sawant AA, Tanpure AA, Srivatsan SG (2012) Posttranscriptional chemical functionalization of azide-modified oligoribonucleotides by bioorthogonal click and Staudinger reactions. Chem Commun 48(4):498–500
23. Rao H, Tanpure AA, Sawant AA, Srivatsan SG (2012) Enzymatic incorporation of an azide-modified UTP analog into oligoribonucleotides for post-transcriptional chemical functionalization. Nat Protoc 7(6):1097–1112
24. Sawant AA, Tanpure AA, Mukherjee PP, Athavale S, Kelkar A, Galande S, Srivatsan SG (2016) A versatile toolbox for posttranscriptional chemical labeling and imaging of RNA. Nucleic Acids Res 44(2):e16

Chapter 25

Detection of the First Round of Translation: The TRICK Assay

Franka Voigt, Jan Eglinger, and Jeffrey A. Chao

Abstract

Quantitative fluorescence microscopy techniques are frequently applied to answer fundamental biological questions. Single-molecule RNA imaging methods have enabled the direct observation of the initial steps of the mRNA life cycle in living cells, however, the dynamic mechanisms that regulate mRNA translation are still poorly understood. We have developed an RNA biosensor that can assess the translational state of individual mRNA transcripts with spatiotemporal resolution in living cells. In this chapter, we describe how to perform a TRICK (*t*ranslating *R*NA *i*maging by *c*oat protein *k*nock-off) experiment and specifically focus on a detailed description of our image processing and data analysis procedure.

Key words mRNA translation, TRICK, MS2/PP7 stem-loops, Coat proteins, Quantitative fluorescence microscopy in live cells, Single-RNA imaging

1 Introduction

Ribosomes translate mRNAs to produce protein. mRNA translation is a highly regulated process that allows appropriate protein production in response to specific cellular needs. Ensemble measurements have contributed to our understanding of translation regulation by assessing ribosome occupancy and protein abundance on a genome-wide scale [1, 2]. These methodologies, however, can provide only limited insight into complex regulatory mechanisms that rely on the spatial or temporal distribution of individual molecules within the same cell. Single-molecule fluorescence microscopy, on the other hand, can image individual mRNAs in living cells [3] and has been used to investigate transcriptional dynamics in both prokaryotes and eukaryotes [4, 5].

In order to deepen our understanding of the spatiotemporal regulation of translation, we have established a dual-color single RNA imaging-based assay that can assess the translational state of individual mRNA molecules in living cells [6]. The TRICK (*t*ranslating *R*NA *i*maging by *c*oat protein *k*nock-off) assay takes advantage of the fact that the translating ribosome removes all proteins

Imre Gaspar (ed.), *RNA Detection: Methods and Protocols*, Methods in Molecular Biology, vol. 1649, DOI 10.1007/978-1-4939-7213-5_25, © Springer Science+Business Media LLC 2018

associated with the coding sequence of an mRNA transcript during the first round of translation (Fig. 1a). A TRICK reporter transcript contains translatable PP7 stem-loops in the coding sequence and MS2 stem loops in the 3′ UTR. Since these orthogonal RNA stem-

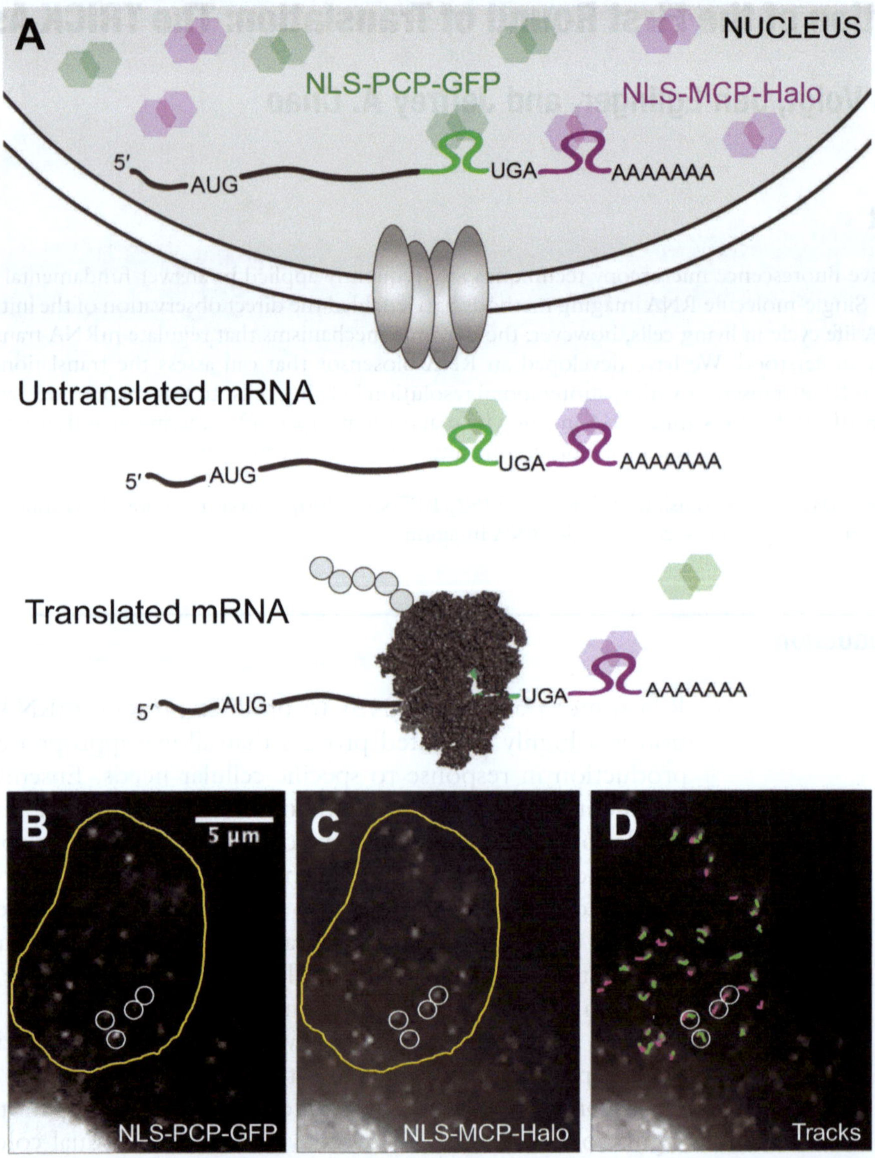

Fig. 1 Detecting the first round of translation. (**a**) Schematic representation of the TRICK assay. The translating ribosome (*black*) removes GFP-labeled PP7 coat proteins (NLS-PCP-GFP, *green*) from the coding region. Halo-labeled MS2 coat proteins (NLS-MCP-Halo, *magenta*) remain bound to the 3′ UTR. Figure modified from [6] and reprinted, with permission, from The American Association for the Advancement of Science © 2015. (**b**, **c**) Representative live cell images of a HeLa cell expressing NLS-PCP-GFP, NLS-MCP-Halo and a TRICK reporter. **b** = GFP, **c** = Halo labeled with JF_{549}. The ROI is shown in *yellow*. Several segmented spots are marked by *white circles*. (**d**) Superposition of tracks acquired for both channels after SPT in five frames

loops can be bound by PP7 and MS2 coat proteins (PCP and MCP) fused to spectrally distinct fluorescent proteins, an untranslated mRNA will be dual-labeled [7, 8]. Upon the first round of translation, the translating ribosome displaces all proteins bound to the coding region of the reporter mRNA including the fluorescent signal arising from the PCP without displacing the MCP signal. Since both MCP and PCP fluorescent fusion proteins contain nuclear localization sequences rebinding of PCP to the transcript after translation does not occur. TRICK measures the translational state of a transcript as a function of the colocalization of the PCP and MCP fluorescent labels. This chapter briefly explains how to perform a TRICK experiment and focuses on the subsequent image and data analysis.

2 Materials

2.1 Sample Preparation

1. A cell line is needed that stably expresses fluorescently labeled MCP and PCP as well as a TRICK mRNA reporter transcript. Specifically, we use the HeLa 11HT cell line [9], which contains a single doxycycline-inducible locus for recombinase-mediated cassette exchange of the TRICK reporter and stably expresses NLS-PCP-GFP and NLS-MCP-Halo. It is available upon request from the Chao laboratory. Alternatively, lentiviral transfer plasmids encoding NLS-PCP-GFP and NLS-MCP-Halo are available from Addgene (Addgene IDs 64539, 64540, 64544, and 84443). TRICK expression and colocalization control plasmids that can serve as cloning templates are also available on Addgene (Addgene IDs 84443, 84444) [6].
2. Cell culture incubator.
3. Gibco™ Dulbecco's Modified Eagle's Medium (DMEM) supplemented with 10 v/v % tetracycline-free FBS and 1 v/v % penicillin and streptomycin (pen/strep).
4. Automated cell counter and counting slides (e.g., Bio-Rad).
5. Imaging dish, 35 mm, high, glass bottom (e.g., ibidi μ-Dish).
6. Fluorobrite™ DMEM imaging medium + 10% FBS.
7. 1 μg/mL doxycycline in Fluorobrite™ DMEM imaging medium + 10% FBS.
8. 100 nM Halo-labeling Janelia Fluor 549 (JF_{549}; HHMI Janelia Research Campus) in DMEM + 10% FBS.
9. Phosphate-buffered saline (PBS).

2.2 Image Acquisition

1. Multipoint confocal spinning disk microscope, e.g., an Olympus IX81 inverted microscope equipped with a CSU-X1 scan-head (Yokogawa) and Borealis modification (Andor) featuring

a dichroic beam-splitter in the scanhead (Semrock Di01-T488/568—13 × 15 × 0.5), 100× 1.45NA PlanApo TIRFM oil immersion objective (Olympus), two back-illuminated EvolveDelta EMCCD cameras (Photometrics), green (Semrock, FF01-617/73-25) and red (Semrock, FF02-525/40-25) emission filters, beam-splitter between cameras (Chroma, 565DCXR), solid-state lasers (100 mW 491 nm and 100 mW 561 nm; Cobolt), and motorized X, Y, Z-Piezo controlled stage (ASI).

2. Incubation chamber around the microscope to provide heating and CO_2 regulation.
3. Multicolor calibration slide for channel alignment (e.g., Argo-light, type SLF-001).

2.3 Data Processing and Analysis

1. FijiImageJ [10] including the TrackMate plugin [11].
2. KNIME Analytics Platform (version 3.2.1, [12]).
3. RStudio (version 0.99.896) with the ggplot2 package installed [13].

3 Methods

TRICK is a single-molecule technique that distinguishes untranslated from translated mRNAs by assessing the presence of two spectrally distinct fluorescent labels bound to the open reading frame (ORF) and 3′ UTR of a reporter transcript. Subheadings 3.1 and 3.2 briefly explain how to perform a TRICK experiment. More detailed instructions on how to design reporter constructs, generate stable cell lines and best acquire images have been described by Halstead et al. [14].

Subsequently, Subheadings 3.3 and 3.4 provide detailed protocols on how to perform the image processing and data analysis part of a TRICK experiment. They focus on the challenges and advantages of multiple-channel single-particle tracking (SPT) using the Fiji plugin TrackMate [10, 11, 15] and provide detailed instructions on how to employ a KNIME [12] data processing pipeline for spot-based track colocalization. The analysis pipeline and a sample data set are available for download on the Chao lab web site (http://www.fmi.ch/research/groupleader/?group=132).

3.1 Performing a TRICK Experiment

1. To generate a TRICK reporter construct, integrate a PP7 cassette into the coding sequence and an MS2 cassette into the 3′UTR of the RNA of interest. This can be done by excision and gel extraction of the complete PP7-stop codon-MS2 fragment via SalI and ClaI restriction enzyme sites from the Renilla TRICK reporter construct (Addgene ID 84443) described by Halstead et al. [6]. Alternatively, the PP7 (SalI/XhoI) and MS2

(BamHI/ClaI) cassettes can be individually cloned into reporter constructs. Conventional PCR amplification of the PP7 and MS2 cassettes described here is not recommended due to the repetitiveness of the DNA sequences. A co-localization control plasmid (*see* Subheading 3.4.3) can be generated by introduction of a stop codon in front of the PP7 cassette using site-directed mutagenesis.

2. Seed 30,000 HeLa cells stably expressing NLS-MCP-Halo, NLS-PCP-GFP and the TRICK reporter construct in 2 mL DMEM + 10% FBS + 1% Pen/Strep into a 35 mm imaging dish ~48 h prior to the experiment. Take care that cells are homogeneously distributed within the dish to improve cytoplasmic imaging of mRNAs (*see* **Note 1**).
3. Incubate cells for 2 days at 37 °C and 5% CO_2. Longer or shorter incubation times are possible depending on the cell line being used.
4. Immediately prior to the experiment, prewarm PBS and Fluorobrite™ DMEM + 10% FBS to 37 °C.
5. Halo-label cells by incubation with 1 mL 100 nM JF_{549} in DMEM + 10% FBS for 30 min at 37 °C and 5% CO_2.
6. Remove medium from cells, wash three times with PBS and add prewarmed Fluorobrite™ DMEM + 10% FBS + 1 μg/mL doxycycline to induce TRICK reporter expression.
7. Since TRICK measures the first round of translation, it is important to consider the timing of induction and the acquisition of images. For the HeLa cell line we use, TRICK transcripts begin to enter the cytoplasm approximately 45 min after doxycycline addition.

3.2 Image Acquisition

1. Equilibrate the incubation chamber of the microscope to 37 °C and 5% CO_2.
2. Align cameras prior to image acquisition using a multicolor calibration slide.
3. Identify suitable cells for image acquisition using the red laser (MCP-Halo channel) at low intensities to reduce photobleaching prior to the experiment. Choose cells with well-resolved (high signal/noise) diffraction-limited mRNA particles (diameter approximately 4 pixels or 0.36 μm for optical settings described above) at densities that facilitate SPT.
4. Image cells in both channels simultaneously using laser powers, camera gain and exposure time compatible with SPT. Exposure times should be no longer than 50 ms to ensure that fast moving particles can be unambiguously tracked between subsequent frames. Laser power should be adapted to yield maximum signal intensity while limiting bleaching the samples,

which would prevent accurate spot detection in the last frames of the image series. Camera gain can be adjusted with respect to the laser power and should result in optimal signal-to-noise ratios without saturation of the detector.

5. Collect 10–50 frame image series per region of interest.
6. To facilitate SPT, take special care to image cells with low to medium spot density in the region of interest. High particle densities can often lead to false assignment of displacements in between neighboring particles and will complicate SPT.

3.3 Single-Particle Tracking Using TrackMate

To assess the translational state of individual mRNAs, it is necessary to reconstruct their trajectories using SPT. For an informative TRICK experiment, it is further essential to track the mRNA movement in both channels independently.

The Fiji plugin TrackMate performs SPT as a two-step process: First, it localizes individual particles (spot detection). Second, it links particle positions in consecutive frames into tracks (tracking). TrackMate outputs all positions assumed by a particle throughout the time course of the experiment as a list of coordinates that serves as input for the track colocalization pipeline described under Subheading 3.4.

3.3.1 Image Preparation in Fiji

1. Before starting TrackMate, make sure that the channels are precisely aligned. To correct potential offset between channels, use a suitable method for channel alignment, e.g., the “Translate” command or the descriptor-based registration plugin. Use the “Search” option within the “Help” menu to find Fiji commands mentioned throughout this chapter.
2. Decide how many frames to include in the analysis and, if needed, reduce the length of the image series accordingly using the “Slice Keeper” function. Accurate tracking is best performed on a small number of frames (typically 3–10) since this limits the number of data points that has to be visually inspected (*see* **Note 2**).
3. Optionally, reduce randomly distributed noise by application of the “Bandpass filter” (filtering small objects below three pixels).
4. Select a single region of interest (ROI, yellow in Fig. 1b and c) via the freehand selection tool (*see* **Notes 3** and **4**). Make sure to use the same ROI in both channels by using the “Restore Selection” function or the ROI Manager, but track both channels individually.

3.3.2 Spot Detection

1. Launch TrackMate while having the first image series selected. Check that image parameters are loaded correctly. If not, calibrate using the image “Properties” function in Fiji and “Refresh source” in the TrackMate graphical user interface

(GUI). Alternatively, edit the calibration settings in the TrackMate GUI. Make sure that both image series are calibrated the same way.

2. Detect spots using the "Laplacian of Gaussian" (LoG) detector. Alternatively, use the "differences of Gaussian" (DoG) detector as a quicker approximation of the LoG detector.
3. Optimize detector settings for the detection of single mRNA particles. Choose an "Estimated blob diameter" that is slightly bigger than the expected spot size (approximately four pixels or 0.36 μm in the imaging set-up described above) and an intensity "Threshold" that detects all, even weak intensity spots throughout the whole image series.
4. Visually inspect detected spots in all frames either using the "Preview" function or after spot detection has been performed. For inspection, choose the "Hyperstack Displayer" to superpose the segmentation results over the images (white circles in Fig. 1b–d).
5. Do not threshold or filter the detected spots. All filtering can be done later as part of the track colocalization pipeline.
6. Make sure that all visible spots are detected in all images. If not, go back and repeat the spot detection using a reduced intensity threshold or larger spot diameter. A slight overdetection is not a problem since all spots that cannot be linked to form tracks will be discarded during tracking.

3.3.3 Tracking

1. Choose the "Simple LAP tracker" as particle-linking algorithm since it allows gaps and prevents track merging or splitting events.
2. Set the "Linking max distance" to match physiological diffusion coefficients observed for individual mRNA particles, i.e., 0.1–0.8 μm to account for diffusion coefficients between 0.009 $\mu m^2/s$ [16] and 3.42 $\mu m^2/s$ [17] that have been observed for different types of transcripts in various subcellular localizations.
3. Set the "Gap-closing max frame gap" to 2 in order to allow single frame gaps in tracks. Gap size is defined as the frame difference, which is 2 if one frame is missing.
4. Set the "Gap-closing max distance" to a multiple of the "Linking max distance" that matches the allowed frame gap if particle density is low and correct tracking likely. Otherwise, reduce the "Gap-closing max distance" to values similar to the "Linking max distance" to allow gaps only for low mobility particles.
5. Perform tracking, do not filter tracks and click through the GUI until the "Display Options" interface is reached. Apply the interface to change track appearance to facilitate visual inspection.

6. Examine all tracks to see whether they describe spot movement accurately. If not, go back and change tracking parameters accordingly.
7. Optimal tracking is always a trade-off between large linking distances that allow tracking of highly mobile particles and false linking of arbitrary spots resulting from spot detection errors. The choice of parameters should be optimized depending on the images being analyzed and their effect should be systematically evaluated.
8. Export the tracking results via the "Analysis" button in the "Display Options" interface. To this aim, save the "Spots in tracks statistics" table that pops up as an .xls file with exactly the same name as the image series used for tracking.
9. Repeat all steps from Subheadings 3.3.2 and 3.3.3 for the second channel using identical tracking but not necessarily identical spot detection parameters since each channel may have different signal-to-noise ratios.

3.4 Spot-Based Track Colocalization in KNIME

TRICK assesses the translational state of individual mRNA transcripts via detection of two spectrally distinct fluorescent labels and relies on the pairwise colocalization of tracks obtained after independent SPT in both fluorescent channels.

Pairwise colocalization analysis is performed by measuring the distances of all spots belonging to the first channel to all spots in the second channel. Using the measured distances, colocalizing pairs of spots are determined by selecting mutual nearest neighbors. To accommodate for cases where two spots in one channel have the same nearest neighbor in the other channel, the mutual nearest neighbor analysis is recursively performed on all unpaired spots until every spot is assigned to a pair. Spot pairs are classified as colocalizing if their distance is below a user-defined cutoff.

Tracks are classified as colocalizing if they contain a user-defined minimum number of colocalizing spot pairs (usually two). All tracks shorter than a user-defined minimum track length (usually three frames) that were not classified as colocalizing are excluded from the analysis. All other tracks are called orphans and represent translated mRNAs.

3.4.1 Data Preparation

1. Assemble both .xls files exported after SPT of the two channels along with both image series used for tracking in a single folder for each cell. Make sure that coordinate files and image series are named identically (except for the file extension).
2. Combine the folders containing tracking and imaging data for each cell in one or several parent folders.

3.4.2 Track Colocalization

1. Open the Track-based colocalization analysis workflow in KNIME.
2. Configure and execute the "Parameter input" node (#391) by choosing the parent folder that combines the tracking and imaging data folders as input directory. Enter the "File Extension" of the coordinate files (xls) and choose "Channel identifiers" that distinguish input files belonging to both channels. Tick "Load images for quality control" to load image files into the pipeline. Identify image files by providing the image file extension (e.g., tif). Define the distance cutoff below which a spot pair is classified as colocalizing (default = 0.3 μm). Enter the pixel size to calibrate imaging data (default = 0.09 μm, depending on microscope set-up). Choose "channel 1 and 2 plotting colors" from the scroll down menus in case defaults (channel 1 = red, channel 2 = green) do not apply.
3. Configure and execute the "Filtering parameters" node (#405). Choose the minimum track length that is to be included in the analysis (default = three frames) and the minimum number of spot colocalizations necessary to define a track as colocalizing (default = two events).
4. Execute and view the "Interactive Segmentation View" node (#454). It provides a control mechanism that allows superposition of the spot coordinates of channel 1 over the image files of the same channel. Open the interactive view panel by double-clicking an image in the (segmented) labeling column. Use the "Bounding Box Renderer" to visually check that spot coordinates match spot positions.
5. Execute and view the "JavaScript Table View" node (#427). The branches leading to this node perform the colocalization analysis. For quality control purposes, they generate line plots showing the tracks belonging to each channel in the colors defined under point 2. Tracks that are classified as colocalizing are depicted in opaque colors while orphan tracks are shown semitransparent (alpha = 0.2).
6. Inspect the line plots (Fig. 1d) that are generated for each cell and included in the output table of node #427. If colocalization analysis was performed satisfactory, continue with point 7. If not, refine the parameters to improve results quality. Specifically, test different cutoff values if track classification does not match visually determined colocalization. Return to Subheadings 3.3.2 or 3.3.3 to improve the tracking of individual cells in case orphan tracks appear systematically or track patterns repeatedly differ between channels.
7. Execute the "XLS Sheet Appender" node (#438) to activate the branch that calculates the statistics of the colocalization analysis.

8. Inspect the interactive table generated by "JavaScript Table View" node (#426). It contains the results of the colocalization analysis for each cell. Table columns depict: (a) the number of tracks discarded from the analysis based on the minimum track length and colocalization criteria defined in node #405 ("Excluded Tracks" for channel 1/2); (b) the number of colocalizing and orphan tracks for each channel; (c) the total number of tracks (= colocalizing + orphan) included in the analysis of each channel; (d) the ratio of colocalizing tracks over total included tracks for both channels ("Ratio Colocalized/Total" for channel 1/2).
9. Inspect the colocalization ratio for the green PCP-GFP channel (channel 2). Since green spots should never be present without red spots, this ratio defines the detection efficiency.
10. Inspect the colocalization ratio for the red MCP-Halo channel (channel 1). This ratio represents the untranslated fraction of mRNAs observed in each cell. Use the detection efficiencies defined above to normalize the translation ratios per cell.

3.4.3 Iterative Tracking and Analysis Cycles

Optimal tracking and colocalization is an iterative process (*see* **Notes 5–7**). Therefore, go back in the analysis pipeline and repeat individual steps as many times as it takes to refine parameters that give the best possible detection efficiency:

1. Before performing a TRICK experiment, it is important to assess the colocalization of a dual-labeled transcript that contains both PP7 and MS2 stem-loops in the 3′ UTR of the reporter construct (addgene ID 84444). Since the colocalization of the fluorescent signals of this transcript is not translation-dependent, this analysis is a measure of the maximum colocalization that can be achieved. Using the HeLa cell line and microscope setup described above, we achieve colocalization of 89 ± 7% in the red and 91 ± 6% in the green channel.
2. If orphan tracks accumulate in a specific area of the cell, this often indicates reduced spot detection efficiency due to high background fluorescence in perinuclear zones. Repeat SPT using a different ROI that excludes the problematic area (*see* **Note 8**).
3. If multiple short tracks localize close to each other, this can indicate too short linking distances. Repeat Subheading 3.3.3 using a larger "linking max distance".
4. If a large number of short tracks are excluded from analysis, this could be due to overdetection of spots that are arbitrarily linked to short tracks. Increase intensity "Threshold" or reduce "Estimated blob diameter" during Subheading 3.3.2.

3.4.4 Output

1. Inspect the results of the translation analysis that are saved in a "KNIME_Results.xlsx" file in the parent folder that served as input directory. The file contains two sheets, one giving a summary and one showing the results for each cell included in the analysis.
2. Find the line plots of colocalizing and orphan tracks for each cell as .png files (superposed to representative cell image in Fig. 1d) named after the channel 1 input files in each subfolder.

4 Notes

1. Consider seeding cells more than 48 h prior to the experiment (at lower densities) to yield larger cells that are widely spread out on the imaging dish. Reduced cell thickness can increase signal/noise via reduction of background fluorescence resulting from fluorescent objects above or below the focal plane.
2. To lessen intensity loss due to photobleaching try the "Bleach Correction" function in Fiji using histogram matching. Note that spot detection can get error-prone if long time series are analyzed.
3. Only analyze those regions in a cell that facilitate SPT, i.e., choose ROIs that exclude low signal/noise areas. However, make sure not to bias the analysis with repeated selection of similar ROIs.
4. Always select an ROI to prevent false-positive spot detection at image boundaries.
5. Good SPT is essential since it generates the data that will be used in the analysis. Refine SPT until the results match the physiological conditions as well as possible.
6. Best SPT results are achieved at low till medium particle densities. There are two options to reduce the average density of labeled particles dependent on the type of experiment performed: (1) use very short induction times, start imaging ≥30 min after induction and do not remove doxycycline from the medium; (2) use longer induction times (1–2 h), remove doxycycline from the medium and start imaging ≥1 h after transcription shut-off.
7. Reduced spot densities allow less stringent tracking parameters, i.e., larger gaps at increased gap closing distances. Set higher "Linking max distances" (Subheading 3.3.3) to allow tracking of highly mobile particles in low particle densities.
8. Overdetect spots to make sure not to miss any particles in noisy data sets.

Acknowledgments

The authors would like to thank all members of the Chao lab for their joint efforts in developing the TRICK analysis pipeline, in particular, Johannes Wilbertz for providing test data and Ivana Horvathova for generation of the colocalization control. We further thank the Facility for Advanced Imaging and Microscopy at FMI for data acquisition and analysis support. Research in the Chao lab is supported by the Novartis Research Foundation and the Swiss National Science Foundation.

References

1. Ingolia NT, Ghaemmaghami S, Newman JRS, Weissman JS (2009) Genome-wide analysis in vivo of translation with nucleotide resolution using ribosome profiling. Science 324:218–223
2. Schwanhäusser B, Busse D, Li N, Dittmar G, Schuchhardt J, Wolf J, Chen W, Selbach M (2011) Global quantification of mammalian gene expression control. Nature 473:337–342
3. Fusco D, Accornero N, Lavoie B, Shenoy SM, Blanchard J-M, Singer RH, Bertrand E (2003) Single mRNA molecules demonstrate probabilistic movement in living mammalian cells. Curr Biol 13:161–167
4. Golding I, Paulsson J, Zawilski SM, Cox EC (2005) Real-time kinetics of gene activity in individual bacteria. Cell 123:1025–1036
5. Larson DR, Zenklusen D, Wu B, Chao JA, Singer RH (2011) Real-time observation of transcription initiation and elongation on an endogenous yeast gene. Science 332 (6028):475–478. doi:10.1126/science.1202142
6. Halstead JM, Lionnet T, Wilbertz JH, Wippich F, Ephrussi A, Singer RH, Chao JA (2015) Translation. An RNA biosensor for imaging the first round of translation from single cells to living animals. Science 347 (6228):1367–1671. doi:10.1126/science.aaa3380
7. Bertrand E, Chartrand P, Schaefer M, Shenoy SM, Singer RH, Long RM (1998) Localization of ASH1 mRNA particles in living yeast. Mol Cell 2:437–445
8. Beach DL, Salmon ED, Bloom K (1999) Localization and anchoring of mRNA in budding yeast. Curr Biol 9:569–578
9. Weidenfeld I, Gossen M, Löw R, Kentner D, Berger S, Görlich D, Bartsch D, Bujard H, Schönig K (2009) Nucleic Acids Res 37(7): e50. doi:10.1093/nar/gkp108
10. Schindelin J, Arganda-Carreras I, Frise E, Kaynig V, Longair M, Pietzsch T, Preibisch S, Rueden C, Saalfeld S, Schmid B, Tinevez J-Y, White DJ, Hartenstein V, Eliceiri K, Tomancak P, Cardona A (2012) Fiji: an open-source platform for biological-image analysis. Nat Methods 9:676–682
11. Tinevez J-Y, Perry N, Schindelin J, Hoopes GM, Reynolds GD, Laplantine E, Bednarek SY, Shorte SL, Eliceiri KW (2017) TrackMate: an open and extensible platform for single-particle tracking. Methods 115:80–90
12. Berthold, MR, Cebron, N, Dill, F, Gabriel, TR, Kötter, T, Meinl, T, et al. (2009) KNIME - the Konstanz information miner: version 2.0 and beyond. ACM SIGKDD Explorations Newsletter 11(1):26–31. doi:10.1145/1656274.1656280
13. Wickham, H. (2009). ggplot2. New York, NY: Springer Science & Business Media. doi:10.1007/978-0-387-98141-3
14. Halstead JM, Wilbertz JH, Wippich F, Lionnet T, Ephrussi A, Chao JA (2016) TRICK: a single-molecule method for imaging the first round of translation in living cells and animals. Methods Enzymol 572:123–157
15. Jaqaman K, Loerke D, Mettlen M, Kuwata H, Grinstein S, Schmid SL, Danuser G (2008) Robust single-particle tracking in live-cell time-lapse sequences. Nat Methods 5:695–702
16. Shav-Tal Y, Darzacq X, Shenoy SM, Fusco D, Janicki SM, Spector DL, Singer RH (2004) Dynamics of single mRNPs in nuclei of living cells. Science 304:1797–1800
17. Ma J, Liu Z, Michelotti N, Pitchiaya S, Veerapaneni R, Androsavich JR, Walter NG, Yang W (2013) High-resolution three-dimensional mapping of mRNA export through the nuclear pore. Nat Commun 4:2414

Chapter 26

Imaging Translation Dynamics of Single mRNA Molecules in Live Cells

Suzan Ruijtenberg, Tim A. Hoek, Xiaowei Yan, and Marvin E. Tanenbaum

Abstract

mRNA translation is a key step in decoding the genetic information stored in DNA. Regulation of translation efficiency contributes to gene expression control and is therefore important for cell fate and function. Here, we describe a recently developed microscopy-based method that allows for visualization of translation of single mRNAs in live cells. The ability to measure translation dynamics of single mRNAs will enable a better understanding of spatiotemporal control of translation, and will provide unique insights into translational heterogeneity of different mRNA molecules in single cells.

Key words mRNA, Translation, Single molecule imaging, Fluorescence, Microscopy, SunTag

1 Introduction

Translation of mRNAs into proteins is a key step in gene expression and is of critical importance for fine-tuning cellular protein levels. In recent years, different methods have provided many new and important insights into the regulation of translation, yet many questions remain. For example, it is still unclear whether all mRNAs transcribed from the same gene are translated with similar efficiencies, or whether translational heterogeneity exists among such mRNAs. Similarly, it is largely unknown how translation efficiencies are controlled in space and time. An important reason for our limited understanding of translational control is that many current methods to assess translation efficiency rely on population-based measurements and frequently require fixation or lysis of the cells to obtain a measurement of translation efficiency. As a consequence, mainly snapshots of average translation efficiencies of thousands of mRNA molecules have been obtained. A major advance in measuring translation efficiency of single mRNAs in live cells was recently achieved by our lab, as well as several other labs [1–5].

Imre Gaspar (ed.), *RNA Detection: Methods and Protocols*, Methods in Molecular Biology, vol. 1649, DOI 10.1007/978-1-4939-7213-5_26, © Springer Science+Business Media LLC 2018

Here, we describe a microscopy-based method, which allows quantitative measurements of ribosome initiation and elongation on individual mRNA molecules in live cells. This approach provides a powerful and widely applicable tool to study dynamics and regulation of translation. In this method, a reporter mRNA is designed which encodes a proteins of interest (POI) fused at its N-terminus to an array of antibody peptide-epitopes, derived from the SunTag system that we previously developed [6]. The SunTag peptide epitopes are recognized by a single chain antibody fragment fused to superfolderGFP (sfGFP) [6], which is coexpressed in cells with the reporter mRNA. When the reporter mRNA is translated, the peptide epitopes emerge from the ribosome while the fused POI is undergoing synthesis (Fig. 1a, upper panel). Binding of the GFP-fused antibodies to the nascent peptide epitopes results in a bright green labeling of the nascent polypeptide, which can be observed under the microscope as a bright fluorescent dot at the site of translation (Fig. 1b), providing a real-time readout of the translation of the reporter mRNA. In addition to fluorescent labeling of the nascent polypeptide, the mRNA molecule is fluorescently labeled in a second color through the MS2- or PP7-based labeling system [7, 8] (Fig. 1a, upper panel, b). In order to improve the long term tracking of mRNA molecules, we have devised an mRNA tethering system, which reduces mRNA mobility and allows tracking of individual mRNA molecules for extended periods of time (>1 h) (Fig. 1a, lower panel, b). In this chapter we provide details on how to design, carry out, and interpret experiments to image translation dynamics of an mRNA of interest and in a cell type of choice.

2 Materials

2.1 Plasmids

1. Translational reporter, containing the 24 SunTag peptides and 24 PP7 binding sites (for example pcDNA4TO-24xGCN4_v4-kif18b-24xPP7, Addgene #74928).
2. scFv-GFP (for example pHR-scFv-GCN4-sfGFP-GB1-dWPRE, Addgene #60907).
3. PCP-mCherry (for example pHR-tdPP7-3xmCherry, #74926, or pHR-PP7-2xmCherry-CAAX, #74925).
4. AgeI, BamHI, EcoRI, HindIII, and MluI restriction enzymes for cloning.

2.2 Cell Culture

1. Glass bottom cell culture dishes suitable for live-cell microscopy. Most high magnification microscope objectives are designed for glass with a thickness of 0.17 mm. We routinely use 96-well glass bottom dishes (Matriplate, Brooks).

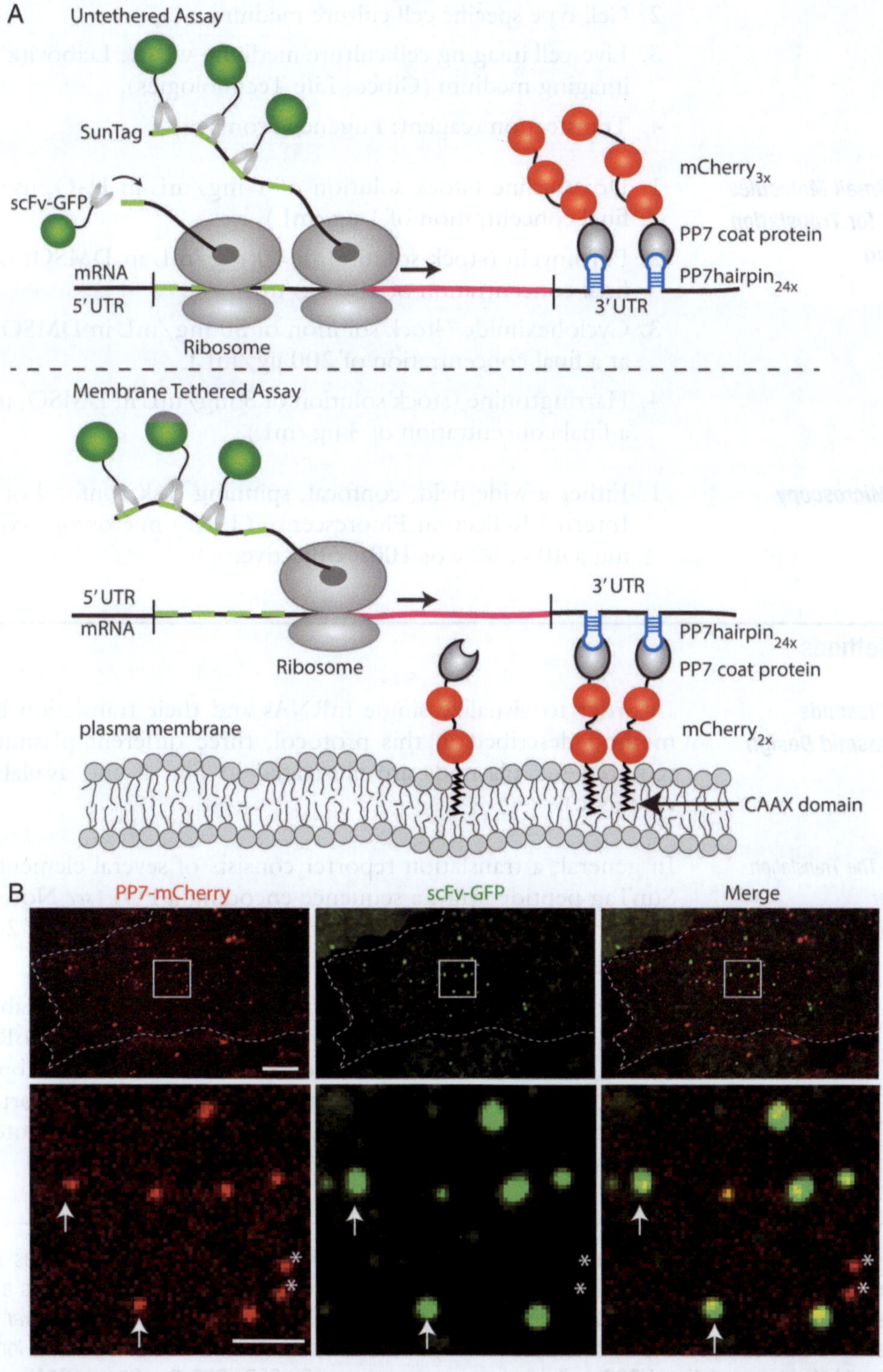

Fig. 1 Imaging translation of single mRNA molecules. (**a**) Schematic representation of the fluorescent labeling of nascent polypeptides using the SunTag system and the labeling of mRNA using the PP7 system. A gene of interest is fused to an array of SunTag peptides. When the mRNA is translated, scFv-GFP antibodies will bind to the SunTag peptides that emerge from the ribosome, resulting in a bright fluorescent spot at the site of translation and thus allowing real-time observation of protein synthesis. To visualize the mRNA, 24 PP7

2. Cell type specific cell culture medium.
3. Live-cell imaging cell culture medium; we use Leibovitz's-L15 imaging medium (Gibco, Life Technologies).
4. Transfection reagent: Fugene (Promega).

2.3 Small Molecules Useful for Translation Imaging

1. Doxycycline (stock solution of 1 mg/mL in H_2O, used at a final concentration of 1 μg/mL).
2. Puromycin (stock solution of 10 mg/mL in DMSO, used at final concentration of 100 μg/mL).
3. Cycloheximide (stock solution of 50 mg/mL in DMSO, used at a final concentration of 200 μg/mL).
4. Harringtonine (stock solution of 3 mg/mL in DMSO, used at a final concentration of 3 μg/mL).

2.4 Microscopy

1. Either a wide-field, confocal, spinning disk confocal or Total Internal Reflection Fluorescence (TIRF) microscope containing a 40×, 60× or 100× objective.

3 Methods

3.1 Plasmids and Plasmid Design

In order to visualize single mRNAs and their translation by the method described in this protocol, three different plasmids are required (all plasmids are illustrated in Fig. 2 and available on Addgene).

3.1.1 The Translation Reporter

In general, a translation reporter consists of several elements; the SunTag peptide array, a sequence encoding a POI (*see* **Note 1** for further discussion on choosing the POI), and an array of 24 PP7 binding motifs.

1. Clone the POI in the translational reporter available on Addgene (#74928). Using the enzymes AgeI and EcoRV the original POI can be removed (a fragment of 2587 bp) and replaced by any POI to create a new translational reporter (*see* **Note 1** for further discussion on choosing the protein of interest).

Fig. 1 (continued) binding sites are inserted in the 3′ UTR of the reporter mRNA. These sites can be recognized by the PP7 bacteriophage coat protein, which is fused to three copies of mCherry. As a result, mRNAs can be observed as mCherry positive foci. By fusing a CAAX-motif to PP7-mCherry (*lower panel*), mRNAs can be tethered to the plasma membrane, which allows tracking of individual mRNAs for long time periods. (**b**) A representative U2OS cell is shown expressing scFv-GFP, PP7-2xmCherry-CAAX and the translational reporter ($SunTag_{24x}$-kif18b-$PP7_{24x}$). mRNAs are visible in *red*, translation in *green*. The *dotted line* indicates the outline of a cell. A zoomed-in view of the *white-boxed area* is shown, containing both mRNAs that are undergoing translation (two examples are indicated with *arrows*) and mRNAs that are not translated (two examples are indicated with *asterisks*). Scale bars, 5 μm (*upper panels*) or 2 μm (*lower panels*)

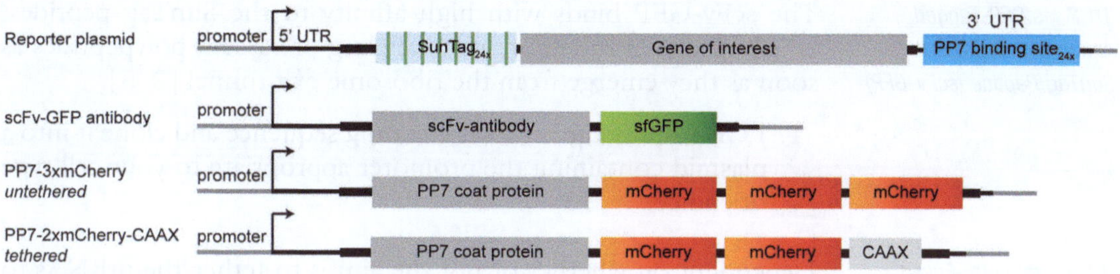

Fig. 2 Schematic overview of the plasmids used for the translation imaging method. Region of plasmid that is transcribed into mRNA is shown in *black*, other DNA in *grey*. *Thin black lines* indicate noncoding parts of the mRNA, *thick black lines* indicate coding sequences. The SunTag peptides are shown in *light blue* with *green stripes* representing individual peptides (eight shown), the PP7 binding sites in *dark blue*, sfGFP in *green*, and mCherry in *red*. All plasmids are available on Addgene (*see* text for catalog number)

2. Alternatively, start with a vector of choice and a POI of choice and clone the individual elements of the translational reporter (i.e., SunTag and PP7 binding sites) into this vector.
   (a) Since the SunTag peptide array forms a repetitive sequence, it is difficult to amplify the array by PCR. We therefore recommend cloning strategies based on enzymatic digestion and ligation (*see* **Note 2** for more information about PCR-based cloning strategies of the SunTag array). To insert the SunTag array (consisting of 5–24 SunTag peptides (*see* **Note 3** about the use of different numbers of SunTag peptides)) at the N-terminus of the POI, use HindIII and AgeI restriction enzymes to digest the translational reporter plasmid available on Addgene (#74928). This results in two fragments (9141 and 1819 bp). The smaller fragment contains the 24× SunTag peptides, which can be cloned into the desired vector.
   (b) Clone the 24 PP7 binding motifs into the 3′ UTR of the mRNA to label the mRNA independently of translation. Use BamHI and EcoRI to digest the translational reporter plasmid available on Addgene (#74928). This results in two fragments (9492 and 1468 bp), the smaller of which contains the PP7 binding motives. Since this array of short hairpin sequences is highly repetitive, we recommend cloning methods based on digestion and ligation rather than through PCR-based cloning.
   (c) Clone a promoter of choice upstream of the reporter coding sequence. Expression of the reporter mRNA is typically driven by a doxycycline-inducible promoter to allow temporal control of reporter mRNA expression (*see* **Note 4** about the use of an inducible promoter). This promoter can be obtained from plasmid (#74928), by using the enzymes MluI and HindIII or by PCR.

3.1.2 sfGFP-Tagged Antibody that Binds the SunTag Peptide (scFv-GFP)

The scFv-GFP binds with high affinity to the SunTag peptides, which results in the fluorescent labeling of nascent polypeptides as soon as they emerge from the ribosome exit tunnel [2–5].

1. PCR amplify the scFv-GFP coding sequence and clone it into a plasmid containing the promoter appropriate to your cell type of choice.

3.1.3 mCherry-Fused PCP (PP7-mCherry)

Dependent on whether or not the aim is to tether the mRNAs to the plasma membrane, plasmid #74926 (untethered) or plasmid #74925 (tethered) can be used as a template. Using PCR-based methods, either PP7 alone, PP7 fused to mCherry or PP7 fused to mCherry and a CAAX domain can be amplified and placed into a vector of choice (*see* **Note 5** on tethering the mRNA to the membrane). Note that plasmids containing multiple copies of mCherry cannot be amplified by PCR and need to be cloned by digestion--ligation methods.

3.1.4 Exchanging Fluorescent Proteins

The three plasmids described above enable visualization of both the reporter mRNA (PP7-mCherry, red) and its translation (scFv-GFP, green) in live cells (Fig. 1b). In principle, the color of the fluorescent proteins (e.g., GFP and mCherry) can be changed, but the functionality of the newly designed constructs needs to be carefully tested, as, for example, addition of sfGFP to the antibody has been shown to be important for preventing scFv aggregation in mammalian cells [6] (*see* **Note 6** for further comments on fluorescent proteins fused to the scFv antibody).

1. Exchange fluorescent protein using PCR-based cloning methods.
2. Test expression level of newly created fusion protein by transient transfection in cell type of choice.
3. Examine aggregation state in cells of fusion protein after transient transfection using widefield or confocal microscopy (bright fluorescent foci in transfected cells indicate protein aggregation).

3.2 Creating a Cell Line for Imaging Translation

3.2.1 Choosing a Suitable Cell Type

We have performed the majority of our experiments in U2OS cells. However, similar SunTag-based translation imaging has been successfully performed in neurons and HeLa cells [2–4], suggesting that the translation imaging approach described here can be performed in most cell types (*see* **Note 7** on how to choose the best cell type for your experiment).

3.2.2 Delivering the Plasmids

To create a cell line in which translation can be imaged, the plasmids described above can be delivered into cells with standard methods such as transfection or viral transduction. Because of the repetitive sequences present in the reporter mRNA plasmid (e.g., the PP7 binding sites and the sequence encoding the SunTag peptides), virus titers may be low, and transfection may be preferred over viral transduction. To transfect the plasmids into U2OS cells in a 6 cm cell culture dish (transient or stable transfection; *see* **Note 8** about transient versus stable transfections) the following Fugene (Promega) transfection protocol can be used. Please note that other cell types may require other transfection protocols.

1. Warm up DMEM medium without serum and without antibiotics (DMEM −/−) to 37 °C.
2. Make a mastermix (number of reactions +1) of Fugene (Promega) containing 100 μL DMEM −/− and 2 μL Fugene per 6 cm dish.
3. Mix by tapping vigorously.
4. Spin down 3 s to collect medium in bottom of the tube.
5. Incubate the master mix for ~5 min at room temperature.
6. Make the DNA mix containing 1 μg total DNA per transfection.
7. Add 100 μL of DMEM/Fugene mastermix to the DNA and mix by pipetting.
8. Incubate for 5–15 min at room temperature.
9. Add 3 mL of fresh cell culture medium to the cells.
10. Add the transfection mix to the cells.
11. After 24 h wash the cells (note: this is not essential).

3.3 Preparing Cells for Imaging

1. Approximately 12–24 h before imaging plate the cells containing the reporter plasmid, PP7-mCherry (−CAAX) and the scFv-GFP, in a glass bottom dish (we routinely use 96-wells glass bottom dishes) at the intended densities (~50% confluency). Depending on the experimental design and cell type, cells can also grow for longer time periods on the glass surface.
2. Immediately before transferring the glass-bottom dish containing the transfected cells to the microscope, replace the culture medium with prewarmed imaging medium per 96-well (*see* **Note 9** about the use of imaging medium).
3. Set the temperature at the microscope to 37 °C for mammalian cells. Changes in temperature may result in cellular stress, which could influence the process of translation. We found temperatures between 36 and 37.5 °C to be acceptable for most human cell lines.
4. When imaging for longer time periods, it is important to prevent evaporation of the cell culture medium, as this may result in changes in medium composition. We recommend

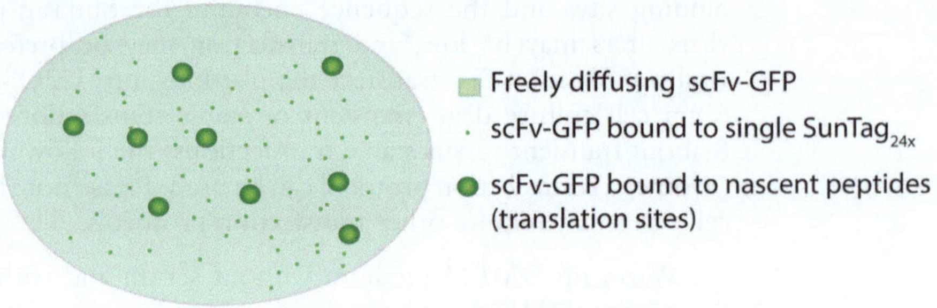

Fig. 3 Different pools of scFv-GFP present in cells. Expression of scFv-GFP in the presence of a reporter mRNA results in the appearance of three different pools of GFP in the cell: (1) a pool of unbound, freely diffusing scFv-GFP, (2) a pool of scFv-GFP bound to mature SunTag proteins, which have 24 SunTag peptides and are thus ~24 times brighter than the single scFv-GFP molecules, and (3) a pool of scFv-GFP bound to the newly synthesized SunTag peptides which represent sites of translation. Translation sites are usually 1–20× brighter than single SunTag proteins, as multiple ribosomes can bind to a single mRNA molecule

keeping a lid on the imaging plate whenever possible to prevent evaporation.

5. When using a doxycycline-inducible promoter to express the translation reporter, add doxycycline (used at a final concentration of 1 μg/mL) approximately 10 min before the start of imaging. Addition of doxycycline to the cells will generally induce expression of the reporter mRNAs within 15–30 min (*see* **Note 10** about how to add doxycycline to the cells). Adding doxycycline immediately before the start of imaging allows imaging the first rounds of translation of newly transcribed mRNAs as well. In addition, adding doxycycline just before the start of imaging prevents cytoplasmic depletion of scFv-GFP, which occurs when high levels of SunTag protein are present in the cell (*see* below and **Note 11** about the levels of scFv-GFP).
6. Select cells for imaging that have the correct levels of GFP- and mCherry-tagged proteins, and in which the translation reporter is expressed. (*see* **Note 12** for more details about the expression levels of mCherry, *see* **Note 11** and Fig. 3 for more details about scFv-GFP levels, *see* **Note 13** for more information on how to select cells with the correct levels of translational reporter). Note that the number of cells that can be imaged at the same time is limited when a high time resolution is required.

3.4 Imaging and Image Acquisition

Different optical imaging techniques, including widefield, point scanning confocal, spinning disc confocal, TIRF and light sheet microscopy can be used for imaging translation (*see* **Note 14** about the advantages and disadvantages of different imaging techniques).

1. Select the appropriate microscope objective. In general, we image with a 100× NA1.49 oil-immersion objective to obtain a high resolution and good sensitivity. Objectives with lower NA or magnification may also be used to image translation sites with multiple ribosomes, but might fail to reliably detect the mCherry-labeled mRNA or single ribosomes translating an mRNA.
2. Set the appropriate laser power and exposure time. Laser power and exposure time settings depend on the objective, microscope, camera and specifics of the experimental design. In the case of the tethered mRNA assay, low laser power in combination with a long exposure time provides the highest image quality and signal-to-noise ratio. Long exposure times (in the range of 500 ms) will cause motion blurring of the highly motile, GFP-labeled mature SunTag proteins and will therefore result in a more homogenous background signal. Since tethered mRNAs diffuse more slowly and are therefore not motion blurred at 500 ms exposure times, long exposure times will help to distinguish translation sites from background signal.
3. Find the correct focal plane to imaging mRNAs and translation sites. When the mRNA is tethered to the plasma membrane, it is important to focus the objective slightly above the plasma membrane of the cells during the imaging, as this is where the fluorescence associated with both the mRNA and the ribosomes translating the tethered mRNAs is located (*see* **Note 15** on how to focus on both the mRNA and translation sites).
4. Set the time interval for time-lapse imaging. In order to image translation dynamics and allow measurements of translation initiation and elongation, we usually acquire an image every 30 s. In general, shorter time intervals make it easier to detect short lived events or to track individual mRNAs over time, while imaging with longer time intervals results in reduced photobleaching and phototoxicity, which allows imaging for longer time periods. In experiments in which mRNAs are tethered, we found a 30 s time interval to be a good compromise between high temporal resolution and photobleaching for most of our experiments. However, when mRNAs are not tethered, a higher temporal resolution may be required to achieve accurate mRNA tracking.
5. Start image acquisition.
6. Add drugs which interfere with translation, as required by experimental setup. In most experiments, we recommend to add drugs after 10–30 min of imaging, when 10–50 translations sites are present per cell. Adding the drugs during imaging allows one to observe their immediate effects on translation. After addition of the drugs (*see* Subheading 3.5), cells should be imaged for another 5–30 min to observe the effect on translation.

3.5 Drugs that Interfere with Translation as Tools to Study Translation Dynamics

Several drugs that are known to interfere with translation can be added to the cells to measure specific aspects of translation dynamics. Note that when adding drugs to the medium, it is advisable to predilute the drug in large volume (~20% of final volume) (*see* **Note 10** on adding drugs to the cells).

1. *Puromycin* (used at 100 μg/mL). Puromycin binds the elongating nascent polypeptide chain, thereby releasing the nascent polypeptide from the ribosome and dissociating the ribosomal subunits from the mRNA. Addition of puromycin to the cells results in the disappearance of bright GFP spots (translation sites) within 1 min after addition. Puromycin can therefore be used as a tool to verify whether the observed GFP spots are active translation sites. Make a 10 mg/mL stock concentration of puromycin in DMSO, which can be diluted to a concentration of 700 μg/mL (7× the final concentration) in imaging medium. Of this dilution, add 50 μL to the 300 μL imaging medium present in a well of a 96 well-plate, to create a final concentration of 100 μg/mL.
2. *Cycloheximide* (used at 200 μg/mL). Cycloheximide (CHX) binds to the E-site of the ribosome, preventing release of the ribosome-bound tRNA and ribosomal translocation along the mRNA. Thus, CHX treatment results in stalling of ribosomes on the mRNA and should lead to the stabilization of GFP signal at translation sites. CHX may therefore be used to address whether changes in GFP intensity at sites of translation are caused by altered ribosomal occupancy. Make a 50 mg/mL stock concentration of cycloheximide in DMSO, which can be diluted to a concentration of 1400 μg/mL (7× the final concentration) in imaging medium. Of this dilution, add 50 μL to the 300 μL imaging medium present in a well of a 96 well-plate, to create a final concentration of 200 μg/mL.
3. *Harringtonine* (used at 3 μg/mL). Harringtonine is a small molecule translation inhibitor that specifically blocks translocation of ribosomes at the initiation codon, without affecting downstream ribosomes. As a consequence, upon harringtonine treatment ribosomes downstream of the start codon will complete translation normally and dissociate from the mRNA one-by-one after translation termination, resulting in a gradual decrease of the translation site GFP signal. Measuring the decay rate of GFP fluorescence from single mRNAs, and fitting the data to a simple mathematical model [5], allows estimation of ribosome translocation rates on a given mRNA transcript. The duration of ribosome run-off is dependent on the length of the POI. In case of the translational reporter Addgene #74928, run-off can be observed within 5–15 min after harringtonine addition. Starting from a 3 mg/mL stock

concentration of harringtonine in DMSO, make a dilution of 21 μg/mL (7× the final concentration) in imaging medium. Of this dilution, add 50 μL to the 300 μL imaging medium present in a well of a 96 well-plate, to create a final concentration of 3 μg/mL.

3.6 Image Analysis

Using the translation imaging method described here, the GFP-intensity of translation sites can be used to determine quantitative features of translation, including ribosome density on mRNAs, translation initiation rates, ribosome translocation rates and ribosome stalling. Determining these characteristics of translation requires (1) precise quantification of GFP intensities and (2) careful interpretation of the GFP fluorescence intensity.

3.6.1 Measuring GFP Intensities and Determining the Number of Ribosomes on an mRNA

The GFP intensity reports on the number of ribosomes on an mRNA. In some cases, it is sufficient to determine the relative number of ribosomes on each mRNA, in which case comparing GFP intensities between translation sites is possible. However, for other experiments it is useful to determine the absolute number of ribosomes on an mRNA. Below, we describe a step-wise protocol on how to measure GFP intensities and how these fluorescence intensity measurements can be used to calculate the number of ribosomes present on an mRNA.

In order to determine the number of ribosomes on an mRNA based on the GFP intensity of the translation site, it is important to compare the observed GFP intensity of a translation site with the GFP intensity of a single mature SunTag protein. Visualizing single mature SunTag proteins requires imaging with a short exposure time (10–30 ms) and a sensitive camera as a single SunTag protein contains at most 24 GFP molecules (in contrast to translation sites, which can contain >100 GFPs). Use a laser power that is sufficiently high to allow detection of single SunTag proteins, but without saturating camera pixel intensities at the much brighter translation sites. Single SunTag molecules are detectable on either EMCCD or sCMOS cameras.

1. Measure the GFP intensity of a single mature protein.
   (a) Draw a region of interest (ROI) around each fluorescent spot that represents a freely diffusing mature SunTag protein (*see* **Note 16** on how to select foci representing mature SunTag proteins). Use an ROI that is as small as possible, but large enough to also accommodate translation sites.
   (b) Measure the mean fluorescence intensity in the ROI.
   (c) Measure the mean background GFP intensity in the cell, by using a large ROI which does not contain any GFP foci.

   (d) Subtract the mean background intensity from the mean spot intensity to obtain the intensity of a single SunTag protein.
2. Measure GFP intensity of a translation site.
   (a) Draw a ROI (of the same size as used for measuring single mature proteins) around each translation site.
   (b) Measure the average fluorescence intensity of each translation site.
   (c) Subtract the mean background intensity (measured in **step 1c**) from the mean translation site intensity to calculate the mean intensity of a translation site.
3. In experiments where substantial photobleaching is observed, correction for photobleaching of the fluorescence intensities is critical (*see* **Note 17** for further discussion about photobleaching).
4. Divide the mean GFP intensity of the translation sites by the mean intensity of the single mature SunTag proteins.
5. The value calculated in **step 4** provides an estimate of the number of SunTag arrays present at a translation site (*see* **Note 18** and Fig. 4 for a further description on how to interpret GFP intensity).

3.6.2 Image Analysis Software to Measure GFP Intensities

In order to analyze the images obtained by microscopy, different image analysis software packages can be used, including Matlab, Python, and ImageJ. The choice for a specific software package mainly depends on the experimenter's previous experience and personal preference. For unique or complex questions, custom analysis software may be required, making Matlab and Python good options. However, for many simple types of analysis, existing ImageJ plugins can be used. Currently, several plugins are available which allow, for example, counting of the number of translation spots per cell, measuring the intensities of individual translation spots, or tracking translation spots over time. One simple ImageJ plugin that is useful for the analysis of translation dynamics is the spot_counter ImageJ plugin (http://fiji.sc/SpotCounter) developed by Nico Stuurman. This plugin counts the number of translation spots in a cell over time, and determines the fluorescence intensity of individual spots.

4 Notes

1. Choosing the gene of interest in the translation reporter. In principle, any gene can be introduced in the translation reporter, and the choice will mainly depend on the goal of the experiment. However, it is important to take into account

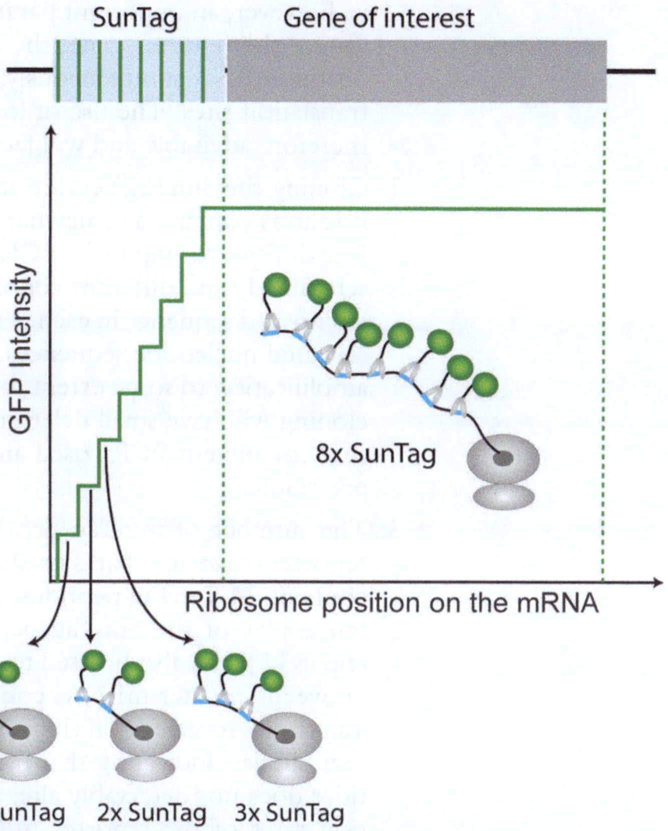

Fig. 4 Interpreting GFP fluorescence intensities of translation sites. The scFv-GFP intensity associated with a single ribosome depends on the location of the ribosome along the mRNA. The GFP intensity will initially increase as the ribosome synthesizes successive SunTag peptides (illustrated in the figure by the binding of 1, 2, or 3 scFv-GFP antibodies). Ribosome-associated fluorescence reaches a maximum once all SunTag peptides have been synthesized and will remain constant while the ribosomes translate the remaining sequence of the mRNA. As a consequence, ribosomes at the 3′ end of the mRNA are labeled brighter, than those at the 5′ end. When the number of ribosomes on an mRNA is calculated based on GFP intensity, these position-dependent effects of GFP intensity need to be taken into account. A simple mathematical model can be used, as described in the text

that other regulatory sequences outside the coding region, such as the 3′ and 5′ UTR, may influence translation efficiency as well and can also be inserted into the reporter construct, either on their own or together with the coding sequence. If the goal of the experiment is not related to the regulation of a specific gene, but rather to study global translational control mechanisms, the specific mRNA sequence inserted downstream of the SunTag sequence may not be critical and different sequences can be inserted. The *length* of the reporter sequence is, however, an important parameter to take into account. The longer the reporter gene, the more ribosomes can be present on the mRNA simultaneously, affecting the brightness of the translation sites. The use of longer reporter genes (1–2 kb) is therefore advisable and will facilitate further analysis.

2. Cloning the SunTag peptide array sequence. The SunTag peptide array contains a somewhat repetitive sequence and is therefore difficult to amplify by PCR. The SunTag sequence is codon scrambled (i.e., different codons are used to encode the same amino acid sequence in each peptide) to minimize the degree of repeated nucleotide sequences. Codon scrambling allows PCR amplification to some extent, but some clones after PCR-based cloning will have small deletions. Therefore, cloning strategies that circumvent PCR-based amplification of this sequence are preferable.
3. The number of SunTag peptides in the reporter gene. The reporter construct that is used by us and others [2–5] generally contains 24 SunTag peptides. However, shorter arrays (five or ten copies of the SunTag peptide [5]) or longer arrays (56 copies [2]) can also be used to image translation). Comparison between reporter mRNAs containing either 5, 10 or 24 SunTag copies revealed that ribosome density on these mRNAs was very similar, indicating that increased number of SunTag peptides does not detectably alter translation initiation or elongation rates of the reporter mRNA. Lowering the number of SunTag peptides will make the imaging of translation sites, especially at low ribosome occupancy, more challenging as it decreases the translation-associated GFP signal. Therefore, having a high copy number of SunTag peptides will be favorable in most situations. However, some specific experimental setups, such as the integration of the SunTag sequence into an endogenous gene locus by CRISPR/Cas9, may benefit from the use of shorter and less repetitive peptide arrays.
4. Using an inducible promoter to express a reporter mRNA. The advantage of using an inducible promoter is that it allows temporal control of mRNA synthesis. Expressing the reporter mRNA only during the imaging experiment has two main advantages; first, it prevents accumulation of high level of relatively bright mature SunTag proteins, which hinders the imaging of translation sites. (*See* also **Note 13** about the expression level of the translational reporter.) Second, limiting the levels of mature SunTag protein will also prevent depletion of the freely diffusible pool of scFv-GFP antibody from the cytoplasm. If mature SunTag protein levels become too high, the majority of scFv-GFP is bound to mature protein and is therefore not available to bind nascent SunTag peptides at translation sites. As a consequence, newly made SunTag

peptides emerging from the ribosome are labeled incompletely, limiting the fluorescence of translation sites.

5. Tethering the mRNA to the plasma membrane. PP7-mCherry binds with high affinity to the hairpin sequence present in the 3′ UTR of the reporter plasmid, resulting in fluorescent labeling of mRNA molecules. However, due to the rapid diffusion of single mRNAs, it is challenging to track single mRNAs over longer periods of time. In order to improve long-term tracking of individual mRNAs, a CAAX prenylation motif can be fused to the PCP. The CAAX motif anchors the PCP to the plasma membrane, which results in tethering of the reporter mRNAs containing the PCP binding sites to the plasma membrane. Membrane tethering of mRNAs reduces their mobility, facilitating long-term tracking of individual mRNA molecules (Fig. 1a, lower panel, b; [5]). Importantly, membrane tethering of mRNAs also allows a specialized form of microscopy, called TIRF microscopy, which significantly improves the signal-to-noise ratio of the image. So far, we have not observed any differences in translation dynamics between tethered and untethered mRNAs ([5] and unpublished results). However, under certain conditions tethering of the mRNA to the membrane could affect translation, for example when translation of the reporter mRNA is spatially regulated. Appropriate controls should therefore be performed in experiments involving mRNA tethering.
6. The importance of fusing sfGFP to the SunTag antibody to create scFv-GFP. The SunTag antibody has a tendency to aggregate at high expression levels in the cytoplasm of mammalian cells. To optimize intracellular expression of the antibody, a variety of N- and C-terminal fusion proteins known to enhance protein solubility were tested. Fusion of one variant of GFP, called sfGFP [9], to the C-terminus of the antibody resulted in soluble expression of the antibody even at high expression levels [5]. Therefore, when changing the fluorophore fused to the antibody, it is important to test whether the newly created antibody–fluorescent protein fusion does not aggregate in cells.
7. Choosing a suitable cell type. In principle, most cell types can be used to study translation using the method described here. However, there are some features which may be worthwhile to take into account when choosing a cell type. The most important aspect of a cell type is whether the plasmids described above can be efficiently delivered into the cells. In addition, it is important to consider whether the cell type can be imaged using high resolution microscopy. For example, cells grown in suspension may be more difficult to image than flat, adherent cells.

8. Transient versus stable transfection. In general, the plasmids required for translation imaging (*see* Subheading 2.1) can be either transiently transfected or used to generate cell lines stably expressing the genes of interest. While the use of a stable cell line will generally make results slightly more reproducible, a transient transfection will save time, as it allows for imaging of translation of a specific reporter 1 day after transfection. Since PP7-mCherry and scFv-GFP are used in every experiment, we recommend making a cell line in which both PP7-mCherry and scFv-GFP are stably expressed. This cell line can then be used to introduce different reporter mRNAs to study their translation. Note that in some cases we observed that stable expression of PP7-mCherry led to lysosomal accumulation of mCherry signal. Lysosomal accumulation was not detected when PP7 was fused to other fluorophores, such as GFP. mCherry-positive lysosomes are readily distinguishable from reporter mRNAs based on size, shape, intensity and motile properties, and therefore do not hinder the imaging of mRNAs, unless their localization overlaps with an mRNA molecule.

9. Using CO_2-independent, phenol red free imaging medium. Replacing the normal cell growth medium (generally CO_2-dependent media containing phenol red) with CO_2-independent, phenol red-free imaging medium ensures a correct pH over the course of the experiment and, in addition prevents that phenol red from interfering with fluorescence imaging. We routinely use Leibovitz-L-15 imaging medium which is CO_2-independent and free of phenol red, and therefore ideal for live-cell microscopy. Note that appropriate levels of serum and antibiotics should still be added to the imaging medium. Alternatively, an optimal pH during the experiment can be achieved by having a CO_2 supply in the microscope imaging chamber.

10. Diluting drugs before addition to the cells. We recommend prediluting drugs that need to be added to the cells during the imaging experiment in a large volume (~20% of final volume) before adding it to the cells. This ensures quick diffusion through the imaging medium so that the drug rapidly reaches the cells without repeated pipetting (which can cause cellular stress). We usually make a 7× dilution, of which we add 50 μL to the 300 μL imaging medium present per 96 well.

11. Expression levels of scFv-GFP. Expression of scFv-GFP in the presence of a reporter mRNA results in the appearance of three different populations of GFP particles in the cell: (1) a pool of freely diffusing scFv-GFP, (2) a pool of scFv-GFP bound to SunTag proteins which have completed translation and have been released from the ribosome (and thus contain a single SunTag peptide array), referred to as "mature proteins," and (3) a pool of scFv-GFP bound to the nascent SunTag peptides

which represent sites of translation (Fig. 3). scFv-GFP expression levels need to be sufficiently high to bind all SunTag peptides present in both the mature SunTag proteins and the newly synthesized nascent SunTag polypeptides as they emerge from the ribosomes. If scFv-GFP levels are too low, all antibody will be bound to mature SunTag protein, and the newly synthesized, nascent SunTag peptides will not be (completely) fluorescently labeled. On the other hand, if scFv-GFP levels are too high, this will give a strong background fluorescence of unbound scFv-GFP that can mask the signal of translation sites.

12. Expression levels of PP7-mCherry. The expression levels of PP7-mCherry should be high enough to saturate binding to the PP7 binding sites, but low enough to prevent high background fluorescence of PP7-mCherry not bound to an mRNA. In addition, high expression of PP7-mCherry can cause accumulation in lysosomes, resulting in bright red dots in the cell (*see* also **Note 8** about lysosomal accumulation of PP7-mCherry).
13. Expression level of the translational reporter. For most experiments it is useful to select cells or a cell line for imaging with a high number of transcripts, as this enables the imaging of many translation events in one experiment. However, very high expression of the reporter also results in rapid antibody depletion (i.e., a situation in which the majority of antibody is bound to mature protein, resulting in weak labeling of translation sites, *see* **Note 11**) and might impair long-term tracking of mRNAs (as moving mRNAs are more likely to cross paths). Optimal expression levels of the reporter mRNA therefore depends on the specific experimental conditions. Inducible expression of the reporter reduces some of the problems of high expression levels of the mRNA reporter (i.e., antibody depletion) and is therefore generally beneficial. An additional approach to prevent high levels of labeled, mature SunTag protein in the cells involves fusion of the SunTag protein to a degron to ensure its rapid degradation after its synthesis is completed. This approach has been successfully used to increase the signal-to-noise ratio during imaging experiments [3, 4]. When selecting cells with the correct level of the translational reporter, it is important to note that in case of very high levels of mature SunTag protein expression, the GFP signal in the cell may appear homogenous throughout the cell without clear SunTag punctae, because each SunTag protein is labeled with so few scFv-GFP molecules that individual SunTag-labeled proteins can no longer be distinguished from unbound scFv-GFP molecules based on their fluorescence intensity.

14. Advantages and disadvantages of different imaging systems. *Widefield microscopy*: Widefield microscopy allows imaging of relatively thick Z-sections, facilitating mRNA tracking in 3D with limited number of Z-slices acquired. The reduced number of Z-slices required for tracking mRNAs in 3D may cause less photobleaching, and increase the time resolution that can be achieved. A drawback of widefield imaging, however, is that the signal-to-noise ratio is lower due to increased out-of-focus light, and therefore only moderately to strongly translating mRNAs (i.e., mRNAs translated by multiple ribosomes) will be detectable. *Point scanning confocal microscopy*: With point scanning confocal microscopy, a higher signal-to-noise ratio can be achieved (compared to widefield microscopy). However, point scanning confocal microscopy is slow and causes relatively high levels of photobleaching and phototoxicity, and is therefore less suitable for long-term live-cell imaging. *Spinning disc confocal microscopy:* Similar to point scanning confocal microscopy, the use of spinning disc confocal microscopy allows imaging with a high signal-to-noise ratio. However, spinning disc confocal imaging is faster than point scanning confocal microscopy and may cause lower levels of phototoxicity to the sample. It is therefore suitable for imaging translation of single mRNA molecules with high sensitivity over longer time periods, and is our system of choice for the majority of experiments involving the translation imaging system. *Total Internal Reflection Fluorescence (TIRF) microscopy*: TIRF microscopy greatly reduces background signal and therefore increases the signal-to-noise ratio. However, only fluorescent molecules that are located close to the glass surface (e.g., near the plasma membrane at the bottom of the cell) can be observed, so tethering of the mRNAs to the membrane is recommended when imaging translation sites using TIRF. An additional disadvantage of TIRF microscopy is that the illumination of cells is generally uneven, which hinders quantitative measurements of fluorescence intensities. *Light sheet microscopy*: Light sheet microscopy can potentially be used to image translation in thick samples, such as tissues and embryos of various model organisms.

15. Finding the correct focal plane for imaging mRNAs and translation sites using the plasma membrane tethering approach. Although red mRNAs and green translation spots largely colocalize, we noted that the mRNA fluorescence signal localizes slightly below (i.e., closer to the plasma membrane) than the translation signal. This is expected as mCherry is fused directly to the CAAX domain, and thus very close to the plasma membrane, whereas the GFP is connected to the membrane through mCherry, the mRNA, and the nascent chain, and

will thus localize slightly further toward the cell interior (Fig. 1a, lower panel). Therefore, care should be taken to ensure both the mRNA and translation signal are in focus.

16. Selecting foci representing mature SunTag proteins to determine their fluorescence intensity. Since mature proteins are not tethered to the plasma membrane and are relatively small compared to translation sites, they diffuse rapidly in 3D throughout the cell. Imaging mature SunTag proteins and measuring their fluorescence intensity is complicated by their fast diffusion, as the rapid movement of single SunTag protein causes motion blurring of the fluorescent foci in the image. In addition, rapid movement of mature SunTag proteins in the *Z*-axis causes many spots to be slightly out of focus when images are acquired. Both issues involving focus and motion blurring affect the fluorescence intensity of mature SunTag proteins. To minimize abovementioned issues, imaging with a short (10–30 ms) exposure time is recommended, which will reduce motion blurring. In addition, manually selecting foci for quantification that appear in focus will help alleviate abovementioned issues.
17. Correcting for photobleaching. As a consequence of exposing fluorophores to excitation light, photobleaching occurs over time, reducing the intensity of GFP measured at translation sites. The rate at which photobleaching occurs can be determined by measuring the GFP signal of a large area in the cell (potentially the whole cell or field of imaging). Choosing an area of the cell lacking translation spots to measure bleaching rates ensures that such measurements are not affected by appearance and disappearance of mRNAs. In experiments where substantial photobleaching is observed, correction for photobleaching of the fluorescence intensities of translation sites is critical.
18. Interpreting the scFv-GFP fluorescence intensity to measure translation dynamics. The GFP signal observed at translation sites is a result of the nascent SunTag peptides bound by scFv-GFP antibodies. Importantly, ribosomes on the 5′ end of the mRNA have not yet translated all the SunTag peptides, and thus have fewer antibodies and fewer GFPs associated with them as compared to ribosomes at the 3′ end of the mRNA (Fig. 4). As a result, the GFP intensity associated with a ribosome at the 5′ end of the mRNA is lower than with a ribosome at the 3′ end of the mRNA (which has synthesized the entire SunTag peptide array). As a consequence, the measured GFP intensity at a translation site is not directly related to the number of ribosomes on the mRNA. To correctly translate GFP intensity to ribosome number, both ribosome density and ribosome location along the mRNA need to be taken

into account. A simple mathematical model can be used to calculate the number of ribosomes from the measured GFP intensities [5].

Acknowledgment

This work was supported by a by a fellowship of the Dutch Cancer Society (KWF) to M.E.T. and a Veni grant from the Netherlands Organization for Scientific Research (NWO) to S.R.

References

1. Morisaki T, Lyon K, DeLuca KF, DeLuca JG, English BP, Zhang Z, Lavis LD, Grimm JB, Viswanathan S, Looger LL, Lionnet T, Stasevich TJ (2016) Real-time quantification of single RNA translation dynamics in living cells. Science (New York, NY) 352(6292):1425–1429. doi:10.1126/science.aaf0899
2. Pichon X, Bastide A, Safieddine A, Chouaib R, Samacoits A, Basyuk E, Peter M, Mueller F, Bertrand E (2016) Visualization of single endogenous polysomes reveals the dynamics of translation in live human cells. J Cell Biol 214 (6):769–781. doi:10.1083/jcb.201605024
3. Wang C, Han B, Zhou R, Zhuang X (2016) Real-time imaging of translation on single mRNA transcripts in live cells. Cell 165 (4):990–1001. doi:10.1016/j.cell.2016.04.040
4. Wu B, Eliscovich C, Yoon YJ, Singer RH (2016) Translation dynamics of single mRNAs in live cells and neurons. Science (New York, NY) 352 (6292):1430–1435. doi:10.1126/science.aaf1084
5. Yan X, Hoek TA, Vale RD, Tanenbaum ME (2016) Dynamics of translation of single mRNA molecules in vivo. Cell 165 (4):976–989. doi:10.1016/j.cell.2016.04.034
6. Tanenbaum ME, Gilbert LA, Qi LS, Weissman JS, Vale RD (2014) A protein-tagging system for signal amplification in gene expression and fluorescence imaging. Cell 159(3):635–646. doi:10.1016/j.cell.2014.09.039
7. Bertrand E, Chartrand P, Schaefer M, Shenoy SM, Singer RH, Long RM (1998) Localization of ASH1 mRNA particles in living yeast. Mol Cell 2(4):437–445
8. Chao JA, Patskovsky Y, Almo SC, Singer RH (2008) Structural basis for the coevolution of a viral RNA-protein complex. Nat Struct Mol Biol 15(1):103–105. doi:10.1038/nsmb1327
9. Pedelacq JD, Cabantous S, Tran T, Terwilliger TC, Waldo GS (2006) Engineering and characterization of a superfolder green fluorescent protein. Nat Biotechnol 24(1):79–88. doi:10.1038/nbt1172

Chapter 27

Systematic Detection of Poly(A)$^+$ RNA-Interacting Proteins and Their Differential Binding

Miha Milek and Markus Landthaler

Abstract

RNA-binding proteins are dynamic posttranscriptional regulators of gene expression. Identification of mRNA-binding proteins in a given experimental setting is thus of great importance. We describe a procedure to enrich for direct poly(A)$^+$ RNA protein binders by 4-thiouridine-enhanced UV cross-linking and oligo(dT) purification. Subsequent nuclease-mediated release of RNA-binding proteins (RBPs) from mRNA allows for detection of eluted proteins by mass spectrometry. In addition, we provide a comparative approach to detect differences in RBP binding activity upon a biological stimulus.

Key words Protein–RNA interactions, RNA-binding proteins, Photoactivatable ribonucleoside, 4-Thiouridine, UV cross-linking, Oligo(dT) affinity purification, Mass spectrometry

1 Introduction

To systematically identify proteins binding to mRNA, a quantitative approach termed mRNA interactome capture was recently published [1, 2]. This method exploits in vivo UV cross-linking, oligo (dT) affinity purification, and mass spectrometry to identify the proteins directly bound to poly(A)$^+$ RNA. Since its initial publication, interactome capture has been successfully applied to other cultured mammalian cells [3–5], *Drosophila* embryos [6, 7], *Caenorhabditis elegans* [8], *Plasmodium falciparum* [9], *Arabidopsis thaliana* [10, 11], and yeast [4, 12]. A common feature of this approach is to stabilize the protein–RNA interactions by either conventional (UV 254 nm) or photoactivatable ribonucleoside (PAR)-enhanced (UV 365 nm) cross-linking. As in cross-linking and immunoprecipitation (CLIP)-related techniques [13–16], covalent bonds are formed at the site of direct contact between a nucleotide in the RNA and an amino acid in the binding protein molecule [17]. This linkage allows for stringent oligo(dT) affinity purification under denaturing conditions and highly selective recovery of protein–mRNA complexes.

Imre Gaspar (ed.), *RNA Detection: Methods and Protocols*, Methods in Molecular Biology, vol. 1649, DOI 10.1007/978-1-4939-7213-5_27,

Compared to other gene expression regulators such as transcription factors, RNA-binding proteins (RBPs) show higher expression [18], are able to act both cotranscriptionally and post-transcriptionally in a concerted manner [19, 20] and are highly responsive to external cues, such as DNA damage [21]. Therefore, they represent an important avenue for future research utilizing systems-wide approaches. An exciting feature of mRNA interactome capture is thus to systematically, quantitatively and differentially analyze the RBPs that change their binding activity upon certain biological stimuli.

Here, we describe our procedure for isolation and differential analysis of poly(A)$^+$ RNA-bound proteins. Initially, mammalian cells in culture are incubated with 4-thiouridine (4SU), resulting in metabolic labeling of cellular RNA. Following the application of biological stimulus to one cell population, all samples are exposed to UV light facilitating the formation of direct protein–RNA cross-links in living cells and stabilization of protein–RNA complexes before cell lysis (Fig. 1). An important feature of our method is the usage of a calibrator "heavy" SILAC-labeled lysate [22] that is spiked into the samples corresponding to both experimental conditions and allows for quantitative comparisons between them. After a stringent oligo(dT) affinity purification and enrichment of poly(A)$^+$ RNA, the cross-linked proteins are identified and quantified using mass spectrometry. Afterward, the "heavy" SILAC intensity allows for the correction of between replicate variability due to separate MS runs, amount of starting material and/or efficacy of oligo(dT) affinity purifications. We present a detailed experimental procedure that may be easily applied to different cell culture systems and stimulus specific responses.

2 Materials

2.1 Equipment

1. General cell culture equipment, e.g., incubator and dishes.
2. UV cross-linker device (Stratalinker or similar) with UV 365 nm light bulbs.
3. 15- and 50-mL Falcon tubes.
4. Heating block at 80 °C.
5. Magnetic stands for Eppendorf tubes, 15- and 50-mL Falcon tubes (e.g., Thermo Fisher Scientific DynaMag-2, -15, and -50 magnets).
6. Sterile cell scrapers.
7. Needles and syringes, 23- and 26-G.
8. Centrifugal filters with 10,000 MW cutoff and 15-mL volume (e.g., Millipore. Amicon Ultra-15 Centrifugal Filter).

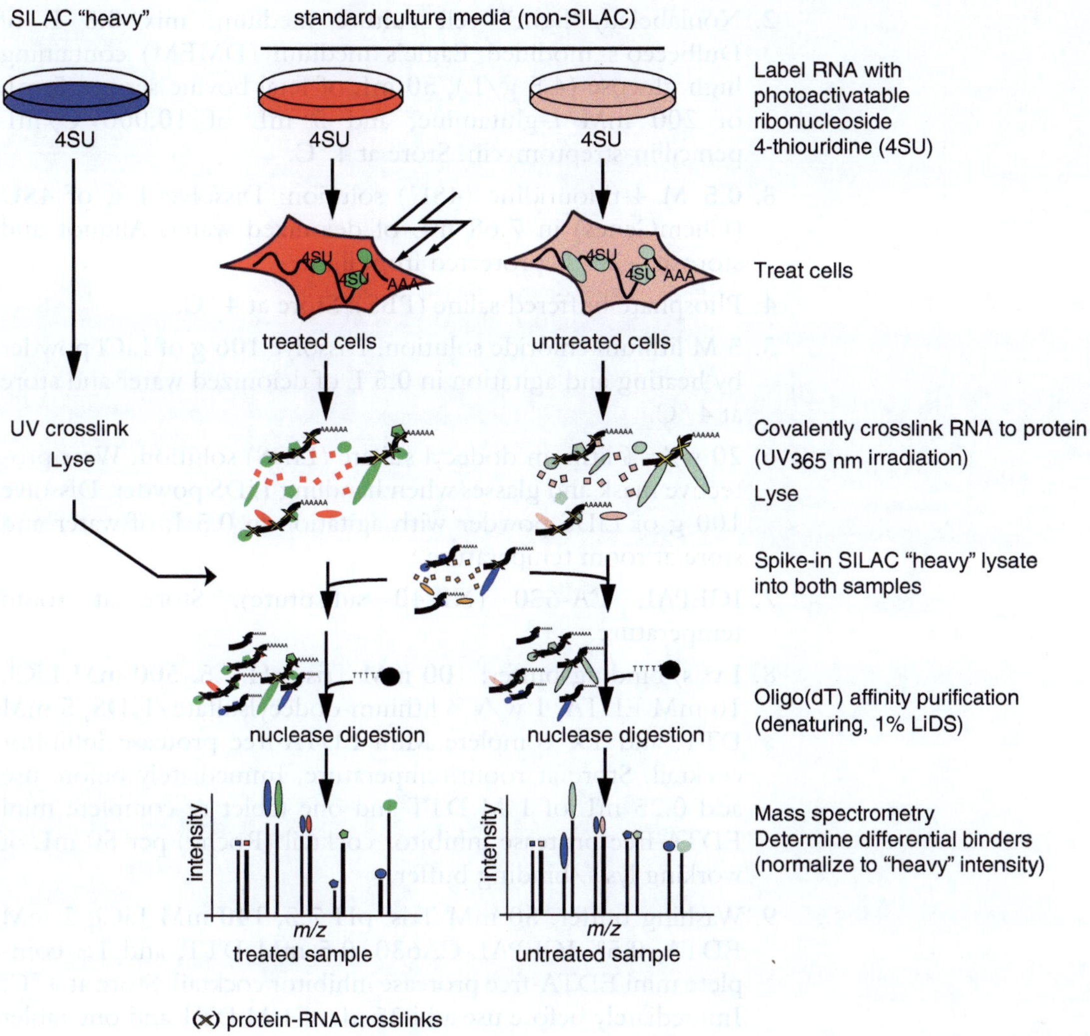

Fig. 1 Graphical representation of experimental procedure to detect differential protein binding to poly(A)$^+$ RNA upon biological stimulus. In order to be able to accurately normalize the data between two experimental conditions, we rely upon the usage of "heavy" SILAC lysate obtained from separately cultured "heavy"-labeled cells. Cell lysates obtained from untreated or treated "light" cells are spiked with the same volume of "heavy" SILAC lysate, followed by oligo(dT) affinity purification. Protein–mRNA-containing eluates are then nuclease-treated, concentrated and used in mass spectrometry runs. The measured light peptide intensities can be normalized to the heavy intensities, which are present due to the spike-in

9. Rotating wheel for falcon tubes.
10. Liquid nitrogen.

2.2 Solutions

We routinely use distilled water (MilliQ grade) to prepare the solutions described below.

1. Adherent cell line, such as HEK293, HeLa, U2OS, and MCF-7 (ATCC).

2. Nonlabeled/"light" cell culture medium: mix 0.5 L of Dulbecco's modified Eagle's medium (DMEM) containing high glucose (4.5 g/L), 50 mL of fetal bovine serum, 5 mL of 200 mM L-glutamine, and 5 mL of 10,000 U/mL penicillin-streptomycin. Store at 4 °C.
3. 0.5 M 4-thiouridine (4SU) solution: Dissolve 1 g of 4SU (ChemGenes) in 7.68 mL of deionized water. Aliquot and store at −20 °C protected from light.
4. Phosphate buffered saline (PBS). Store at 4 °C.
5. 5 M lithium chloride solution. Dissolve 106 g of LiCl powder by heating and agitation in 0.5 L of deionized water and store at 4 °C.
6. 20 w/v % lithium dodecyl sulfate (LiDS) solution. Wear protective mask and glasses when handling LiDS powder. Dissolve 100 g of LiDS powder with agitation in 0.5 L of water and store at room temperature.
7. IGEPAL CA-630 (NP-40 substitute). Store at room temperature.
8. Lysis/binding buffer: 100 mM Tris, pH 7.5, 500 mM LiCl, 10 mM EDTA, 1 w/v % lithium-dodecylsulfate/LiDS, 5 mM DTT, and 1× complete Mini EDTA-free protease inhibitor cocktail. Store at room temperature. Immediately before use add 0.25 mL of 1 M DTT and one tablet of complete mini EDTA-free protease inhibitor cocktail (Roche) per 50 mL of working lysis/binding buffer.
9. Washing buffer: 50 mM Tris, pH 7.5, 140 mM LiCl, 2 mM EDTA, 0.5% IGEPAL-CA630, 0.5 mM DTT, and 1× complete mini EDTA-free protease inhibitor cocktail. Store at 4 °C. Immediately before use add 25 μL of 1 M DTT and one tablet of complete Mini EDTA-free protease inhibitor cocktail (Roche) per 50 mL of working buffer.
10. Magnetic oligo(dT) beads (e.g., Thermo Fisher Scientific Oligo$(dT)_{25}$ Dynabeads, component of the mRNA DIRECT kit).
11. Low-salt elution buffer: 10 mM Tris, pH 7.5. Store at 4 °C. The elution buffer provided in the mRNA DIRECT kit (Thermo Fisher Scientific) may also be used.
12. Concentrating buffer: 10 mM Tris, pH 7.5, 50 mM NaCl. Store at 4 °C.
13. 100 U/μL RNase I solution. Store at −20 °C.
14. 125 U/μL benzonase solution. Store at −20 °C.
15. 1 M $MgCl_2$. Dissolve 4.76 g of $MgCl_2$ in 50 mL of deionized water. Store at 4 °C.

16. 0.4 M Arg10 solution: Reconstitute 62.1 mg of $^{13}C_6$,$^{15}N_4$ L-arginine HCl (Cambridge Isotope Laboratories) in 700 μL deionized water. Store at −20 °C.
17. 0.8 M Lys8 solution. Reconstitute 128.1 mg of $^{13}C_6$,$^{15}N_4$ L-lysine 2HCl (Cambridge Isotope Laboratories) in 700 μL deionized water. Store at −20 °C.
18. SILAC "heavy" cell culture medium: Mix 0.5 L of high glucose SILAC DMEM (PAA or Thermo Fisher Scientific), 50 mL of dialyzed fetal bovine serum, 5 mL of 200 mM L-glutamine, 5 mL of 10,000 U/mL penicillin-streptomycin, 0.5 mL of 0.4 M Arg10 and 0.5 mL of 0.8 M Lys8. Filter the medium through 0.2 μm filter membrane and store at 4 °C.

3 Methods

3.1 Cell Culture and UV Cross-Linking

1. Culture adherent cells in desired medium until 40–70 % confluence and incubate them in the presence of 100–200 μM 4-thiouridine (4SU) for 6–10 h. Duration and concentration of 4SU treatment may have to be optimized (*see* **Note 1**).
2. (Optional: Apply an additional labeling pulse with 100 μM 4SU 1–2 h prior to UV cross-linking to ensure the labeling of short-lived transcripts.)
3. Decant culture medium and place the dishes on ice. Wash once with 5–10 mL ice-cold PBS to remove the residual culture medium. Make sure you remove most of the liquid before exposure to UV light (*see* **Note 2**). Next, cross-link cells with UV light with a wavelength of 365 nm and energy of 0.2 J/cm^2 using a UV cross-linker. Repeat this process for all remaining dishes in culture.
4. Immediately harvest UV-exposed cells by scraping them off in 5 mL of ice-cold PBS using a cell scraper. Process a second batch of cells without exposing them to UV light. This sample serves as a background control.
5. Transfer cell suspensions into 50-mL falcon tubes and pellet cells by centrifugation with 400 × *g* at 4 °C for 5 min. Wash the cell pellet once with ice-cold PBS. Collect cells and freeze pellets in liquid nitrogen or continue with the cell lysis.

3.2 Cell Lysis and Oligo(dT) Purification

1. Lyse cells in approximately five times the cell pellet volume of lysis/binding buffer containing 5 mM DTT and Complete Mini EDTA-free protease inhibitor cocktail by pipetting up and down and incubate the extract for 15 min on ice (*see* **Note 3**). If sample volumes exceed 50 mL, aliquot in several 50-mL falcon tubes.

2. Pass the lysate with a syringe several times through a 23- and 26-G needle (*see* **Note 4**). Afterward, save 0.2 mL of input lysate for downstream analyses (e.g., Western analysis and RNA extraction) and store it at −80 °C.
3. Pre-wash the magnetic oligo(dT) beads in 5 mL of lysis/binding buffer in a 15 mL Falcon tube and concentrate them on a magnet. Afterward, resuspend beads into the cell extract by first taking them up in 1 mL of the prepared lysate and transferring them to the tube containing the remaining lysate (*see* **Note 5**).
4. Incubate for 1 h at room temperature on a rotating wheel (*see* **Note 6**).
5. Concentrate beads on magnet. The initial concentration of the beads on the magnet may take from 10 min to 1 h.
6. Carefully collect supernatant and place it into a new tube for additional rounds of oligo(dT) precipitation (*see* **Note 7**).
7. Wash beads two times in one sample volume of room-temperature lysis/binding buffer. For each washing step, rotate the tubes 2–5 min on a rotating wheel at room temperature. Afterward, concentrate the beads on magnet and carefully remove the supernatant (*see* **Note 8**).
8. Wash beads two times in one sample volume of room-temperature washing buffer containing 0.5 mM DTT and complete EDTA-free protease inhibitor cocktail. For each washing step, rotate the sample 2–5 min on a rotating wheel at room temperature. Afterward, concentrate the beads on magnet and carefully remove the supernatant.
9. For the third washing step, resuspend beads in 1 mL of washing buffer and transfer to a fresh 1.5-mL tube. Afterward, concentrate the beads on magnet and remove the supernatant.
10. Wash beads once with 1 mL of ice-cold low-salt elution buffer to remove traces of IGEPAL CA-630 (*see* **Note 9**). Concentrate beads with magnet and remove supernatant.
11. Resuspend beads in 0.5–1 mL of low-salt elution buffer and incubate them for 2 min at 80 °C. Remove supernatant into a new precooled 15-mL falcon tube and place it on ice as quickly as possible. Pool the eluates from different aliquots. Once the eluate is removed, immediately cool down the beads by placing them on ice (*see* **Note 10**).
12. Repeat the elution **step 11** once more.
13. Perform several additional rounds of oligo(dT) affinity purifications. We typically perform two additional rounds with only one elution per round for a total of three affinity purifications (*see* **Note 11**).

14. Keep the eluates on ice or store them at −80 °C until further use. Wash and store oligo(dT) beads in lysis/binding buffer at 4 °C. Reuse them up to three times.

3.3 Analyses of Oligo (dT) Eluates

1. Save an aliquot of the eluate (100 μL) for further RNA analyses and store it at −80 °C (*see* **Note 12**).
2. To prepare the remaining eluate for mass spectrometry analysis, digest RNA by incubation with RNAse I and benzonase. Dilute the RNAse I, benzonase and 1 M $MgCl_2$ 1:1000 in the eluate to obtain the final concentrations of 100 U/mL, 125 U/mL and 1 mM, respectively. Incubate for 1 h at 37 °C in a water bath.
3. Concentrate the eluate by the 15 mL centrifugal filter. Load your sample into the filter unit and fill up the filter device with concentrating buffer to a total volume of 14 mL. Centrifuge the filter unit at 4000 × *g* for 45 min at 4 °C (*see* **Note 13**). Recover the retentate from the filter device (approximately 150 μL).
4. Analyze the protein sample by silver staining (Fig. 2a) and Western analysis.
5. Submit the sample for mass spectrometry analysis and analyze the data to obtain the identity of enriched RNA-binding proteins (*see* **Note 14**).

3.4 Detection of Differences in Binding Activity Upon a Biological Stimulus

General consideration: If studying stress responses (e.g., oxidative stress, ER stress, and DNA damage), it is important to evaluate the effect of perturbations (4SU treatment, UV exposure, SILAC media incubation) on stress induction and cell growth, as well as 4SU labeling efficacy (*see* **Note 15**).

1. Grow a population of cells in SILAC heavy media formulation and passage them at least six times before proceeding with the experiment (*see* **Note 16**).
2. Once the SILAC cells are expanded and around 70–80% confluent, grow additional two populations of cells in “light” (i.e., non-SILAC) media until 80% confluency.
3. Incubate all three cell populations with 200 μM 4SU for 6–10 h (*see* **Note 17**).
4. Perform stimulation for one “light” cell population.
5. After the desired incubation, UV-cross-link the cells as described in Subheading 3.1 (*see* **Note 18**).
6. Immediately harvest UV-exposed cells by scraping them off in 5 mL ice-cold PBS using a rubber policeman. Transfer cell suspensions into 50-mL falcon tubes and pellet them by centrifugation (400 × *g*, 4 °C, 5 min). Wash the cell pellet once with

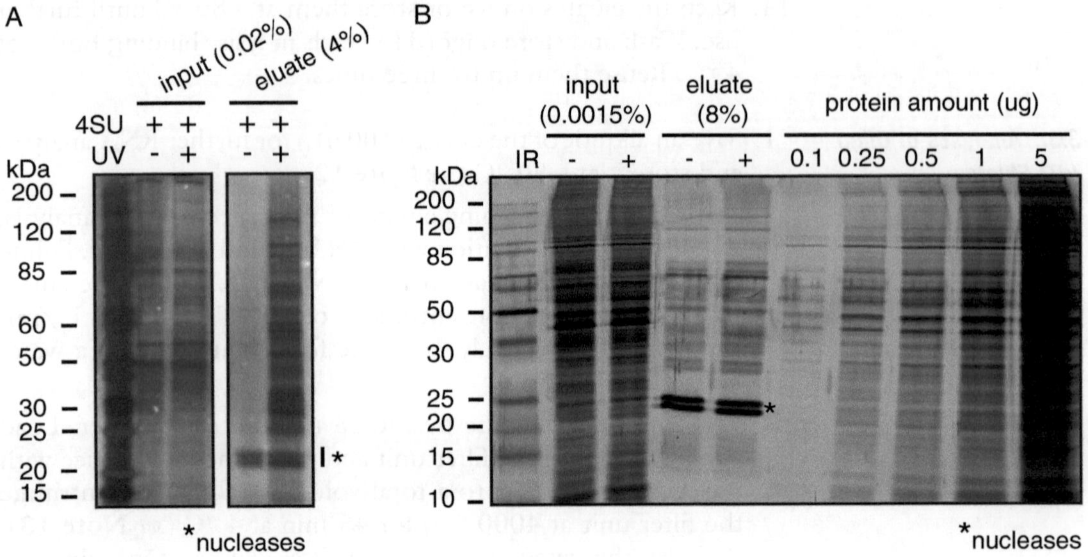

Fig. 2 Quality control and analysis of oligo(dT) eluates. (**a**) Isolation of poly(A)$^+$ RNA binding proteins. MCF-7 cells were incubated in the presence of 200 μM 4SU for 16 h, UV cross-linked (0.2 J/cm^2) or left untreated. After lysis and oligo(dT) purification, mRNA complexes were eluted and RNase treated. Indicated fractions of total input and eluate volumes were resolved with SDS-PAGE. Proteins were visualized using silver staining. (**b**) Example of determination of differential protein binding to poly(A)$^+$ RNA upon DNA damage induction. MCF-7 cells were incubated in the presence of 200 μM 4SU for 16 h, followed by ionizing radiation (IR) exposure (10 Gy) to induce double strand breaks, and UV cross-linking (0.2 J/cm^2). After oligo(dT) affinity purification, protein–mRNA complexes were eluted and RNase treated. Indicated fractions of total input and eluate volumes were resolved with SDS-PAGE. Proteins were visualized using silver staining. To judge the amount of proteins in oligo(dT) eluates, increasing amounts (0.1–5 μg) of an MCF-7 NP40 lysate with a known concentration were loaded on the same gel

ice-cold PBS. Collect cells and freeze pellets in liquid nitrogen or continue with lysis (*see* **Note 19**).

7. Lyse 'light" and "heavy" cells in approximately five times the cell pellet volume of lysis/binding buffer and incubate the extract for 15 min on ice.
8. Pass the lysate several times through a 23- and 26-G needle. Save an aliquot (0.2 mL) for downstream analyses and store it at −80 °C.
9. Spike-in the same volume of the heavy lysate into untreated and treated "light" cell lysates. We typically mix "heavy" to "light" lysates in a 1:5 (v/v) ratio (*see* **Note 20**).
10. Prewash the magnetic oligo(dT) beads in 5 mL of lysis/binding buffer. Afterward, resuspend them into the cell extract by first taking them up in 1 mL of the prepared lysate and transferring them to the tube containing the remaining lysate.
11. Perform the incubation, washing, and elution steps as described in Subheading 3.2.

12. Analyze the oligo(dT) eluates as described in Subheading 3.3. Representative results are shown in Fig. 2b.
13. Submit the sample for mass spectrometry analysis and analyze the data to obtain RNA-binding proteins that show differential binding upon stimulation (*see* **Notes 21** and **22**).

4 Notes

1. 4SU incorporation into RNA may be easily followed by dot blot analysis as described elsewhere [23]. Determine the conditions of incubations that give you a similar signal as 4SU-labeled RNA extracted from HEK293/HeLa/MCF-7 cells that were labeled for 6–10 h in the presence of 100–200 μM 4SU.
2. It is critical to remove the culture medium to allow the maximal penetration of UV light into the cells. Five 15-cm culture dishes will fit into the Stratalinker and can be UV irradiated at the same time.
3. We typically lyse one cell pellet obtained from an 80–90% confluent 15-cm culture dish in 1–3 mL of lysis/binding buffer, depending on cell type. For HEK293 cells the amount of poly(A)$^+$ RNA obtained from one 15-cm culture dish is high, requiring 3 mL of lysis buffer. Since the amount obtained from MCF-7 cells is lower, we use 1 mL for lysis. In the experiments shown in Fig. 2b, we used 60 dishes per sample and lysed the cells in 60 mL of lysis/binding buffer. Note that such a high amount of starting material can yield 50–150 μg of eluted proteins, and this can be beneficial when a high number of quantified proteins is needed. However, starting material may also be decreased to about 5–10 dishes per sample (yield 0.5–5 μg) provided the mass spectrometry experiments are carried out with the current state-of-the art sensitive MS equipment.
4. Due to the presence of strong detergent and high salt concentration, the lysate will be very viscous. Viscous material normally floats at the top of the solution. For efficacy of oligo(dT) affinity purification it is critical to decrease viscosity and remove insoluble material. Typically, we pass the lysate four times through a 23-G needle, followed by 2 passes through a 26-G needle. Note that it takes around 20–30 min per sample to perform this step. Be careful to keep the lysate cold at all times. If insoluble/viscous material is still visible at the top after this step, try to remove it by pipetting the bottom layer of the solution into a new precooled 50-mL falcon tube thus

separating it from the top viscous layer. To avoid RNA fragmentation, sonication is not recommended.

5. We suggest to use 200 μL of oligo(dT)$_{25}$ Dynabeads (bed volume equal to approximately 15 mL of original oligo(dT) Dynabeads) per 60 mL of cell lysate.
6. Care should be taken to dry the outside of the tubes before closing them, in order to prevent leakage during the incubation.
7. Keep the supernatant tube on ice until you are ready for the second incubation round with oligo(dT) beads. Alternatively, the supernatant may also be flash frozen in liquid nitrogen and kept at −80 °C until further use. If possible, prepare a second batch of oligo(dT) beads to start the second round of purifications while performing the washing steps after the first round. This may significantly decrease the total processing time.
8. Avoid excessive foaming of the buffer as this may lead to retention of beads on the top of the solution. If this occurs, pipette the top of the solution several times up and down using a 1000-μL pipette to suspend the floating beads.
9. These washing steps are critical for performing shotgun proteomics that is incompatible with contaminating detergents (LiDS, IGEPAL-CA630) in the sample. Ideally, the compatibility of the sample with downstream analyses should be discussed with the scientists performing the mass spectrometry.
10. After the incubation at 80 °C, the beads must be concentrated and supernatant removed as quickly as possible. If processing many tubes, perform the 2-min incubations at 80 °C for each tube separately in order to avoid cooling down the sample and reannealing of the RNA to oligo(dT) beads.
11. We suggest elution in 0.5–1 mL of elution buffer per 100 μL (bed volume) of oligo(dT) beads. For 60 mL of starting cell lysate, this results in 4–8 mL of eluate after completing all three affinity purification rounds. Typical yields can range from 0.5–5 μg (low input amount) to 50–150 μg of total protein (high input amount). In our experience, some proteins (e.g., AGO2) are readily detected in oligo(dT) eluates by Western analysis only after 2–3 rounds of affinity purification.
12. To analyze RNA in eluted samples by qRT-PCR and/or RNA-seq, RNA can be first treated with proteinase K to digest the proteins, followed by TRIzol extraction and further downstream analyses.
13. Alternatively, precipitate proteins with 0.25× eluate volume of trichloroacetic acid and incubation on ice for 1 h. Centrifuge at maximum speed (16,000 × *g*, 30 min) and wash the pellet twice with 0.20× eluate volume of cold acetone. Remove supernatant, air-dry the pellet and resuspend it in the desired

volume of water. Note that due to denaturation some proteins cannot be effectively resuspended after TCA precipitation.

14. After receiving the dataset, compare the intensity values against those obtained from the non-cross-linked control sample. Compare log2-fold enrichments and rank the proteins accordingly. To validate novel RNA-binding protein candidates we stably or transiently express FLAG/HA-tagged version of proteins of interest in cells, which are labeled for 6–10 h with 100–200 μM 4SU. After cross-linking the cells with exposure to 365 nm UV light and the energy of 0.15 J/cm^2 we follow Subheading 3.3, **steps 2**–7 of a recently published PAR-CLIP protocol [24] to examine the formation of protein–RNA complexes which can be detected by radiolabeling RNA in these complexes.
15. Use a biological marker that should be induced upon stimulus and evaluate the effect of perturbations that are applied to both unstimulated and stimulated cells, i.e., 4SU treatment, UV exposure and presence of SILAC media components. For example, in the case of DNA damage induction, a general marker for double strand break formation is phosphorylation of histone γ-H2AX. Quantity of γ-H2AX can be followed by Western analysis or immunofluorescence microscopy.
16. In order to prevent the usage of large amounts of SILAC heavy culture media, grow the cells in small culture dishes for the initial passages. Since cells generally grow slower in SILAC media, we do not recommend diluting them lower than threefold during passaging. Incorporation of “heavy” amino acids into proteins may be tested by performing a whole proteome analysis of nonlabeled/“light” and “heavy” cells. For successful incorporation, one would expect the distribution of SILAC heavy-to-light ratios to be centered at 1. Although we have not observed SILAC incorporation to be a critical issue in HEK293, HeLa, and MCF-7 cells, it may be of great importance to some specialized cell cultures.
17. SILAC media may significantly affect the efficacy of RNA metabolic labeling by 4SU. Perform optimizations as described in **Note 1**.
18. If you have many culture dishes for treated cells, make sure that the incubation time will be the same for all of them by stimulating them at slightly delayed time points. We typically complete the cross-linking and harvesting of five 15-cm dishes in 5–7 min.
19. It is a good idea to collect a separate aliquot of cells (1 mL) at this point to test if the biological marker is present in treated but not in untreated cells before proceeding with the large-scale experiment.

20. The amount of spiked-in volume should be discussed with the scientist performing the mass spectrometry experiments.

21. After receiving the dataset, normalize the "light" peptide intensity data points by normalization factors obtained from "heavy" peptide intensity. Linear regression between the "heavy" intensity from both conditions may be applied and the slope of the linear fit used as normalization factor. Alternatively, normalization may also be carried out by quantile normalization (preprocessCore, baySeq Bioconductor packages) [25] or DESeq2-style normalization which is based on geometric means [26]. Identify differential binders by computing log2-transformed fold changes between treated and untreated sample.

22. Differential amount of proteins in oligo(dT) eluates may arise due to differences in amounts of total poly(A)$^+$ RNA between treated and untreated cells as well as differences in total protein abundance. Therefore, mRNA and protein levels should be quantified in input lysates by qRT-PCR, RNA-seq, Western, and whole proteome analysis. If changes can only be observed on the RBPome level, but not on the mRNA and protein level the differential binding activity for RNA-binding proteins was successfully detected.

Acknowledgment

We thank Koshi Imami and Matthias Selbach (Max Delbrück Center for Molecular Medicine, Berlin) for their expertise in mass spectrometry. This work was supported by an International European Fellowship (Maria Sklodowska Actions FP7-PEOPLE-2011-IEF) and DFG grant LA 2941/5-1.

References

1. Baltz AG, Munschauer M, Schwanhausser B, Vasile A, Murakawa Y, Schueler M, Youngs N, Penfold-Brown D, Drew K, Milek M, Wyler E, Bonneau R, Selbach M, Dieterich C, Landthaler M (2012) The mRNA-bound proteome and its global occupancy profile on protein-coding transcripts. Mol Cell 46(5):674–690. doi:10.1016/j.molcel.2012.05.021. S1097-2765(12)00437-6 [pii]
2. Castello A, Fischer B, Eichelbaum K, Horos R, Beckmann BM, Strein C, Davey NE, Humphreys DT, Preiss T, Steinmetz LM, Krijgsveld J, Hentze MW (2012) Insights into RNA biology from an atlas of mammalian mRNA-binding proteins. Cell 149(6):1393–1406. doi:10.1016/j.cell.2012.04.031. S0092-8674(12)00576-4 [pii]
3. Kwon SC, Yi H, Eichelbaum K, Fohr S, Fischer B, You KT, Castello A, Krijgsveld J, Hentze MW, Kim VN (2013) The RNA-binding protein repertoire of embryonic stem cells. Nat Struct Mol Biol 20(9):1122–1130. doi:10.1038/nsmb.2638. nsmb.2638 [pii]
4. Beckmann BM, Horos R, Fischer B, Castello A, Eichelbaum K, Alleaume AM, Schwarzl T, Curk T, Foehr S, Huber W, Krijgsveld J, Hentze MW (2015) The RNA-binding proteomes from yeast to man harbour conserved enigmRBPs. Nat Commun 6:10127. doi:10.1038/ncomms10127

5. Liepelt A, Naarmann-de Vries IS, Simons N, Eichelbaum K, Fohr S, Archer SK, Castello A, Usadel B, Krijgsveld J, Preiss T, Marx G, Hentze MW, Ostareck DH, Ostareck-Lederer A (2016) Identification of RNA-binding proteins in macrophages by interactome capture. Mol Cell Proteomics 15(8):2699–2714. doi:10.1074/mcp.M115.056564
6. Wessels HH, Imami K, Baltz AG, Kolinski M, Beldovskaya A, Selbach M, Small S, Ohler U, Landthaler M (2016) The mRNA-bound proteome of the early fly embryo. Genome Res 26 (7):1000–1009. doi:10.1101/gr.200386.115
7. Sysoev VO, Fischer B, Frese CK, Gupta I, Krijgsveld J, Hentze MW, Castello A, Ephrussi A (2016) Global changes of the RNA-bound proteome during the maternal-to-zygotic transition in Drosophila. Nat Commun 7:12128. doi:10.1038/ncomms12128
8. Matia-Gonzalez AM, Laing EE, Gerber AP (2015) Conserved mRNA-binding proteomes in eukaryotic organisms. Nat Struct Mol Biol 22(12):1027–1033. doi:10.1038/nsmb.3128
9. Bunnik EM, Batugedara G, Saraf A, Prudhomme J, Florens L, Le Roch KG (2016) The mRNA-bound proteome of the human malaria parasite Plasmodium falciparum. Genome Biol 17(1):147. doi:10.1186/s13059-016-1014-0
10. Marondedze C, Thomas L, Serrano NL, Lilley KS, Gehring C (2016) The RNA-binding protein repertoire of Arabidopsis thaliana. Sci Rep 6:29766. doi:10.1038/srep29766
11. Reichel M, Liao Y, Rettel M, Ragan C, Evers M, Alleaume AM, Horos R, Hentze MW, Preiss T, Millar AA (2016) In planta determination of the mRNA-binding proteome of Arabidopsis etiolated seedlings. Plant Cell. doi:10.1105/tpc.16.00562
12. Mitchell SF, Jain S, She M, Parker R (2013) Global analysis of yeast mRNPs. Nat Struct Mol Biol 20(1):127–133. doi:10.1038/nsmb.2468
13. Ule J, Jensen KB, Ruggiu M, Mele A, Ule A, Darnell RB (2003) CLIP identifies Nova-regulated RNA networks in the brain. Science 302(5648):1212–1215. doi:10.1126/science.1090095. 302/5648/1212 [pii]
14. Konig J, Zarnack K, Rot G, Curk T, Kayikci M, Zupan B, Turner DJ, Luscombe NM, Ule J (2010) iCLIP reveals the function of hnRNP particles in splicing at individual nucleotide resolution. Nat Struct Mol Biol 17 (7):909–915. doi:10.1038/nsmb.1838
15. Licatalosi DD, Mele A, Fak JJ, Ule J, Kayikci M, Chi SW, Clark TA, Schweitzer AC, Blume JE, Wang X, Darnell JC, Darnell RB (2008) HITS-CLIP yields genome-wide insights into brain alternative RNA processing. Nature 456 (7221):464–469. doi:10.1038/nature07488
16. Hafner M, Landthaler M, Burger L, Khorshid M, Hausser J, Berninger P, Rothballer A, Ascano M Jr, Jungkamp AC, Munschauer M, Ulrich A, Wardle GS, Dewell S, Zavolan M, Tuschl T (2010) Transcriptome-wide identification of RNA-binding protein and microRNA target sites by PAR-CLIP. Cell 141 (1):129–141. doi:10.1016/j.cell.2010.03.009
17. Kramer K, Sachsenberg T, Beckmann BM, Qamar S, Boon KL, Hentze MW, Kohlbacher O, Urlaub H (2014) Photo-cross-linking and high-resolution mass spectrometry for assignment of RNA-binding sites in RNA-binding proteins. Nat Methods 11(10):1064–1070. doi:10.1038/nmeth.3092
18. Gerstberger S, Hafner M, Tuschl T (2014) A census of human RNA-binding proteins. Nat Rev Genet 15(12):829–845. doi:10.1038/nrg3813
19. Keene JD (2007) RNA regulons: coordination of post-transcriptional events. Nat Rev Genet 8 (7):533–543. doi:10.1038/nrg2111
20. Keene JD (2007) Biological clocks and the coordination theory of RNA operons and regulons. Cold Spring Harb Symp Quant Biol 72:157–165. doi:10.1101/sqb.2007.72.013
21. Dutertre M, Lambert S, Carreira A, Amor-Gueret M, Vagner S (2014) DNA damage: RNA-binding proteins protect from near and far. Trends Biochem Sci 39(3):141–149. doi:10.1016/j.tibs.2014.01.003
22. Ong SE, Blagoev B, Kratchmarova I, Kristensen DB, Steen H, Pandey A, Mann M (2002) Stable isotope labeling by amino acids in cell culture, SILAC, as a simple and accurate approach to expression proteomics. Mol Cell Proteomics 1(5):376–386
23. Radle B, Rutkowski AJ, Ruzsics Z, Friedel CC, Koszinowski UH, Dolken L (2013) Metabolic labeling of newly transcribed RNA for high resolution gene expression profiling of RNA synthesis, processing and decay in cell culture. J Vis Exp 78. doi:10.3791/50195
24. Garzia A, Meyer C, Morozov P, Sajek M, Tuschl T (2016) Optimization of PAR-CLIP for transcriptome-wide identification of binding sites of RNA-binding proteins. Methods. doi:10.1016/j.ymeth.2016.10.007
25. Hardcastle TJ, Kelly KA (2010) baySeq: empirical Bayesian methods for identifying differential expression in sequence count data. BMC Bioinformatics 11:422. doi:10.1186/1471-2105-11-422
26. Love MI, Huber W, Anders S (2014) Moderated estimation of fold change and dispersion for RNA-seq data with DESeq2. Genome Biol 15 (12):550. doi:10.1186/s13059-014-0550-8

Chapter 28

Isolation and Characterization of Endogenous RNPs from Brain Tissues

Rico Schieweck, Foong yee Ang, Renate Fritzsche, and Michael A. Kiebler

Abstract

Identification of physiological target RNAs and protein interactors bound to RNA-binding proteins is a key prerequisite to understand the underlying mechanisms of posttranscriptional expression control and RNA granule assembly. Here, we describe a multistep biochemical approach to isolate endogenous ribonucleoprotein particles from brain tissues by exploiting differential centrifugation and gradient fractionation followed by immunoprecipitation with monospecific, affinity-purified antibodies directed against selected RNA-binding proteins. This protocol results in highly enriched endogenous ribonucleoprotein particles that then can be analyzed by mass spectrometry (for proteins composition) and microarray or RNA sequencing technologies (for target mRNAs).

Key words Neuronal RNA granules, Staufen2, Barentsz, Differential centrifugation, Gradient fractionation, Immunoprecipitation

1 Introduction

Ribonucleoprotein particles (RNPs) are highly diverse multimolecular complexes consisting of RNA binding proteins (RNPs), RNA and other protein interactors [1]. Those RNA granules are essential to transport selected transcripts to specific compartments in the cell, thereby controlling local protein expression in mammalian neurons [2]. Research in the last decades has revealed their importance for regulating neuronal signaling cascades necessary for excitability and the establishment of neuronal circuits [2]. In the last few years, a serious effort was taken by several groups to unravel the composition of these granules [3, 4].

To identify protein interactors and RNA targets of RBPs, immunoprecipitation is the method of choice. By definition, immunoprecipitation is the isolation of protein-protein or protein–RNA complexes from lysates using antibodies directed against either an endogenous protein or peptide. Alternatively, exogenous RNPs

Imre Gaspar (ed.), *RNA Detection: Methods and Protocols*, Methods in Molecular Biology, vol. 1649, DOI 10.1007/978-1-4939-7213-5_28,

containing tagged RBPs can be immunoprecipitated using antibodies directed against selected protein tags. Brain tissues are highly heterogeneous and complex tissues challenging immunoprecipitation from crude lysates. Therefore, we developed a combinatorial approach to (1) enrich for RNPs and then (2) to perform immunoprecipitation using monospecific, affinity-purified antibodies [5, 6].

2 Materials

An essential prerequisite for immunoprecipitation is the availability of highly specific and purified antibodies. Therefore, we purified overexpressed full-length Staufen2 (Stau2) and Barentsz (Btz) from *E. coli* cells and injected them into rabbits to generate polyclonal antibodies. For immunoprecipitation, we exclusively used affinity-purified antibodies. As a control for immunoprecipitation, we routinely chose the preimmune sera of the same rabbit.

To isolate RNA granules, we prepared all solutions with RNase-free DEPC water and worked under RNase-free conditions.

2.1 Antibody Coupling and Cross-Linking to Protein A Sepharose Beads

1. Phosphate buffered saline (PBS).
2. Antibody (Ab) (in PBS, affinity-purified).
3. Preimmune sera (PIS).
4. Protein A sepharose.
5. RNase inhibitor (e.g., Thermo Fisher Scientific Ribolock).
6. Ab and PIS solutions: 100 μg protein (antibody or PIS) in 250 μL PBS.
7. 4× brain extraction buffer (BEB): 100 mM HEPES pH 7.3, 600 mM KCl, 32% glycerol, and 0.4% NP-40.
8. Full BEB: 1× BEB supplemented with 1 mM DTT, EDTA-free protease inhibitor (e.g., Roche complete) and RNase inhibitor.
9. 0.2 M triethanolamine in PBS, pH 9.5 (TEA). Titrate with NaOH or KOH.
10. 40 mM dimethylpimelimidate dihydrochloride (DMP) in TEA (prepare always freshly for every single reaction step).
11. 200 mM ethanolamine in water pH 8.
12. 0.02% NaN_3 in PBS.
13. Motor-driven Dounce homogenizer or a hand-driven Douncer.

2.2 OptiPrep Gradient Fractionation

1. Gradient mixer.
2. Polyallomer tubes (e.g., Beckmann 14 × 89 mm, for 11–12 mL).

3. Ultracentrifuge.
4. Swing-out rotor (e.g., Beckmann SW41).
5. OptiPrep: 60 w/v % solution of iodixanol in dH_2O (sterile) (e.g., Sigma Aldrich).
6. OptiPrep gradient solution: 15% and 30% OptiPrep in 1× BEB, 1 mM DTT, protease inhibitor.

2.3 Immuno-precipitation

1. Blocking solution: 1.25 mg/mL BSA in 1× BEB + 0.125 mg/mL yeast tRNA.
2. 10 mg/mL yeast tRNA.
3. 1 mM $MgCl_2$ in PBS.
4. RNase A + T1 mix.
5. 0.2 M glycine pH 2.5.
6. 1 M Tris–HCl pH 8.5.
7. Wash solutions: 1× BEB.
8. RNase elution buffer: 200 μg/mL RNase A/T1 and 1 mM $MgCl_2$ in PBS, pH 7.1.

3 Methods

All steps for immunoprecipitation should be carried out on ice. Proteins and RNAs are eluted at room temperature (RT).

3.1 Antibody Cross-Linking to Protein A Sepharose Beads

1. Hydrate beads in 10 mL PBS at RT.
2. Incubate them on a shaker for 30 min.
3. Spin beads at 500 × *g* for 2 min.
4. Remove supernatant and resuspend in equal volume PBS (= 1:1 slurry).
5. Beads can be stored hydrated at 4 °C in PBS supplemented with 0.02% NaN_3.
6. Prepare Ab and PIS solution by diluting 100 μg protein in 250 μL PBS, respectively.
7. Pipette 100 μL 1:1 slurry protein A sepharose beads (*see* **Note 1**) into 1.5 mL tube (= 50 μL protein A sepharose beads), spin at 500 × *g* for 1 min, wash shortly three times with PBS.
8. Add Ab and PIS dilutions to the beads.
9. Incubate at 4 °C for 1 h constantly rotating.
10. Store supernatant for coupling test using a Coomassie gel (*see* **Note 2**).
11. Wash beads at 4 °C by centrifuging at 500 × *g* for 1 min, three times with PBS, two times with TEA.

12. Add 1 mL 40 mM DMP (*see* **Note 3**) in TEA to the beads (cross-linking step).
13. Incubate at RT for 30 min, constantly rotating.
14. Wash beads 1× with TEA.
15. Repeat **steps 12–14** three times.
16. Wash with 200 mM ethanolamine.
17. Incubate 2× with 200 mM ethanolamine for 10 min at RT.
18. Wash 2× with PBS.
19. Store beads in 1 mL PBS supplemented with 0.02% NaN_3.

3.2 Brain Homogenization and Differential Centrifugation

1. Brains are homogenized on ice in 5 mL full BEB using a motor-driven Dounce homogenizer or using a hand-driven Douncer (*see* **Note 4**).
2. Centrifuge at 20,000 × *g* for 15 min at 4 °C (resulting supernatant S20).
3. Chill the S20 supernatant (lysate) on ice.

3.3 OptiPrep Gradient Fractionation

1. To prepare OptiPrep gradients (Fig. 1), we are using a gradient mixer. Prepare gradient solutions always freshly.
2. Rinse polyallomer tubes and gradient mixer with RNase-free water, and fill the 15% chamber with 5 mL of 15% OptiPrep solution. Then open the valve to remove air bubbles from the chamber connection and pipet the solution from the 30% chamber (front chamber) back. Remove the bubbles from the outlet tubing of 30% chamber as well by filling with 30% OptiPrep solution and sucking the air from the outlet by a 100 μL pipette. Make sure to have an equal volume (5 mL) of corresponding OptiPrep solutions in both chambers (*see* **Note 5**).
3. Turn on the magnetic stirrer and open the two taps carefully, so that the gradient solutions can mix and flow into the tube.
4. Upon pouring the gradients, turn off the magnetic stirrer and put the tube on ice.
5. Wash the gradient maker once with RNase-free water and repeat the whole procedure for the second gradient.
6. Then, carefully add ~1.7 mL of the soluble brain lysate (S20) to each gradient and centrifuge at 280,000 × *g* (40,000 rpm in a SW41 swing-out rotor) for 2.5 h at 4 °C (max. acceleration, slow deceleration).
7. Collect 11× 0.9 mL fractions from the top to the bottom of the gradient. Check every fraction on a Western blot (*see* **Note 6**). Fractions can be flash-frozen in liquid nitrogen and stored at −80 °C until further use.

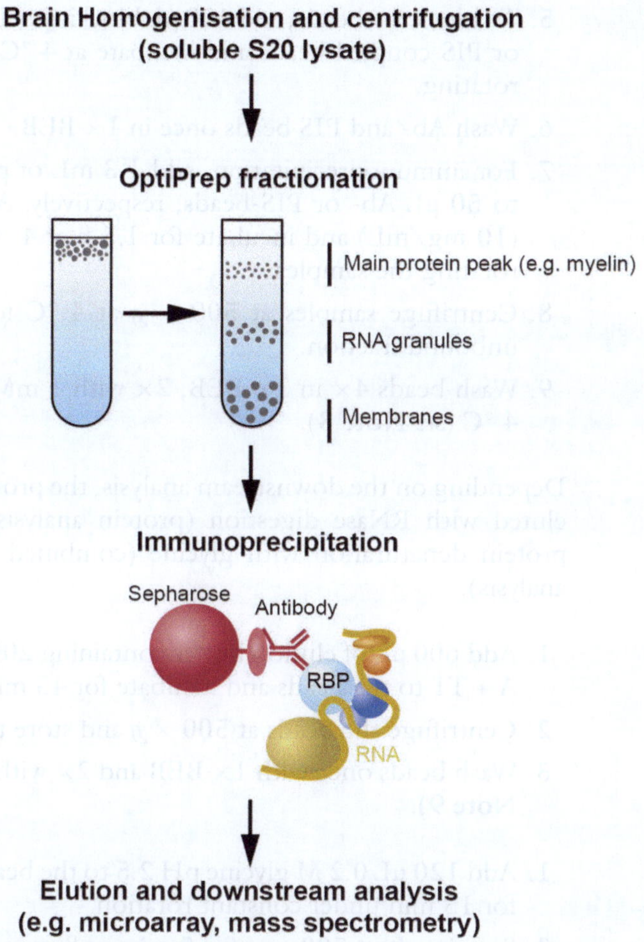

Fig. 1 Workflow for neuronal RNP isolation and subsequent downstream analysis. The S20 of brain homogenates is biochemically separated on an OptiPrep gradient (ranging from 15% to 30% of OptiPrep). Fractions enriched for Stau2 or Btz proteins are pooled and used for immunoprecipitation with highly specific and affinity purified antibodies (scheme modified from [6])

3.4 Immuno-precipitation of RNPs

1. Pool OptiPrep fractions, which show an enrichment of the protein of interest (in our case Stau2 and Btz) on a Western blot. Add 2.22 μL Ribolock/mL of pooled fractions (*see* **Note 7**).
2. Wash 100 μL protein A sepharose beads in PBS, then wash them twice in 1× BEB.
3. Add pooled fractions to the washed beads and incubate at 4 °C for 1 h constantly rotating (*preclearing step*).
4. In the meantime, wash Ab- and PIS-coupled beads 1× in PBS and 2× in 1× BEB.

5. To block the beads, add 400 μL blocking solution to 50 μL Ab- or PIS-coupled beads and incubate at 4 °C for 1 h constantly rotating.
6. Wash Ab- and PIS beads once in 1× BEB.
7. For immunoprecipitation, add 1.3 mL of precleared fractions to 50 μL Ab- or PIS-beads, respectively. Add 5 μL of tRNA (10 mg/mL) and incubate for 1.5 h at 4 °C while constantly rotating the samples.
8. Centrifuge samples at 500 × *g* at 4 °C for 5 min and save unbound fraction.
9. Wash beads 4× in 1× BEB, 2× with 1 mM $MgCl_2$ in PBS at 4 °C (*see* **Note 8**).

3.5 Elution of the Beads

Depending on the downstream analysis, the protein A beads can be eluted with RNase digestion (protein analysis) or by reversible protein denaturation with glycine (combined protein and RNA analysis).

3.5.1 RNase Elution

1. Add 600 μL of elution buffer containing 200 μg/mL of RNase A + T1 to the beads and incubate for 45 min at RT.
2. Centrifuge the beads at 500 × *g* and store the supernatant.
3. Wash beads once with 1× BEB and 2× with ice-cold water (*see* **Note 9**).

3.5.2 Glycine Elution

1. Add 120 μL 0.2 M glycine pH 2.5 to the beads. Incubate at RT for 15 min under constant rotation.
2. Spin beads at 500 × *g* for 5 min, collect supernatant and mix immediately with 30 μL 1 M Tris–HCl pH 8.5.
3. Wash beads 2× with PBS.
4. Store beads in PBS supplemented with 0.02% NaN_3 at 4 °C (*see* **Note 10**).

3.6 Analysis of RNPs

To identify protein interactors of RBPs, elution fractions were processed for mass spectrometry (*see* **Note 11**).

4 Notes

1. Depending on the antibody, sepharose A or G show different cross-linking efficiencies. Beads for cross-linking have to be tested.
2. Successful coupling of antibodies to protein A or G beads results in much reduced or almost absent IgG bands around 55 and 25 kDa on a Coomassie gel.
3. DMP is unstable in aqueous solutions. Prepare DMP solutions always freshly before using.

4. Different RBPs behave differently during or upon lysis. For every protein, lysis conditions have to be optimized. Importantly, Mili and Steitz raised the important experimental issue that there might be reassociation of molecules, especially for RBPs, after cell lysis [7].
5. OptiPrep gradients have to be optimized for each RNP of choice. The migration of the respective RBP into the gradient strongly depends on the granule density that is influenced by the number of proteins and/or RNAs embedded in those particles.
6. Check gradient fractions on Western blot. Generally, Stau2 containing RNA granules are expected to accumulate in fraction 4–6 while Btz granules are predominately found in fraction 5–7 [6].
7. We recommend to compare Western blots with Coomassie gels of the respective fractions. Pool those fractions that show a signal in the Western blot for your protein of interest and less staining intensity of the total protein.
8. RNase A + T1 requires Mg^{2+} to work. For RNA extraction, add TRIzol to the beads after washing. For RNA isolation, it is recommended to increase the stringency of washing steps by increasing NP-40 concentration.
9. The efficiency of RNA digestion strongly depends on the chosen RNase. This step has to be optimized. Keep in mind, some RNases cut only single stranded or double stranded RNA while others cut both.
10. Cross-linked beads can be reused several times. Importantly, the yield of proteins in the elution fraction is decreasing with frequency of usage.
11. For analysis of protein interactors or RNA targets, we used the PIS from the same rabbit as negative control. For analytical immunoprecipitation, proteins eluted from antibody and PIS beads are precipitated by trichloroacetic acid or methanol chloroform extraction [8].

References

1. Kiebler MA, Bassell GJ (2006) Neuronal RNA granules: movers and makers. Neuron 51:685–690. doi:10.1016/j.neuron.2006.08.021
2. Jung H, Gkogkas CG, Sonenberg N, Holt CE (2014) Remote control of gene function by local translation. Cell 157:26–40. doi:10.1016/j.cell.2014.03.005
3. Doyle M, Kiebler MA (2012) A zipcode unzipped. Genes Dev 26:110–113. doi:10.1101/gad.184945.111
4. Buxbaum AR, Haimovich G, Singer RH (2014) In the right place at the right time: visualizing and understanding mRNA localization. Nat Rev Mol Cell Biol 16:95–109. doi:10.1038/nrm3918
5. Heraud-Farlow JE, Sharangdhar T, Li X et al (2013) Staufen2 regulates neuronal target

RNAs. Cell Rep 5:1511–1518. doi:10.1016/j.celrep.2013.11.039
6. Fritzsche R, Karra D, Bennett KL et al (2013) Interactome of two diverse RNA granules links mRNA localization to translational repression in neurons. Cell Rep 5:1749–1762. doi:10.1016/j.celrep.2013.11.023
7. Mili S, JA S (2004) Evidence for reassociation of RNA-binding proteins after cell lysis : implications for the interpretation of immunoprecipitation analyses. RNA 10:1692–1694. doi:10.1261/rna.7151404.mRNA
8. Wessel D, Flügge UI (1984) A method for the quantitative recovery of protein in dilute solution in the presence of detergents and lipids. Anal Biochem 138:141–143. doi:10.1016/0003-2697(84)90782-6

Chapter 29

Individual Nucleotide Resolution UV Cross-Linking and Immunoprecipitation (iCLIP) to Determine Protein–RNA Interactions

Christopher R. Sibley

Abstract

RNA-binding proteins (RBPs) interact with and determine the fate of many cellular RNA transcripts. In doing so they help direct many essential roles in cellular physiology, while their perturbed activity can contribute to disease etiology. In this chapter we detail a functional genomics approach, termed individual nucleotide resolution UV cross-linking and immunoprecipitation (iCLIP), that can determine the interactions of RBPs with their RNA targets in high throughput and at nucleotide resolution. iCLIP achieves this by exploiting UV-induced covalent cross-links formed between RBPs and their target RNAs to both purify the RBP–RNA complexes under stringent conditions, and to cause reverse transcription stalling that then identifies the direct cross-link sites in the high throughput sequenced cDNA libraries.

Key words iCLIP, CLIP, RNA-binding protein, RNA, Protein–RNA interactions, Post-transcriptional regulation

1 Introduction

The fate of a transcribed RNA is largely determined by interactions with RNA-binding proteins (RBPs) [1, 2]. For example, RBPs can regulate the posttranscriptional processing, stability, localization, and translation of RNA transcripts as they progresses from transcription to degradation. According to a recent census, ~1542 RBPs [2] are found in humans, with each RBP expected to interact with hundreds-to-thousands of RNA targets. Conversely, individual RNA transcripts can interact with hundreds of RBPs during their lifetime [3, 4]. These RBP–RNA interactions occur because the RBPs are recruited to their specific target loci through recognition of specific features (e.g., sequence motifs, secondary/tertiary structures, and protein–protein interactions), with these features widely dispersed across the transcriptome. Unsurprisingly, this one-to-many activity means that numerous RBPs have fundamental roles in cell biology [5, 6], while perturbed activity of several

Imre Gaspar (ed.), *RNA Detection: Methods and Protocols*, Methods in Molecular Biology, vol. 1649, DOI 10.1007/978-1-4939-7213-5_29, © Springer Science+Business Media LLC 2018

RBPs contributes to various disease etiologies [1, 7–9]. Accordingly, RBPs have attracted considerable interest in recent years, while specialized techniques have been established in order to determine the sites of RBP occupancy in a transcriptome-wide manner.

RBP–RNA interactions are primarily studied using variants of the RNA immunoprecipitation (RIP) and UV cross-linking and immunoprecipitation (CLIP) techniques. RIP involves the immunoprecipitation of a RBP together with its bound RNA that is converted to cDNA then sequenced. Although native immunoprecipitations are most commonly used [10], stability of interactions is in some cases strengthened through use of reversible paraformaldehyde cross-linking. In contrast, CLIP-based approaches initially use UV cross-linking to form an irreversible covalent bond directly between RBPs and their target RNAs. Unlike the use of paraformaldehyde that can cross-link protein–protein, protein–DNA, and protein–RNA interactions, this UV-induced cross-linking is specific to protein–RNA interactions over zero-length distances [11, 12]. This allows CLIP to use more stringent biochemical purification of the bound RNAs that reduces the signal-to-noise ratio, and which also helps eliminate non-specific interactions [12–16]. Accordingly, several CLIP protocols have now been developed (Table 1), as they have become the methods of choice for studying protein–RNA interactions.

In this chapter we discuss the rationale and application of one CLIP method in detail; individual nucleotide resolution UV-cross-linking and immunoprecipitation (iCLIP) [13, 19]. Like other CLIP approaches, iCLIP involves cross-linking RBPs to their RNA targets, immunoprecipitating the RBP-of-interest, digesting away the RBP, and converting the bound RNA into a cDNA library that can be high-throughput sequenced. The advantage of iCLIP over other methods is that it specifically exploits the fact that ~80–100% of cDNAs produced during the reverse transcription step of the protocol truncate at the protein–RNA cross-link site (Fig. 1a). This truncation has been both experimentally and computationally validated [20, 27], and implies that CLIP protocols requiring cross-link read-through by the reverse transcriptase (e.g., HITS-CLIP, CLIP-Seq, and PAR-CLIP) are losing information during library production. In contrast, by using an adapter configuration that captures both truncation events and read-through events (Fig. 1b), iCLIP allows identification of the cross-link site with nucleotide resolution and permits quantitative assessment of RBP-binding activity [19, 28]. The eCLIP protocol additionally exploits this truncation event using a different adapter configuration, but fundamentally differs from iCLIP in its absence of a protein–RNA complex visualization step [24]. We always recommend complex visualization to help identify or exclude contaminating complexes contributing to the signal, reveal dimers/trimers

Table 1
Variants of UV cross-linking and immunoprecipitation

Method	Difference	Advantage	Disadvantages	References
CLIP	Initial CLIP experimental design, based on Sanger sequencing	Stringent purification of RBP–RNA complexes and identification of RNA binding sites	Low throughput	[12]
HITS-CLIP/ CLIP-seq	CLIP with longer PCR primers that allow next-generation sequencing	High throughput, cross-link sites can be determined at nucleotide resolution from the deletions in some reads	Requires cDNAs to read-through the cross-link site, deletions usual only present in <10% of reads	[16–18]
PAR-CLIP	Incorporation of photoreactive ribonucleosides into RNA transcripts allows UV-A cross-linking	UV-A enhances cross-linking efficiency for some RBPs, allows pulse-labeling analysis of nascent RNA, cross-link site can be determined by T–C transitions at ribonucleoside incorporation sites	Presently restricted to cell lines, 4SU can cause toxicity, requires cDNAs to read-through the cross-link site, most reads either lack T–C transitions or have more than one transition	[15]
iCLIP	Adapter design and library preparation targets dominant reverse transcription truncation events	Captures both read-through and dominant truncation events to identify cross-link sites with nucleotide resolution. Unique molecular identifiers allow quantitative study	Data interpretation is sensitive to sub-optimal RNase conditions and insert size	[19, 20]
iCLAP	Use of epitope tags that allow denaturing purification. This can be either via peptide tags and their associated antibodies (e.g., 3X FLAG), or via epitope tags that allow direct binding to Strep and/or His beads	Allows analysis of RBPs with no/poor antibodies for iCLIP, purify RBP–RNA complexes under highly denaturing conditions, now possible to add tags with CRISPR/Cas9 and combine with other protocols	Requires RBP tagging so not endogenous protein (i.e., functional testing of tagged protein required)	[21, 22]

(continued)

Table 1 (continued)

Method	Difference	Advantage	Disadvantages	References
CRAC	Dual purification through use of cleavable bipartite tag that can allow purification without antibodies	Allows analysis of RBPs with no/poor antibodies for iCLIP, purify RBP–RNA complexes under highly denaturing conditions	Requires RBP tagging so not endogenous (i.e., functional testing of tagged protein required)	[23]
eCLIP	Modified iCLIP workflow includes 3′ adapter ligated to cDNA to avoid circularization step, and no visualization of the purified protein–RNA complex	Reduced protocol duration and improved efficiency of ligation steps allows high-throughput analysis of many RBPs	No complex visualization can mean non-antigen contamination of library. Blind cut means library may be dominated by short RNAs around the antigen Mw that may reduce efficiency in protocol	[24]
Fast-iCLIP	Modified library preparation to reduce iCLIP protocol duration	Quick biotin purification of CLIP'd material instead of precipitations, improved efficiency of circLigase step	Insert size selection only post-PCR may bias library toward short products	[25]
irCLIP	Further optimization of fast-iCLIP protocol and introduction of an adapter for non-radioactive visualization of the purified protein–RNA complex	Near-infrared adapter is as sensitive as radioactive labeling, dot blots allow monitoring of RNA inputs into library preparation steps, increased recovery of protein–RNA complexes from membrane, less tube transfers	Less stringent size selection of cDNA inserts due to bead-based cleanups and PCR	[26]

that may be of interest, allow accurate isolation of RNAs that are of suitable length for library construction, and to provide an additional purification of free RNA that may stick to the beads used in the immunoprecipitation.

The iCLIP protocol has been extensively optimized in recent years, and we refer the reader to two closely related publications from the Ule [13] and Konig [29] research groups that will assist

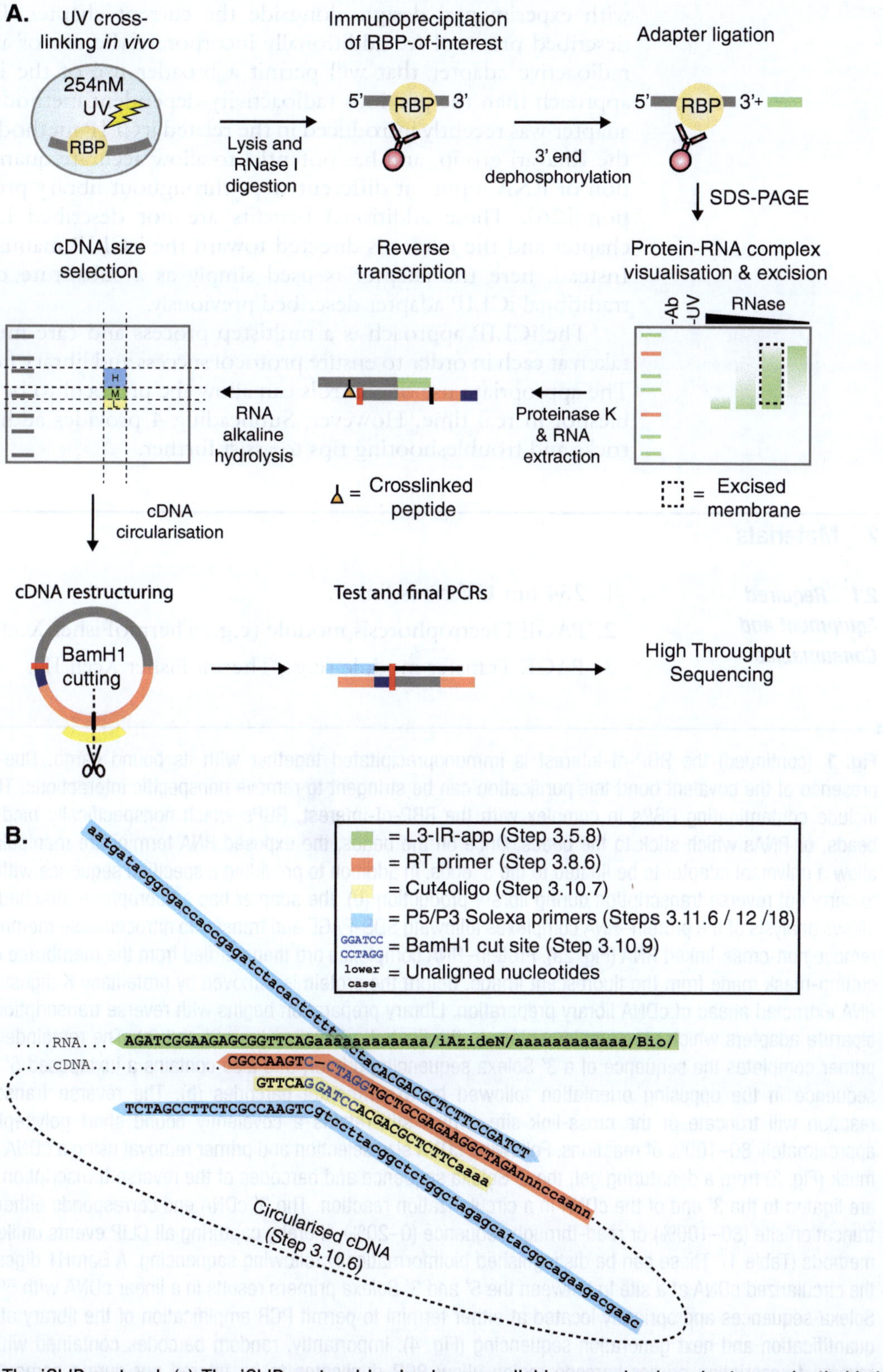

Fig. 1 The iCLIP protocol. (**a**) The iCLIP protocol begins by cross-linking samples in order to covalently bind RBPs and their RNA targets together. Samples are then lysed before a controlled RNase I digestion is carried out in order to shorten cross-linked RNA fragments to lengths compatible with RNA sequencing. At this point

with experimental design alongside the current chapter. In the described protocol we additionally incorporate the use of a non-radioactive adapter that will permit a broader use of the iCLIP approach than the previous radioactivity-dependent method. This adapter was recently introduced in the related irCLIP method from the Khavari group, and has potential to allow accurate quantification of RNA inputs at different steps throughout library preparation [26]. These additional benefits are not described in this chapter and the reader is directed toward the irCLIP manuscript. Instead, here the adapter is used simply as a substitute of the traditional iCLIP adapter described previously.

The iCLIP approach is a multistep process and care must be taken at each in order to ensure protocol success and library fidelity. The appropriate use of controls can allow the protocol to be troubleshot in real time. However, Subheading 4 provides additional tricks and troubleshooting tips to assist further.

2 Materials

2.1 Required Equipment and Consumables

1. 254 nm UV cross-linker.
2. PAGE Electrophoresis module (e.g., ThermoFisher Xcell II).
3. PAGE Transfer module (e.g., ThermoFisher Xcell II).

Fig. 1 (continued) the RBP-of-interest is immunoprecipitated together with its bound cargo. Due to the presence of the covalent bond this purification can be stringent to remove nonspecific interactions. This can include contaminating RBPs in complex with the RBP-of-interest, RBPs which nonspecifically bind to the beads, or RNAs which stick to the beads. Once on the beads, the exposed RNA termini are manipulated to allow a universal adapter to be ligated to the 3′ ends. In addition to providing a specified sequence with which to carry out reverse transcription during library production (**b**), the adapter has a fluorophore attached which allows analysis of the protein–RNA complexes following SDS-PAGE and transfer to nitrocellulose membrane to remove non-cross-linked RNA (Fig. 2a). Protein–RNA complexes are then purified from the membrane using a cutting-mask made from the fluorescent image, before the protein is removed by proteinase K digestion and RNA extracted ahead of cDNA library preparation. Library preparation begins with reverse transcription using bipartite adapters which are complementary to the ligated adapter at their 3′ end (**b**). The remainder of the primer completes the sequence of a 3′ Solexa sequencing primer, and also contains a juxtaposed 5′ Solexa sequence in the opposing orientation followed by experimental barcodes (**b**). The reverse transcription reaction will truncate at the cross-link site, which still retains a covalently bound short polypeptide, in approximately 80–100% of reactions. Following cDNA size selection and primer removal using a cDNA cutting mask (Fig. 3) from a denaturing gel, the 5′ Solexa sequence and barcodes of the reverse transcription primer are ligated to the 3′ end of the cDNA in a circularization reaction. The 3′ cDNA end corresponds either to the truncation site (80–100%) or read-through sequence (0–20%), thereby capturing all CLIP events unlike other methods (Table 1). These can be distinguished bioinformatically following sequencing. A BamH1 digestion of the circularized cDNA at a site in between the 5′ and 3′ Solexa primers results in a linear cDNA with 5′ and 3′ Solexa sequences appropriately located at either termini to permit PCR amplification of the library ahead of quantification and next generation sequencing (Fig. 4). Importantly, random barcodes contained within the reverse transcription primer barcode region allow PCR duplicates to be filtered out during computational analysis to ensure quantitative nature of the approach is maintained. (**b**) Adapter and oligonucleotide alignments for the iCLIP protocol. *Colours* are matched to the adapters and oligonucleotides shown in (**a**)

4. Thermocycler.
5. qPCR machine.
6. 1.5/2 mL tube thermomixer.
7. 1.5 mL centrifuge.
8. Vacuum pump.
9. Sonicator.
10. Magnetic rack.
11. Acetate printing film.
12. Printer.
13. Near infrared imager (*see* **Note 1**).
14. Cell scrapers.
15. 15 mL centrifuge tubes.
16. 2 mL microcentrifuge tubes.
17. 1.5 mL microcentrifuge tubes.
18. Low-binding 0.5 mL microcentrifuge tubes (e.g., ThermoFisher AM12350).
19. Low-binding 1.5 mL microcentrifuge tubes (e.g., ThermoFisher AM12450).
20. PCR tubes.
21. qPCR tubes/microplates.
22. Sterile filters.
23. Protran 0.45 nitrocellulose membrane.
24. Whatman filter paper.
25. Razor blades.
26. 30 G syringe needles.
27. 16 G syringe needles.
28. Phase-lock heavy columns.
29. Costar-X filter spin columns (e.g., VWR International).
30. Proteus Clarification columns (e.g., Generon).
31. Glass prefilters.

2.2 Required Reagents and Solutions

1. 1× PBS.
2. Tris–HCl, pH 7.4.
3. Tris–HCl, pH 6.5.
4. Tris–HCl, pH 7.8.
5. NaCl solution.
6. Igepal CA-630.
7. SDS.

8. Sodium deoxycholate.
9. Urea.
10. EDTA pH 8.0.
11. $MgCl_2$.
12. Tween 20.
13. Dithiothreitol.
14. PEG400.
15. LDS-4× sample buffer.
16. Methanol.
17. 20× MOPS-SDS running buffer.
18. 20× transfer buffer.
19. Neutral phenol–chloroform .
20. TE buffer pH 7.0.
21. TE buffer pH 8.0.
22. 3 M sodium acetate, pH 5.5.
23. 80% and 100% ethanol (molecular biology grade).
24. 1 M NaOH.
25. HEPES pH 7.3.
26. 2× TBE-UREA loading buffer.
27. TBE buffer.
28. Magnetic protein G/A beads.
29. RBP antibodies.
30. Anti-hnRNP C (positive control) (e.g., Santa Cruz sc-32308).
31. Protease inhibitor cocktail.
32. Anti-RNase antibody (e.g., Thermo Fisher AM 2690).
33. RNase I (e.g., Thermo Fisher).
34. Turbo DNase (e.g., Thermo Fisher).
35. T4 PNK (e.g., NEB).
36. RNase inhibitor (e.g., Promega RNasin).
37. T4 RNA ligase I.
38. Near infrared protein marker (e.g., Li-Cor Chameleon).
39. Antioxidant (e.g., Life Technologies, NP0005).
40. Reducing agent (e.g., Life Technologies, NP0004).
41. Proteinase K (e.g., NEB).
42. GlycoBlue.
43. 10 mM dNTPs.
44. Reverse transcriptase (e.g., ThermoFisher Superscript III).

45. Low molecular DNA weight marker (e.g., NEB).
46. SYBR safe.
47. Circligase II, $MnCl_2$, and 10× buffer.
48. Fast Digest BamH1.
49. PCR mastermix (e.g., Thermo Fisher Accuprime supermix I).
50. Illumina qPCR library quantification kit (e.g., Kapa Biosystems).
51. 4–12% protein denaturing precast gels (e.g., Thermo Fisher NuPAGE) (*see* **Note 2**).
52. 6% TBE-UREA precast gels (e.g., Thermo Fisher).
53. 6% TBE precast gels (e.g., Thermo Fisher).

2.3 Required Oligonucleotides

1. L3-IR-app adapter: /5Phos/AGATCGGAAGAGCGGTTCA GAAAAAAAAAAAA/iAzideN/AAAAAAAAAAAA/3Bio/

   Adapter requires adenylation then click chemistry conjugation of the IRDye 800CW DBCO infrared dye (Li-Cor) before use.
2. RT-primers: /5Phos/NNAACCNNNAGATCGGAAGAGCG TCGTGgatcCTGAACCGC

   underlined nucleotides can be replaced with different index barcodes e.g., RT1: AACC, RT2: ACAA, RT3: ATTG, RT4: AGGT, RT6: CCGG, RT7: CTAA, RT8: CATT.
3. Cut4oligo: 5′GTTCAGGATCCACGACGCTCT TCaaaa.
4. P5 primer: 5′AATGATACGGCGACCACCGAGATCTACA CTCTTTCCCTACACGACGCTCTTCCGATCT.
5. P3 primer: 5′CAAGCAGAAGACGGCATACGAGATCGGT CTCGGCATTCCTGCTGAACCGCTCTTCCGATCT.

2.4 Buffers

1. Lysis Buffer: 50 mM Tris–HCl, pH 7.4, 100 mM NaCl, 1% Igepal CA-630, 0.1% SDS, and 0.5% sodium deoxycholate. Sterile filter and store at 4 °C.
2. High-Salt Wash: 50 mM Tris–HCl, pH 7.4, 1 M NaCl, 1 mM EDTA, 1% Igepal CA-630, 0.1% SDS, and 0.5% sodium deoxycholate. Sterile filter and store at 4 °C.
3. PNK Buffer: 20 mM Tris–HCl, pH 7.4, 10 mM $MgCl_2$, 0.2% Tween-20. Sterile filter and store at 4 °C.
4. 5× PNK pH 6.5 Buffer: 350 mM Tris–HCl, pH 6.5, 50 mM $MgCl_2$, and 5 mM dithiothreitol. Use nuclease-free H_2O. Dispense to 50 μL aliquots and store at −20 °C. Only use aliquots once.
5. 4× Ligation Buffer: 200 mM Tris–HCl, pH 7.8, 40 mM $MgCl_2$, and 4 mM dithiothreitol. Use nuclease-free H_2O. Dispense to 50 μL aliquots and store at −20 °C. Only use aliquots once.

6. PK Buffer: 100 mM Tris–HCl, pH 7.4, 50 mM NaCl, and 10 mM EDTA. Sterile filter and store at 4 °C.
7. PK Buffer +7 M Urea: 100 mM Tris–HCl, pH 7.4, 50 mM NaCl, 10 mM EDTA, and 7 M Urea. Sterile filter and store at 4 °C.
8. 1× NuPAGE MOPS-SDS buffer: Add 25 mL of 20× MOPS-SDS buffer to 475 mL of water.
9. 1× NuPAGE loading buffer (per sample): 5 μL 4× NuPAGE Loading Buffer, 2 μL sample reducing reagent, and 13 μL nuclease-free H_2O.

2.5 Reaction Mixes

1. PNK de-phosphorylation mix (per sample): 15 μL nuclease-free H_2O, 4 μL of 5× PNK pH 6.5 Buffer, 0.5 μL PNK, and 0.5 μL RNAsin.
2. Ligation mix (add reagents in the indicated order): 8.5 μL nuclease-free H_2O, 5 μL of 4× ligation buffer, 0.5 μL T4 RNA ligase, 0.5 μL RNAsin, 1.5 μL 1 μM L3-IR-app oligo, and 4 μL PEG 400.
3. Reverse transcription mix (per sample): 7 μL nuclease-free H_2O, 4 μL 5× RT buffer, 1 μL 0.1 M DTT, 0.5 μL RNasin, 0.5 μL Superscript III.
4. Circularization mix (per sample): 6.5 μL nuclease-free H_2O, 0.8 μL 10× CircLigase II buffer, 0.4 μL 50 mM $MnCl_2$, and 0.3 μL CircLigase II.
5. Cut oligo mix (per sample): 25 μL nuclease-free H_2O, 4 μL FastDigest buffer, and 1 μL 10 μM cut oligo.
6. Test-PCR mix (per sample): 3.75 μL ddH_2O, 5 μL AccuPrime Supermix 1, 0.25 μL 10 μM P5/P3 Solexa primer mix.
7. Final-PCR mix (per sample): 9 μL ddH_2O, 20 μL AccuPrime Supermix 1, and 1 μL 10 μM P5/P3 Solexa primer mix.
8. Universal qPCR Master mix (per sample): 12.4 μL qPCR Master mix (Primer Premix and ROX High/Low should be added to Master Mix prior to first use as per manufacturer's instructions), 3.6 μL ddH_2O.

3 Methods

3.1 Sample Collection

1. Remove media from cells growing at 80–90% confluency on a 10 cm dish and replace with 6 mL of ice-cold PBS (*see* **Note 3**).

   Important: Samples should remain at 4 °C from this point forward unless indicated.

   Optional: To test if the signal is dependent on expected antigen and not a contaminant RBP, use a control cell-line in which the RBP of interested is knocked down or knocked out (*see* Note 4).

2. Place plate on ice-filled tray which has dimensions suitable for UV cross-linker (*see* **Note 5**).
3. Ensure that cell culture dish lids are removed before irradiating cells at 150 mJ/cm^2 at 254 nm.

   Important: Remember to proceed with some plates to step 4 with no cross-linking. These will be used as a no UV control (*see* Note 6).
4. Immediately harvest cells through gentle use of a cell scraper. Cells should be aliquoted into 2 mL samples (*see* **Note 7**).
5. Spin cell suspensions at 376 × *g* for 1 min at 4 °C to pellet cells.
6. Remove supernatant and snap-freeze cells on dry ice. Store cell pellets at −80 °C until use.

3.2 Bead Preparation

1. Add 100 μL of protein G or protein A magnetic beads to a microcentrifuge tube (*see* **Note 8**).
2. Place microcentrifuge tube on a magnetic rack, remove supernatant, then wash beads twice in 900 μL of lysis buffer. Beads should be resuspended by rotation for each wash (*see* **Note 9**).
3. Resuspend in 100 μL per sample of lysis buffer.
4. Transfer 100 μL of bead suspension to a second microcentrifuge tube to act as no antibody control in order to confirm that the signal is dependent on antigen and not unspecific RBPs or RNA binding to beads (*see* **Note 10**).
5. To the first microcentrifuge tube add 2–10 μg antibody per sample (*see* **Note 11**).
6. Rotate tubes in a cold room for 30–60 min while proceeding with sample preparation (Subheading 3.3).
7. When lysate is ready wash beads in 1× high salt wash and 2× lysis buffer. After final wash resuspend beads in 100 μL lysis buffer per sample. Proceed Subheading 3.4.

3.3 Sample Preparation

1. Prepare 2.1 mL of ice-cold lysis buffer per sample by adding 21 μL (1:100) proteinase inhibitor cocktail. For starting material with high RNase activity add an additional 2.1 μL anti-RNase per sample.
2. Resuspend cell pellets in 1 mL of prepared lysis buffer and place on ice.

   Recommended: At this stage the protein content of the lysates can be assessed with a Bradford assay. Samples using single pellets should be ~ 2 mg/mL and standardized by removing lysate from samples with excess. Removed volumes of lysate should be replaced with fresh lysis buffer to ensure identical 1 mL volumes.

3. Prepare a 1/1000 dilution of RNase I in 500 μL of ice-cold lysis buffer. This dilution is used for low RNase samples (*see* **Note 12**).
4. Prepare a 1/10 dilution of RNase I in 15 μL of ice-cold lysis buffer for high RNase condition (*see* **Note 13**).
5. Add 10 μL of low RNase dilution to all lysates together with 2 μL Turbo DNase. To high RNase sample add additional 10 μL of high RNase dilution.
6. Incubate samples for exactly 3 min at 37 °C while shaking at 1100 rpm. After incubation immediately transfer samples to ice for 3 min.
7. Centrifuge lysates at >18,000 × *g* for 10 min at 4 °C and transfer lysate to new 1.5 mL microcentrifuge tube. Take care not to disturb pelleted debris during transfer.
8. Load 500 μL of lysate per sample into a Proteus Clarification spin column and spin at >18,000 × *g* for 1 min. Transfer flow-through to a new 2 mL microcentrifuge tube. Repeat step with remaining lysate and combine accordingly.
9. Add 1 mL of prepared ice-cold lysis buffer and place on ice until beads are washed and ready for immunoprecipitation.

   Optional: At this stage 20 μL of lysate can be retained as a pre-immunoprecipitation sample and used to evaluate immunoprecipitation efficiency through western analysis.

3.4 Immunoprecipitation

1. Ensure beads are uniformly resuspended before adding 100 μL of beads to cell lysates.
2. Rotate beads for 1 h at 4 °C in cold room.

   Optional: Once beads have been separated from lysate on a magnetic rack, 20 μL of lysate can be retained as a post-immunoprecipitation sample. This can be used with previously collected sample to evaluate immunoprecipitation efficiency through western analysis.
3. Discard supernatant and wash beads in 2× high salt wash. Rotate the second wash for 5 min in the cold room.
4. Wash beads 2× PNK wash buffer and re-suspend in 1 mL PNK wash buffer until ready to proceed with next step.

3.5 Adapter Ligation

1. Remove PNK wash buffer from beads and remove microcentrifuge tubes from magnet for 30 s. Return microcentrifuge tubes to magnet and remove any small volumes of buffer still retained (*see* **Note 14**).
2. Remove beads from magnet and resuspend in 20 μL of dephosphorylation mix.
3. Incubate samples for exactly 20 min at 37 °C while shaking at 1100 rpm.

4. Wash with 1× PNK buffer.
5. Wash with 1× high salt wash buffer. Rotate the wash for 5 min in the cold room.
6. Wash with 2× PNK buffer and leave in final wash until ready for ligation.
7. Remove PNK buffer from beads and remove microcentrifuge tubes from magnet for 30 s. Return microcentrifuge tubes to magnet and remove any small volumes of buffer still retained (*see* **Note 14**).
8. Remove beads from magnet and resuspend in 20 μL of ligation mix.
9. Incubate samples overnight at 16 °C while shaking at 100 rpm.
10. Wash with 1× PNK buffer.
11. Wash with 2× high salt wash buffer. Rotate the first wash for 5 min in the cold room (*see* **Note 15**).
12. Wash with 1× PNK buffer and transfer to new 1.5 mL microcentrifuge tube.
13. Wash with 1× PNK buffer and leave in final wash until ready for SDS-PAGE.

3.6 Protein–RNA Complex Visualization

1. Assemble the XCell III gel system according to manufacturer's instructions using a 4–12% NuPAGE Bis-Tris gel (*see* **Note 2**) and 1× MOPS-SDS buffer.
2. Add 500 μL of antioxidant to the upper chamber (*see* **Note 16**).
3. Remove PNK buffer from beads and remove microcentrifuge tubes from magnet for 30 s. Return microcentrifuge tubes to magnet and remove any small volumes of buffer still retained (*see* **Note 14**).
4. Re-suspend beads in 20 μL of 1× NuPage sample loading buffer.
5. Incubate samples at 80 °C for 5 min to dissociate sample from beads. The chameleon protein ladder does not need to be heated.
6. Spin down any precipitation of the samples using a desktop microcentrifuge then place on magnet to separate beads.
7. Load 20 μL of sample supernatant to gel using gel loading tips, and 5 μL of chameleon ladder. It is suggested that gaps are left between samples to facilitate extraction of protein–RNA complexes from the membrane.
8. Run the gel for the 50 min at 180 V.

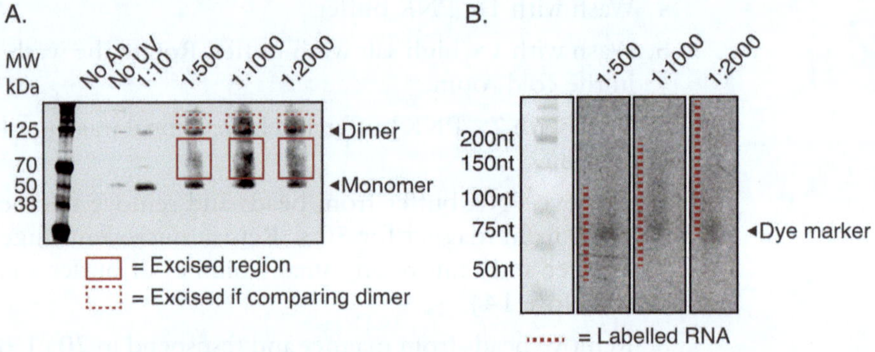

Fig. 2 Protein–RNA complex visualization and RNase digestion analysis. (**a**) Fluorescent adapter labeled RNA–protein complexes visualized following SDS-PAGE. The fluorescent image has been converted to a grey-scale image. This is best suited to create the cutting mask that aids protein–RNA complex excision from the membrane (*see* **step 14** of Subheading 3.6). Samples include no UV and no antibody negative controls, and a gradient of RNase I conditions from which suitable digestions patterns can be determined. The immunoprecipitated RBP is hnRNP C, and the *arrowhead* indicates both monomers and dimers particularly discernible in the high RNase condition. Note that once optimal digestion conditions have been determined for RNase I on the desired sample batch, only the high RNase and determined concentration (in this case 1:1000) need to be carried out for each new iCLIP experiment alongside the no UV and no antibody negative controls. The *solid red markers* indicate regions excised from the membrane using a cutting mask made from the gel image. The *dotted red markers* indicate region that would be cut if comparisons between monomer and dimer were to be made. (**b**) RNA size distributions from the different regions and different RNase concentrations indicated in cut regions from (**a**). Extracted RNA was denatured and run for 40 min at 180 V on a 6% TBE-UREA gel. *Dotted red line* indicates detectable size distributions of each RNase condition. Signal intensity can be limited relative to alternative radiolabeling methods due to maximum exposure time limits on some imaging machines

9. Prepare fresh transfer buffer by adding 25 mL of 20× transfer buffer and 50 mL of methanol to 425 mL of water.
10. Following SDS-PAGE remove gel cassette and carefully open. Assemble blotting "sandwich" and XCell III blotting module as per manufacturer's instructions.
11. Transfer protein–RNA complexes for 1.5 h at 30 V.
12. Remove nitrocellulose transfer membrane and transfer to a light protected PBS containing box.
13. Visualize protein–RNA complexes using appropriate fluorescent imager and return to PBS until protein–RNA complex isolation (*see* Fig. 2a).
14. Print image at 100% scale on acetate film. It is recommended that a grey scale image be printed to facilitate alignment of ladder markers between the film and the membrane (*see* Fig. 2a).

3.7 Protein–RNA Complex Isolation and RNA Purification

1. Make a cutting mask to guide protein–RNA complex excision by drawing a box around protein–RNA signal that starts just above the Mw of the RBP-of-interest (*see* **Note 17**). The high RNase condition can be used to assist accurate assessment of

the antigens Mw. Cut out these boxes from the acetate film to create windows that locate protein–RNA signal (Fig. 2a).

2. Remove transfer membrane from PBS and wrap in Saran Wrap. Secure firmly to fixed cutting surface.
3. Align cutting mask with membrane by matching protein ladder markers. Use windows to guide removal of nitrocellulose membrane segments containing protein–RNA complex. Remove coexcised saran wrap and cut membrane sections into small pieces. Transfer to 1.5 mL microcentrifuge tube with the assistance of a 30 G syringe needle (*see* **Note 18**).

   Recommended: The cut membrane can be reanalyzed to assess cutting accuracy.
4. Prepare proteinase K digest mix by adding 10 μL of proteinase K to 200 μL of PK buffer per sample.
5. Add 200 μL of proteinase K digest mix to nitrocellulose membrane pieces and incubate for 20 min at 37 °C while shaking at 1100 rpm.
6. Add 200 μL of PK-urea buffer and incubate for an additional 20 min at 37 °C while shaking at 1100 rpm.
7. Transfer supernatant to Phase Lock Heavy gel columns together with 400 μL of neutral phenol–chloroform.
8. Incubate for 5 min at 30 °C while shaking at 1100 rpm.
9. Separate phases by centrifuging at >18,000 × *g* at room temperature for 5 min.
10. Taking care not to touch the gel matrix, transfer the aqueous upper phase to a new low-binding 1.5 mL microcentrifuge tube.
11. Spin at >18,000 × *g* for 1 min then transfer to a new low-binding 1.5 mL microcentrifuge tube.
12. Purify RNA by adding 0.75 μL of GlycoBlue and 40 μL of 3 M sodium acetate pH 5.5, then mixing. Add 1 mL of ice-cold 100% ethanol, mix again, then precipitate overnight at −20 °C.

3.8 Reverse Transcription

1. Spin samples for 20 min at 4 °C and >18,000 × *g*.
2. Remove supernatant leaving ~50 μL around blue pellet. Add 1 mL of ice-cold 80% ethanol (do not re-suspend) and spin samples for additional 10 min at 4 °C and >18,000 × *g*.
3. Carefully remove all supernatant. Use a P10 pipette tip for last few μL of supernatant around pellet.
4. Air-dry pellet at room temperature for 5 min with lid opened.
5. Resuspend in 5 μL nuclease-free water and transfer to PCR tube (*see* **Note 19**, Fig. 2b).

6. Add 1 μL of 10 mM dNTPs and 0.5 μL of 0.5 pmol/μL RT# primer to each sample (*see* **Note 20**).
7. Denature RNA using the following thermocycler program:

(a)	70 °C	5 min
(b)	25 °C	Forever

8. Add 13 μL of freshly prepared reverse transcription mix to each sample and mix by pipetting. Carry out reverse transcription using the following settings:

(a)	25 °C	5 min
(b)	42 °C	20 min
(c)	50 °C	40 min
(d)	80 °C	5 min
(e)	4 °C	Forever

9. Add 1.65 μL of 1 M NaOH to each sample and incubate at 98 °C for 20 min in order to cause alkaline hydrolysis of RNA that can interfere with the CircLigase reaction.
10. Neutralize samples by adding 20 μL of HEPES pH 7.3.

    Optional: At this stage samples to be multiplexed can be combined to reduce the number of samples to work with. It is recommended that negative controls are kept separate from working samples.
11. Add 350 μL TE buffer, 0.75 μL GlycoBlue, and 40 μL 3 sodium acetate pH 5.5, and mix. Add 1 mL of ice-cold 100% ethanol, mix again, then precipitate overnight at −20 °C.

3.9 cDNA Purification and Size Selection

1. Spin samples for 20 min at 4 °C and >18,000 × *g*.
2. Remove supernatant leaving ~50 μL around blue pellet. Add 1 mL of ice-cold 80% ethanol (do not re-suspend) and spin samples for additional 10 min at 4 °C and >18,000 × *g*.
3. Carefully remove all supernatant. Use a P10 pipette tip for last few μL of supernatant around pellet.
4. Air-dry pellet at room temperature for 5 min with lid opened.
5. Resuspend pellet in 12 μL of 1× TBE-UREA loading buffer prepared using nuclease free water. Also add 6 μL of 2× TBE-UREA loading buffer to 6 μL of a 1:30 dilution of low molecular weight marker.
6. Heat samples for 80 °C for 5 min directly before gel loading.

7. Assemble the XCell III gel system according to manufacturer's instructions using a 6% TBE-UREA gel (*see* **Note 21**) and fill chambers with TBE buffer.
8. While samples are heating flush UREA out of gel wells.
9. Load 12 μL of samples per well, allowing 1 well gaps between samples to facilitate gel cutting. Load ladder into a well at one end of the gel.
10. Run the gel for 40 min at 180 V (*see* **Note 22**).
11. While gel is running, prepare 0.5 mL microcentrifuge tubes for gel crushing. Carefully using a 16 G syringe needle to pierce a clean hole in the bottom of 0.5 mL microcentrifuge tubes then place these inside new low-bind 1.5 mL microcentrifuge tubes.
12. Open the cassette and cut off the last lane containing the ladder. Stain the ladder using SYBR Safe and an appropriate imager, and use to construct a cutting guide that is printed out at 100% scale and wrapped in saran wrap (*see* **Note 23**). Alternatively, a pre-prepared cutting guide can be used (*see* **Note 22**, Fig. 3).
13. Place the opened cassette upon the cutting guide and use markers to determine gel cut sites for each sample. The adapter and primer account for 52 nt. Cut sizes and corresponding cDNA insert sizes are below (*see* **Note 24**):

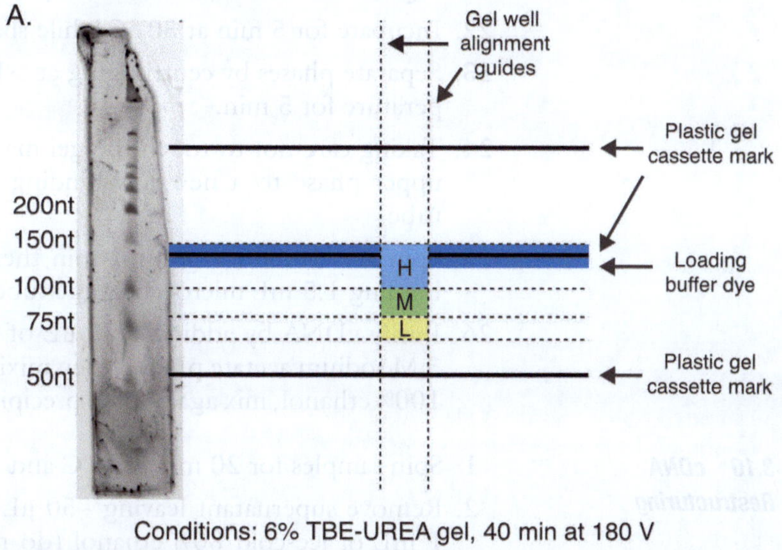

Fig. 3 cDNA size selection. Example cutting mask determined from low molecular weight marker used for excising cDNA from denaturing TBE-UREA gel. Low, medium, and high cut sites determined by denatured ladder are indicated, while gel cassette marks of the recommended TBE-UREA gel setup act as additional alignment guides

		Cut size:	Insert size:
(a)	Low band	70–80 nt	18–28 nt
(b)	Medium band	80–100 nt	28–48 nt
(c)	High band	100–150 nt	48–98 nt

14. Transfer gel pieces to pierced 0.5 mL tubes sitting inside new low-binding 1.5 mL microcentrifuge tubes.
15. Spin samples at 10,000 × *g* to crush gel pieces.
16. Add 400 μL TE buffer and incubate for 1 h at 37 °C shaking at 1100 rpm.
17. Place samples on dry ice to snap-freeze samples ahead of rapid gel expansion.
18. Incubate samples for 1 h at 37 °C shaking at 1100 rpm.
19. While incubating, place two glass filters per sample into a Costar Spin-X column.
20. Cut end of P1000 tip to allow suction of gel pieces. Transfer buffer and gel pieces to prepared coster spin-X column placed in low-bind 1.5 mL microcentrifuge tube and spin at >18,000 × *g* for 1 min.
21. Transfer flow-through to Phase Lock Heavy gel columns together with 400 μL of neutral phenol–chloroform.
22. Incubate for 5 min at 30 °C while shaking at 1100 rpm.
23. Separate phases by centrifuging at >18,000 × *g* at room temperature for 5 min.
24. Taking care not to touch the gel matrix, transfer the aqueous upper phase to a new low-binding 1.5 mL microcentrifuge tube.
25. Spin at >18,000 × *g* for 1 min then transfer to a new low-binding 1.5 mL microcentrifuge tube.
26. Purify cDNA by adding 0.75 μL of GlycoBlue and 40 μL of 3 M sodium acetate pH 5.5 then mixing. Add 1 mL of ice-cold 100% ethanol, mix again, then precipitate overnight at −20 °C.

3.10 cDNA Restructuring

1. Spin samples for 20 min at 4 °C and >18,000 × *g*.
2. Remove supernatant leaving ~50 μL around blue pellet. Add 1 mL of ice-cold 80% ethanol (do not re-suspend) and spin samples for additional 10 min at 4 °C and >18,000 × *g*.
3. Carefully remove all supernatant. Use a P10 pipette tip for last few μL of supernatant around pellet.
4. Air-dry pellet at room temperature for 5 min with lid opened.

5. Resuspend cDNA pellet in 8 μL of freshly prepared circularization mix and transfer to 0.2 mL PCR tubes.
6. Incubate for 1 h at 60 °C.
7. Add 30 μL of freshly prepared cut oligo mix to each PCR tube
8. Anneal the cut oligo with the following program:

(a)	95 °C	2 min
(b)	95 °C to 25 °C ramp	20 s for each °C
(c)	25 °C	Forever

9. Add 2 μL of BamH1 to each PCR tube and incubate with the following program:

(a)	37 °C	30 min
(b)	80 °C	5 min
(c)	25 °C	Forever

10. Transfer samples to new low-bind 1.5 mL microcentrifuge tubes.
11. Add 350 μL TE buffer, 0.75 μL GlycoBlue, and 40 μL 3 sodium acetate pH 5.5, and mix. Add 1 mL of ice-cold 100% ethanol, mix again, then precipitate overnight at −20 °C.

3.11 cDNA Library PCR

1. Spin samples for 20 min at 4 °C and >18,000 × *g*.
2. Remove supernatant leaving ~50 μL around blue pellet. Add 1 mL of ice-cold 80% Ethanol (do not resuspend) and spin samples for additional 10 min at 4 °C and >18,000 × *g*.
3. Carefully remove all supernatant. Use a P10 pipette tip for last few μL of supernatant around pellet.
4. Air-dry pellet at room temperature for 5 min with lid opened.
5. Resuspend the cDNA pellet in 22 μL nuclease-free water
6. Add 1 μL of resuspended cDNA to 9 μL of test-PCR mix.
7. Perform test PCR with the following program:

		Time:	Cycles:
(a)	94 °C	2 min	1
(b)	94 °C	15 s	
	65 °C	30 s	20/25
	68 °C	30 s	
(c)	68 °C	3 min	1
(d)	25 °C	Hold	Hold

Important: Following the PCR, do not open PCR tubes in same room where iCLIP library preparations (i.e., all steps before step 8 of Subheading 3.11) were carried out. All post-PCR analysis and storage needs to be carried out in separate room to avoid amplicon contamination.

8. Assemble the XCell III gel system according to manufacturer's instructions using a 6% TBE gel and TBE running buffer.
9. Add 2 μL of 6× loading dye to each PCR sample and 10 μL of 1:30 low molecular weight marker. Load 12 μL of samples/ladder into gel wells.
10. Run gel for 30 min at 180 V
11. Remove gel from cassette and stain for 5 min in SYBR Safe in TBE buffer. Visualize using appropriate imager (Fig. 4a). The P5/P3 primers account for 128 nt of PCR product. Appropriate PCR band sizes are therefore:

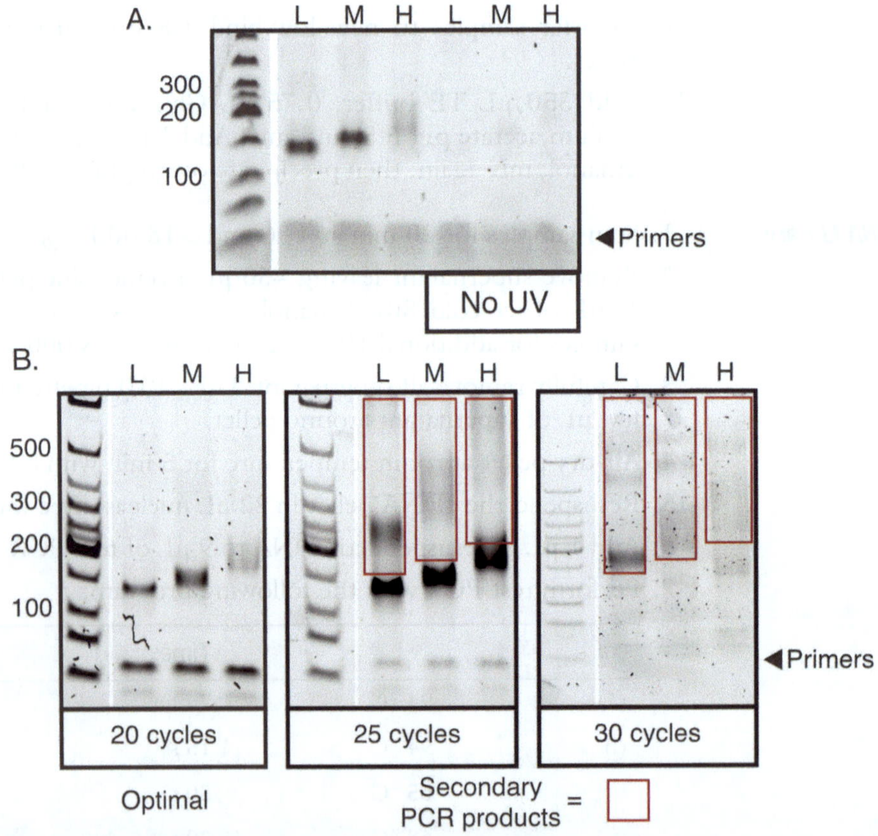

Fig. 4 cDNA library PCRs. (**a**) Example of final library PCR (**steps 13–17** of Subheading 3.11) showing low, medium, and high bands. Both a low RNase sample and a no UV control sample are shown. (**b**) Example of a cDNA library amplified with different PCR cycle numbers. *Red boxes* indicate secondary spurious products that migrate at a higher molecular weight to the expected products

		Band size:	Insert size:
(a)	Low	145–155 nt	18–28 nt
(b)	Medium	155–175 nt	28–48 nt
(c)	High	175–225 nt	48–98 nt

Optional: Steps 6–11 of Subheading 3.11 can be repeated with adjusted cycling conditions to optimize cDNA library amplification (*see* Notes 25 and 26).

12. Once optimal PCR conditions are determined, add 10 μL of resuspended cDNA to 30 μL of final-PCR mix:
13. Perform final PCR with the following program using one cycle less than optimal test-PCR (see **Note 27**):

		Time:	Cycles:
(a)	94 °C	2 min	1
(b)	94 °C	15 s	
	65 °C	30 s	Test PCR-1
	68 °C	30 s	
(c)	68 °C	3 min	1
(d)	25 °C	Hold	Hold

14. Assemble the XCell III gel system according to manufacturer's instructions using a 6% TBE gel and TBE running buffer.
15. Add 2 μL of 6× loading dye to 10 μL aliquots of each final PCR sample and 10 μL of 1:30 low molecular weight marker. Load 12 μL of samples/ladder into gel wells.
16. Run gel for 30 min at 180 V.
17. Remove gel from cassette and stain for 5 min in SYBR Safe in TBE buffer. Visualize using appropriate imager.

    Optional: If sample is under-amplified then remaining PCR samples can be returned to thermocycler for 1–2 additional cycles before proceeding.
18. Once optimal final PCR cycling number is confirmed then repeat final PCR (**steps 13** and **14** of Subheading 3.11) using appropriate cycle number. Pool final PCRs together.
19. If PCR products are of correct size then aliquots of different sized PCR products can be combined in ratio of 1:5:5 (Low–Medium–High), or 1:1 (Medium–High) if no low band is included.

3.12 Library Quantification

1. Prepare a 1:10 dilution of the final library, and then make serial dilutions to obtain 1:100, 1:1000, and 1:10,000 samples.

2. Prepare qPCR library quantification standards and nontemplate controls as per manufacturer guidelines.
3. In technical triplicate, add 16 μL of universal qPCR Master mix per qPCR plate/strip well.
4. Dispense 4 μL of non-template control, standards and serial library dilutions into reaction wells accordingly.
5. Seal plate, spin down and transfer to qPCR instrument. Run qPCR with the following cycling protocol:

		Time:	Cycles:
(a)	95 °C	5 min	1
(b)	95 °C	30 s	35
	60 °C	45 s	
(c)	Melt curve analysis (65–95 °C)		

6. Determine concentration of library from qPCR standards and then correct for library type and insert size using predetermined factor (*see* **Notes 28** and **29**).
7. Dilute library to 10 nM and submit >10 μL to sequencing facility for sequencing. Provide facility with details about correction factor from **step 6** of Subheading 3.12 so that it can be factored into any repeat quantification. Sequencing can be carried out in single-end 50-nucleotide runs (*see* **Notes 30** and **31**).

4 Notes

1. The adapter described in this protocol has the Li-Cor IRDye 800CW dye conjugated. An imager capable of detecting the 794 nm wavelength is required. The high wavelength reduces background and increases sensitivity to levels that are required in iCLIP. It is expected that additional fluorophores with high wavelength characteristics may be viable alternatives with appropriate imagers e.g., >600 nm.
2. It is critical that the pH is stably maintained during gel electrophoresis in order to prevent alkaline hydrolysis of the RNA. A pour-your-own gel is not suitable, while the Novex NuPage gels and system are strongly recommended when using the MOPS-SDS running buffer.
3. The use of 1/3rd of a 10 cm dish for each sample is a guideline suitable for standard cell lines. Combining pellets, or using smaller/larger vessels can increase sample size for hard to

iCLIP proteins. It is also recommended that inputs are normalized by carrying out protein quantification following lysis and diluting to standard amounts.

4. Silencing of the RBP-of-interest should be confirmed through both qRT-PCR and western blotting to ensure specificity of knockdown. Appropriate silencing will reduce intensity of protein–RNA smear in **step 13** of Subheading 3.6 rather than reducing size of smear.
5. It is important to keep the cross-linking conditions identical between experiments. This includes the height between UV bulbs and exposed samples. Finding a suitable tray that fits inside the cross-linker without interfering with the automatic cutoff switch will facilitate this.
6. The no UV control is critical to confirm that the signal is UV dependent and derived from cross-linked complexes. This control requires samples to be prepared in identical manner to experimental samples with the exception of UV cross-linking. Samples can be carried through to final PCR stage to assess library preparation efficiency and assess background levels of contamination contributing to library signal. In some cases where the RBP is both abundant and a particularly strong binder of RNA the no UV control may reveal some signal at the expected molecular weight. However, this should appear much weaker relative to the low RNase conditions (Fig. 2a).
7. Less variable pellet sizes can be achieved by transferring 6 mL cell suspensions to a 15 mL falcon tube, mixing by pipetting, then separating cells into 2 mL aliquots.
8. Protein G beads work well with most antibodies except sometimes rabbit. Protein A may work better with rabbit antibodies but protein A beads tend to stick to microcentrifuge tubes.
9. Wash steps are in 900 μL of indicated buffer unless specified otherwise. Washing steps involved magnetic separation of beads, supernatant removal, then resuspension of beads in new buffer.
10. The no antibody negative control can use half the input to a normal sample and be carried through to final PCR stage as in case of the no-UV control (*see* **Note 6**).
11. A positive control, such as anti-hnRNP C antibody [13, 19, 28], may be used to confirm that iCLIP is working in the hands of the experimenter, and provide a comparator of antigen size against RBP-of-interest. It is strongly recommended that a new iCLIP setup is first tested with a positive control antibody before moving on to a new RBP.
12. The integrity of the iCLIP protocol and the subsequent bioinformatics analysis strongly depends on optimal RNase

digestion conditions [20]. In order to ensure samples are not over or under digested, it is critical that each new batch of RNase I is tested for activity on each batch of sample preparations used. In test experiments, dilutions of RNase I should span from 1:10–1:2000 (Fig. 2a). In addition to fluorescent probe analysis following SDS-PAGE and nitrocellulose transfer, small aliquots of RNA extracted following proteinase-K digestions should be run on denaturing UREA gels in order to accurately assess RNA sizes corresponding to each digestion. Suitable digestions will produce an enrichment of RNA fragments in the size range of 50–300 nt (Fig. 2b).

13. A high RNase is used to confirm that the signal is sensitive to availability of RNase. It may also identify contaminating sources of signal or dimer/trimer complexes. Half the input to a normal sample should be used for the high RNase sample. High RNase will lead to a less heterogeneous pool of RNA lengths that will cause an intense signal above Mw of cross-linked antigen.

14. Enzymatic steps in Subheading 3.5 and gel loading step in Subheading 3.6 involve small volumes of 20 μL. Care should be taken to remove all carryover of wash buffers that can increase this volume. This can be achieved by removing the microcentrifuge tubes from the magnet for 15 s to allow beads to settle, then returning microcentrifuge tubes to the magnet and removing excess wash buffer.

15. Un-ligated adapter carried through into library preparation steps can be processed and lead to sample contamination. Various steps are carried out to ensure unligated adapter is removed. This includes stringent washing, transferring of washes to new tubes, nitrocellulose transfer following SDS-PAGE, and size selection following cDNA synthesis.

16. The use of the antioxidant and reducing reagent maintains proteins in a reduced state during gel electrophoresis. Although use is optional, it is strongly recommended if the antibody bands run at a similar Mw to the RBP-of-interest. Without antioxidant and reducing reagent this will cause a band of reduced signal intensity across the protein–RNA complex smear in **step 13** of Subheading 3.6.

17. Cross-linking efficiency is ~1%. The Mw of the RBP-of-interest is therefore excluded since this position will include the majority of immunoprecipitated protein with no cross-linked and labeled RNA. Exclusion avoids potential saturation of proteinase K. Note that each extra 20 nt of RNA will add ~7 kDa to the molecular weight of the protein–RNA complex. Accordingly, high molecular weight RBPs or higher order complexes (e.g., dimers in Fig. 2a) may benefit from being run for

extended periods on higher percentage precast gels in order to provide better resolution at these molecular weights.

18. The high RNase sample does not need to be excised. However, it is recommended that the no antibody and no UV negative controls are excised in order to assess library preparation integrity.

19. At this stage an aliquot of RNA can be run on a TBE-UREA denaturing gel to assess the size of the RNA fragments that have been extracted. The optimal size of RNA is between 50 and 300 nt (Fig. 2b). Since the cross-link site will be located at variable positions within RNA fragments, this size range leads to optimal cDNA insert sizes between 25 and 150 nt.

20. Variation will exist between different batches of reverse transcription primer [13]. It is therefore recommended to test each new batch of reverse transcription primers to identify those that are producing optimal libraries. This can be achieved by taking a single low RNase sample up to **step 5** of Subheading 3.8, then splitting the sample into equal aliquots for each reverse transcription primer. The remainder of the iCLIP protocol should be completed up to the test PCR stage (**step 11** of Subheading 3.11) for each sample with no multiplexing. This will allow evaluation of the performance of each RT primer against one another under identical conditions.

21. TBE-UREA gels are required to fully denature cDNA for accurate size determination. Precast TBE-UREA gels have short shelf-lives and should be not used outside of indicated dates.

22. Consistent conditions for cDNA gels allow the user to decide whether to use a previously made cutting guide or generate a new guide each time. Once it has been established that the denaturing conditions used can fully separate the low molecular weight marker into single stranded nucleotides (i.e., no ladder double bands due to partial denaturing), then it is possible to use a previous cutting guide if conditions have been kept identical (Fig. 3).

23. Ladder staining should be carried out as quickly as possible as the TBE-UREA gels will start to swell once the cassette is open. A practice run on a ladder is recommended before the first iCLIP experiment, while a premade cutting guide can also be used as indicated in **Note 22**.

24. Three bands are cut from each sample. This is to avoid PCR bias toward shorter fragments at later steps. The lowest band contains short insert sizes that are less mappable than the medium and high bands. However, they may contain short RNAs including miRNAs. Unless short RNAs are desirable, the crushed low band can be stored at −20 °C for processing at

a later date if required. Should PCR analysis reveal inaccurate and high cutting, then the stored low band may also be processed to produce optimal cDNA libraries.

25. Over amplification of cDNA can lead to secondary products appearing above the expected size (Fig. 4b) [13]. At higher cycle numbers these products are too large to migrate on the 6% TBE gels. These secondary products carry the necessary sequences required for flow-cell hybridization and so should be removed through PCR optimization.
26. A low cycle final PCR (~15–21 cycles) will produce well-diversified libraries, while high PCR cycles (>25 cycles) will be produce libraries where limited products are sequenced many times over. It is suggested that an initial test PCR of 25 cycles is carried out to ascertain where the amplification stands relative to Fig. 4b libraries. Appropriate cycle number changes can then be made for a second test PCR to finalize conditions. For example, if the test PCR has a similar appearance to the 25 PCR cycles in Fig. 4b, then the next test PCR should be reduced by 4–5 cycles in order to get optimal amplification at 20–21 cycles.
27. The final PCR uses 2.5-fold more concentrated input cDNA than the test PCR. Accordingly, an adjustment of 1 reduced cycle results in optimal final PCR cycling number.
28. In our hands we find that the calculated concentration divided by two results in accurate values for Illumina sequencing. However, it is recommended that relationships between qPCR and sequencing cluster density are closely monitored over initial sequencing runs in order to optimize in a new iCLIP setup.
29. Alternatives to qPCR quantification can include Agilent Bioanalyser/TapeStation analysis together with Qubit quantification, e.g., [29].
30. Following high-throughput sequencing the bioinformatics analysis of iCLIP libraries requires demultiplexing, mapping, collapsing PCR duplicates using random barcodes, and determination of the cross-link site. Note that the nucleotide immediately preceding the first nucleotide of a mapped read is considered the site of cross-linking. After replicate comparisons, biological and technical replicates may be merged before subsequent analysis of the RBPs binding landscape.
31. Following bioinformatics removal of PCR duplicates based on random barcode evaluation and the merging of technical/biological replicates, an iCLIP library will ideally have ~1×10^6 unique reads for analysis. However, libraries with >1×10^5 can allow sufficient information to perform a basic analysis.

Acknowledgments

The iCLIP protocol described is based on previous versions developed in the Ule and Konig labs by many individuals. I would like to extend thanks to all those who have contributed to the method development over the years. I would also like to thank Prof. Jernej Ule for critical reading of the chapter, and Prof. Jernej Ule and Flora Lee for providing the adapter oligo used in this protocol. This work is supported by an Edmond Lily Safra Fellowship to C.R.S.

References

1. Modic M, Ule J, Sibley CR (2013) CLIPing the brain: studies of protein-RNA interactions important for neurodegenerative disorders. Mol Cell Neurosci 56:429–435. doi:10.1016/j.mcn.2013.04.002
2. Gerstberger S, Hafner M, Tuschl T (2014) A census of human RNA-binding proteins. Nat Rev Genet 15(12):829–845. doi:10.1038/nrg3813
3. Baltz AG, Munschauer M, Schwanhausser B, Vasile A, Murakawa Y, Schueler M, Youngs N, Penfold-Brown D, Drew K, Milek M, Wyler E, Bonneau R, Selbach M, Dieterich C, Landthaler M (2012) The mRNA-bound proteome and its global occupancy profile on protein-coding transcripts. Mol Cell 46(5):674–690. doi:10.1016/j.molcel.2012.05.021
4. Castello A, Horos R, Strein C, Fischer B, Eichelbaum K, Steinmetz LM, Krijgsveld J, Hentze MW (2013) System-wide identification of RNA-binding proteins by interactome capture. Nat Protoc 8(3):491–500. doi:10.1038/nprot.2013.020
5. Jangi M, Sharp PA (2014) Building robust transcriptomes with master splicing factors. Cell 159(3):487–498. doi:10.1016/j.cell.2014.09.054
6. Lunde BM, Moore C, Varani G (2007) RNA-binding proteins: modular design for efficient function. Nat Rev Mol Cell Biol 8 (6):479–490. doi:10.1038/nrm2178
7. Nussbacher JK, Batra R, Lagier-Tourenne C, Yeo GW (2015) RNA-binding proteins in neurodegeneration: Seq and you shall receive. Trends Neurosci 38(4):226–236. doi:10.1016/j.tins.2015.02.003
8. Cookson MR (2017) RNA-binding proteins implicated in neurodegenerative diseases. Wiley Interdiscip Rev RNA 8(1). doi:10.1002/wrna.1397
9. Wurth L, Gebauer F (2015) RNA-binding proteins, multifaceted translational regulators in cancer. Biochim Biophys Acta 1849 (7):881–886. doi:10.1016/j.bbagrm.2014.10.001
10. Jayaseelan S, Doyle F, Tenenbaum SA (2014) Profiling post-transcriptionally networked mRNA subsets using RIP-Chip and RIP-Seq. Methods 67(1):13–19. doi:10.1016/j.ymeth.2013.11.001
11. Greenberg JR (1979) Ultraviolet light-induced crosslinking of mRNA to proteins. Nucleic Acids Res 6(2):715–732
12. Ule J, Jensen K, Mele A, Darnell RB (2005) CLIP: a method for identifying protein-RNA interaction sites in living cells. Methods 37(4): 376–386. doi:10.1016/j.ymeth.2005.07.018
13. Huppertz I, Attig J, D'Ambrogio A, Easton LE, Sibley CR, Sugimoto Y, Tajnik M, Konig J, Ule J (2014) iCLIP: protein-RNA interactions at nucleotide resolution. Methods 65(3): 274–287. doi:10.1016/j.ymeth.2013.10.011
14. Gonzalez-Buendia E, Saldana-Meyer R, Meier K, Recillas-Targa F (2015) Transcriptome-wide identification of in vivo interactions between RNAs and RNA-binding proteins by RIP and PAR-CLIP assays. Methods Mol Biol 1288:413–428. doi:10.1007/978-1-4939-2474-5_24
15. Hafner M, Landthaler M, Burger L, Khorshid M, Hausser J, Berninger P, Rothballer A, Ascano M Jr, Jungkamp AC, Munschauer M, Ulrich A, Wardle GS, Dewell S, Zavolan M, Tuschl T (2010) Transcriptome-wide identification of RNA-binding protein and microRNA target sites by PAR-CLIP. Cell 141 (1):129–141. doi:10.1016/j.cell.2010.03.009
16. Licatalosi DD, Mele A, Fak JJ, Ule J, Kayikci M, Chi SW, Clark TA, Schweitzer AC, Blume JE, Wang X, Darnell JC, Darnell RB (2008) HITS-CLIP yields genome-wide insights into brain alternative RNA processing. Nature 456 (7221):464–469. doi:10.1038/nature07488
17. Zhang C, Darnell RB (2011) Mapping in vivo protein-RNA interactions at single-nucleotide resolution from HITS-CLIP data. Nat

Biotechnol 29(7):607–614. doi:10.1038/nbt.1873

18. Yeo GW, Coufal NG, Liang TY, Peng GE, Fu XD, Gage FH (2009) An RNA code for the FOX2 splicing regulator revealed by mapping RNA-protein interactions in stem cells. Nat Struct Mol Biol 16(2):130–137. doi:10.1038/nsmb.1545
19. Konig J, Zarnack K, Rot G, Curk T, Kayikci M, Zupan B, Turner DJ, Luscombe NM, Ule J (2010) iCLIP reveals the function of hnRNP particles in splicing at individual nucleotide resolution. Nat Struct Mol Biol 17(7):909–915. doi:10.1038/nsmb.1838
20. Haberman N, Huppertz I, Attig J, Konig J, Wang Z, Hauer C, Hentze MW, Kulozik AE, Le Hir H, Curk T, Sibley CR, Zarnack K, Ule J (2017) Insights into the design and interpretation of iCLIP experiments. Genome Biol 18(1):7. doi:10.1186/s13059-016-1130-x
21. Wang Z, Kayikci M, Briese M, Zarnack K, Luscombe NM, Rot G, Zupan B, Curk T, Ule J (2010) iCLIP predicts the dual splicing effects of TIA-RNA interactions. PLoS Biol 8(10):e1000530. doi:10.1371/journal.pbio.1000530
22. Van Nostrand EL, Gelboin-Burkhart C, Wang R, Pratt GA, Blue SM, Yeo GW (2016) CRISPR/Cas9-mediated integration enables TAG-eCLIP of endogenously tagged RNA binding proteins. Methods. doi:10.1016/j.ymeth.2016.12.007
23. Bohnsack MT, Tollervey D, Granneman S (2012) Identification of RNA helicase target sites by UV cross-linking and analysis of cDNA. Methods Enzymol 511:275–288. doi:10.1016/B978-0-12-396546-2.00013-9
24. Van Nostrand EL, Pratt GA, Shishkin AA, Gelboin-Burkhart C, Fang MY, Sundararaman B, Blue SM, Nguyen TB, Surka C, Elkins K, Stanton R, Rigo F, Guttman M, Yeo GW (2016) Robust transcriptome-wide discovery of RNA-binding protein binding sites with enhanced CLIP (eCLIP). Nat Methods 13(6):508–514. doi:10.1038/nmeth.3810
25. Flynn RA, Martin L, Spitale RC, Do BT, Sagan SM, Zarnegar B, Qu K, Khavari PA, Quake SR, Sarnow P, Chang HY (2015) Dissecting noncoding and pathogen RNA-protein interactomes. RNA 21(1):135–143. doi:10.1261/rna.047803.114
26. Zarnegar BJ, Flynn RA, Shen Y, Do BT, Chang HY, Khavari PA (2016) irCLIP platform for efficient characterization of protein-RNA interactions. Nat Methods 13(6):489–492. doi:10.1038/nmeth.3840
27. Sugimoto Y, Konig J, Hussain S, Zupan B, Curk T, Frye M, Ule J (2012) Analysis of CLIP and iCLIP methods for nucleotide-resolution studies of protein-RNA interactions. Genome Biol 13(8):R67. doi:10.1186/gb-2012-13-8-r67
28. Zarnack K, Konig J, Tajnik M, Martincorena I, Eustermann S, Stevant I, Reyes A, Anders S, Luscombe NM, Ule J (2013) Direct competition between hnRNP C and U2AF65 protects the transcriptome from the exonization of Alu elements. Cell 152(3):453–466. doi:10.1016/j.cell.2012.12.023
29. Sutandy FX, Hildebrandt A, Konig J (2016) Profiling the binding sites of RNA-binding proteins with nucleotide resolution using iCLIP. Methods Mol Biol 1358:175–195. doi:10.1007/978-1-4939-3067-8_11

Chapter 30

RNA Tagging: Preparation of High-Throughput Sequencing Libraries

Christopher P. Lapointe and Marvin Wickens

Abstract

Protein–RNA networks, in which a single protein binds and controls multiple mRNAs, are central in biological control. As a result, methods to identify protein–RNA interactions that occur in vivo are valuable. The "RNA Tagging" approach enables the investigator to unambiguously identify global protein–RNA interactions in vivo and is independent of protein purification, cross-linking, and radioactive labeling steps. Here, we provide a protocol to prepare high-throughput sequencing libraries for RNA Tagging experiments.

Key words RNA Tagging, High-throughput sequencing, Protein–RNA interactions, Protein–RNA networks, RNA regulatory networks, Poly(U) polymerases, RNA, RNA-binding proteins

1 Introduction

Protein–RNA interactions underlie fundamental cellular functions. Single proteins often bind to hundreds of RNAs inside the cell to control when the RNAs are made, where they are located, when they are degraded, and what they do [1–3]. These "protein–RNA networks" have important roles in the activity of protein-coding genes, and thus underlie a diverse range of biological processes. As a result, methods to identify transcriptome-wide protein–RNA interactions in vivo are valuable and multiple strategies have been described, including several forms of CLIP (cross-linking and immunopurification) [4–9].

We developed a facile and unambiguous approach, called RNA Tagging, which enables an investigator to identify and analyze global protein–RNA interactions in vivo [10]. In the version of RNA Tagging described here, we express an RNA-binding protein (RBP) of interest fused to a poly(U) polymerase (PUP) . The RBP–PUP chimera covalently marks RNAs it binds in vivo with 3′ terminal uridines, which we refer to as a "U-tag." The U-tagged RNAs are then identified from a pool of total RNA extracted from

Imre Gaspar (ed.), *RNA Detection: Methods and Protocols*, Methods in Molecular Biology, vol. 1649, DOI 10.1007/978-1-4939-7213-5_30, © Springer Science+Business Media LLC 2018

the cell using high-throughput sequencing. For applications involving mRNA networks, we commonly purify poly(A)-containing RNAs to obtain an enriched pool of mRNAs, but this can be dispensed with to allow nonadenylated RNAs to be captured. The RNA Tagging strategy does not require cross-linking, protein purification, or radioactive-labeling steps. The data are unambiguous because the U-tags are directly detected via sequencing and cells are lysed under denaturing conditions, which ensures RNAs are U-tagged in vivo and not in the cell lysate. Furthermore, the approach highlights mRNAs that likely are regulated from those that likely are not. The TRIBE approach developed by McMahon et al. is conceptually similar and uses ADAR as the tagging agent [11].

We describe here a detailed protocol to prepare RNA Tagging high-throughput sequencing libraries. We first outline our protocol to isolate total RNA from *Saccharomyces cerevisiae* in denaturing conditions (Subheading 3.1) (Fig. 1). We next describe our protocols to enrich for polyadenylated RNAs using poly(A) selection (Subheading 3.2) and rRNA depletion (Subheading 3.3) steps. To enable transcriptome-wide analyses, we then G–I tail all RNAs (Subheading 3.4), which both captures U-tags and ensures all RNAs have a common sequence at their 3′ termini. Next, we selectively reverse transcribe U-tagged RNAs using a U-select oligo that preferentially hybridizes to U-tags and the G–I tail (Subheading 3.5), synthesize the second strand of DNA (Subheading 3.6), and PCR amplify and purify the DNA libraries (Subheading 3.7). At the conclusion of this protocol, users will have RNA Tagging libraries that are ready for analysis by high-throughput sequencing.

2 Materials

Use nuclease-free water to make all buffers. Test buffers to ensure they are RNase-free before use.

2.1 Total RNA Isolation from Yeast

1. Vortex.
2. Nuclease-free 1.75 mL microcentrifuge tubes.
3. Refrigerated general purpose centrifuge.
4. Refrigerated microcentrifuge.
5. UV-Vis spectrophotometer (e.g., NanoDrop).
6. Agarose gel electrophoresis setup.
7. −20 to −50 °C freezer, and −80 °C freezer.
8. Optional: bioanalyzer.
9. Ice cold water.

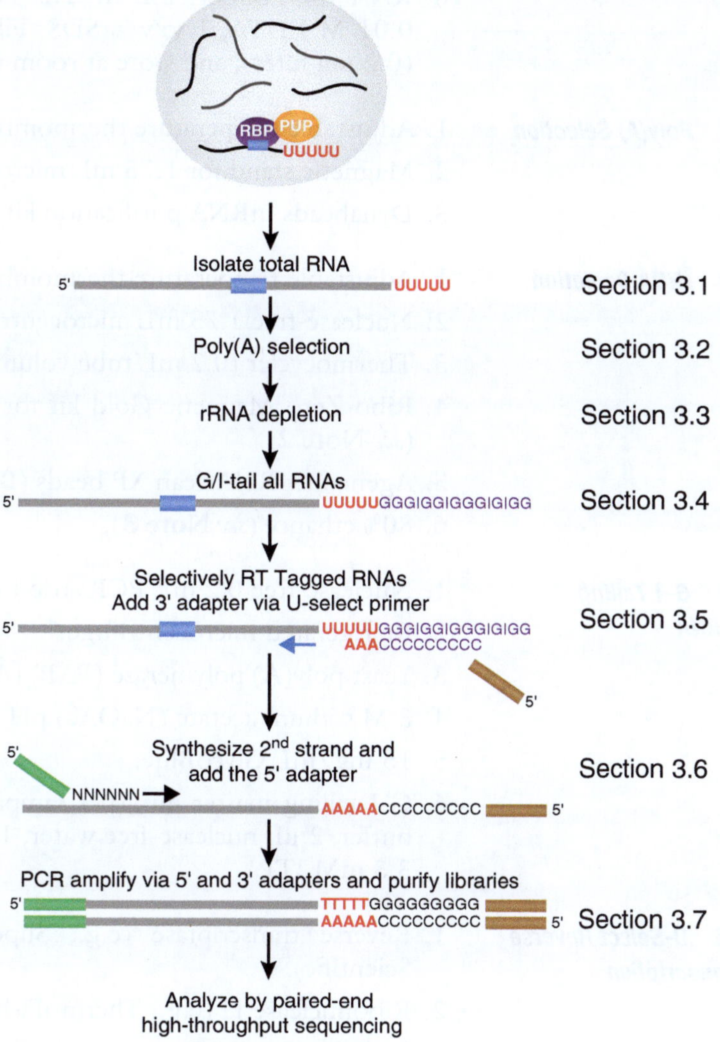

Fig. 1 Schematic of the RNA Tagging library preparation workflow. Adapted from Fig. 1 in Lapointe et al. [10]

10. 425–600 μm acid-washed glass beads.
11. Phenol–chloroform–isoamyl alcohol (25:24:1) pH 6.7.
12. Chloroform.
13. 100% ethanol.
14. 80% ethanol.
15. TURBO DNaseI.
16. Nuclease-free water.
17. RNA Purification kit (e.g., GeneJET, ThermoFisher Scientific).

18. RNA ISO Buffer: 0.2 M Tris–HCl pH 7.5, 0.5 M NaCl, 0.01 M EDTA, 1 v/v % SDS. Filter-sterilize before first use (0.2 μm filter), and store at room temperature (*see* **Note 1**).

2.2 Poly(A) Selection

1. Adjustable temperature thermomixer.
2. Magnetic stand for 1.75 mL microcentrifuge tubes.
3. Dynabeads mRNA purification kit (ThermoFisher Scientific).

2.3 rRNA Depletion

1. Adjustable temperature thermomixer.
2. Nuclease-free 1.75 mL microcentrifuge tubes.
3. Thermocycler (0.2 mL tube volume).
4. Ribo-Zero Magnetic Gold kit for yeast (Epicentre/Illumina) (*see* **Note 2**).
5. Agencourt RNAclean XP beads (Beckman Coulter).
6. 80% ethanol (*see* **Note 3**).

2.4 G–I Tailing of RNA

1. Nuclease-free 0.2 mL PCR strip tubes.
2. Refrigerated microcentrifuge.
3. Yeast poly(A) polymerase (PAP) (Affymetrix, 74225Y).
4. 3 M sodium acetate (NaOAc) pH 5.5.
5. 15 mg/mL GlycoBlue.
6. G-I tailing master mix (per sample): 4 μL 5× PAP Reaction buffer, 2 μL nuclease-free water, 1 μL 10 mM GTP, and 1 μL 3.3 mM ITP.

2.5 U-Select Reverse Transcription

1. Reverse transcriptase (e.g., SuperScript III ThermoFisher Scientific).
2. Ribonuclease H (e.g., ThermoFisher Scientific).
3. PCR Purification kit (e.g., GeneJET, ThermoFisher Scientific).
4. 1 μM U-select oligo 5′- *GCCTTGGCACCCGAGAATTC-CACCCCCCCCCAAA*-3′.

   The three adenosines on the 3′ end of the U-select oligo preferentially anneal to RNAs that end in uridines (prior to the G/I-tailing), thus selectively enriching for U-Tagged RNAs. The nine cytosines anneal to the G–I tail. The underlined portion corresponds to Illumina 3′ adapter sequence vital for PCR amplification.
5. RT master mix (per sample): 4 μL 5× SuperScript III Reaction Buffer, 1 μL 100 mM DTT, and 1 μL RNaseOUT (e.g., ThermoFisher Scientific).

2.6 Second Strand Synthesis of DNA

1. 5 U/μL Exo-Klenow Fragment DNA Polymerase I.
2. 80% ethanol (*see* **Note 3**).

3. Second strand synthesis oligo 5′-GTTCAGAGTTCTACA GTCCGACGATCNNNNNN-3′.

   The underlined portion corresponds to Illumina 5′ adapter sequence vital for PCR amplification.

4. 10× Klenow Buffer: 500 mM Tris–HCl pH 7.5, 100 mM $MgCl_2$, 10 mM DTT, and 0.5 mg/mL BSA. Store buffer at −20 °C.

5. S3 master mix (per sample):12 μL nuclease-free water, 10 μL 10× Klenow Buffer 5 μL 10 mM dNTPs, and 10 μL 10 μM second strand synthesis oligo.

2.7 PCR Amplification and Size-Selection of Libraries

1. GoTaq Green 2× PCR Master Mix (Promega).
2. 10 μM 5′ PCR primer

   5′-*AATGATACGGCGACCACCGAGATCTACACGTTCA GAGTTCTACAGTCCGA*-3′
3. 10 μM 3′ Barcoded PCR Primer

   5-*CAAGCAGAAGACGGCATACGAGATNNNNNNGTG ACTGGAGTTCCTTGGCACCCGAGAATTCCA*-3′. The *NNNNNN* represents the unique barcode. Use a unique barcode for each sample.
4. 80% ethanol (*see* **Note 3**).

3 Methods

This protocol requires that the relevant RBP–PUP fusion proteins already be engineered and introduced into cells. Upon request, we provide a kit that includes plasmids that encode the PUP-2 open-reading frame, free of charge to any academic lab.

3.1 Isolate Total RNA from S. cerevisiae

Isolate total RNA from yeast in completely denaturing conditions (*see* **Note 4**).

Timing: **steps 3–21**: 1.5–2 h; **steps 22–31**: 3–4 h.

1. Place 25 mL of A_{660} ~ 0.5–0.8 cultures on ice for 5 min.
2. Harvest cultures by centrifugation at 1900 rcf for 5 min at 4 °C.
3. Wash yeast pellets once with 40 mL of ice-cold water.
4. Resuspend yeast in 500 μL RNA ISO buffer.
5. Add ~200 μL of acid-washed beads.
6. Add 500 μL of phenol–chloroform–isoamyl alcohol.
7. Vortex for 20 s at room temp., then incubate for 20 s on ice. Repeat for a total of ten cycles.
8. Split into two tubes. Each sample is now in two tubes.

9. Add 375 μL RNA ISO buffer and 375 μL PCA.
10. Mix gently by several inversions.
11. Spin at max speed (16,100 × *g*) at 4 °C for 10 min.
12. Transfer aqueous phase to new tube.
13. Add equal volume PCA and mix gently.
14. Spin at max speed at 4 °C for 10 min.
15. Remove aqueous phase to new tube.
16. Add equal volume chloroform and mix gently.
17. Spin at max speed at 4 °C for 10 min.
18. Remove aqueous phase to new tube.
19. Add 1 mL of 100% ethanol, mix gently. Incubate for at least 1 h at −50 °C (*see* **Note 5**).
20. Spin at max speed at 4 °C for 20 min. Remove supernatant.
21. Wash pellet at least once in 80% ethanol (*see* **Note 6**).
22. Resuspend RNA pellets in 43 μL of nuclease-free water.
23. Combine the two tubes for each sample (back to one tube per sample), add 10 μL of 10× TURBO DNase Buffer and 4 μL (8 U) of TURBO DNase.
24. Incubate for 1 h at 37 °C (*see* **Note 7**).
25. Clean reactions with an RNA Purification kit (e.g., using the "RNA cleanup protocol" of the GeneJET kit).
26. Elute in 30 μL of nuclease-free water.
27. Determine the concentration of total RNA (*see* **Note 8**).
28. Analyze isolated RNA by agarose gel electrophoresis to assess RNA quality. Load 500 ng of total RNA (*see* **Note 9**).
29. Store RNA at −80 °C until next step.

3.2 Poly(A) Selection

Many small, noncoding RNAs end in several uridine residues. To decrease the number of these "background" RNAs in our libraries, we enrich for polyadenylated RNAs (*see* **Note 10**). This step can be omitted, but the resulting libraries will be primarily composed of small noncoding RNAs, such as SCR1 (≈90% of the sample). Thus deeper sequencing (larger numbers of reads) will be required to identify the complete collection of U-tagged RNAs.

Timing: ~1 h.

1. Prior to starting, let the aliquots of beads and buffers warm to room temp (~15 min), slow thaw RNAs on ice (15–20 min), and set a thermomixer to 65 °C.
2. Use 75 μg of total RNA per sample and adjust volume to 100 μL using nuclease-free water.

3. Heat RNA solution to 65 °C in the thermomixer for 2 min to disrupt secondary structures.
4. Place samples on ice until use in **step 9**.
5. Set the thermomixer to 70 °C.
6. Transfer 200 μL of well-resuspended Dynabeads to a microcentrifuge tube.
7. Place tube on magnet. Discard supernatant.
8. Add 100 μL of Binding Buffer to calibrate the beads. Mix well.
9. Place tube back on magnet and discard the supernatant.
10. Add 100 μL of Binding Buffer to the beads. Mix well.
11. Add the RNA (from **step 3**) to the bead solution. It's important to have a 1:1 ratio of RNA volume to Binding Buffer volume.
12. Mix beads thoroughly.
13. Gently vortex samples briefly every 30 s for 5 min at room temp.
14. Place tube on magnet to separate beads. Remove the supernatant.
15. Wash with 200 μL Washing Buffer B. Mix thoroughly.
16. Place tube on magnet. Remove all of the supernatant.
17. Repeat **steps 15–16**.
18. After removing all of the supernatant from the second wash, add 28 μL of water.
19. Resuspend the beads well.
20. Heat the samples to 70 °C in the thermomixer for 2 min.
21. Place the samples on the magnet *immediately.*
22. Collect the supernatant. This is your poly(A)+ mRNA.

3.3 rRNA Depletion

The poly(A) selection efficiently removes small, noncoding RNAs (such as snRNAs and SCR1) but there is bleed through of rRNAs. Thus, we use this step to remove the remaining rRNAs (*see* **Note 10**). This step can be omitted, but the resulting libraries will be primarily composed of rRNAs (90–95% of the sample).

Timing: ~2 h.

1. *Prior to starting:* let the magnetic beads and associated buffers warm to room temperature (~15 min), slow thaw reagents stored at −80 °C on ice (15–20 min), set a thermocycler to 68 °C, set a thermomixer to 50 °C, and aliquot the needed volume of RNA Clean XP beads (*see* Subheading 3.3.4) and store them at room temp until use in Subheading 3.3.4 (*see* **Note 11**).

3.3.1 Bead Washing

1. For each reaction, pipet 225 μL of Ribo-Zero magnetic beads into a 1.75 mL centrifuge tube. Pipet slowly to avoid air bubbles. Store the unused beads at 4 °C.
2. Place tubes on magnetic stand for >1 min.
3. Remove and discard the supernatant.
4. Remove the tube from the stand and add 225 μL of water to each tube. Mix well by repeated pipetting or vortexing at medium speed.
5. Repeat **steps 2–4**. Remove the tube from the stand. Add 65 μL of Magnetic Bead Resuspension Solution to each tube. Mix well by pipetting or vortexing on medium speed.
6. Add 1 μL of RiboGuard RNase Inhibitor and mix briefly by vortexing.
7. Keep the tubes at room temp until needed in Subheading 3.3.3.

3.3.2 Treatment of Total RNA with Ribo-Zero rRNA Removal Solution

1. For each sample, combine the following in an RNase-free 0.2 mL PCR strip-tube (volumes reflect 1 reaction):

4 μL	Ribo-Zero Reaction Buffer
26 μL	poly(A)+ RNA
10 μL	Ribo-Zero rRNA Removal Solution

2. Gently mix the reaction by pipetting and incubate at 68 °C for 10 min in a thermocycler.
3. Store the rest of the unused kit at −80 °C.
4. Remove the reaction tube and incubate at room temperature for 5 min.

3.3.3 Magnetic Bead Reaction and rRNA Removal (See Note 12)

1. Using a pipette, add the treated RNA sample to the washed magnetic beads and *immediately* mix by pipetting at least ten times to thoroughly mix the sample. Then, vortex the tube at medium setting for 10 s and place at room temperature.
2. Incubate the samples at room temperature for 5 min.
3. Then, mix the reactions by vortexing at medium speed for 10 s and place at 50 °C for 5 min. Avoid any significant condensation during this step (e.g., make sure the cover for the thermomixer is on the instrument during the incubation to keep the lids of the tubes exposed to warm air).
4. Remove the tubes from the 50 °C heat block and place on a magnetic stand for >1 min.
5. While on the stand, carefully remove the supernatant (*this is your rRNA-free RNA!*) and place in a labeled, RNase-free tube.

6. If there are residual beads in the supernatant, repeat the magnetic separation.
7. Place the RNA on ice and immediately proceed to RNA cleanup.

3.3.4 Agencourt RNAClean XP Bead Mediated RNA Cleanup

1. Mix the Agencourt RNAClean XP beads well by vortexing.
2. Add 160 μL of the mixed beads to each reaction containing 85–90 μL of rRNA-depleted sample. Mix thoroughly by pipetting >10 times. Vortex gently.
3. Incubate at room temperature for 15 min. During the incubation prepare a fresh 80% ethanol solution.
4. Place the tube on a magnetic stand for >5 min.
5. Remove the supernatant without disturbing the beads.
6. With the tube still on the stand, add 400 μL of fresh 80% ethanol without disturbing the beads.
7. Incubate at room temperature for 1 min.
8. Remove the ethanol supernatant.
9. Repeat the 80% ethanol wash for a total of two wash steps.
10. Allow the tube to air dry on the magnetic stand (*see* **Note 13**).
11. Add 12 μL of RNase-free water to the tube and immediately and thoroughly mix.
12. Incubate the tubes at room temperature for 2 min.
13. Place the tubes on the magnetic stand for at least 5 min. Transfer the clear supernatant to a new tube, always leaving 1–2 μL behind to prevent carryover of the beads to the next steps.

3.4 G-I Tailing of RNA

This protocol adds a known sequence to the 3′ end of all RNAs that can be exploited to reverse transcribe the RNA (*see* **Note 14**).

Timing: **steps 1–22**: ~2.5 h; **steps 23–28**: ~1 h.

1. For each sample, aliquot 8 μL of the G-I tailing master mix into nuclease-free 0.2 mL PCR strip-tubes.
2. For each sample, add 10 μL of the appropriate poly(A)+/rRNA-depleted RNA and mix.
3. Add 2 μL of 600 U/μL Yeast PAP to each reaction.
4. Incubate at 37 °C for 90 min.
5. Add an additional 2 μL of 600 U/μL Yeast PAP to each reaction.
6. Incubate at 37 °C for 30 min.
7. Add 80 μL of nuclease-free water to each reaction (volume should now be ~100 μL).

8. Transfer reactions to 1.75 mL microcentrifuge tubes.
9. Add 100 μL of phenol–chloroform–isoamyl alcohol (25:24:1) pH 6.7 to each reaction.
10. Gently vortex or mix thoroughly.
11. Spin at max speed for 5 min at 4 °C.
12. Collect the aqueous phase (top layer, aim for 95–100 μL) and transfer to a new tube (*see* **Note 15**).
13. Add an equal volume of chloroform.
14. Gently vortex or mix thoroughly.
15. Spin at max speed for 5 min at 4 °C.
16. Collect the aqueous phase (top layer, aim for 90 μL) and transfer to a new tube.
17. Add 10 μL of 3 M NaOAc, 1 μL of 15 mg/mL GlycoBlue, and 500 μL of 100% ethanol.
18. Mix thoroughly.
19. Incubate for at least 1 h at −50 °C (*see* **Note 5**).
20. Spin at max speed for 25 min at 4 °C.
21. Remove supernatant. Wash pellet with 70–80% ethanol.
22. Spin at max speed for 25 min at 4 °C.
23. Remove supernatant.
24. Pulse-spin and remove residual ethanol.
25. Resuspend pelleted RNA in 10 μL of nuclease-free water.

3.5 U-select Reverse Transcription

Use the G-I nucleotides on the 3′ end of the RNA to prime reverse transcription via the U-select oligo. The three adenosines on the 3′ end of the U-select oligo preferentially anneal to RNAs that end in uridines (prior to the G/I-tailing), thus selectively enriching the U-Tagged RNAs. The U-select oligo also contains Illumina 3′ adapter sequence (underlined).

Timing: ~2.5 h

1. *Recommended*: Also prepare –RT reactions for each sample as a comparison.
2. Assemble the following reaction in strip tubes (volumes reflect one reaction):

1 μL	1 μM U-select oligo
5 μL	G-I-tailed RNA
1 μL	10 mM dNTP mix
6 μL	nuclease-free water

3. Heat the reactions and the RT master mix to 65 °C for 5 min.

4. Cool the reactions and the RT master mix to 50 °C for 5 min.

   Important: Perform steps 5 and 6 while the RNA/primer mix and the RT master mix are in the thermomixer. It's important to keep the reactions at 50 °C to maintain the U-selection.

5. Add 6 μL of the preheated (50 °C) RT master mix in the thermocycler to each reaction.
6. Add 1 μL of 200 U/μL SuperScript III reverse transcriptase to each reaction.
7. Incubate at 50 °C for 60 min.
8. Incubate at 85 °C for 5 min.
9. Cool reactions to 4 °C.
10. Add 1 μL RNase H to each reaction.
11. Incubate at 37 °C for 20 min.
12. Add 80 μL of water to increase reaction volume to ~100 μL.
13. Clean cDNA using the GeneJET PCR Purification kit. (We do not add isopropanol.)
14. Add 32 μL nuclease-free water to the dry column.
15. Incubate the water on the column for a least 2 min at room temperature.
16. Centrifuge at max speed for 2 min.
17. Repeat **steps 14–16**.
18. Combine elution fraction to get ~60 μL of cDNA for each reaction.

3.6 Second Strand Synthesis

Randomly synthesize the second strand of DNA that is complementary to the cDNA sequence, while at the same time adding the Illumina 5′ adapter sequence.

Timing: ~2 h

1. *Prior to starting:* aliquot the required volume of Agencourt RNAClean XP beads and keep them at room temperature until use.
2. Aliquot 37 μL of S3 master mix for each reaction into 0.2 mL nuclease-free PCR strip tubes.
3. Add 60 μL purified cDNA to the S3 master mix for each sample.
4. Add 3 μL of 5 U/μL Exo-Klenow fragment DNA polymerase I to each reaction.
5. Incubate at 37 °C for 30 min.
6. Cool to 4 °C.
7. Warm reactions to room temperature (*see* **Note 16**).
8. Mix the Agencourt RNAClean XP beads well by vortexing.

9. Add 100 μL of the mixed beads to each 100 μL of sample (*see* **Note 17**).
10. Mix thoroughly by pipetting >10 times. Vortex gently.
11. Incubate at room temperature for 15 min. During the incubation prepare a fresh 80% ethanol solution.
12. Place the tube on a magnetic stand for >5 min.
13. Remove the supernatant without disturbing the beads.
14. With the tube still on the stand, add 400 μL of fresh 80% ethanol without disturbing the beads.
15. Incubate at room temperature for 1 min while still on the magnetic stand.
16. Remove the ethanol supernatant.
17. Repeat the 80% ethanol wash for a total of two wash steps.
18. Allow the tube to air dry on the magnetic stand (*see* **Note 13**).
19. Add 100 μL of nuclease-free water to the tube and immediately and thoroughly mix.
20. Incubate the tubes at room temperature for 2 min.
21. Place the tubes on the magnetic stand for 5 min. Transfer the clear supernatant to a new tube, always leaving 1–2 μL behind to prevent carryover of the beads to the next steps.
22. Repeat **steps 8–18**.
23. Add 50 μL of nuclease-free water to the tube and immediately and thoroughly mix.
24. Incubate the tubes at room temperature for 2 min.
25. Place the tubes on the magnetic stand for 5 min. Transfer the clear supernatant to a new tube, always leaving 1–2 μL behind to prevent carryover of beads to the next steps.

3.7 PCR Amplification and Purification

Amplify the dsDNA and add the remaining 5′ and 3′ Illumina adapter sequences for subsequent high-throughput sequencing. The PCR purification step efficiently removes adapter–adapter products (5′ Illumina adapter–3′ Illumina adapter with no RNA insert) that will preferentially sequence.

Timing: ~2.5 h.

1. Assemble the following reaction for each sample:

83.3 μL	2× GoTaq Master Mix
6.7 μL	10 μM 5′ PCR primer
6.7 μL	10 μM 3′ barcoded PCR primer
20 μL	Nuclease-free water
50 μL	Purified cDNA

2. Aliquot 20 μL of the PCR reaction mix into eight separate 0.2 mL nuclease-free PCR tubes.
3. Amplify via the following protocol (*see* **Note 18**):
   (a) 94 °C: 2 min.
   (b) 94 °C: 10 s.
   (c) 40 °C: 2 min.
   (d) 72 °C: 1 min.
   (e) Go to **step b** once.
   (f) 94 °C: 10 s.
   (g) 55 °C: 30 s.
   (h) 72 °C: 1 min.
   (i) Go to **step f** seven times.
   (j) 94 °C: 15 s.
   (k) 55 °C: 30 s.
   (l) 72 °C: 1 min.
   (m) Go to **step j** fourteen times.
   (n) 72 °C: 5 min.
   (o) 4 °C: pause.
4. Warm Agencourt RNAClean XP beads and PCR samples to room temperature before proceeding (≈15 min).
5. Combine individual PCRs for each sample into a single 1.75 mL microcentrifuge tube (*see* **Note 19**).
6. Measure the volume of each sample using a pipette.
7. Mix the Agencourt RNAClean XP beads well by vortexing.
8. Add 0.8 volumes (relative to sample volume) of the pre-warmed, mixed beads to each sample (*see* **Note 20**).
9. Mix thoroughly by pipetting >10 times. Vortex gently.
10. Incubate at room temperature for 15 min. During the incubation prepare a fresh 80% ethanol solution.
11. Place the tube on a magnetic stand for >5 min.
12. Remove the supernatant without disturbing the beads.
13. With the tube still on the stand, add 400 μL of fresh 80% ethanol without disturbing the beads.
14. Incubate at room temperature for 1 min while still on the magnetic stand.
15. Remove the ethanol supernatant.
16. Repeat the 80% ethanol wash for a total of two wash steps.
17. Allow the tube to air dry on the magnetic stand (*see* **Note 13**).
18. Add 100 μL of nuclease-free water to the tube and immediately and thoroughly mix.

19. Incubate the tubes at room temperature for 2 min.
20. Place the tubes on the magnetic stand for 5 min. Transfer the clear supernatant to a new tube, always leaving 1–2 μL behind to prevent carryover of the beads to the next steps.
21. Repeat **steps 6–17**.
22. Add 15 μL of nuclease-free water to the dried beads and immediately and thoroughly mix.
23. Incubate the tubes at room temperature for 2 min.
24. Place the tubes on the magnetic stand for 5 min.
25. Transfer the clear supernatant to a new tube, always leaving 1–2 μL behind to prevent carryover of the beads to the next steps.
26. Store reactions at −80 °C until submission for high-throughput sequencing.
27. Analyze 2–4 μL of the libraries on a 1% agarose gel to check library quality. *See* Fig. 2 for an image of our analysis with an example library.
28. Optional: Topo–Cloning analysis of initial library preps is recommended to ensure libraries are constructed correctly.
29. Analyze libraries by paired-end high-throughput sequencing.

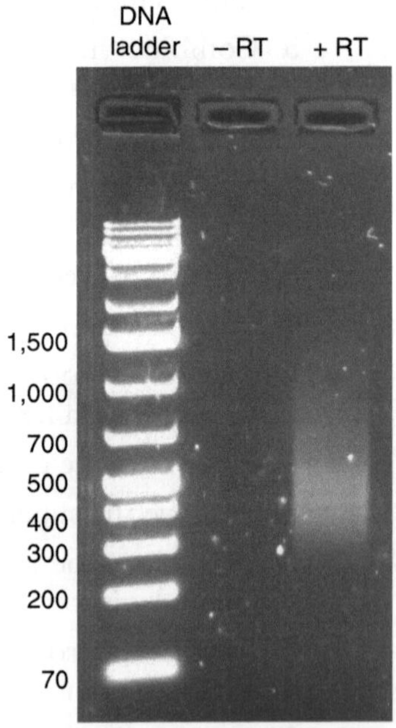

Fig. 2 Image of ethidium bromide-stained agarose gel that shows an RNA Tagging library ready for analysis via high-throughput sequencing. The "+RT" lane is the sample, and the "−RT" lane is the negative control that lacked reverse transcriptase. The numbers to the left of the DNA ladder lane indicate the size of the band (nucleotides)

4 Notes

1. The SDS in the buffer may precipitate at room temperature. If it does, heat the buffer very briefly in a 37 °C water buffer until the solution is clear.
2. For non-yeast experiments, substitute the yeast rRNA depletion kit with an rRNA depletion kit appropriate for the desired organism.
3. Prepare fresh 80% ethanol solution prior to each use.
4. If not using *S. cerevisiae*, proceedYeast directly to the poly(A) selection Subheading 3.2. Ensure total RNA was isolated in denaturing conditions.
5. If desired, an overnight incubation at −50 °C works well as a stopping point.
6. Two washes with 80% ethanol work well.
7. We do this in 1.75 mL microcentrifuge tubes in a 37 °C incubator.
8. We analyze 1:10 dilutions of total RNA using a NanoDrop spectrophotometer.
9. Optionally, total RNA can be analyzed using a BioAnalyzer.
10. The poly(A) selection protocol is essentially done as recommended by the manufacturer.
11. Mix the Agencourt RNAClean XP beads very well prior to aliquoting. Also, pipette slowly to help avoid air bubbles.
12. Always add the treated RNA sample to the washed magnetic beads and immediately mix by pipetting. Immediate and thorough mixing prevents the beads from forming clumps that can significantly impact rRNA removal efficiency.
13. The beads shouldn't over dry, but make sure there is no ethanol remaining in the tube. The elution in the next step is only 12 μL, so even very small volumes of ethanol could negatively impact the downstream enzymatic reactions. We typically add the water once the beads go from glossy to more matte-like appearance, just before they start cracking (thin white lines). It typically takes about 5 min. According to the manufacturer, overdrying reduces elution efficiency.
14. The G/I-tailing step can be replaced by a 3′ ligation step.
15. The aqueous phase will likely have a white, cloudy precipitate. Do not worry about it. Just collect as much of the aqueous phase as possible, even if it includes the precipitate. The precipitate will disappear once the second extraction is complete.
16. We let the cooled reactions sit on a lab bench for 5 min.

17. Use a 1:1 bead-to-reaction ratio (by volume) to efficiently remove the second strand synthesis oligo.
18. We most often run this PCR protocol overnight. A standard cycling protocol (94 °C for 10 s; 55 °C for 30 s; 72 °C 1 min; repeated 25 times) would likely work fine, too.
19. Add the eight PCR reactions per sample into a single 1.75 mL tube, but keep each of the samples in separate 1.75 mL tubes. There should be 140–150 μL per sample.
20. For example, add 120 μL of beads to 150 μL of sample.

Acknowledgments

We would like to thank Melanie Preston, Shruti Waghray, Hugo Medina, Brian Carrick, Harriet Saunders, Daniel Wilinski, and other members of the Wickens lab for their feedback on this protocol. We also thank Laura Vanderploeg and the Biochemistry Department Media Lab at UW-Madison for help with preparing figures. We appreciate the excellent assistance of the University of Wisconsin DNA Sequencing Facility. We are grateful for funding from NIH (GM R01 50942), and for Genentech, Wharton, and Biochemistry Scholar fellowships to C.P.L.

References

1. Keene JD (2007) RNA regulons: coordination of post-transcriptional events. Nat Rev Genet 8 (7):533–543. doi:10.1038/nrg2111
2. Konig J, Zarnack K, Luscombe NM, Ule J (2011) Protein-RNA interactions: new genomic technologies and perspectives. Nat Rev Genet 13(2):77–83. doi:10.1038/nrg3141
3. Gerstberger S, Hafner M, Tuschl T (2014) A census of human RNA-binding proteins. Nat Rev Genet 15(12):829–845. doi:10.1038/nrg3813
4. Ule J, Jensen KB, Ruggiu M, Mele A, Ule A, Darnell RB (2003) CLIP identifies Nova-regulated RNA networks in the brain. Science 302(5648):1212–1215. doi:10.1126/science.1090095
5. Licatalosi DD, Mele A, Fak JJ, Ule J, Kayikci M, Chi SW, Clark TA, Schweitzer AC, Blume JE, Wang X, Darnell JC, Darnell RB (2008) HITS-CLIP yields genome-wide insights into brain alternative RNA processing. Nature 456 (7221):464–469. doi:10.1038/nature07488
6. Hafner M, Landthaler M, Burger L, Khorshid M, Hausser J, Berninger P, Rothballer A, Ascano M Jr, Jungkamp AC, Munschauer M, Ulrich A, Wardle GS, Dewell S, Zavolan M, Tuschl T (2010) Transcriptome-wide identification of RNA-binding protein and microRNA target sites by PAR-CLIP. Cell 141(1):129–141. doi:10.1016/j.cell.2010.03.009
7. Konig J, Zarnack K, Rot G, Curk T, Kayikci M, Zupan B, Turner DJ, Luscombe NM, Ule J (2010) iCLIP reveals the function of hnRNP particles in splicing at individual nucleotide resolution. Nat Struct Mol Biol 17 (7):909–915. doi:10.1038/nsmb.1838
8. Van Nostrand EL, Pratt GA, Shishkin AA, Gelboin-Burkhart C, Fang MY, Sundararaman B, Blue SM, Nguyen TB, Surka C, Elkins K, Stanton R, Rigo F, Guttman M, Yeo GW (2016) Robust transcriptome-wide discovery of RNA-binding protein binding sites with enhanced CLIP (eCLIP). Nat Methods 13 (6):508–514. doi:10.1038/nmeth.3810
9. Zarnegar BJ, Flynn RA, Shen Y, Do BT, Chang HY, Khavari PA (2016) irCLIP platform for efficient characterization of protein-RNA interactions. Nat Methods 13(6):489–492. doi:10.1038/nmeth.3840

10. Lapointe CP, Wilinski D, Saunders HA, Wickens M (2015) Protein-RNA networks revealed through covalent RNA marks. Nat Methods 12(12):1163–1170. doi:10.1038/nmeth.3651

11. McMahon AC, Rahman R, Jin H, Shen JL, Fieldsend A, Luo W, Rosbash M (2016) TRIBE: hijacking an RNA-editing enzyme to identify cell-specific targets of RNA-binding proteins. Cell 165(3):742–753. doi:10.1016/j.cell.2016.03.007

Chapter 31

RAP-MS: A Method to Identify Proteins that Interact Directly with a Specific RNA Molecule in Cells

Colleen A. McHugh and Mitchell Guttman

Abstract

RNA molecules interact with proteins to perform a variety of functions in living cells. The binding partners of many RNAs, in particular the newly discovered class of long noncoding RNAs (lncRNAs), remain largely unknown. RNA antisense purification coupled with mass spectrometry (RAP-MS) is a method that enables the identification of direct and specific protein interaction partners of a specific RNA molecule. Because RAP-MS uses direct RNA–protein cross-linking methods coupled along with highly denaturing purification conditions, RAP-MS provides a short list of high confidence protein interactors.

Key words RNA–protein interactions, UV cross-linking, Mass spectrometry, RNA purification, Protein purification, Antisense nucleic acid capture, RNA-binding proteins

1 Introduction

Long noncoding RNAs (lncRNAs) are emerging as a new class of cellular regulators that play important roles in gene regulation, chromatin structure, and cell fate during development [1], yet the mechanisms by which most lncRNAs work remain unknown. Addressing this question requires knowledge of the protein interaction partners that these RNA molecules use to achieve their functions.

There have been technical challenges in addressing this goal because of a lack of available methods that can successfully isolate direct RNA interacting proteins that occur in vivo (for a review of methods *see* McHugh et al. [1]). Briefly, methods that measure in vitro association of an RNA with cellular proteins fail to separate interactions that occur in vivo from those that occur in solution. Purifications of RNA-protein complexes from formaldehyde cross-linked samples identify both direct and indirect protein interactors, leading to a potentially long list of proteins making functional characterization of these interactions challenging [1].

Imre Gaspar (ed.), *RNA Detection: Methods and Protocols*, Methods in Molecular Biology, vol. 1649, DOI 10.1007/978-1-4939-7213-5_31, © Springer Science+Business Media LLC 2018

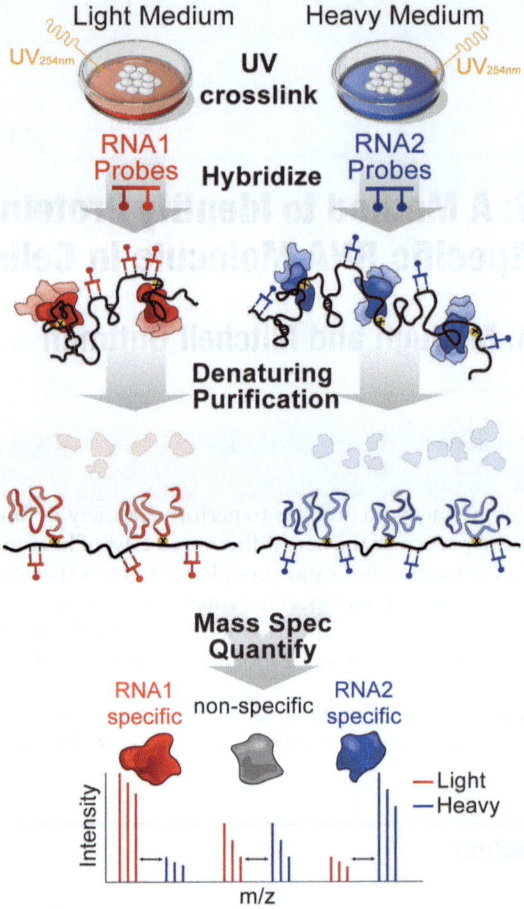

Fig. 1 Schematic of RAP-MS purification procedure from SILAC labeled mouse ES cells. Target RNA and control RNA are captured from cross-linked SILAC labeled cell lysates and purified under denaturing conditions. The resulting protein preparations are mixed and analyzed by mass spectrometry to identify proteins that bind specifically and directly to the target RNA versus the control RNA. Reproduced from [4]

In recent years, methods such as cross-linking and immunoprecipitation (CLIP) have been developed that have become the gold standard for studying direct RNA–protein interactions in vivo [2]. These methods utilize UV cross-linking to create a covalent link between RNA and protein interactions and are coupled with stringent purification conditions to enable the precise genome-wide mapping of the RNA binding sites of a specific protein [2]. Despite their success, CLIP methods are of more limited utility for identifying new protein interaction partners for a specific lncRNA.

To address this goal, we developed the RNA antisense purification coupled with mass spectrometry (RAP-MS) method to identify proteins that directly and specifically interact with a target RNA molecule (Fig. 1). The RAP-MS protocol uses ultraviolet light to

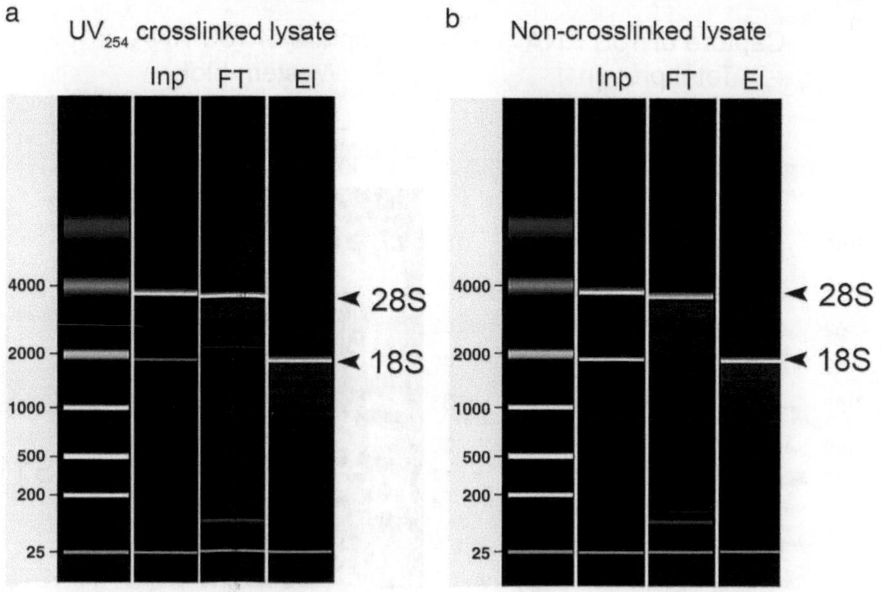

Fig. 2 Example of purified RNA captured by RAP-MS for 18S rRNA. Agilent Bioanalyzer gel-like images from RNA 6000 Pico chip for RNA Input (Inp), Flow-Through (FT), and Elution (El) samples from RAP-MS captures of 18S performed in (**a**) UV_{254} cross-linked or (**b**) non-cross-linked control (20 million cells each). The target RNA is efficiently captured and recovered from both UV cross-linked and non-cross-linked lysates

cross-link zero-distance interacting RNA and protein partners, followed by capture of the RNA of interest through hybridization with biotin-labeled DNA probes on streptavidin beads (Fig. 2). RAP-MS incorporates stringent washing with buffers containing high concentrations of denaturing and reducing agents to isolate only direct and specific proteins that are covalently cross-linked in vivo to the target RNA (Fig. 3). Stable isotope labeling of amino acids in cell culture (SILAC) tagging [3] is used to compare multiple protein capture samples in a single mass spectrometry experiment (Fig. 4), reducing instrument time and resulting in highly accurate quantitation of relative protein levels in each sample. At the end of the experiment, a short list of high-confidence direct protein interactors can be identified for a target RNA, simplifying follow-up analysis and subsequent functional assays.

The RAP-MS method has been validated on several well-characterized cellular RNAs (18S rRNA, U1 snRNA, and 45S preribosomal RNA) and was successfully used to identify the key functional proteins that interact with the Xist lncRNA during the initiation of X chromosome inactivation during development [4].

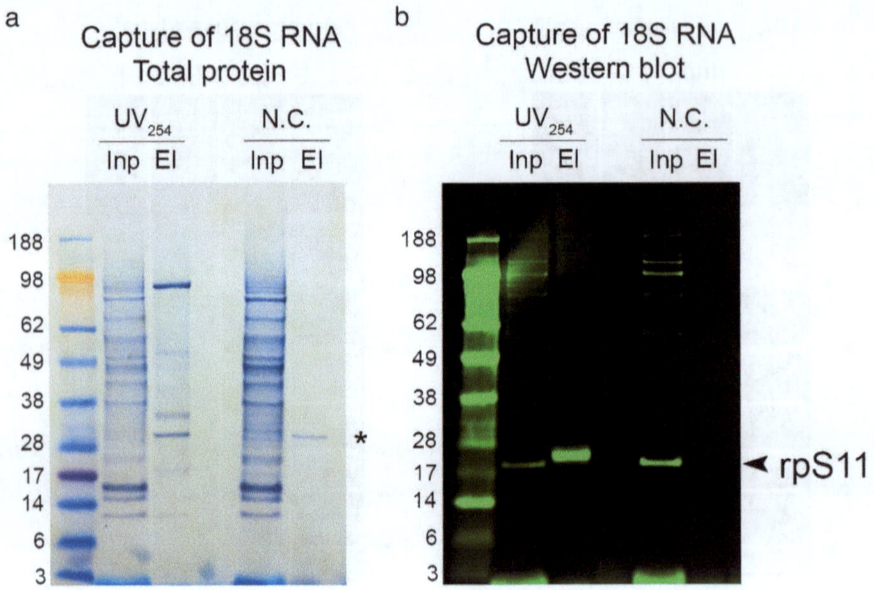

Fig. 3 Example of interacting proteins captured by RAP-MS for 18S rRNA. Proteins from RAP-MS captures of 18S in UV_{254} cross-linked and non-cross-linked N.C. control were separated by SDS-PAGE and transferred to nitrocellulose membrane. Each capture was performed from 20 million cells lysate. (**a**) Total protein staining of input (Inp) and elution (El) samples was performed with Blot FastStain (G Biosciences). *Asterisk* (*) indicates benzonase enzyme that was added to the elution sample. (**b**) For Western blotting, the same membrane was probed with an antibody (Abcam ab175213) against the rpS11 protein that is known to interact with 18s rRNA. Direct and specific 18S interacting proteins are recovered from the UV cross-linked sample, but not from the non-cross-linked lysate

2 Materials

Prepare all solutions using ultrapure water and the highest possible purity reagents. Whenever possible, use certified RNase- and DNase-free tubes and water for preparing samples. Please follow all safety and waste disposal guidelines when disposing of waste materials.

2.1 Specialized Equipment and Reagents

1. Custom-designed 90-mer DNA oligonucleotides with 5′-biotin modification (*see* **Note 1**).
2. Sonicator with microtip (e.g., Branson).
3. Thermomixer with heating and shaking functions.
4. Magnetic separation rack for Eppendorf tubes, 15 mL tubes, and 96-well plates (e.g., Life Technologies DynaMag).
5. Peptide HPLC column (e.g., Michrom Bioresources peptide MicroTrap column).
6. HPLC system (e.g., Agilent 1200 HPLC system), with Buffer A (0.2% formic acid) and Buffer B (100% acetonitrile).

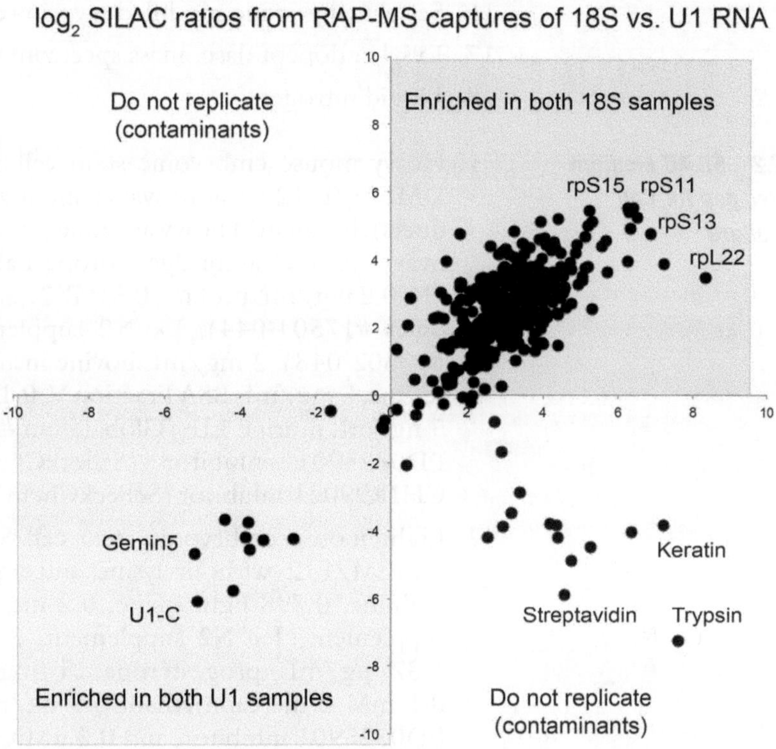

Fig. 4 Example of SILAC ratio plot for proteins from RAP-MS captures. Captures from more than one RNA target can be mixed and quantitated by mass spectrometry. RAP-MS captures for U1 and 18S were performed in both heavy and light lysates and resulting proteins were mixed together. Proteins identified in two replicates with label swap (18S light label vs. U1 heavy label, 18S heavy label vs. U1 light label) are plotted by their $\log_2$ SILAC ratio. Proteins that replicate in label-swap captures are high-confidence interactors for the target RNA molecule. Known contaminants (keratins, trypsin, benzonase, and streptavidin) are always purified in the light labeled sample and can be excluded from the final list of interactors. Adapted from [4]

7. q-PCR and/or Agilent Bioanalyzer.
8. TurboDNase with high salt tolerance.
9. Glass dounce homogenizer, 2 mL size.
10. UV cross-linker with 254 nm wavelength bulbs (e.g., Spectrolinker).
11. Vacuum lyophilizer.
12. Streptavidin-coated magnetic beads (e.g., Life Technologies Dynabeads).
13. Protease inhibitor cocktail set III, EDTA free.
14. Benzonase nuclease.
15. Detergent removal columns (e.g., Life Technologies HiPPR).

16. Sequencing grade modified trypsin (e.g., Promega).
17. Lysyl endopeptidase, mass spectrometry grade (e.g., Wako).
18. Liquid nitrogen.

2.2 SILAC Medium Recipes for Cell Culture

1. Heavy mouse embryonic stem cell SILAC medium: custom DMEM/F-12 without lysine and arginine (Dundee Cell Products), 0.398 mM heavy arginine (Sigma #608033), 0.798 mM heavy lysine (Cambridge Isotope Laboratories #CNLM-291-H), 0.2 mg/mL proline, 0.5× B-27 supplement (Life Technologies #17504-044), 1× N2 supplement (Life Technologies #17502-048), 2 mg/mL bovine insulin, 1.37 μg/mL progesterone, 5 mg/mL BSA Fraction V, 0.1 mM 2-mercaptoethanol, 5 ng/mL murine LIF (GlobalStem #GSR-7001), and 0.1 μM PD0325901 inhibitor (SelleckChem #S1036), 0.3 μM CHIR99021 inhibitor (SelleckChem #S2924).
2. Light mouse embryonic stem cell SILAC medium: Custom DMEM/F12 without lysine and arginine, 0.398 mM light arginine, 0.798 light lysine, 0.2 mg/mL proline, 0.5× B-27 supplement, 1× N2 supplement, 2 mg/mL bovine insulin, 1.37 μg/mL progesterone, 5 mg/mL BSA Fraction V, 0.1 mM 2-mercaptoethanol, 5 ng/mL murine LIF, 0.1 μM PD0325901 inhibitor, and 0.3 μM CHIR99021 inhibitor.

2.3 Buffer Recipes for RAP-MS Captures

1. Phosphate-buffered saline (1× PBS): 1 mM monobasic potassium phosphate, 155 mM sodium chloride, and 3 mM dibasic sodium phosphate.
2. 10 mM Tris–HCl pH 7.5.
3. 100 mM Tris–HCl pH 8.5.
4. Tris(2-carboxyethyl)phosphine (TCEP).
5. 8 M urea dissolved in 100 mM Tris–HCl pH 8.5.
6. 100 mM $CaCl_2$.
7. 500 mM iodoacetamide.
8. 80 w/v % aqueous solution tricholoroacetic acid (TCA).
9. 98% formic acid.
10. 200× DNase salt solution: 500 mM $MgCl_2$, 100 mM $CaCl_2$.
11. Cell lysis buffer I: 10 mM HEPES pH 7.4, 20 mM KCl, 1.5 mM $MgCl_2$, 0.5 mM EDTA, 1 mM TCEP, and 0.5 mM phenylmethylsulfonyl fluoride (PMSF).
12. Cell lysis buffer I with dodecyl maltoside (DDM): 10 mM HEPES pH 7.4, 20 mM KCl, 1.5 mM $MgCl_2$, 0.5 mM EDTA, 1 mM TCEP, 0.5 mM PMSF, and 0.1% DDM.
13. Cell lysis buffer II: 20 mM Tris–HCl pH 7.5, 50 mM KCl, 1.5 mM $MgCl_2$, 2 mM TCEP, 0.5 mM PMSF, 0.4% sodium deoxycholate, 1% DDM, and 0.1% *N*-lauroylsarcosine (NLS).

14. 1× Hybridization buffer: 10 mM Tris–HCl pH 7.5, 5 mM EDTA, 500 mM LiCl, 0.5% DDM, 0.2% SDS, 0.1% sodium deoxycholate, 4 M urea, and 2.5 mM TCEP.
15. Total cell lysis buffer: 10 mM Tris–HCl pH 7.5, 500 mM LiCl, 0.5% DDM, 0.2% SDS, 0.1% sodium deoxycholate, 1× protease inhibitor cocktail, and 1000 U/mL murine RNase inhibitor.
16. Benzonase elution buffer: 20 mM Tris–HCl pH 8.0, 0.05% NLS, 2 mM $MgCl_2$, and 0.5 mM TCEP.
17. NLS elution buffer: 20 mM Tris–HCl pH 8.0, 10 mM EDTA, 2% NLS, and 2.5 mM TCEP.
18. Bead based nucleic acid purification kit (We use Life Technologies Dynabeads® MyOne™ SILANE).

3 Methods

Maintain samples on ice and perform all centrifugation steps at 4 °C unless otherwise noted. All buffers should be prepared before starting the protocol. Buffers containing urea should be freshly prepared, or alternatively, can be prepared then immediately aliquoted and frozen at −20 °C. Do not store urea buffers at room temperature for extended periods of time. Before beginning experiments, select the target RNA of interest and controls as needed (*see* **Note 2**) and determine number of input cells to be used for each capture (*see* **Note 3**). If the cellular localization of the RNA is already known, select the appropriate lysis method for the target RNA; otherwise, use the whole cell lysis method (*see* **Note 4**).

3.1 Preparation of SILAC Labeled Cell Pellets

1. Initiate culture of cell line of interest. Once cells are growing well, split into two parallel cultures in SILAC Heavy and SILAC Light medium and grow for at least three passages to incorporate SILAC labels (*see* **Note 5**).
2. Seed adherent cells on 15 cm tissue culture plate. Grow to 70–90% confluence then remove medium from plate and replace with 10 mL ice-cold 1× PBS. Rock gently for 10 s then remove PBS wash. Add another 10 mL ice-cold PBS to plate to prevent cells from drying during cross-linking.
3. Place plates on a shallow tray of ice and UV cross-link at 254 nm for a total energy of 0.8 J/cm^2. Remove plates from cross-linker and keep on ice for the remainder of the procedure.
4. Scrape cells from the plate using a cell lifter and transfer to a sterile 15 mL tube. Cell suspensions from multiple plates can be pooled at this point.
5. Pellet cells by centrifugation at 1000 × *g* for 5 min.

6. Remove supernatant and resuspend cells in cold PBS to a final concentration of 50 million cells per 1 mL of buffer, pipetting gently to break up pellet.
7. Aliquot 1 mL of PBS/cell mixture into microcentrifuge tubes and centrifuge at 1000 × *g* for 5 min. Remove supernatant by aspirating or pipetting gently. At this point, pellets may be flash frozen in liquid nitrogen and stored at −80 °C.

3.2 Preparation of Nuclear Lysates

1. Resuspend each cell pellet in 1 mL of Cell Lysis Buffer I.
2. Centrifuge at 3300 × *g* for 10 min. Discard supernatant and resuspend cell pellet in 1 mL of Cell Lysis Buffer I with 0.01% DDM.
3. Incubate for 10 min on ice, then transfer sample to a tissue homogenizer and dounce with B (small clearance) pestle 20 times to break cells (*see* **Note 6**).
4. Transfer sample to microcentrifuge tube, and pellet nuclei by centrifugation at 3300 × *g* for 10 min.
5. Discard supernatant and resuspend pellet in 580 μL of Cell Lysis Buffer II.
6. Incubate for 10 min on ice, then sonicate on ice with microtip using 5 W of power (25% duty) for 60 s total in pulses of 0.7 s on, followed by 3.3 s off.
7. Add 1× DNase salt solution (3.75 μL) and 330 U TurboDNase (165 μL).
8. Incubate for 12 min at 37 °C.
9. Mix lysate with equal volume of 2× Hybridization Buffer (750 μL).
10. Centrifuge at 16,000 × *g* for 10 min at 4 °C.
11. Transfer supernatant to fresh tube and flash freeze in liquid nitrogen.
12. Store lysate at −80 °C until ready to perform RAP captures.

3.3 Preparation of Whole Cell Lysates

1. Resuspend each cell pellet in 900 μL Total Cell Lysis Buffer and incubate on ice for 10 min.
2. Pass cell suspension 3–5 times through a 26-G needle, then sonicate with a microtip at 5 W power for 30 s in pulses of 0.7 s on followed by 1.3 s off.
3. Perform DNase treatment as described for nuclear lysate (**steps** 7 and **8** of Subheading 3.2), then add salt and detergents to adjust sample buffer to match 1× Hybridization Buffer.
4. Centrifuge at 16,000 × *g* for 10 min to pellet insoluble material.
5. Transfer the supernatant to a fresh tube and flash freeze in liquid nitrogen.
6. Store lysate at −80 °C until ready to perform RAP captures.

3.4 Preclearing of Lysate

1. On the day of the experiment, warm frozen aliquots of either whole cell lysate or nuclear lysate to 37 °C using a thermomixer. Pool samples in a single tube.
2. Transfer 1.2 mL of streptavidin-coated magnetic beads per 200 million cell sample into a fresh microfuge tube (*see* **Note** 7).
3. Separate on magnetic rack and remove storage buffer from beads.
4. Resuspend beads in 1 mL of 10 mM Tris–HCl pH 7.5 with gentle pipetting, then separate on magnetic rack and remove supernatant.
5. Repeat bead washes (**step 4**) for a total of four washes in 10 mM Tris pH 7.5, and two washes in 1× Hybridization Buffer (*see* **Note 8**).
6. Magnetically separate and remove last wash from beads, then transfer lysate to beads and resuspend by pipetting gently.
7. Incubate for 30 min at 37 °C with intermittent mixing at 1100 rpm on thermomixer (30 s shaking, 30 s off).
8. Magnetically separate beads and transfer supernatant to fresh tubes. Repeat this step to transfer lysate to fresh tubes a second time, to remove all traces of beads from sample.
9. Remove sample of 100,000 cells worth of lysate and transfer to PCR strip tube. This is the "RNA Input" sample.

3.5 RAP Captures of Target RNA–Protein Complexes

1. Denature appropriate quantity of probe by heating at 85 °C for 3 min, then place on ice (*see* **Note** 7).
2. Mix lysate and probe, then incubate for 2 h at 67 °C with intermittent mixing at 1100 rpm on thermomixer (30 s shaking, 30 s off).
3. During the 2 h incubation, prepare streptavidin beads as previously described (**steps 4** and **5** of Subheading 3.4).
4. Magnetically separate beads and remove final wash from beads.
5. At the end of the 2 h incubation, remove sample of 100,000 cells worth of lysate and transfer to PCR strip tube. This is the "RNA Input + Probe" sample.
6. Resuspend beads in lysate (*see* **Note 8**).
7. Incubate for 30 min at 67 °C with intermittent mixing at 1100 rpm on thermomixer (30 s shaking, 30 s off).
8. Magnetically separate beads and remove supernatant. Take sample of 100,000 cells worth of supernatant and transfer to PCR strip tube. This is the "RNA Flow-Through" sample.

9. Wash beads 3–6 times, with at least one bead volume of 1× Hybridization Buffer per wash. Incubate each wash for 5 min at 67 °C.
10. Remove a sample of beads between 0.5% and 1% of the total volume and transfer to a PCR strip tube. This is the "RNA elution" sample.

3.6 Elution of Captured Protein

1. Magnetically separate remaining beads and remove supernatant.
2. Resuspend beads in 1 mL of Benzonase Elution Buffer.
3. Add 125 U of benzonase nuclease to sample.
4. Incubate for 2 h at 37 °C with intermittent mixing at 1100 rpm on thermomixer (30 s shaking, 30 s off) to digest nucleic acids and release proteins from beads.
5. Magnetically separate beads and transfer supernatant to a fresh microcentrifuge tube. Repeat this step for a total of six transfers to fresh tubes to remove all traces of streptavidin beads. The last supernatant is the "protein elution" sample.
6. If desired, a second nonspecific elution can be performed by boiling the bead sample in NLS elution buffer for 2 min at 95 °C. This elution will remove all remaining RNA as well as the streptavidin and bound proteins from the beads and can be used to test for remaining RNA or protein after the benzonase elution.

3.7 Elution of Captured RNA

1. Take the "RNA elution" sample of beads from the final step of Subheading 3.5 and separate on magnetic rack.
2. Remove and discard supernatant. Resuspend beads in 20 μL of NLS Elution Buffer.
3. Heat samples for 2 min at 95 °C.
4. Magnetically separate and transfer supernatant containing eluted RNA to a fresh PCR strip tube.
5. Take the previously collected samples (RNA Input, Input + Probe, and Flow-Through) and dilute each sample to 20 μL total volume with NLS Elution Buffer.
6. Add 1 mg/mL Proteinase K to each sample.
7. Incubate for 1 h at 52–55 °C to digest proteins.
8. Store samples at −20 °C for short term or −80 °C for long term.

3.8 Quantitation of Captured RNA

1. Perform SILANE bead based nucleic acid cleanup using the following steps (for a 20 μL sample):
2. Aliquot 20 μL of beads per RNA sample into clean PCR strip tubes. Magnetically separate beads and remove storage buffer.

3. Resuspend beads in 60 μL RLT Buffer (3× original sample volume).
4. Transfer beads in RLT to 20 μL RNA sample and mix well.
5. Add 120 μL of 100% ethanol (6× original sample volume). Wait for 2 min for sample to bind beads.
6. Wash beads two times with 150 μL of 70% ethanol.
7. Remove supernatant and allow to air-dry for approximately 5 min.
8. Elute RNA in water or desired buffer. In this case, elute RNA samples by adding 26 μL of 1× TURBO DNase Buffer (dilute from 10× stock supplied by manufacturer).
9. Leave beads in tube. Perform DNase treatment to remove background DNA by adding 1 μL of murine RNase inhibitor and 3 μL of TURBO DNase to each sample (30 μL total reaction volume). Incubate for 20 min at 37 °C.
10. Perform a second SILANE cleanup using beads already in the tube:
11. Add 90 μL RLT Buffer to each 30 μL sample.
12. Add 180 μL of 100% ethanol and mix well.
13. Wait 2 min for sample to bind beads.
14. Wash beads two times with 70% ethanol.
15. Remove supernatant and allow beads to air dry approximately 5 min.
16. Elute in 10 μL of UltraPure water.
17. Analyze RNA samples using Agilent Bioanalyzer or by RT-qPCR (*see* **Note 9**).

3.9 Mass Spectrometry of Captured Proteins

3.9.1 Protein Precipitation

1. Add 10% final concentration of TCA to protein elution sample. Incubate at 4 °C overnight.
2. Centrifuge at 16,000 × *g* for 30 min to pellet protein.
3. Remove supernatant and replace with 1 mL of cold acetone.
4. Centrifuge at 16,000 × *g* for 15 min.
5. Remove supernatant and allow pellet to dry in open tube in fume hood or on bench. Store protein elution samples at −20 °C.

3.9.2 In-Solution Digest of Protein Samples for Mass Spectrometry

1. Resuspend protein elution sample in 40 μL of freshly prepared 8 M urea dissolved in 100 mM Tris–HCl pH 8.5.
2. Add 3 mM TCEP and incubate 20 min at room temperature.
3. Add 11 mM freshly prepared iodoacetamide and incubate for 15 min at room temperature in the dark.

4. Digest samples with 0.1 μg Lysyl endopeptidase for 4 h at room temperature.
5. Dilute samples to final concentration of 2 M urea by adding appropriate volume of 100 mM Tris–HCl pH 8.5.
6. Add 1 mM $CaCl_2$ to sample.
7. Digest with 0.1–0.5 g of trypsin overnight at room temperature.

3.9.3 Purify Peptides to Remove Detergent

1. Use HiPPR resin spin columns to remove detergent according to manufacturer's instructions.
2. Add 5% formic acid and centrifuge for 1 min at 16,000 × *g*.
3. Desalt by HPLC (*see* **Note 10**).
4. Collect fractions containing peptides.
5. Lyophilize peptides in SpeedVac.
6. Store samples at −20 °C until ready for mass spectrometry.
7. Resuspend samples in 0.2% formic acid and 5% acetonitrile.
8. Mix SILAC heavy labeled target RNA capture sample with SILAC light labeled control RNA capture sample, or vice versa (*see* **Note 11**).
9. Analyze mixed samples by mass spectrometry and quantify peptide ratios using MaxQuant or similar analysis software (*see* **Note 12**).

4 Notes

1. *Oligonucleotide probe design for RAP-MS captures.* Design 90-nucleotide oligos that tile across the target RNA sequence of interest without overlapping. Probe design software is available at www.lncRNA.caltech.edu/software.php. To avoid off-target hybridization, use BLAST, or similar alignment programs, to remove sequences that contain a perfect 30 base pair match or an imperfect (90%) identity 60 base pair match with another transcript or genomic region. Compare the oligos to Repeat-Masker annotations and remove probes that contain more than 30 bases that overlap with a repeat annotation. Order oligos with 5′ biotin standard modification from an oligonucleotide synthesis company such as Integrated DNA Technologies. Individual probes should be resuspended at 500 μM or 1 mM concentration depending on the synthesis scale. Dilute probe stocks 1:100 from 96-well plates or individual tubes into Ultra-Pure water. Mix all individual probes together to create a probe stock to cover the length of the target RNA. We usually make several aliquots of probe stock mixtures and store at −20 °C,

avoiding multiple freeze–thaw cycles. Sequences for control capture oligonucleotide probes for mouse RNA targets (U1, 18S, and 45S) are available upon request.

2. *Selection of appropriate controls.* The U1 snRNA control and non-cross-linked capture control are generally used as our standard negative controls. Antisense probes that do not bind the target RNA, probes targeting a known mRNA, or other controls could also be used to evaluate the level and identify of background or nonspecific proteins in the cell type of interest. Negative controls like antisense probes and non-cross-linked samples are important to perform to ascertain the level of background but generally do not yield enough protein for useful quantitation in mixed samples. We use U1 as a standard nuclear control for calculating SILAC ratios in mixing experiments, but 18S or other RNA capture controls could also be used.
3. *Number of cells per capture.* Volumes indicated are for 200 million cells per capture. Increase or decrease the cell number used for each capture depending on the abundance of the target RNA. For high abundance RNAs (U1, 18S, and 45S), between 20 and 200 million mouse embryonic stem cells are usually sufficient per capture. For lower abundance RNAs (Xist or other lncRNAs), we used between 200 and 800 million mouse embryonic stem cells per capture. Other RNA targets may require different cell input numbers to reliably obtain sufficient quantities of captured protein for mass spectrometry analysis.
4. *Cell lysis method.* Select either the nuclear lysis or whole cell lysis method as needed for the target RNA. For nuclear RNAs that are chromatin associated, like Xist, the nuclear lysis method is optimal. If the lncRNA is not chromatin associated or a whole cell extraction method is preferred, the whole cell lysis procedure can be followed. For 18S rRNA captures shown here, the whole cell lysis method was used. For Xist lncRNA captures shown in McHugh et al. 2015 [4], the nuclear lysis method was used. Alternative methods for nuclear extraction, particularly the method described in [5], are also compatible with RAP-MS as long as the guidelines for removing detergents and salts from proteins before mass spectrometry analysis are followed.
5. *Initiating SILAC cultures.* Sample SILAC medium recipes for mouse ES cell culture are provided above. SILAC medium should be adapted to fit the requirements of the desired cell line. The most important factor is to use base medium without lysine and arginine amino acids. Serum that has been dialyzed to remove unlabeled amino acids may be required to support the growth of some cell types. Cells should be grown in SILAC

medium for at least three generations and the incorporation should be tested by mass spectrometry to ensure >95% labeling of peptides. It is also possible to perform experiments in label-free samples and identify and quantitate captured proteins by intensity of their peptides only. However, this approach requires more controls and experimental replicates to obtain a list of high-confidence interactors with a target RNA, since nonspecific contaminant proteins will likely dominate the total mass spectrum by intensity.

6. *Douncing for cell lysis.* Number of strokes for douncing to lyse cells but retain nuclear integrity should be optimized for the cell type so that the cell membrane is broken but the nuclei remain intact. When establishing any cell lysis procedure, check the cells under a microscope to confirm appropriate lysis.

7. *Selection of appropriate probe to lysate and bead to lysate ratios.* The exact concentration of probe and streptavidin beads needed for a particular experiment will vary depending on the target concentration. Many targets will require a lower probe and bead quantity than is suggested here. Optimization of the RNA capture in small-scale lysates is advisable before scaling up for protein identification experiments. The quantity of streptavidin beads and probe can be adjusted up or down relative to the amounts given here, and capture temperatures ranging from 45 to 67 °C can be evaluated to determine the best combination for efficient capture of a particular target RNA molecule.

8. *Sample handling during magnetic bead captures.* The buffers used for hybridization and washes contain detergents that may create bubbles and make magnetic separation of beads challenging. Pipette gently to avoid creating bubbles when washing beads, and ensure magnetic beads have separated sufficiently from the liquid phase before removing and discarding supernatants.

9. *Samples for analysis of RNA yield.* In the first experiment for a new target RNA, do a small-scale capture in 1–10 million cells lysate to test that probes and washes are performing as expected. Collect and evaluate the RNA Input lysate, Input plus Probe, Flowthrough, and Elution samples to evaluate where the target RNA is located at each step of the procedure. Captured RNA is usually of sufficient quantity to detect by Agilent Bioanalyzer or RT-qPCR analysis but may not be enough to detect on a standard agarose gel. If performing qPCR analysis, ensure that primers are designed such that they do not amplify a region contained within a single probe sequence. Highly stable DNA:RNA hybrids formed during RAP capture are not always completely removed during SILANE cleanups, even after DNase treatment. Inclusion of

the Input plus Probe sample during RT-qPCR controls for any DNA probe that was not removed during RNA cleanup.

10. *Cleaning of peptides for mass spectrometry.* It is imperative that peptide samples be fully cleaned to remove all detergents and salts that may interfere with analysis. The method described here uses a combination of detergent-binding resin and a desalting column, but Stop-and-Go extraction tips [6] or other methods may also be used as an alternative.
11. *Mixing samples for mass spectrometry.* Accurate mixing is important to achieve correct SILAC peptide ratios. Take a small amount of unmixed peptide sample (about one tenth of the total sample) and perform a short quantitation run on the mass spectrometer to get an accurate measurement of the amount of protein present in each sample. Perform peptide searches for each sample. Filter to remove common contaminants using the database provided by MaxQuant or other analysis software. Finally, calculate the median peptide intensity for each sample and mix heavy and light SILAC samples based on the median intensity of peptides in the sample. Protein concentration measurement methods like Bradford or Coomassie assays could be used to estimate sample concentration instead of a quantitation MS run, but these methods may not be accurate enough to precisely measure the concentration of small amounts of captured protein in the final RAP-MS sample.
12. *Mass spectrometry data analysis and identification of final RNA–protein interactors list.* At the end of this protocol, samples are ready for LC-MS measurements using the desired instrument. For example, a nanoflow LC system (Proxeon EASy-nLC1000) coupled to a hybrid linear ion trap Orbitrap Elite mass spectrometer (Thermo Fisher Scientific) was used for the experiments described in McHugh et al. [4]. MaxQuant analysis software was used for peptide searches and to calculate SILAC median peptide ratios for proteins with at least two matched SILAC pairs. Other software may be used to identify and quantitate peptides after SILAC labeling. Judicious use of background controls including non-cross-linked samples and a variety of nontarget RNAs (U1, 18S, or others) can help distinguish real and specific interactors from proteins that interact with many RNA molecules in the cell. We exclude known contaminants from the final protein list, including keratins and proteins introduced during the sample purification and preparation process (such as streptavidin, benzonase, and trypsin, Fig. 4), as well as naturally biotinylated proteins like histones that can contaminate the preparation by binding to streptavidin beads. The threshold SILAC ratio was set at ≥ 3.0 for Xist vs. U1 experiments but high or lower cutoff may be appropriate for other combinations of target RNA and control samples.

Acknowledgments

We thank Christine Surka, Alexander Shishkin, and Jesse Engreitz for their help in developing the RAP-MS method and Parham Peyda, Tony Szempruch, and Ward Walkup IV for their helpful comments on the manuscript. This work was supported by a Caltech Division of Biology and Biological Engineering Fellowship and a National Institute of General Medical Sciences of the National Institutes of Health Pathway to Independence Award (K99GM120494) to C.A.M., as well as the New York Stem Cell Foundation, an NIH Director's Early Independence Award (DP5OD012190), the Edward Mallinckrodt Foundation, Sontag Foundation, Searle Scholars Program, Pew-Steward Scholars program, and funds from the California Institute of Technology to M.G.

References

1. McHugh CA, Russell P, Guttman M (2014) Methods for comprehensive experimental identification of RNA-protein interactions. Genome Biol 15(1):203
2. Ule J, Jensen KB, Ruggiu M, Mele A, Ule A, Darnell RB (2003) CLIP identifies Nova-regulated RNA networks in the brain. Science 302(5648):1212–1215
3. Ong SE, Blagoev B, Kratchmarova I, Kristensen DB, Steen H, Pandey A, Mann M (2002) Stable isotope labeling by amino acids in cell culture, SILAC, as a simple and accurate approach to expression proteomics. Mol Cell Proteomics 1 (5):376–386
4. McHugh CA, Chen C, Chow A, Surka CF, Tran C, McDonel P, Pandya-Jones A, Blanco M, Burghard C, Moradian A, Sweredoski MJ, Shishkin AA, Su J, Lander ES, Hess S, Plath K, Guttman M (2015) The Xist lncRNA directly interacts with SHARP to silence transcription through HDAC3. Nature 521 (7551):232–236
5. Gagnon KT, Li L, Janowski BA, Corey DR (2014) Analysis of nuclear RNA interference (RNAi) in human cells by subcellular fractionation and Argonaute loading. Nat Protoc 9 (9):2045–2060
6. Rappsilber J, Ishihama Y, Mann M (2003) Stop and Go extraction tips for matrix-assisted laser desorption/ionization, nanoelectrospray, and LC/MS sample pretreatment in proteomics. Anal Chem 75(3):663–670

Erratum to: Super-Resolution Single Molecule FISH at the *Drosophila* Neuromuscular Junction

Joshua S. Titlow, Lu Yang, Richard M. Parton, Ana Palanca, and Ilan Davis

Erratum to:
Chapter 10 in: Imre Gaspar (ed.), *RNA Detection: Methods and Protocols*, Methods in Molecular Biology, vol. 1649, https://doi.org/10.1007/978-1-4939-7213-5_10

This chapter was originally published under © Springer Science+Business Media LLC 2018, but has now been republished © The Author(s).

The updated online version of this chapter can be found at
https://doi.org/10.1007/978-1-4939-7213-5_10

Imre Gaspar (ed.), *RNA Detection: Methods and Protocols*, Methods in Molecular Biology, vol. 1649, https://doi.org/10.1007/978-1-4939-7213-5_32, © Springer Science+Business Media LLC 2018

Index

A

Affinity purifications ...49–57, 85–93, 405–407, 410, 412, 414, 420, 423
Alkaline fragmentation ... 30
4'-Aminomethyltrioxsalen hydrochloride (AMT) ...63, 67, 68, 70, 81
Antibody staining ...102, 103, 160, 164
Aptamer ...11, 305–316
Automated cell segmentation ... 144, 157
Axons ...11, 85–93, 167, 169
Azide-alkyne cycloaddition reaction ... 359, 362
Azide-modified UTP analogs (AMUTP) ... 359–370

B

Bacteria ...11, 16, 47, 289–296, 298, 299, 301–303, 355
Bioinformatics ... 50, 113, 432, 449, 452
Bioorthogonal reaction ...359
Biotin ...4, 62, 177, 179, 180, 183, 430, 474, 476, 484
Blood sample ... 191, 193, 194
BoxB-λN system ... 11, 321, 324, 327–331, 333
Branched DNA (bDNA) ...6, 209

C

Cell culture
Gag293 cells ... 339, 341
HEK-293FT cells ... 233
HeLa cells ...7, 50, 244, 251, 256, 314, 315, 364, 366, 369, 389
HT1080 cells ... 233, 236
MCF-7 ... 407, 412, 415
U2OS cells ... 339, 343, 389
Chemical synthesis ... 306
Click chemistry ...434
Contact quenching ...290, 291, 299, 301, 302
Cre-recombination ...86, 87, 92
Cresyl violet ...97, 101, 102, 108
Cycloheximide ...88–90, 93, 388, 393

D

Danio rerio, *see* Zebrafish
Differential centrifugation ... 422
Digoxigenin (DIG) ...177
2-Dimensional HCR ... 189, 193
3D Structured Illumination Microscopy (3D-SIM) ...165, 168, 169, 173, 174
Direct lysis without RNA purification ...81, 96
Doxycycline ... 375, 377, 383, 388, 389, 392
Drosophila melanogaster, *see* Fruitfly

E

Electron microscopy (EM) ... 18, 129, 177, 259, 276
ELISA-like assay ...189
Embryos ... 3, 87, 96, 127, 144, 164, 221, 231, 349, 402, 477
Endosomal mRNA transport ...320, 327–329
Escherichia Coli (E.coli), *see* Bacteria
Exciton-controlled hybridization-sensitive fluorescent oligonucleotide probe (ECHO) ... 11, 259
Expression atlases ...5, 111–113, 115, 122
Expression plasmids ... 269, 294, 295, 298, 302, 326, 328, 342, 345

F

FIJI, *see* ImageJ
FISHQuant ... 166, 171, 172
Fixation
chemical ...177
freezing ...96, 106, 110, 158, 159
Fluorescence imaging ...262, 265, 338–345, 400
Fluorescent in situ hybridization (FISH) ... 276, 321
Fluorescent oligonucleotides ...260, 273
Fluorescent protein
split GFP ... 227, 303, 338
superfolder GFP (sfGFP) ...386, 388, 389, 399
Fluorogenic detection ... 10, 273
Fluorogenic oligonucleotides ... 5
Forced intercalation of thiazole orange probe (FIT) ...11, 273–286
Freezing ... 159, 224
Fruitfly ...349

Imre Gaspar (ed.), *RNA Detection: Methods and Protocols*, Methods in Molecular Biology, vol. 1649, DOI 10.1007/978-1-4939-7213-5, © Springer Science+Business Media LLC 2018

G

Gel electrophoresis
  agarose gel electrophoresis 32, 38, 53, 92, 187, 189, 194, 456, 460
  2D PAGE gel electrophoresis 432
  urea PAGE electrophoresis 432
G–I tailing of RNA 457, 463
Gradient fractionation 421, 422, 425

H

Halo-tag 374, 377
Harringtonine 88, 91, 93, 388, 393
High-throughput sequencing 49, 59, 60, 452, 455, 466, 468
Histogenes 97, 101, 102, 106, 108
Hybridization chain reaction (HCR) 6, 10, 187–196, 209
Hybridizations 5, 35, 129, 146, 165, 178, 200, 209, 232, 243, 260, 273, 316, 321, 452, 475

I

Image registration 113, 114, 168
ImageJ 148, 166, 169, 220, 240, 254, 298, 333, 396
Immunoelectron microscopy (IEM) 178, 183
Immunohistochemistry, *see* Antibody staining
Immunoprecipitation 15, 51, 88, 90, 91, 420, 421, 423, 425, 428–430, 432, 438
Individual nucleotide resolution UV cross-linking and immunoprecipitation (iCLIP) 428, 431, 432, 434, 436–439, 441, 442, 444, 445, 447–452
In situ hybridization (ISH) 5, 18, 112, 127, 129, 132, 133, 143, 177, 225, 260
In situ hybridization (ISH)–Immuno-electron microscopy (IEM) 177, 178, 180, 181, 184, 185
In vivo electroporation 262, 264, 266, 268, 269
In vivo hybridization 10, 473
Interactomes 15, 16, 405

K

KNIME 144, 148, 149, 152, 153, 160, 376, 380, 381, 383

L

Larval neuromuscular junction 164
Laser capture microscopy/microdissection (LCM) 96–100, 103, 105, 106, 108–110
LCM-Seq 96–100, 103, 105, 106, 108–110
Live-cell RNA detection
  of endogenous RNA 12
LiveFISH 259
Locked nucleic acid (LNA) 99, 198, 210–213, 215, 219, 223, 228, 274, 278, 284

M

m^6A sequencing, *see* N^6-methyladenosine (m^6A) sequencing
Mass spectrometry (MS) 16, 278, 405–407, 410, 412–414, 416, 424, 474, 475, 478, 483–485, 487
Methylome 49–51, 53–57
Microinjection 245, 251, 252, 254, 256, 273, 274, 276, 280–283, 285, 286, 351
Microporation 233–236, 238–240, 245, 247, 249, 253–256
MicroRNA (miRNA) 10, 55, 59, 79, 197–201, 203–207, 269, 451
Microscopy-photon microscopy
  brightfield microscopy 206, 217, 354
  point scanning confocal microscopy 402
  spinning disc confocal microscopy 402
  TIRF microscopy 402
  wide-field epifluorescence microscopy 148, 152, 220, 227, 294
Molecular beacon (MB) 10, 231, 243–250, 252, 254–256, 273, 305
Mouse
  brain 5, 106, 260, 264–266
  spinal cord 101, 106
mRNA genotyping 211, 221
mRNA localization 177, 303, 319
mRNA transport 11, 13, 320, 325, 327–329, 349
MS2-MCP system 284, 355

N

N^6-methyladenosine (m^6A) sequencing 2, 30, 49–51, 53–57
Nascent mRNA 6, 13, 136, 141, 349, 389, 393
Nascent polypeptide 386, 387, 389, 393
Neurons 2, 4, 9, 11, 14, 16, 17, 85–93, 95, 102, 110, 164, 231, 389, 419, 423
Non coding RNA
  long non-coding RNA (lncRNA) 16, 473

O

2'-O-methyl RNA 233, 244, 245
Oocytes 11, 13, 18, 164, 177–179, 184, 197, 274, 276, 282–284, 286, 349
Optimal cutting temperature (OCT) compound 97, 100, 108, 144, 146, 149, 150, 158, 159

P

Padlock probes 7, 209–211, 213, 215–220, 222–228
PAR-CLIP 415, 428, 429
Phi29 DNA polymerase 213, 215
Phosphorothioate (PS) 244, 250, 251
Plasmodium falciparum 405
Poly(U) polymerase (PUP) 16, 455, 459
Poly-A$^+$ RNA 122
PP7-PCP system 11–14, 321
Protein–RNA interactions 16, 405, 427–453, 455
Proximity ligation 60, 64, 70, 81
Psoralen 59–65, 67, 68, 70–78, 80–82
Psoralen analysis of RNA interactions and structures (PARIS) 59–83
Pumilio homology domain (PUM-HD) 12, 338, 340–342
Puromycin 269, 388, 393

Q

Quantitation 165, 209, 220, 475, 482, 485, 487
Quencher 11, 232–234, 243, 254, 289–296, 298, 299, 301–303

R

Ratiometric bimolecular beacon (RBMB) 232, 233, 235, 236, 238–240
Ratiometric imaging 244, 254
Reporter constructs 324, 351–353, 376, 382, 397, 398
Retinal ganglion cell (RGC) 86, 87, 89, 92
Reverse transcription (RT) 70, 98, 99, 105, 115, 122, 210, 211, 219, 307, 313, 428, 429, 432, 436, 441, 451, 464, 465
RiboMethSeq 29–33, 35, 37–39, 41, 42, 44, 46, 47
Ribose methylation 30
Ribosomal RNA (rRNA) 30, 55, 59, 475
Ribosomes 3, 85, 320, 373, 386
RiboTag 85–93
RNA affinity purification (RAP) 474–477, 480, 481, 484–487
RNA binding proteins (RBPs) 11, 12, 14–16, 320, 324–331, 333, 338, 340, 410, 412, 413, 415, 416, 419, 425, 427, 429, 431, 434, 436, 440, 449, 455, 473–488
RNA capture 5, 16, 195, 338, 406, 456, 474, 475, 482, 484–486
RNA modifications 2, 29, 59, 245
RNA polymerase 31, 35, 37, 295, 302, 360, 369
RNA structures 3, 59, 60, 74, 77, 79, 81, 82, 263
RNA structurome 3
RNA tagging 10–12, 15, 16, 455–470
RNA-RNA interaction 74, 77–79, 298
RNAscope 197–201, 203–207
Rolling circle amplification (RCA) 210, 211, 213, 215, 216, 219, 223, 227

S

Saccharomices cerevisae, *see* Yeast
Sandwich hybridization 189, 190, 440
SELEX 292
Single molecule 127, 136, 143, 163–170, 173, 174, 238
Single nucleotide polymorphism (SNP) 209–211, 213, 215–220, 222–228
Single-cell mRNA-seq 111, 112, 123
Single-molecules 3, 6, 8, 9, 11, 13, 14, 18, 337–346, 355, 373, 376
Smart-seq2 .... 95, 122
Spatial transcriptomics 5, 95–124
Stable isotope labelling by amino acids (SILAC) 406, 407, 409, 410, 415, 474, 475, 477, 479, 484, 485, 487
Structured illumination 169
SunTag 13, 14, 386, 388, 389, 392, 393, 395, 397, 398, 400, 401, 403
Super-resolution imaging 7
Synapse 51, 164
Systematic Evolution of Ligands by EXponential enrichment (SELEX) 290

T

Telomeric-repeat containing RNA (TERRA) 338, 340, 342, 343, 345
Thiazole orange (TO) 260, 273, 277, 290
4-Thiouridine (4SU) 406, 408, 410, 412, 415, 429
Tissue sectioning
  cryosectioning 144, 146, 150, 181, 199

Transcription 4, 31, 70, 79, 82, 98, 115, 141, 144, 210, 267, 273, 295, 307, 349, 360, 373, 427
Transcriptomes 1, 3, 4, 6, 8, 9, 16, 30, 49, 50, 111–113, 115–119, 121–124, 197, 427, 456
Transfection 234, 238, 254, 255, 305, 308, 315, 339, 341, 342, 345, 362, 364, 369, 388, 389, 391, 400
Transgenesis
  stable 351
  transient 351
Translating ribosome affinity purification (TRAP) 4, 85–93
Translating RNA imaging by coat protein knock-off assay (TRICK) 14, 324, 373–384
Translation 2, 29, 49, 85, 114, 177, 198, 269, 273, 289, 338, 373, 385, 427
Translatomes 87

U

Ultrastructure 177, 184
Ustilago maydis (U.maydis) 319–333
UV crosslinking 405, 408, 428–430, 474

W

Wash-free FISH 280, 281, 285

Y

Yeast 6, 30, 128, 284, 293, 322, 351, 421, 456

Z

Zebrafish
  embryo 5, 143–146, 148–153, 156–162
  heart 200, 201, 205

MIX
Papier aus verantwortungsvollen Quellen
Paper from responsible sources
FSC® C105338

If you have any concerns about our products,
you can contact us on
ProductSafety@springernature.com

In case Publisher is established outside the EU,
the EU authorized representative is:
Springer Nature Customer Service Center GmbH
Europaplatz 3, 69115 Heidelberg, Germany

Printed by Libri Plureos GmbH
in Hamburg, Germany